# Springer Hydrogeology

**Series Editor**

Juan Carlos Santamarta Cerezal, San Cristóbal de la Laguna, Sta. Cruz Tenerife, Spain

The *Springer Hydrogeology* series seeks to publish a broad portfolio of scientific books, aiming at researchers, students, and everyone interested in hydrogeology. The series includes peer-reviewed monographs, edited volumes, textbooks, and conference proceedings. It covers the entire area of hydrogeology including, but not limited to, isotope hydrology, groundwater models, water resources and systems, and related subjects.

More information about this series at http://www.springer.com/series/10174

Uri Kafri · Yoseph Yechieli
Editors

# The Many Facets of Israel's Hydrogeology

Springer

*Editors*
Uri Kafri
Water and Natural Resources
Geological Survey of Israel
Jerusalem, Israel

Yoseph Yechieli
Water and Natural Resources
Geological Survey of Israel
Jerusalem, Israel

Ben Gurion University
Sede Boker Campus
Sede Boker, Israel

ISSN 2364-6454 ISSN 2364-6462 (electronic)
Springer Hydrogeology
ISBN 978-3-030-51147-0 ISBN 978-3-030-51148-7 (eBook)
https://doi.org/10.1007/978-3-030-51148-7



This Springer imprint is published by the registered company Springer Nature Switzerland AG
The registered company address is: Gewerbestrasse 11, 6330 Cham, Switzerland

# Acknowledgements

We acknowledge the support of the management of the Geological Survey, which allowed us to spend the needed time for this project. We also wish to thank all our colleagues for their cooperation throughout the many years of studies of the Israeli hydrogeological systems. This book could not be written without their help. Thanks are also due to Batsheva Cohen and Chana Netzer from the graphic department of the Geological Survey of Israel for their assistance in the design and production of part of the figures in the book.

# Contents

# About the Editors

**Dr. Uri Kafri** was born in 1935 in Israel. He received his Ph.D. degree in geology and hydrogeology in 1969 from the Hebrew University in Jerusalem. He is with the Geological Survey of Israel since 1960, serving as a senior geologist and hydrogeologist, and as a director of the Survey between 1979 and 1983. Currently, he is active as an Emeritus in the Geological Survey. Along with his career, he was involved in part-time teaching in the Ben Gurion University in Israel and was involved in various geological and hydrological studies in Israel. His professional activities abroad included hydrogeological surveys and studies in Nepal, Central Africa, South Africa, Tonga, Germany and the USA. He served as a member of board of directors of Earth Science research institutes, as well as a member of several professional committees. He has been serving for a long time in the editorial board of Environmental Earth Sciences.

**Prof. Yoseph Yechieli** was born in Israel in 1955. He received his B.Sc. and M.Sc. degrees in Geology and Hydrogeology from the Hebrew University in Jerusalem and Ph.D. in hydrogeology and geochemistry of groundwater from the Weizmann Institute of Science. He is a senior researcher at the Geological Survey of Israel and an adjunct professor at Ben Gurion University in Sede Boker since the last 10 years. He is currently the director of the Geological Survey of Israel. He was the president of the Israeli Geological Society in 2006–2007 and chairman of the Israeli Association of Water Resources in 2010–2012. He served as an associate editor of the journals *Groundwater* and *Hydrogeology Journal*. Throughout his career, he has been involved in hydrogeological and geochemical research of coastal aquifers, including the issues of seawater intrusion and dating of groundwater near the Mediterranean Sea and the Dead Sea and in research of sinkholes formation.

# Introduction

**Uri Kafri and Yoseph Yechieli**

This chapter is a concise overview of the main hydrogeological (groundwater) issues of Israel. The reader will find a broad general picture of the hydrogeological setup prior to delving into specific hydrogeological issues that are discussed and detailed in the following chapters.

Israel, as well as the entire region, suffers from a water shortage due to a population growth and an increase in the standard of living that have led to increased water demand. Due to Israel's climate, the groundwater component is the country's major natural water resource.

## 1 Climate

Israel is located between latitude 30 and 33 in the transition between the southern semiarid and the northern subtropical climatic zones. Thus, the northern part of the country is influenced by a Mediterranean climate and the south by the arid deserts (Diskin 1970). The rainy season extends mainly from October through March, and the rest of the year is dry. The annual average rainfall map (Fig. 1) shows rainfall distribution, which is determined by distance from the Mediterranean Sea, latitude, topography and elevation above or below sea level (Diskin 1970). The total average annual volume of rainfall is close to $8 \times 10^9$ cubic meters. Out of this total, some 70% is lost to evapotranspiration, around 5% is drained by runoff and the small remainder is naturally recharged to the groundwater system (Stanhill and Rapaport 1988).

U. Kafri (✉) · Y. Yechieli
Geological Survey of Israel, 32 Leibovitz Str., Jerusalem 9692100, Israel
e-mail: uri.kafri@gsi.gov.il

Y. Yechieli
e-mail: yechieli@gsi.gov.il



U. Kafri and Y. Yechieli (eds.), *The Many Facets of Israel's Hydrogeology*,
Springer Hydrogeology, https://doi.org/10.1007/978-3-030-51148-7_1

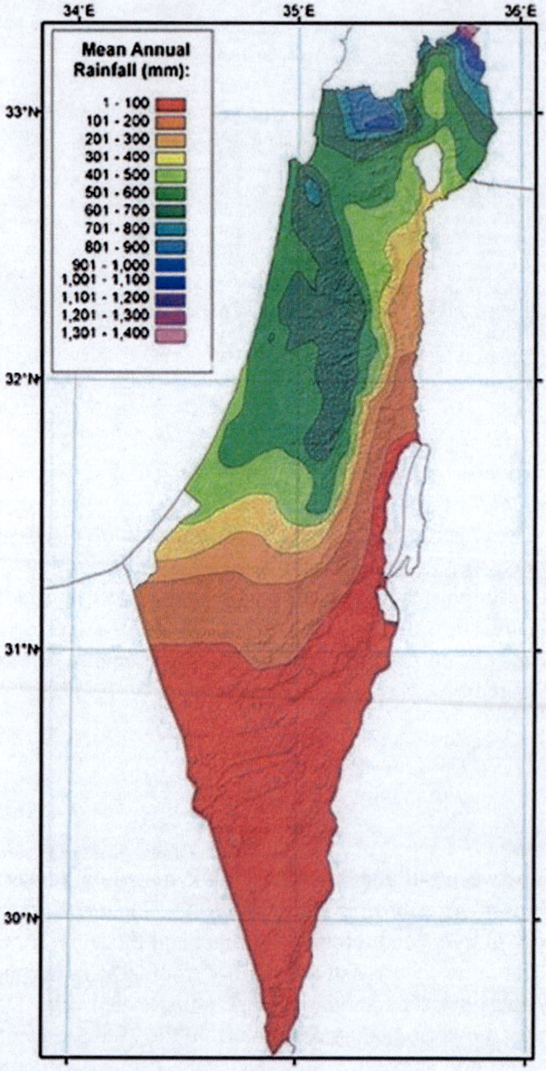


**Fig. 1** Annual average rainfall map of Israel (after the Israeli Meteorological Service)

## 2 Groundwater Base Levels

The country is bounded by three groundwater base levels as follows.

The Mediterranean Sea in the west serves as a base level to regional groundwater flows from the east, from both the mountainous aquifers, consisting mostly of carbonate rocks, and from the coastal detrital aquifer, consisting mostly of calcareous sandstone.

The Gulf of Elat (Aqaba), in the south, which is a tongue of the Red Sea, serves as a base level to convergent flows from two mountainous aquifers: carbonate and detrital. In addition, the southern Arava Valley (part of the Dead Sea Rift, hereafter DSR) detrital graben fill aquifer is drained into The Gulf of Elat.

The DSR in the east serves as an endorheic base level to convergent flows into it from the mountainous aquifers from north, west and east. Also, younger detrital aquifers, as well as basalt aquifers, associated with the DSR and its neighborhood drain into it. The Sea of Galilee and the Dead Sea are part of the discharge zones (detailed in Chapter "The Eastern Dead Sea Rift Continental Groundwater Base Level").

The location of these base levels, coupled with the fact that the Dead Sea is endorheic (terminal) and below sea level (~435 m bsl in 2019) control the interrelationship between them. They also play an important role on the groundwater flow regime as well as on the salinization of the groundwater systems.

## 3 Groundwater Resources

The main water resources of the country, as detailed by Gvirtzman (2002), among others, are described in brief below. The first three are considered herein as the primary resources, followed by resources of secondary importance. The different hydro-stratigraphic units relevant to the above sources are listed in Table 1. The extent of the exposed lithostratigraphic units that host the above is shown in Fig. 2.

**Primary resources:**

1. The Judea Group carbonate regional aquifer (JGA) of Cretaceous age is ca 600 m thick. It is exposed to natural recharge mainly along the anticlinorium of the

**Table 1** Main lithostratigraphic units in Israel and their hosted main aquifers

| Age | Main lithostratigraphic unit | Main lithology | Hydrogeological unit |
|---|---|---|---|
| Neogene to Quaternary | Kurkar Group, Hazeva Formation | Sandstone, gravel | Aquifer |
| | Cover Basalt | Basalts | Aquifer |
| Eocene | Avedat Group | Limestones, chalky limestones | Aquifer, aquitard |
| Early to Upper Cretaceous | Judea Group | Limestones, dolomites | Aquifer |
| Early Cretaceous | Kurnub Group, Nubian Sandstone | Sandstones | Aquifer |
| Jurassic | Arad Group | Limestones | Aquifer |

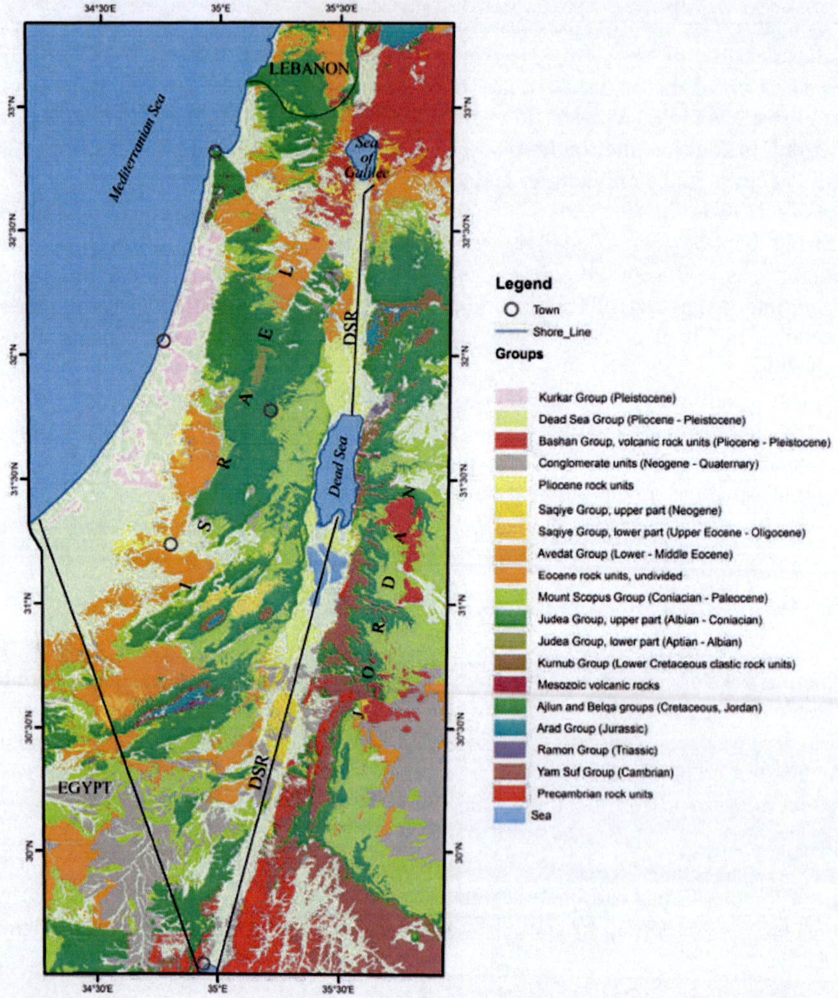


**Fig. 2** Extent of the different lithostratigraphic units in Israel (modified after groups map, Geological Survey of Israel 1998)

mountainous regions of Israel from the Galilee in the north, through the Samariya and Judea mountains in the center to the Negev in the south. The aquifer generally dips to the Mediterranean and the DSR base levels and as a result is confined under an upper Cretaceous to Neogene confining sequence. In places, the aquifer is subdivided into lower and upper sub-aquifers. The aquifer drains to both base levels either directly into the sea or through several springs, most of which are presently managed and exploited.

2. The coastal (Mediterranean) aquifer of the Kurkar Group of Pleistocene age extends from the Gaza strip in the south to the Lebanese border in the north. Its thickness varies from a maximum of roughly 160 m near the shore in the south to a few tens of m in the north. The aquifer also wedges out in the eastern foothill regions. The aquifer consists mainly of detrital calcareous sandstones, subdivided by clay and loam confining layers into four known sub-aquifers in its western portion. Its hydrological potential attains on the average around 300 million cubic meters (MCM) per year. Due to a high water demand, the aquifer has been over-exploited in the last decades and as a result subjected in most of its regions to seawater encroachment.
3. The Sea of Galilee has been until recently a major water resource in the northern part of the DSR. It is fed mostly by the Jordan River and its tributaries, namely the Hermon, Dan and Snir streams. The latter drain groundwater that issues via large springs from the Jurassic Arad Group aquifer is exposed and recharged in Mount Hermon. The average yearly inflow to the Sea of Galilee, which is in origin mainly groundwater, is around 500 MCM. The Sea of Galilee is basically a flow-through freshwater lake. However, some salinity is contributed to the lake water from saline springs that issue on the margins of the lake and at its bottom. The lake water was pumped and diverted via a major national conduit system to the more southern parts of the country at an annual amount of a several hundred MCM.

**Secondary resources**:

1. The Kurnub Group, Nubian Sandstone aquifer of Early Cretaceous age, is known from the subsurface of the southern arid part of the country. Most of its intake area is outside Israel in the Sinai Peninsula. Due to the present-day prevailing arid climate, it was found that most of its water content is paleo groundwater, recharged in the past during more humid climates. The aquifer drains to the Dead Sea and the Red Sea, as well as to the Gulf of Suez in Egypt.
2. The Avedat Group aquitard–aquifer system of Eocene age consists mainly of chalky limestone and chert horizons that constitute an aquitard over most of the country. In certain areas, due to facies changes, this unit passes laterally into a reefal and jointed limestone sequence of a high permeability, essentially forming an aquifer. The Avedat Group aquifer is known mostly from the northern and eastern parts of the country and close to the DSR. The aquifer is perched above the Senonian aquicludal sequence, whereby it is being drained by medium flow size perched springs. Elsewhere, the Avedat Group aquitard is also perched, being drained by several small flow size springs.
3. The basalt aquifers of Neogene to Quaternary age are known in the northeastern part of the country and the Golan Heights close to the DSR. They consist mainly of basalts and agglomerates and are drained by several perched springs. The water contribution of these aquifers to the northern part of the DSR amounts to several tens of MCM, annually.

4. The DSR graben fill aquifers of the Dead Sea Group include the detrital, coarse clastic and sandy aquifers aligned along and adjacent to the DSR. Among these is the Neogene Hazeva sandy aquifer in the south and the Quaternary alluvial, mostly gravel aquifers that were formed as alluvial fans along the DSR, mainly in its southern portion, the Arava Valley. Due to the arid climate that prevails in the south, these aquifers are recharged mainly laterally from the mountainous aquifers on both sides of the DSR as well as by downward percolating seasonal flows along streams that drain to the DSR.

## 4 Salinization and Pollution of the Groundwater System

The natural salinization processes of the different freshwater aquifers in Israel are controlled mainly by its bordering saline base levels. The western Mediterranean and the southern Gulf of Elat marine base levels contribute salinity to their adjoined coastal aquifers, whereby the latter are encroached by seawater. In addition, in places where the JGA is adjacent and hydraulically connected to the Mediterranean base level, it is also encroached upon by seawater.

The DSR base level in the east, however, occupies concentrated brines, since it was intruded in the past by Mediterranean seawater due to the elevation difference between both base levels. Due to the endorheic nature of the DSR, concentrated brines were and are formed which are encroached laterally in the neighboring mountainous and graben fill aquifers. All the described salinization processes are enhanced by the exploitation or over-exploitation of the relevant aquifer systems.

In addition to the above, some of the shallow exposed aquifers are subjected to anthropogenic salinization and contamination processes. These include domestic, industrial and agricultural pollution.

## References

Diskin MH (1970) Factors affecting variation of mean annual rainfall in Israel. Bull Int Assoc Hydrol Sci 15:41–49

Gvirtzman H (2002) Israel water resources. Yad Ben Zvi Press, Jerusalem, 287p (In Hebrew)

Stanhill G, Rapaport C (1988) Temporal and spatial variation in the volume of rain falling annually in Israel. Isr J Earth Sci 37:211–221

# The Yarkon-Taninim Basin—An Example of a Major Carbonate Aquifers in Israel

**Joseph Guttman**

## 1 Background and Purpose of the Article

Carbonate aquifers constitute a considerable portion of the sedimentary sequence in Israel. Among these, the main aquifers are the Jurassic, the Cretaceous and the Eocene aquifers. Their functioning as potential active freshwater aquifers depends on their exploitation depth or in addition on their exposures to natural recharge through precipitation. Regarding the latter, it depends also on whether their recharge area is located in an arid area or a humid one. In the case of a humid regime, the amount of natural recharge to the aquifers depends also on the altitude of their recharge areas which affects the amount of precipitation. In addition, those aquifers do not play a role as freshwater aquifers where and when they host, or are intruded by saline or hypersaline water sources.

The main and most important carbonate aquifers in Israel are as follows.

### *1.1 The Jurassic Arad Group Aquifer*

This aquifer is a deep-seated aquifer found over almost the entire country, which consists mostly of limestones that are partly karstic. The aquifer contains mostly concentrated brines. Only in the northeastern part of the country, in Mount Hermon, the aquifer, some 2000 m thick, was uplifted to elevations up to close to 3 km above sea level. As a result, the aquifer through its recharge area is exposed to considerable natural recharge via rainfall and snowmelt (i.e., Gila'd and Bonne 1990). The upper sequence of this freshwater aquifer feeds the main three Hermon, Dan and Snir springs and in turn the tributaries of the Jordan River at an average annual amount

J. Guttman (✉)
Mekorot, National Water Company of Israel, 9 Lincoln Str., 6713402 Tel Aviv, Israel
e-mail: yosgut@gmail.com



U. Kafri and Y. Yechieli (eds.), *The Many Facets of Israel's Hydrogeology*,
Springer Hydrogeology, https://doi.org/10.1007/978-3-030-51148-7_2

of some 500 million cubic meters (MCM). The latter flow downstream to the Sea of Galilee that serves as the biggest natural water reservoir in Israel. The lower part of this aquifer is assumed to be flowing to deep outlets in Syria, Northern Israel and Lebanon (Burg 2011; Guttman et al. 2012, 2017; Babad 2018).

## *1.2 The Eocene Avedat Group Aquifer or Aquitard*

This hydrogeological unit is of secondary importance regarding its groundwater potential. It extends along the western foothills of the central mountain crest of Israel and close to the margins of the Dead Sea Rift system in the east, as well as in the northern Negev in the south. The aquifer is missing in the central mountain crest of the country. The Avedat Group exhibits lithologically, in general, two lithological facies: (a) chalk and chalky limestone facies throughout most of its occurrences in the west, and thus, is regarded as an aquitard due to its relatively poor hydrological properties. It is, thus, regarded as of secondary importance. (b) a limestone facies, reefal in places, mostly in the eastern part of the country, and considerably jointed, exhibiting a rather high hydrological conductivity and regarded as a good aquifer. However, due to its limited exposed recharge area, mostly close to the Dead Sea Rift in the east and northeastern parts of the country, it is also secondary of importance. The same goes for the northern Negev Avedat Group aquifer due to the arid regime that prevails there.

## *1.3 The Judea Group Carbonate Aquifer*

The Judea Group carbonate aquifer of late Cretaceous age is the most important regional aquifer in Israel. The aquifer consists mainly of dolomites and limestones, in places interbedded by chert-bearing chalk and marl units that subdivide the sequence to sub-aquifers. The total thickness of the aquifer is around 700–800 m all over most of the country except for its southernmost parts where the thickness is reduced approaching the Arabo-Nubian massive. The aquifer extends over most of the country but it is being exposed mainly in its north–south directed-central mountain crest from the Negev in the southward and northward along the Judea and Samaria Mountains as well as in Mount Carmel and in the Galilee Mountains. Its exposure, and being subjected to natural recharge, is related to the uplift of the mountainous region along the Syrian Arc since the Senonian (i.e., Shahar 1994) forming northeast–southwest-directed synclines and anticlines an accompanied by the erosion of the overlying formations. Two later uplift phases, two of which predated the formation of the Dead Sea Rift system in the east (Bar et al. 2016), enhanced the process of exposure. As a result, the Judea Group aquifer vast exposures serve as recharge areas to natural recharge to the aquifer that except of the arid south, serves as the main groundwater reservoir of the country. The aquifer was naturally drained by several springs that are currently managed and artificially exploited by a vast network of exploitation wells. The general flow pattern of groundwater from the elevated recharge areas is being, basically controlled by the existence and location of the western, Mediterranean and the eastern Dead Sea Rift base levels. In addition, it is also influenced by the tectonic system, which includes faults, grabens, horsts, and secondary structures of anticlines and synclines. The tectonic system also determines the location of the natural outlets of the aquifer and affects the flow path and the flow gradients.

The structural axis of the mountain crest, in general, dictates the location of the regional groundwater divide between the so-called Western Aquifer which drains to the Mediterranean base level and the "Eastern Aquifer""which drains to the Dead Sea Rift base level. The "Western Aquifer" in Western Galilee and in the central Judea and Samaria Mountains is the main fresh groundwater resources of Israel.

The Judea Group aquifer was intruded from the west since the Neogene and later on during Pleistocene transgressions by seawater. During subsequent regression periods, the system was flushed by naturally recharged meteoric water. The meteoric water replaced the salt water in the carbonate formation. Nevertheless, there are few areas in the western foothills and close to the sea where the aquifer still contains unflushed or currently intruded seawater. Similarly, the Judea Group aquifer was intruded in the past and is still intruded by concentrated brines that occupy the Dead Sea Rift base level. Thus, proper management of water resources in each of the aquifer basins is based on a deep understanding of the geological, hydrological and salinization hazards that may be caused by different operating regimes.

This chapter focuses on part of the main carbonate aquifer of Israel, namely the Western Mountain Basin of the central mountain ridge, named as the "Yarkon-Taninin Basin," which constitutes the major part of the Judea Group aquifer of the western basin. It was chosen, herein, as an example for the others carbonate aquifers in Israel due to its importance to the water supply system, to its hydrogeological complexity and to the considerable abundance of geological and hydrological data.

The "Yarkon-Taninin Basin" is an important component of the Israel National Water supply system, and the pumping from this basin is influenced by its hydrological conditions as well as by the abstraction from the other water supply sources of the Israel National Water system (desalinated seawater, the Coastal Aquifer and Lake Kinneret-Sea of Galilee).

## 2 Hydrogeology of the Yarkon-Taninim Basin

This chapter summarizes the conceptual hydroglogical model of the "Yarkon-Taninin Basin" and its groundwater management policy of this basin. It is a unique example how to manage a large carbonate and karstic aquifer by combining deep understanding of the geology, hydrology and geochemistry of water sources with local and regional water supply needs.

A schematic geological cross section (Fig. 1) across the central mountain region in Israel exhibits the western and eastern basins and the anticlinal crest in between.

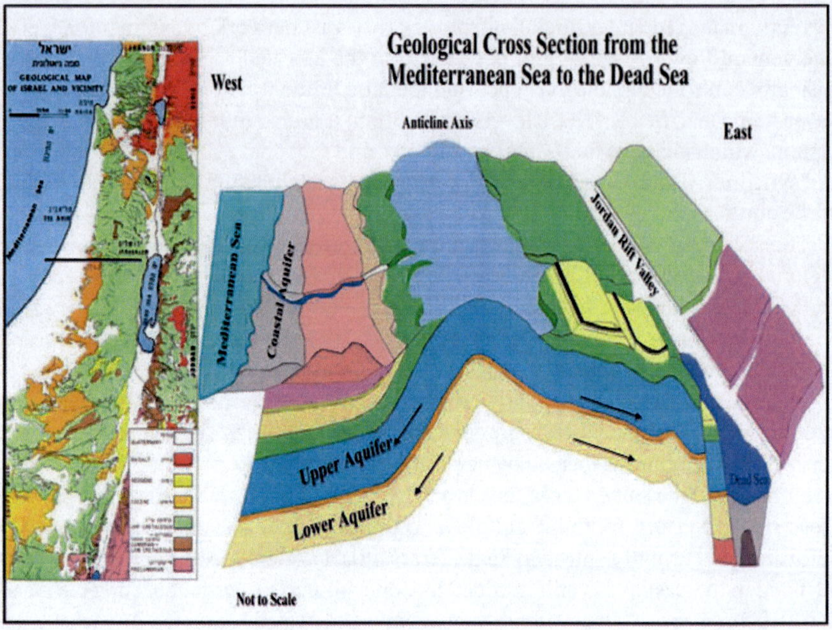


**Fig. 1** Schematic geological cross section through the central mountain region in Israel showing the western and eastern basins and the anticlinal crest

The groundwater divide that generally coincides with the crest of the anticlinal axis divides the system to the separate western and eastern hydrological basins.

The Yarkon-Taninim (Western Mountain Aquifer) Basin extends over the western flank of the central mountain ridge of Israel. This aquifer is one of the three largest water reservoirs in Israel and has the highest water quality.

The boundaries of the basin are from Nitzana in the south to Yoqne'am area in the north and from the coastal area in the west to the mountain crest in the east (Fig. 2). The size of the basin is around 10,000 $km^2$.

The hydrogeology of the basin was studied and described in several articles (i.e., Weinberger et al. 1994) and reports. A summary of the latter is found in Dafny (2009) and Dafny et al. (2010). Therefore, only elements that are important for the understanding of the general hydrological conceptual model and to the management of this aquifer are mentioned herein.

According to the above-mentioned studies, the aquifer consists mainly of limestone and dolomite layers with interbedded layers of chalk, marl and clay. The total thickness of the aquifer varies from 800 m in the northern and central part, to about 500–550 m in the southern part of the basin. The aquifer is divided into three hydro-stratigraphic units:

Lower unit (KUJ1), consisting of dolomite and limestone, partly karstic of high permeability. In its southern part, it is also interbedded by thin layers of shale. The unit is characterized by a moderate to high hydraulic conductivity.

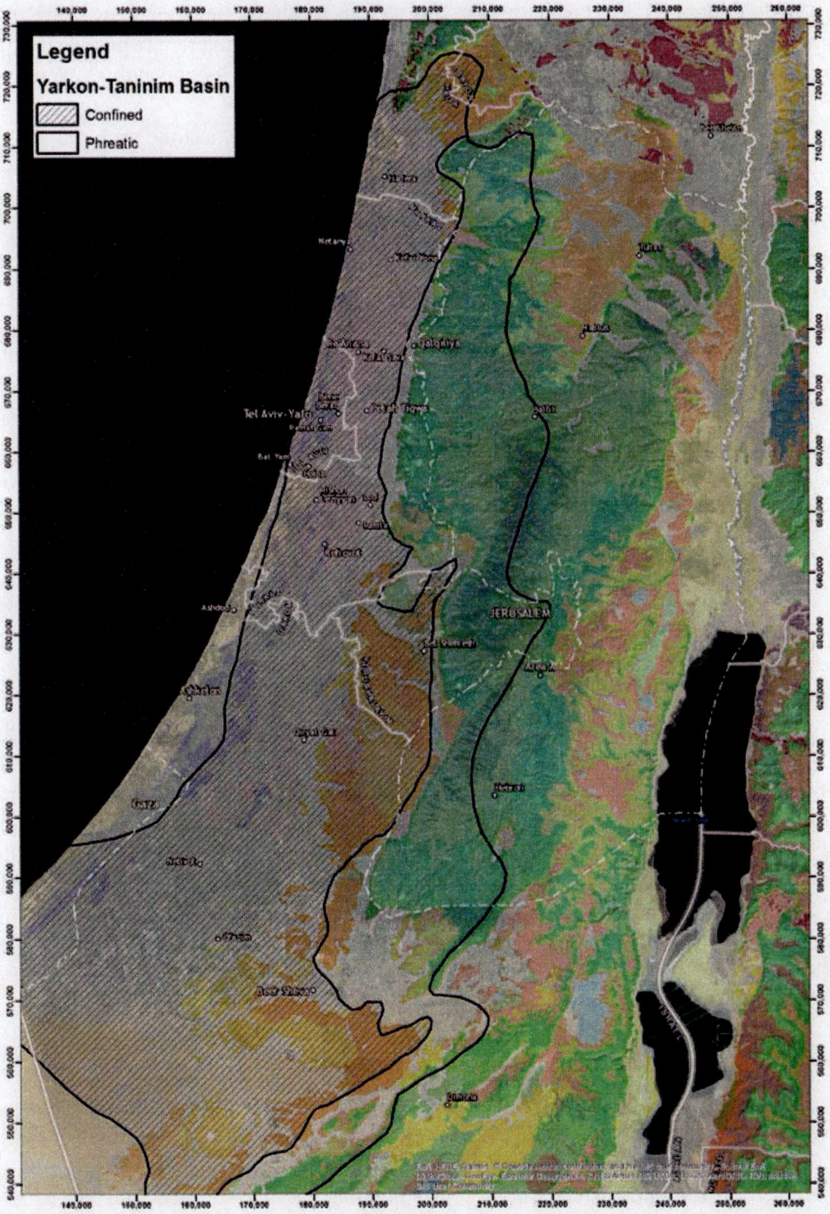


**Fig. 2** Boundaries of the “Yarkon-Taninim Basin“

Middle unit (KUJ2), consisting of alternations of chalk, limestone and marl and sometimes dolomite. The unit is characterized by a low to moderate conductivity. In most of the basin, it is practically an aquiclude that separates between the lower unit and the upper one.

Upper unit (KUJ3, KUJ4), consisting of dolomite and limestone with abundant karst phenomena and thus characterized by a high conductivity.

Drilling results accompanied by analyses of geophysical logs from many wells and television profiles in some of them, show that the aquifer, whose thickness reaches several hundred meters, contains a large number of joints, cracks and karst cavities. All the above revealed karstic cavities of several meters height together with a denser network of joints and cracks.

The locations of the cracks and the significant karstic caves vary between adjacent wells and along the aquifer section. It was found, as expected, that there is a direct relationship between the quantity and the size of the cracks and the karstic caverns along specific well profiles and the hydraulic properties (transmissivity and specific discharge) obtained from their pumping tests. Wells typical of a dense system of cracks and karstic caves indeed reveal a relatively high hourly yield with a small drawdown and a high transmissivity at the range of thousands to tens of thousands of $m^2$/day and vice versa (Table 1 and Fig. 3).

Yechieli et al. (2009) claimed that in some regions, the lower sub-aquifer in this basin is less conductive than the upper one as was simulated in their model. Their results revealed, in general, a higher specific discharge and transmissivity values typical to the upper sub-aquifer as compared to the lower one.

As mentioned before, the transmissivity value and the specific discharge in each well are combination result of the aquifer properties (the amount and the dimension of the karst caves and cracks and the thickness of saturated layers) together with the well construction (diameter, length of the open aquifer section and the percentage of the screen open area) in each well. It is usually difficult to determine based on joints and caverns abundance in each well, its preferred conductive portion of the sequence. Therefore, as a common rule in Israel, the length of the production section in each well (open screen) in carbonate aquifers is planned to be chosen between 100–200 m.

The distribution of the transmissivity values from selected wells in the aquifer area, based on the data of Table 1, is exhibited in Fig. 4. It is evident that the transmissivities are, in general, lower at the mountainous areas in the east and are considerably higher toward the foothills region in the west. In some cases, local high transmissivity values are found also in the high elevated areas. The same transmissivity distribution was modeled and described by Yechieli et al. (2009).

Based on the hydro-stratigraphic classification, the water levels and the water quality, the entire aquifer, as mentioned before, are sub-divided into two sub-aquifers, namely an upper sub-aquifer and a lower one, whereby the middle unit separates between the two sub-aquifers (Guttman 1991, 1998, 2002). However, based on water balance considerations, it is quite clear that the middle unit is acting in several parts of the basin as an aquiferous unit that forms a hydraulic connection between the two sub-aquifers (Guttman 1988).

**Table 1** Hydrological data from pumping tests in selected wells

| Name | Sub-aquifer | Q (max) ($m^3$/h) | Drawdown (m') | Specific discharge ($m^3$/h/m') | Transmissivity ($m^2$/day) |
|---|---|---|---|---|---|
| M-Sade 1a | Upper | 324 | 8 | 40.5 | 11,535 |
| Omer 1 | Upper | 210 | 2.4 | 87.5 | 15,400 |
| B-Sheva 1a | Upper | 269 | 1.1 | 244.5 | |
| Nechusha 1a | Upper | 720 | 11 | 65 | 2100 |
| Ein Karem 12 | Lower | 58 | 95.5 | 0.61 | 9.1 |
| Ein Karem 9 | Lower | 350 | 3.75 | 96 | 6160 |
| Ein Karem 6 | Lower | 84 | 54.8 | 1.5 | 105.6 |
| Eshtahol 5 | Lower | 405 | 11.3 | 35.8 | 4245 |
| Eshtahol 8 | Lower | 665 | 8.67 | 76.7 | 17,150 |
| Achisemech 1 | Lower | 1148 | 5.5 | 215 | 11,842 |
| Lod 32 | Upper | 1718 | 4.77 | 360.2 | 492,500 |
| Lod 17a | Upper | 648 | 0.22 | 2945 | |
| Yarkon 9 | Upper | 1056 | 29.2 | 36.2 | 18,586 |
| R. Haayin 10 | Upper | 1825 | 7.19 | 253.8 | 88,815 |
| R. Haayin 12 | Lower | 1501 | 11.09 | 135.3 | 21,821 |
| Neve Yarak 2 | Upper | 2458 | 12.75 | 192.8 | 107,665 |
| S-Deromi 103 | Upper | 2370 | 19.05 | 124.4 | |
| S-Zefoni 203 | Upper | 2028 | 11.34 | 178.8 | 425,300 |
| Kakun 5 | Upper | 280 | 1 | 280 | 5500 |
| K-Shomrom 1 | Lower | 195 | 31.7 | 6.15 | 238 |
| Ariel 1a | Lower | 215 | 79.6 | 2.7 | 80 |
| Tut 4 | Upper | 638 | 8 | 80 | 12,848 |
| S-Menashe 1 | Upper | 1300 | 23.5 | 55.2 | 10,000 |
| Maanit 5 | Upper | 1440 | 3.9 | 370 | |

*Source* Mekorot Water Company archive

When dividing the aquifer's outcrops and recharge areas according to their belonging to the lower or upper sub-aquifers, it turns out that most of the natural recharge (about 70%) finds its way to the lower sub-aquifer. It takes place in the mountainous region through the exposures of the lower sub-aquifer as well as through the exposures of both saturated and unsaturated upper sub-aquifer via the connection areas (Fig. 5). At the same time, most of the pumping is carried out through wells that penetrated the upper sub-aquifer. Since the considerable exploitation of the upper sub-aquifer is not expressed by a parallel drop of the water level, it demands a compensation mechanism which contributes groundwater from the lower, into the upper sub-aquifer at the rate of at least tens of million cubic meters (MCM) in order to balance the pumping and springs flow of the upper sub-aquifer.

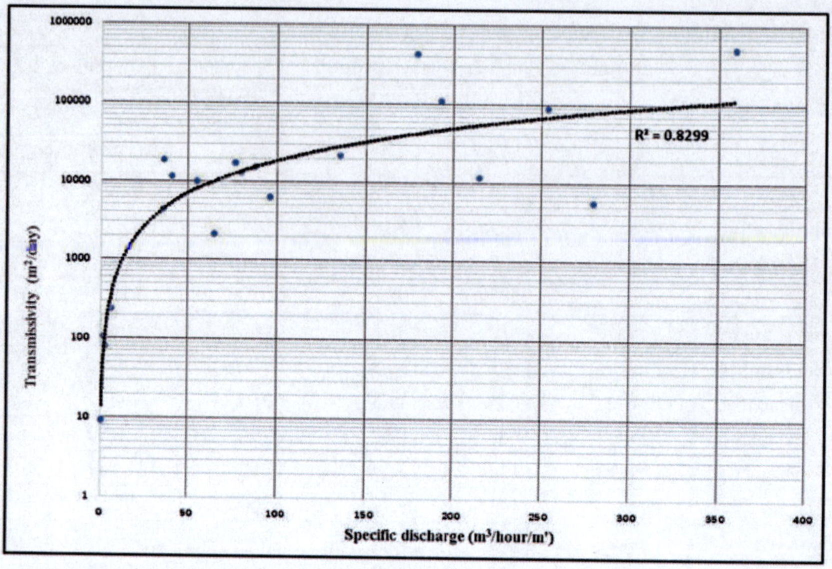


**Fig. 3** Correlation between the specific discharge and transmissivity

The massive amount of water that potentially flows from the lower sub-aquifer into the upper one requires large connection areas. The delineation of the connection areas in the entire basin is essential to any 3D flow model that attempts to simulate the water passage from the lower to the upper sub-aquifer (Guttman 1988), and the results (Fig. 5) are valid until today.

According to the above described setup, and due to regulatory reasons aimed to maintain protection zones, the current policy is to drill the new wells west to the foothill, in the phreatic part of the aquifer, mainly into the deep sections of the upper sub-aquifer and\ or to the lower sub-aquifer.

The phreatic region extends from the groundwater divide in the east to about 1–3 km west of the foothills region (Figs. 4 and 5). More to the west, the aquifer is overlain by Senonian confining layers, forming confined conditions, whereby the water level is above the top of the aquifer. Most of the pumping takes place through wells located near the foothill and a few kilometers west of the confining line, which delimits the passage from phreatic to confined conditions.

The first wells were drilled close to the foothills as early as in the 1940s. Accelerated activity of drilling and development of the pumping capability from the aquifer had taken place during the fifties and sixties of the twentieth century. The massive pumping in the central part of the basin and close to the Yarkon springs dried up these springs around the mid-sixties.

Most of the wells that were drilled 50–60 years ago penetrated only a few tens meters into the upper sub-aquifer. Most of those oldest wells were drilled in percussion and in old rotary methods. In those wells, the uppermost unsaturated zone, up to

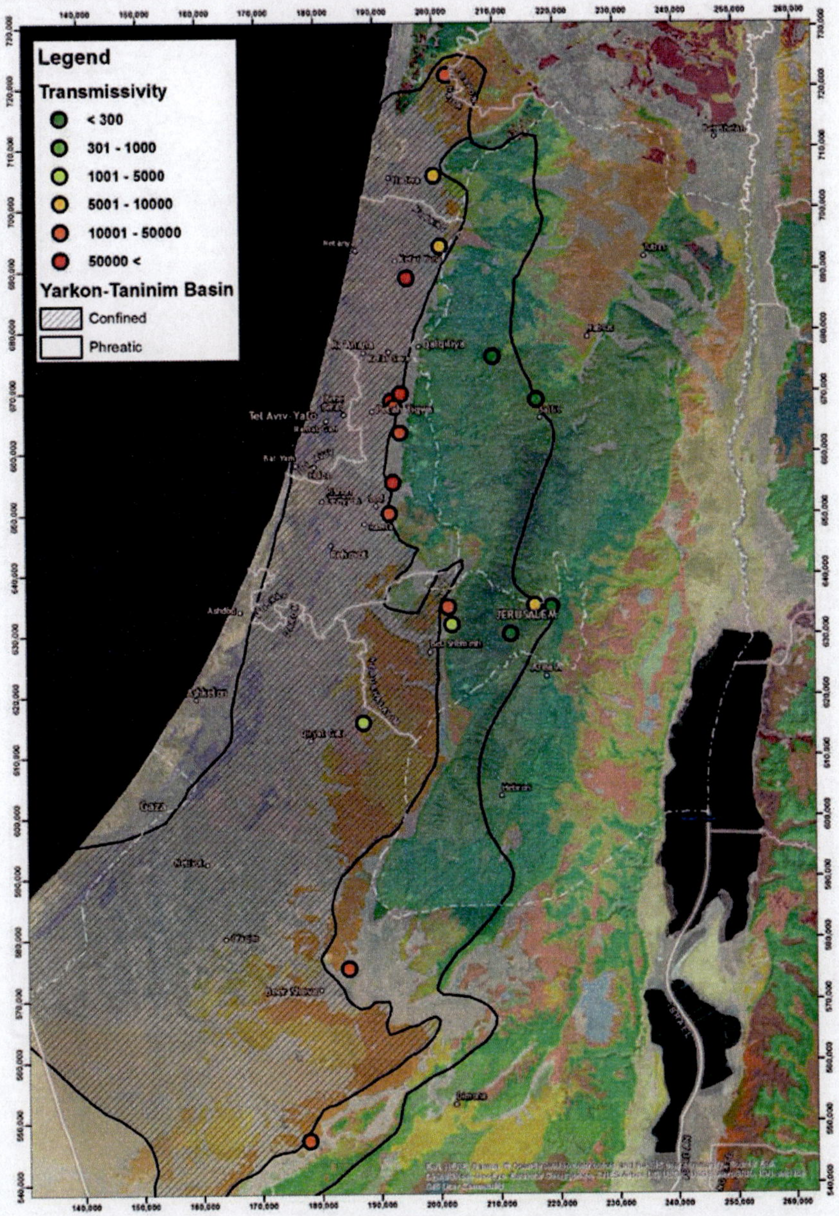


**Fig. 4** Transmissivity values in the "Yarkon-Taninim Basin" (based on data in Table 1)

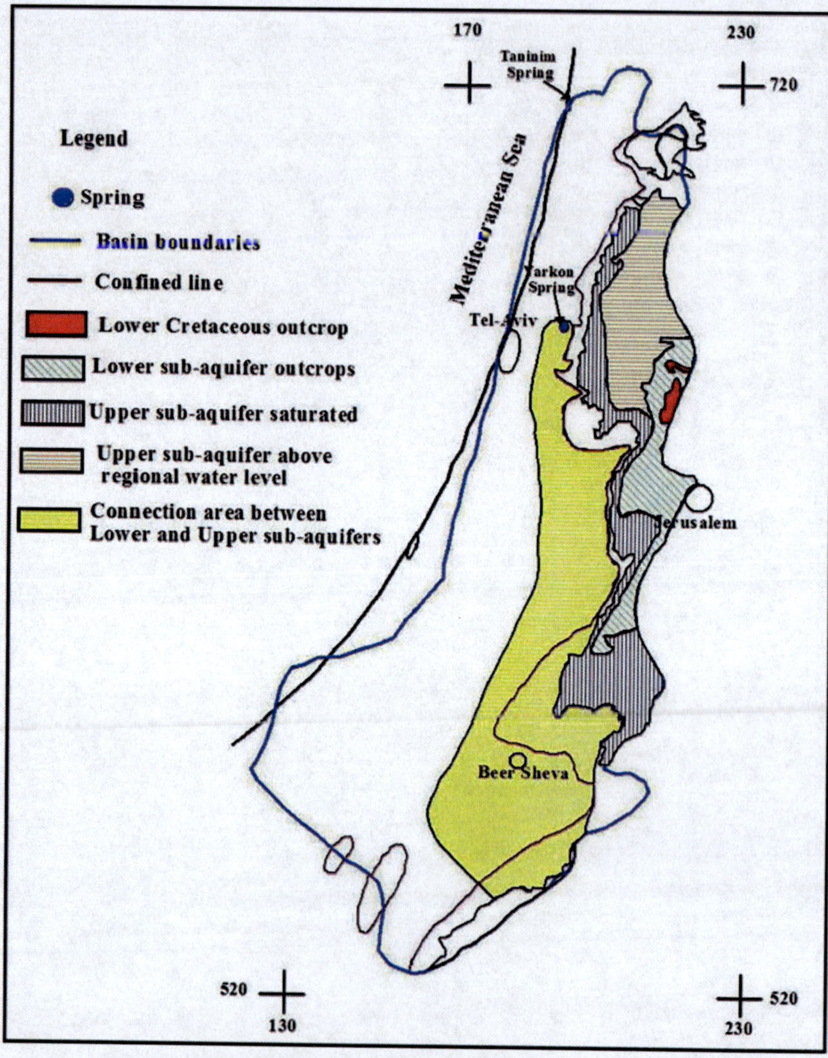


**Fig. 5** Location of the "connection areas", after Guttman (1988)

few meters beneath the static water level, was cased by steel casing without cementation and proper isolation of the aquifer from the surface. Some of the oldest wells are, thus, not properly protected from surficial potential contamination. However, contamination events, resulting from leaking of pollutants from the surface, are found to be quite random and explained by the high hydraulic conductivity of the aquifer, and the resultant rapid dilution of the contaminants.

Most of the natural recharge to the aquifers occurs in the mountainous area. The water levels along the high mountain ridge range from 250 m asl, in the Ariel area, to 400–450 m asl in the Ein Karem well field near Jerusalem. Near the foothills, in the west, water level elevation levels are currently between 10 and 15 m asl.

The groundwater flow is from the mountainous area downslope, being controlled by the major geological structures, bypassing the structural obstacles. The water level differences between the mountainous area and the foothills area are some 250–450 m, over a distance of 15–20 km only. In order to maintain the higher water table in the mountainous area and to account for such a steep gradient, it is required to have in between some low hydraulic conductivity strips that act as a hydrological barrier. The location of few potential hydrological barriers is known, based on geology and drilling results. Still, there are large areas in the high mountainous area, where water level data and the locations of the hydraulic parameters are scarce or nonexistent. In this regard, Meiri and Guttman (1984) and Yechieli et al. (2007) have modeled and delimited such a low conductive "barrier" between the Ein Karem well field (close to Jerusalem) and the foothills, which coincides with the western regional steep flexure of the anticlines structure. Across this "barrier," indeed, the water level drops from around 300 m to 15 m asl in the foothill region along a few kilometers distance, exhibiting a very steep groundwater flow gradient.

Upon reaching the foothills and the confined part of the basin, the westward groundwater flow paths are diverted, flowing northward to the natural outlets of the Yarkon springs in the central part and the Taninim springs in the northern part of the basin, as shown in Fig. 6 (Dafny 2009; Dafny et al. 2010).

From the foothills region, to the west and farther to the north, the water level in the phreatic and in the confined parts exhibits a very mild gradient. The drop of 3–5 m over about 150 km is indicative of a very high transmissivity (Fig. 4). In the confined part of the system, the storativity, as expected, is very small, typical to confined conditions. The natural outlets prior to the exploitation of the aquifer were mainly the Yarkon and Taninim springs (Fig. 5). The Yarkon springs in the center of the basin yielded an estimated yearly flow between 220 and 240 MCM/year, and the Taninim springs in the northwestern margin of the basin yielded a yearly discharge around 110 MCM/year.

The Yarkon springs are in practice overflow springs, which resemble an upper drainage tank, sensitive to changes in its water head. In the 1960s, due to the massive pumping and continuous droughts, the water level in the aquifer declined beneath elevation of the spring's outlet and the springs flow ceased. The flow returned after the extreme rainy winter of 1991/92, when the water levels in the entire basin raised by more than 10 m within 3 months.

The Taninim springs, located on the northwestern edge of the basin, serve as the lower outlet of the aquifer. Despite the intensive pumping upstream, they are still flowing, but with a very low discharge.

The forcing head of the Taninim springs was determined during the calibration of several models (Guttman and Zukerman 1995; Guttman and Zeitoun 1996) at an elevation height of 3.6-4.0 m asl. The water level in the nearby pumping fields located some 7–10 km from the springs (Pardes Hana area), vary between 9 and 14 m asl, depending on the natural recharge and the pumping regime. Figure 7 shows the water level in Karkur Pardes Hana well fields and the parallel Taninim springs discharge.

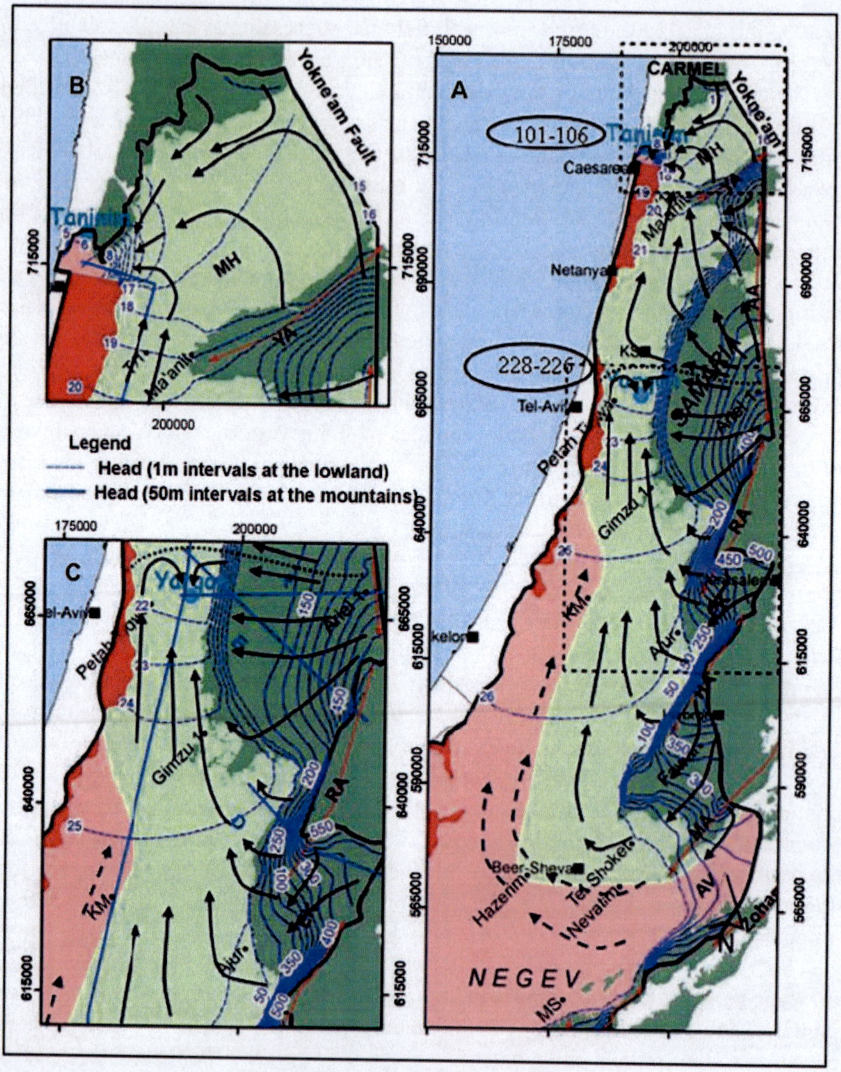


Fig. 6 Regional flow path pattern (used by permission of Elsevier from Dafny et al. 2010, J Hydrol 389:260–275, Figs. 1, 5)

The discharge of the Taninim springs will assumingly continue to drop but will not be completely dry up, like the case of the Yarkon springs.

The average calculated annual recharge into the aquifer for the period 1952–2017 is 346 MCM, whereby the average for 1995–2017 is only 288 MCM (Fig. 8). The reduction of the average annual recharge during the latter period is due to several cycles of dry years. The standard deviation is about 30% and the drop is related to

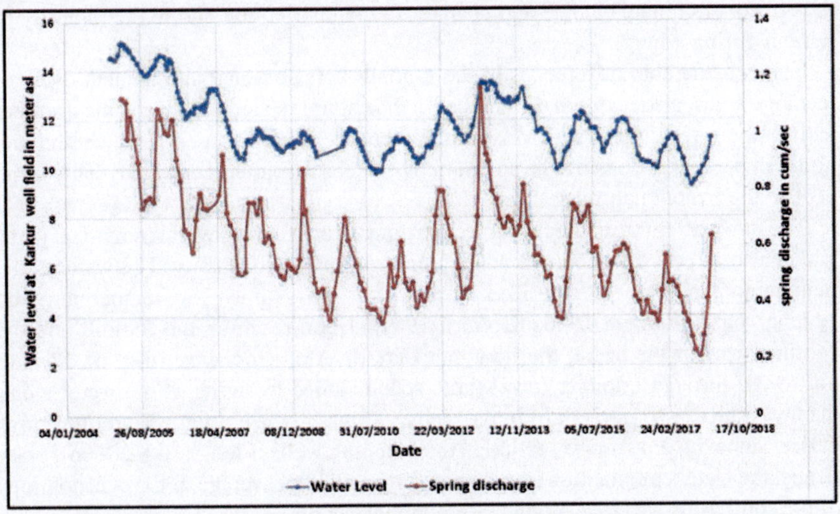


**Fig. 7** Water level in the Karkur-Pardes Hana well field and the Taninim spring discharge

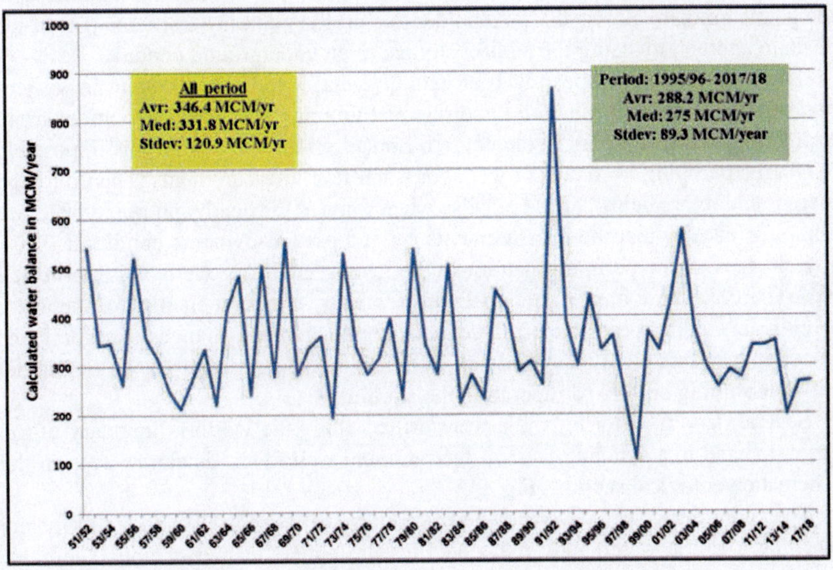


**Fig. 8** Calculated natural recharge (Guttman and Zukerman 1995; Guttman, personal calculation)

the effects of climate change and changes in the distribution and in the intensity of rainfall during winter.

The considerable differences in the rainfall, both seasonal and annual, over the recharge areas, also result in differences in the natural recharge between the northern and southern recharge areas. Most of the recharge (90%) occurs in the central and northern parts of the basin. The southern part (Hebron region) contributes only some 10% to the overall natural recharge.

The flow pattern from the recharge area downstream is via several individual paths which also affects the chemical water composition and its quality. The water at the mountainous area is characterized by relatively low salinity, low temperature and "young" age (Guttman 1980; Kroitoru 1987). Upon reaching the foothill and the confined part of the basin, the flow paths are diverted clockwise (Fig. 6), flowing northward through a longer travel time, accompanied by aging of the water along the flow path down-gradient from the recharge areas to the northern Taninim springs outlet. Since the production section in the pumping wells is long (100–200 m), practically, the water composition in each pumping well is an average of the contributions of the aquiferous horizons along the production section.

The aquifer is acting like a huge reservoir. The water level gradient is about 5 m over approximately 150 km. The high transmissivity of the aquifer together with the gentle gradient allows the stakeholder to shift the pumping from one part of the basin to another, by using the aquifer storage as an underground conduit.

The high transmissivities together with the gentle gradients all over the western part of the basin result in similar and regional groundwater level fluctuations as an expression of the pumping in the entire basin (Fig. 12). Practically, it is impossible to halt the pumping from tens of wells for water level measurements. Therefore, the measurements are being carried out also when some of the nearby pumping wells are pumping, namely that the measurements are at a pseudo-dynamic condition. Thus, any changes of the pumping pattern in the nearest pumping wells adjacent to the observation well are immediately been expressed by temporal changes of the water level by few tenth of centimeters. Due to the gentle gradient, changes of several tens of centimeters are significant and can create errors when calibrating a flow model and determining criteria for operating the aquifer.

Several deep monitoring wells were drilled along the western boundary of the basin, located in a syncline west of the pumping well fields, as can be seen in the schematic geological section (Fig. 9).

The monitoring wells were drilled to depths between 1000 and 1400 meters reaching a saline water body that underlies the upper freshwater one. The chlorinity of the saline water was found to be between 17,000 and 20,000 mg/lit. Based on chemical and isotopical considerations, the saline water is interpreted as ancient seawater that had penetrated the aquifer and remained stored in the deep western syncline (Kafri and Arad 1979; Guttman 1980; Guttman et al. 1988; Paster et al. 2005; Burg and Talhami 2014).

The Israel Hydrology Service (IHS) is continuously monitoring the level of the interface between the freshwater and the saline water that is located beneath the

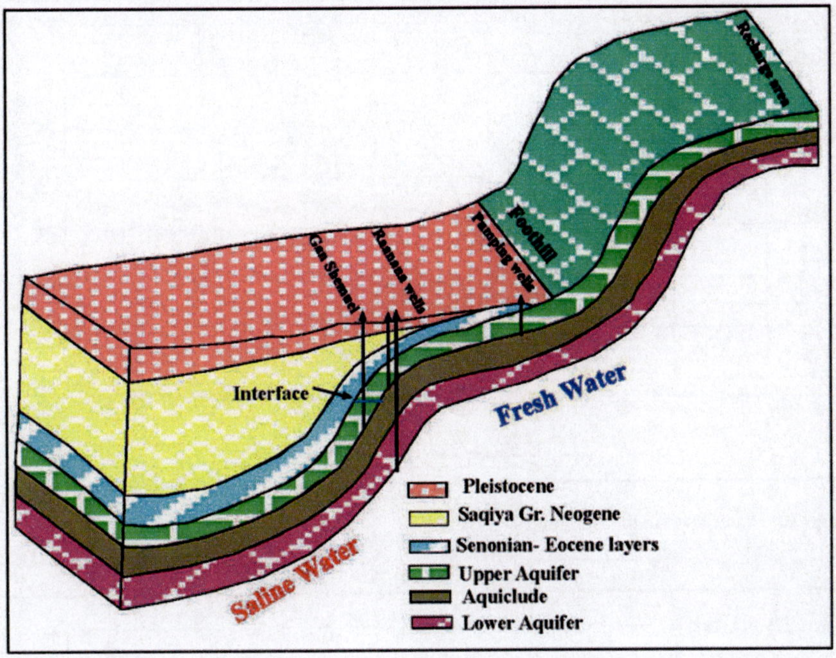


**Fig. 9** 3D schematic geological section with the locations of the pumping well fields and the deep monitoring wells

freshwater body. Movement of the interface level is part of the management criteria of the aquifer.

The water level fluctuations of the fresh and saline water bodies are identical (Fig. 10), indicating a hydraulic connection between them. The brackish water of Taninim springs is a mixture of the fresh and the saline water. During the last decades, the salinity of the Taninim springs increased parallel to the drop of the spring's flow due to intensive pumpage upstream (Fig. 11). Moreover, a sharp rise of salinity was evident after the extreme rainy year of 1991/92. Namely, high precipitation and spring flow resulted in a mechanism that raised the saline water by the driving head of the overlying fresh water body, and as a result increased the spring water salinity. This same phenomenon and mechanism was described for the Taninim springs also by Smoler (1975) and by Kafri and Arad (1979). The same phenomenon was described for the nearby Na'aman spring and basin by Arad et al. (1975) and Kessler and Kafri (2007).

The long-term decline in the spring's flow and the salinity rise caused by the pumping of the freshwater upstream thus reduces the amount of freshwater that feeds the Taninim springs.

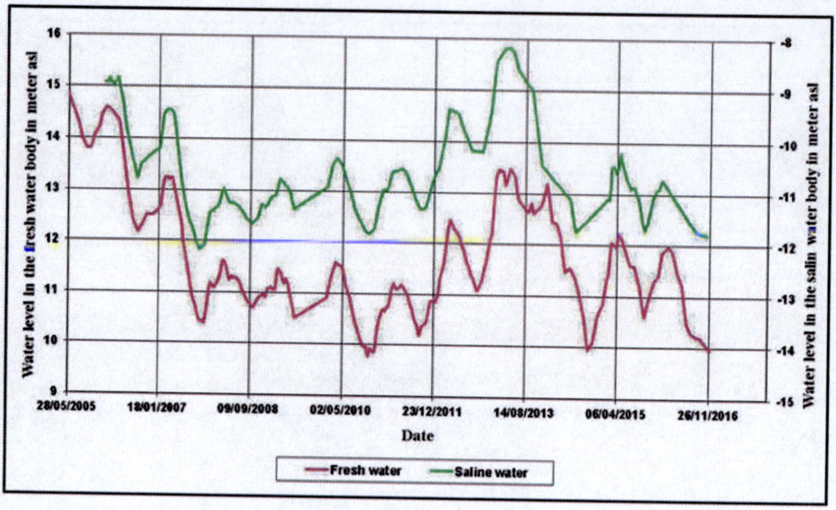


**Fig. 10** Water level in the fresh and saline bodies

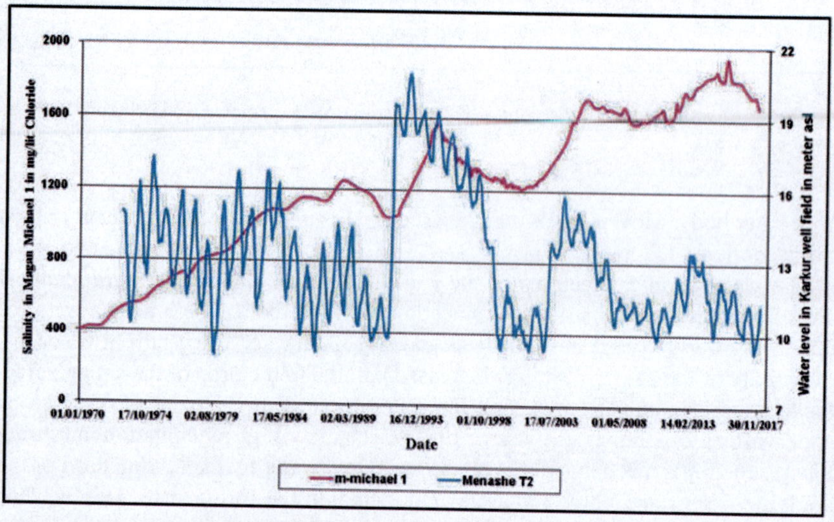


**Fig. 11** Salinity of the Taninim spring and the water level in the pumping well field. The salinity is measured in the nearby Ma'agan Michael 1 well

## 3 Management Issues and Policy

The current main resources of the Israel National Water (INW) system are desalinated seawater and groundwater from natural resources (Yarkon-Taninim Basin, Coastal Aquifer and Lake Kinneret). Prior to the entry of desalinated seawater to the system in 2005, the Yarkon-Taninim Basin served as the most significant source of water for the national system.

The operational policy of the Yarkon-Taninim Basin benefits from its high hydraulic properties (Table 1 and Fig. 4) coupled with its gentle water level gradient, which prevail in its western part. It enables to operate the aquifer as a wide and regional underground system and transfer water by shifting the pumping from one part of the basin to other parts according to the water supply demands, while changing the direction of the groundwater flow and using the aquifer as a reservoir and as an underground water carrier.

In order to reach these capabilities, large amounts of wells were drilled with a total hourly pumping capacity almost twice that of the average natural recharge, enabling the flexibility in the pumping regime.

The addition of desalinated seawater into the INW changed the management policy of the pumpage from the "Yarkon-Taninin Basin." This additional contribution is supplied currently to the INW system continuously throughout the year in quantities determined in the agreements signed by the State of Israel and each sea desalination plant. During the peak high demand in summer, the completion comes mainly from the Coastal Aquifer and the "Yarkon-Taninin Basin." According to the current operational policy, the total pumpage from the aquifer occurs in a short time span mainly during the summer time.

Recently, most of the pumpage from the basin exploits the upper sub-aquifer, and the over-pumping from the upper sub-aquifer is compensated by the overflow from the lower sub-aquifer. Currently, the optimal management is to sub-divide the pumpage between the two sub-aquifers, namely to increase the pumpage from the lower one which in many areas has a better water quality, since it is protected from potential pollution from above. This is planned to be done gradually, accompanied by the shifting of part of the pumpage to areas close to the foothills. This, in turn, may lead to some decline in water levels, but in the long-run will enable to utilize water also from the phreatic zone located east of the main well fields in the west and near the foothills.

Decades ago, two minimal operations' water threshold levels (red lines) were determined. At the northern well field, the red line was determined at an elevation of 9 m asl and in the central well field at 12 m asl (Fig. 12). The red lines were set so as to preserve a water level gradient from the central part northward as well as to ensure a certain volume of fresh groundwater flow from the central area to the northern outlet of the Taninim springs.

The new outcomes from the deep monitoring wells that were drilled since 2002 along the western boundary of the basin, show that the water level fluctuations of the saline body coincide with those of the freshwater body (Fig. 10). This behavior differs from the regular "mirror shape" setup of water level and salinity found in coastal aquifers. Here, the head of the saline water body is being affected by the head of the freshwater body as described before. Therefore, the correct red line elevations are required to be at an elevation that will maintain a dynamic equilibrium between the freshwater body and the saline water one and thus prevent the emergence of the saline water body toward the freshwater one.

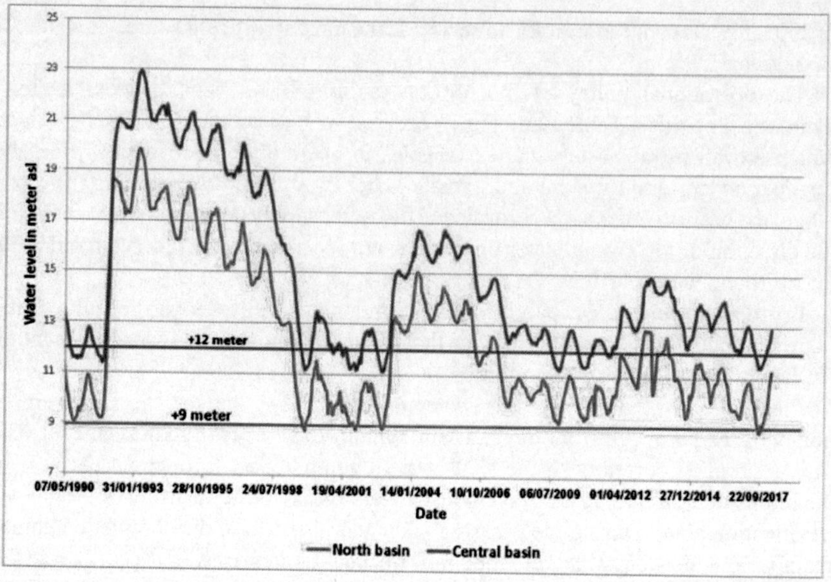


**Fig. 12** Water level in few observation wells in relation to the red lines

Due to climate changes and the decline of the natural recharge in recent years, the water levels in the center and north of the basin are most of the time close or lower than the red lines (Fig. 12). Based on gathered knowledge and experience regarding the effect of the pumpage on the water level changes and water quality, it is assumed that it is possible to lower temporarily the current red lines by 0.5–1.0 m during drought cycles only, which will add some 50–100 MCM to the system during such time spans.

## References


Arad A, Kafri U, Fleisher E (1975) The Na'aman springs, northern Israel, salination mechanism of an irregular freshwater-seawater interface. J Hydrol 25:81–104

Babad A (2018) Regional groundwater flow regime into Hula Valley. Ph.D. thesis, Ben-Gurion University (in Hebrew)

Bar O, Zilberman E, Feinstein S, Calvo R, Gvirtzman Z (2016) The uplift history of the Arabian Plateau as inferred from geomorphologic analysis of its northwestern edge. Tectonophysics 671:9–23

Burg A (2011) Prelaminary geochemistry survey in the water resources in North-East Hula Valley. Geol Surv Israel Rep. GSI/04/2011

Burg A, Talhami P (2014) Yarkon Taninim monitoring project: water quality and isotopes compositions in monitoring wells. Progress report 8th year. Geol Srv Israel GSI/17/2014 (in Hebrew)

Dafny E (2009) Groundwater flow and solute transport within the Yarqon-Taninim Aquifer, Israel. Ph.D. thesis, Hebrew Univ. Jerusalem (in Hebrew)
Dafny E, Burg A, Shffer N, Weinberger G, Gvirtzman H (2009) The dynamic groundwater flow field at the central Yarqon-Taninim basin, Israel. A 3D geological based numerical model. Isr J Earth Sci 58:99–111
Dafny E, Burg A, Gvirtzman H (2010) Effects of Karst and geological structure on groundwater flow: The case of Yarqon-Taninim Aquifer, Israel. J Hydrol 389:260–275
Gila'd D, Bonne J (1990) The snowmelt of Mt. Hermon and its contribution to the sources of the Jordan River. J Hydrol 114:1–15
Guttman J (1980) Study of Uranium Isotopes in the Cenomanian Turonian aquifer in Yarkon-Tanninim Basin. M.Sc. thesis, Tel-Aviv University (in Hebrew)
Guttman J (1988) Simulation of groundwater flow and salinity at the Yarkon-Taninim basin using dual-layer model. 1st progress report. Tahal Ltd., Rep. 01/88/23, 17p (in Hebrew)
Guttman J (1991) Availability to increase pumping at the Beer-Sheva basin. Tahal Ltd., Rep. 01/91/08, 1p (in Hebrew)
Guttman J (1998) Defining flow systems and groundwater interactions in the multi-aquifer system of the Carmel Coast region, Israel. Ph.D. thesis, Tel-Aviv University
Guttman J (2002) Exploitation of brackish water for desalination at Or-'Akiva-Caesarea area. Mekorot Ltd., Rep. 756, 9p (in Hebrew)
Guttman J, Zeitoun D (1996) Flow and salinity model in Northern part of Yarkon-Taninim and Taninim springs. Tahal Ltd. Rep. 6106-96.382, 65p (in Hebrew)
Guttman J, Zukerman H (1995) Yarkon-Taninim—Beer Sheva groundwater basin: setting and calibrating flow model. Tahal Ltd. Rep. 01/95/72, 37p (in Hebrew)
Guttman J, Mercado A, Michaeli A, Baida U (1988) Analysis of the findings in Menashe T/3 well and influence on the monitoring and management of the Yarkon Taninim aquifer. Tahal Ltd., Rep. 01/88/08, 25p (in Hebrew)
Guttman J, Berger D, Burg A, Gev I (2012) Aquifers and groundwater flow pattern in Upper Galilee-Conceptual model and calibration of cell model. Mekorot Ltd. Rep 1582 and Geol Surv Israel Rep GSI/25/2011 (in Hebrew)
Guttman (2017) Gome 2 well. Summary of drilling process. Mekorot report no. 1700 (in Hebrew)
Kafri U, Arad A (1979) Current subsurface intrusion of Mediterranean seawater - a possible source of groundwater salinity in the Rift Valley system, Israel. J Hydrol 44:267--287
Kessler A, Kafri U (2007) Application of a Cell Model for operational management of the Na'aman groundwater basin, Isr. Isr J Earth Sci 56:29–46
Kroitoru L (1987) The characterization of flow systems in carbonatic rocks defined by the ground water parameters: Central Israel. Ph.D. thesis, Weizmann Institute of Science, Rehovot, Israel
Meiri D, Guttman J (1984) Reconstruction by flow model of the flow pattern of the Jerusalem effluents that leak into the Judea Group aquifer in the Soreq wadi. Tahal Ltd., Rep 01/84/52 (in Hebrew)
Paster A, Dagan G, Guttman J (2005) The salt-water body in the northern part of Yarqon-Taninim aquifer: field data analysis, conceptual model and prediction. J Hydrol 323(1):154–167
Shahar J (1994) The Syrian arc system: an overview. Palaeo Palaeo Palaeo 112:125–142
Smoler B (1975) A study of the salination mechanism in the Taninim springs. M.Sc. thesis. Hebrew University, Jerusalem (In Hebrew)
Weinberger G, Rosenthal E, Ben-Zvi A, Zeitoun DG (1994) The Yarkon-Taninim groundwater basin, Israel: case study and critical review. J Hydrol 161:227–255
Yechieli Y, Kafri U, Wollman S, Lyakhovsky V, Weinberger R (2007) On the relation between steep monoclonal flexure zones and steep hydraulic gradients. Ground Water 45:616–626
Yechieli Y, Kafri U, Wollman S, Shalev E, Lyakhovsky V (2009) The effect of base level changes and geological structures on the location of the groundwater divide, as exhibited in the hydrological system between the Dead Sea and the Mediterranean Sea. J Hydrol 378:218–229

# The Coastal Aquifers of Israel

# Introduction to Studies of Coastal Aquifers of Israel

**Yoseph Yechieli**

This chapter deals with several aspects of coastal aquifers, including methods of monitoring (levels and salinities) and processes that determine the chemical and isotopic composition of groundwater. The chapter describes the interrelation between the sea and the aquifer, mostly in the direction from the sea toward the land.

This part is divided to several chapters according to several relevant topics. Chapter "General Information and Hydrogeology of the Mediterranean and Dead Sea Coastal Aquifers and Their Relation with Their Base Level" deals with the general information and hydrogeology of the Mediterranean and Dead Sea coastal aquifers. This sub-chapter also deals with several methods of monitoring of the variation in salinity of the aquifer from different sources, mainly regarding seawater intrusion. Chapter "Dynamic Relationship Between the Sea and the Aquifer" deals with the dynamics relation between these two coastal aquifers and the sea. Chapter "Geochemical Aspects of Seawater Intrusion into the Mediterranean Coastal Aquifer" deals with the geochemical evidences for seawater intrusion into the Mediterranean coastal aquifer and the process that occurs within the fresh–saline water interface zone. Chapter "Geochemical Aspects of Groundwater in the Dead Sea Coastal Aquifer" deals with the geochemical processes in the Dead Sea coastal aquifer and their significance with regard to the groundwater flow regime and velocity. Chapter "Submarine Groundwater Discharge Along the Israeli Eastern Mediterranean Coast and in Inland Basins" deals with the component of submarine groundwater discharge (SGD) to the sea.

Y. Yechieli (✉)
Geological Survey of Israel, Jerusalem, Israel
e-mail: yechieli@gsi.gov.il



U. Kafri and Y. Yechieli (eds.), *The Many Facets of Israel's Hydrogeology*,
Springer Hydrogeology, https://doi.org/10.1007/978-3-030-51148-7_3

# General Information and Hydrogeology of the Mediterranean and Dead Sea Coastal Aquifers and Their Relation with Their Base Level

**Yoseph Yechieli, Itay J. Reznik, Adi Tal, Lior Netzer, Yaakov Livshitz, and Shaked Stein**

## 1 General Background

Coastal aquifers are a very important water source, as large populations are concentrated in coastal areas around the world, depending on their water as their only source. As a result of an increasing demand along with poor management, over exploitation has led to a decrease in groundwater levels which in turn has resulted in intrusion of seawater and salinization of potable groundwater reservoirs (Melloul and Zeitoun 1999). The location and geometry of the fresh–saline water interface depend on the interplay of many parameters such as recharge versus discharge, production rates, hydraulic gradients, hydraulic conductivity and water density. Determination of seawater intrusion/regression rates was inferred by examining the changes in salinity (Acworth and Dasey 2003; Linderfelt and Turner 2001), chemical composition (Russak et al. 2016; Mercado 1985) or mapping the location of the fresh–saline water interface (Melloul and Zeitoun 1999). Direct determination of the seawater

Y. Yechieli (✉) · I. J. Reznik
Geological Survey of Israel, Jerusalem, Israel
e-mail: yechieli@gsi.gov.il

I. J. Reznik
e-mail: reznik@gsi.gov.il

A. Tal · L. Netzer · Y. Livshitz
Hydrological Survey of Israel, Jerusalem, Israel
e-mail: adit20@water.gov.il

L. Netzer
e-mail: Liorn@water.gov.il

Y. Livshitz
e-mail: YakovL20@water.gov.il

S. Stein
Ben Gurion University, Beer Sheva, Israel
e-mail: shakedstein@gmail.com


U. Kafri and Y. Yechieli (eds.), *The Many Facets of Israel's Hydrogeology*,
Springer Hydrogeology, https://doi.org/10.1007/978-3-030-51148-7_4

intrusion rate was previously preformed using radiocarbon and tritium tracers (Sivan et al. 2005; Yechieli et al. 2009) (see also Chapter "Dating of Groundwater in Israeli Aquifers and Determination of Groundwater Flow Velocities").

Two examples of coastal aquifers are provided in this chapter: (a) The Mediterranean coastal aquifer, representing typical coastal aquifers in many parts of the world that drain to oceans and seas and (b) the Dead Sea coastal aquifer, an extremely unique and interesting hydrological system, which is affected by both the hyper-salinity of the Dead Sea water and its extremely rapid receding water level (~1 m/year). This example resembles the situation in other continental endorheic and hyper-saline base levels around the world (Yechieli et al. 2010).

## 2 The Israeli Mediterranean Coastal Aquifer

The Israeli Mediterranean coastal aquifer extends ~120 km along the eastern part of the Mediterranean Sea (Fig. 1). Its thickness decreases eastwards from ~180 meters

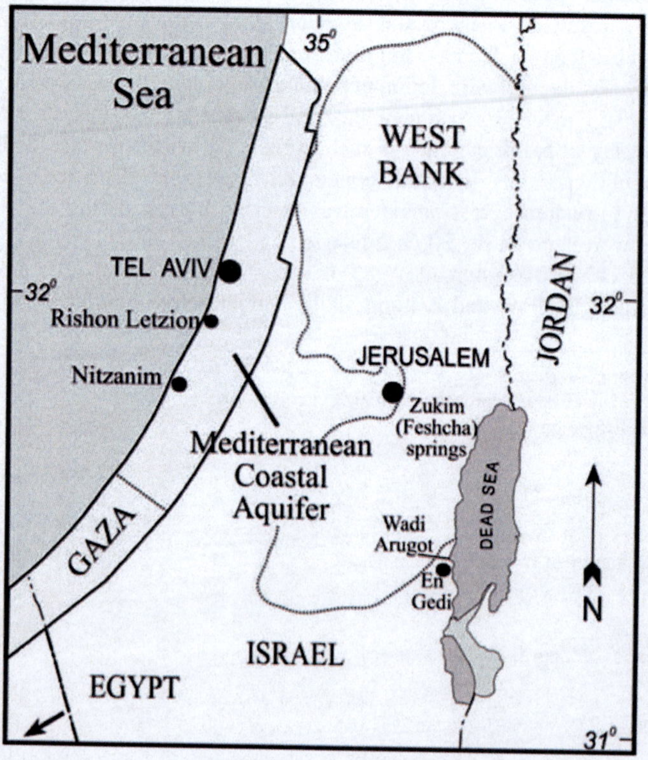


Fig. 1 Map of the coastal aquifers in Israel (after Yechieli and Sivan 2011)

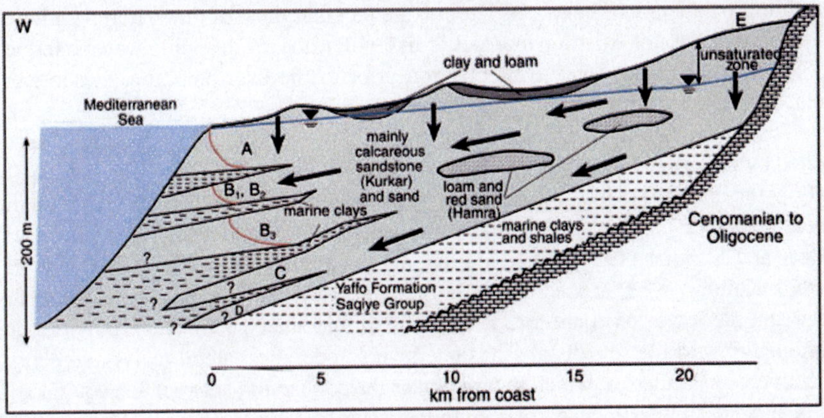


**Fig. 2** Schematic cross section of the Mediterranean coastal aquifer, showing the subdivision to several sub-aquifers (after Yechieli and Sivan 2011)

near the shoreline to less than a few meters near the eastern boundary (Fig. 2) (Ecker 1999). The aquifer, belonging to the Pleistocene Kurkar Group, consists of interlayered sandstone, calcareous sandstone, siltstone and red loam which alternate with continental and marine clays. The aquifer overlays impervious marine clays of the Saqiye Group of Pliocene age (Issar 1968). East of the shoreline, up to a distance of ~5 km, clayey inter-layers subdivide the aquifer into four sub-aquifers. Some of the lower sub-aquifers are confined, while the upper ones are phreatic (Nativ and Weisbrod 1994). The precipitation over most of the coastal aquifer is ~550 mm/yr, and the recharge coefficient is estimated to be ~0.3 (Hydrological-Service-of-Israel 2018; Gvirtzman 2002). Thus, the main direct recharge from rain to most parts of the aquifer is ~200 mm/yr. The total natural recharge from rain is estimated to be ~240 to 290 MCM/yr.

Information regarding the salinity and major ion chemistry of the coastal aquifer groundwater is available from hundreds of monitoring wells since 1933 (Hydrological-Service-of-Israel 2018; Vengosh et al. 1991; Yechieli et al. 1997). In general, the most saline groundwater samples in the coastal aquifer have a salinity and composition of the eastern Mediterranean water (39‰). However, in the southern part of this aquifer, i.e., Gaza Strip and Sinai Peninsula, water bodies that are even more saline than seawater can be found, mostly in the lower sub-aquifers (Vengosh et al. 1991; Fink 1986), implying evaporation of seawater in ancient saltpans (sabkhas).

In several areas, the fresh–saline water interface (FSI) has shifted inland over a distance of more than 1 km, as a result of over-pumping during the last decades (Melloul and Zeitoun 1999). This over-pumping reached a maximum in the 1960s, which created hydrological depressions where the fresh groundwater table declined below sea level as compared to the pristine condition recorded in the 1930s (Zilberbrand et al. 2001). This in turn enhanced the seawater intrusion. Mitigation programs along with cautious management have led to restoration of water levels in some areas and improvement in other areas. Despite the fact that most of these aforementioned depressions do not exist anymore, it is likely that the fresh–saline water interface (FSI) has not receded yet to its position prior to the over-pumping, due to slow response times of the groundwater systems.

The coastal aquifer is divided, as mentioned above, into several sub-aquifers separated by impermeable clayey layers (Fig. 2). Such a geological setup can potentially allow for seawater to intrude only if a connection of each sub-aquifer exists with the sea. The mode of connection of each sub-aquifer has been debated for many years. Bear and Kapuler (1981) suggested that all sub-aquifers are connected to the sea, while Kolton (1988) argued that only the upper sub-aquifers are connected to the sea and the lower ones are not. It was argued that since sediments are more fine-grained toward the sea, distal sandstone layers undergoes lateral facies into clayey sediments near the sea, which in turn blocks the connection between the aquifer and sea. Indeed, some evidence has been found to support this hypothesis (Kolton 1988) demonstrating that the lower sub-aquifer is disconnected from the sea since it hosts freshwater in proximity to the shoreline. Furthermore, groundwater dating studies have shown that old freshwater is present near the shoreline (Kafri and Goldman 2006; Yechieli et al. 2009). This fresh groundwater was found to extend offshore to a distance of 3 km in places, using geoelectric measurements (Levi et al. 2018) (see also Chapter "Geoelectric, Geoelectromagnetic and Combined Geophysical Methods in Groundwater Exploration in Israel").

The configuration of the seawater intrusion seems to be more complex in cases where the clay layers are discontinuous and with some vertical pathways between them, that connect the different sub-aquifers. Such a case was examined by numerical simulations (Feflow), showing that it allows saline water from the upper aquifer to flow down to the lower sub-aquifer via these pathways. Alternatively, when the lower sub-aquifers are blocked to the sea, their freshwater may be drained via these pathways to the upper sub-aquifers (Amir et al. 2013).

A few more salinization sources were observed in the eastern and central parts of the coastal aquifer (Vengosh and Rosenthal 1994), where seawater intrusion was overruled as a potential source due to the distance from the sea. These salinization sources were suggested to include (a) upwelling of diluted brines from the Saqiye Group aquiclude (Vengosh and Ben-Zvi 1994; Shavit and Furman 2001; Livshitz et al. 2002; Livshitz and Issar 2010). Through this process, the salinity in the Be'er Tuvia well increased from ~200 to ~1000 mg/l Cl within a few decades. (b) Irrigation, leading to a simultaneous increase in Cl and $NO_3$ (Baram et al. 2012). (c) Dairy farms in this area. (d) Intrusion of brackish water from the eastern part from the on-lapping Eocene aquitard. This lateral inflow is estimated to be ~38 mcm/yr, which contributes a salt flux of ~30,000 ton Cl/yr (Livshitz et al. 2002; Hydrological-Service-of-Israel 2018; Livshitz and Issar 2010). Recently, several wells were drilled into the brackish Eocene aquifer in order to prevent the water from flowing downstream (westward) toward the fresher coastal aquifer as well as to use the water as for desalination (see also Chapter "Groundwater Management: An Example of a Project Related to Rehabilitation of the Southern Coastal Aquifer, As Part of the National Water System in Israel").

Another issue related to desalination is the examination of the option of pumping saline groundwater below the interface as a source of water for desalination. The different sources of saline water for desalination were examined by Mercado (2000) and Kafri (2000), comparing between open seawater, seawater from the bottom sediment and saline groundwater near the shoreline. The advantage of using saline groundwater over seawater directly from the sea is that groundwater undergoes natural filtration in the aquifer and requires less expensive pretreatment. Furthermore, it has been recently shown that desalination of saline groundwater from the coastal aquifer performs less fouling on the reverse osmosis membranes compared to seawater desalination (Stein et al. 2016). The constant temperature in groundwater (~20 to 22 °C), compared to the variable temperature in the sea (16–30 °C), is another advantage when attempting to prevent fouling/scaling on RO membranes. Moreover, hydrological simulations and field studies showed that pumping saline water may have a positive effect on the salinity field of the aquifer, keeping the seawater from penetrating deep inland into the aquifer and rehabilitating parts of the aquifer that had been salinized (Stein et al. 2019).

Contamination of the coastal aquifer includes mainly industrial point sources and widespread agricultural sources. The industrial pollution usually starts as a point source; however, it disperses with time to form a plume as it flows and disperses down gradient. The agricultural pollution tends to cover vast areas due to the application of fertilizers and other material over large fields. Irrigation of agricultural fields by treated sewage water has also led to elevated nitrate concentrations as well as additional elements (Ronen et al. 1983; Kass et al. 2005).

A smaller area of the main coastal aquifer is the Mount Carmel coastal aquifer, where a more complicated setup exists. In this area, the upper sub-aquifer and to lesser degree in the middle sub-aquifer (unit B) (Fig. 3), are polluted by fishponds (Tal et al. 2017, 2018). It is interesting to note that the extensive pumping in this area does not result in a significant movement of the fresh–saline water interface as expected. This is explained by the connection of the shallow alluvial coastal aquifer with a deeper and much larger Judea Group Aquifer (JGA) which replenishes most the pumped water (Tal et al. 2018), while keeping the water level steady even in this dynamic situation. Using a CHIRP geophysical survey in the shallow sea and the FEFLOW hydrological simulation, Tal et al. (2018) showed that the salinity in the different sub-aquifers is affected by the seaward extent of the confining clayey roof and thus the resultant connection of the aquifer with the sea. In places where the roof terminates ~100 m offshore, the FSI reached the shore line just a few years after the pumping has begun, while no seawater intrusion occurred in the area where the roof is continuous farther offshore.

The northern portion of the coastal aquifer extends to the Zevulun Plain and the Western Galilee coastal Plain (Kafri and Ecker 1964; Kessler and Kafri 2007). The lithological sequence and geological structure resemble that of the southern coastal aquifer. It consists mainly of Pliocene to Pleistocene calcareous sandstone units with clay inter-layers that are overlain in the Zevulun Plain by a perched dune sand unit. In the Zevulun Plain, the aquifer is fed by brackish JGA water and is being intruded by seawater from the west. The sand dune unit hosts freshwater recharged directly

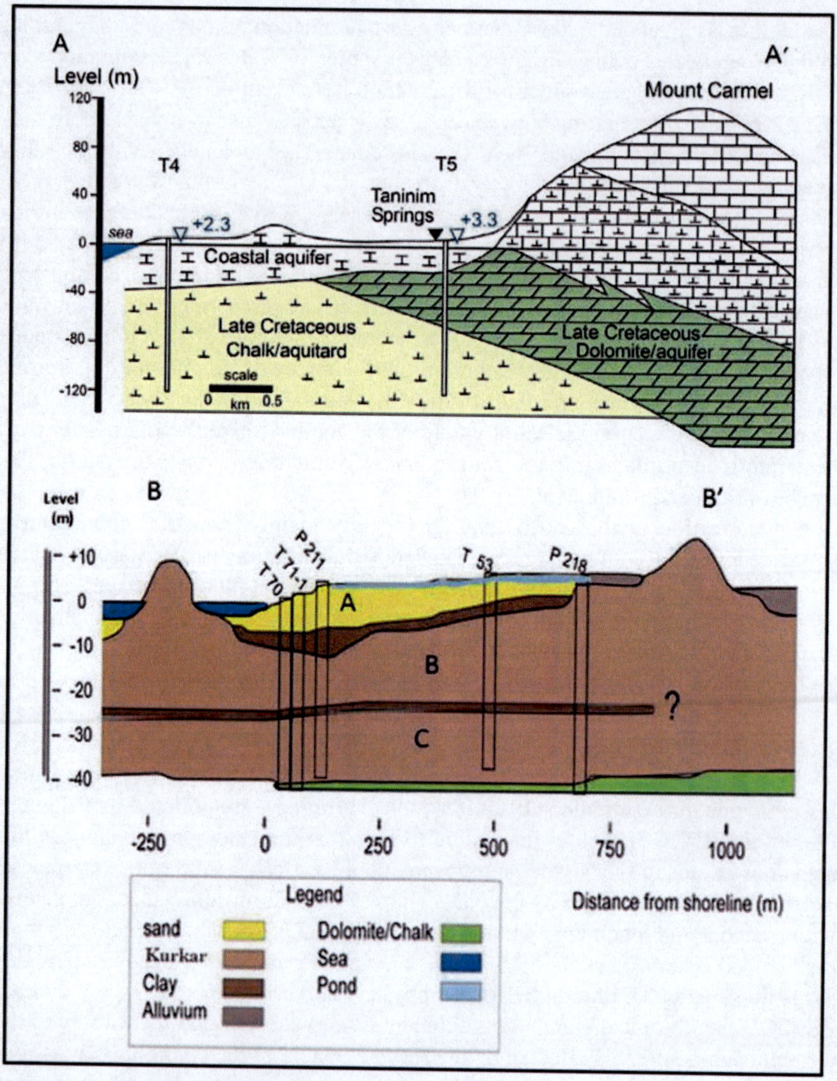


**Fig. 3** Schematic cross section of the Mount Carmel coastal aquifer (used by permission of Elsevier from Tal et al. (2017), J Hydrol 551:768–783, Figs. 2, 3)

by precipitation. The Western Galilee coastal Plain is thin (about 10–20 m), limiting the distance of seawater intrusion. It is mainly fed laterally by JGA water from the east but is being polluted down gradient by pollutants and irrigation return flows as indicated by increasing Cl and $NO_3$ concentrations.

A somewhat different hydrological setup exists near the main rivers next to the shoreline (Smith and Turner 2001), such as the Alexander Stream (Shalem et al.

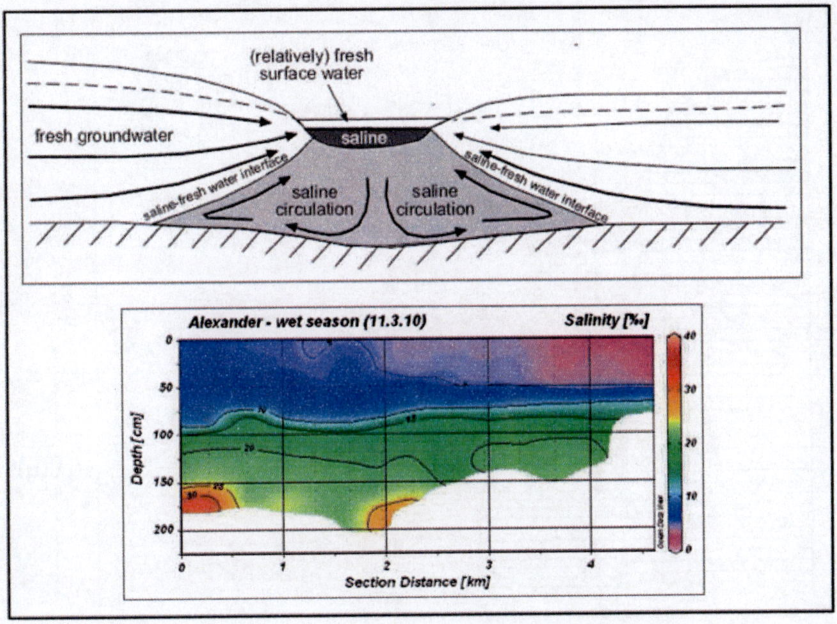


**Fig. 4** General hydrological relationships of sea–river–aquifer (after Shalem et al. 2015)

2015). This setup consists of three water bodies, namely the sea, the saline river and the aquifer next to the river (Shalem et al. 2015). The water column of the river itself is stratified, where the lower water mass is much denser and more saline than the upper column (Fig. 4). During stormy events, when the sea bar is breached, the river can be flooded by seawater. The response of the groundwater system to the flooding by the sea storms' events depends on the specific hydraulic characteristic of the sediments next to the river (Shalem et al. 2019).

## 3 The Dead Sea Coastal Aquifer

The Dead Sea (DS) is a terminal endorheic lake and a base level to converging flows, situated in the deepest part of the Dead Sea Rift System. The DS salinity and density are 340 g/L and 1.24 kg/L, respectively. The DS has a unique chemical composition whereby the main anion is Cl with very little $SO_4$ and $HCO_3$, while the main cations are Mg and Na. The source of salts was attributed mainly to ingression of seawater into the DS rift in the Neogene times and to several water–rock interactions that have occurred since then, together with high evaporation (Starinsky 1974). The DS groundwater system consists of two main aquifers (Fig. 5): The upper Cretaceous JGA and the quaternary alluvial coastal aquifer (Arad and Michaeli 1967; Yechieli

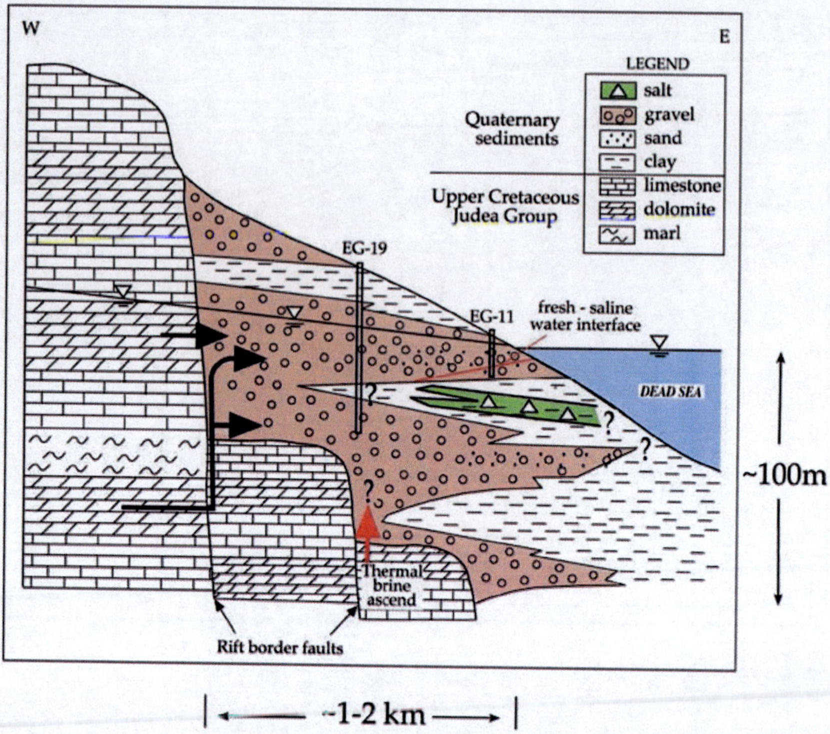


**Fig. 5** Schematic cross section of the Dead Sea coastal aquifer, showing the subdivision to sub-aquifers (after Yechieli and Sivan 2011)

et al. 1995). The DS coastal aquifer consists mainly of clastic sediments, such as gravel, sand and clay deposited in fan deltas and lacustrine sediments such as clays, aragonite, gypsum and salts. The alternations between gravel and clay divide the aquifer into several sub-aquifers that differ in their groundwater level and chemical composition. This aquifer is bound by normal faults, which set the carbonate JGA against quaternary alluvial and lacustrine sediments. The recharge of the aquifer is mainly through lateral flow from the JGA, which is replenished in the highlands 10–30 km to the west and by flash floods. Direct recharge is negligible because of the arid climate and high evaporation in the Dead Sea region.

The extremely high density of the DS induces a very shallow interface between the fresh groundwater and brine. According to the Ghyben–Herzberg approximation, the depth of the interface is 4.35 times that of the groundwater head above the lake level, as compared to 40 times in the case of normal ocean water (Fig. 6) (Yechieli 2000).

Beside the DS brine, there are also thermal brines (Kedem-Shalem) with temperature of up to 45 °C, whose chemistry is somewhat different from that of DS (e.g., equivalent Na/Cl ratio of 0.33–0.40 as compared ~0.25 in the DS in 1980). The brines

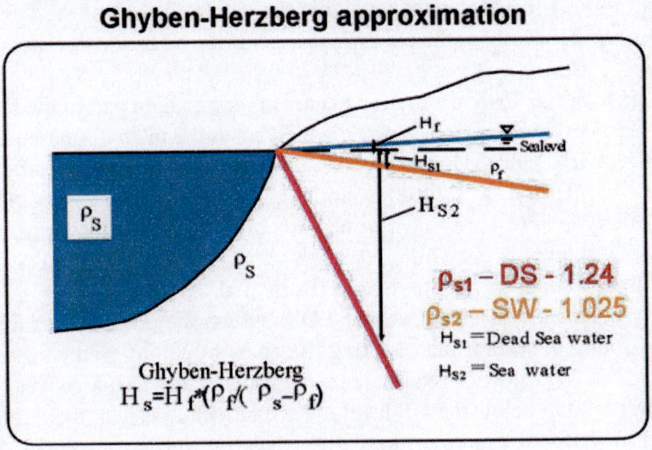


**Fig. 6** Configuration of the fresh–saline water interface in the Dead Sea area, compared to the interface next to the ocean (modified after Yechieli 2000)

were considered to be remnants of previous lakes that existed in this area prior to the present DS, which penetrated into the subsurface in eras of higher lakes levels (Gavrieli et al. 2001; Shalev and Yechieli 2007; Weber et al. 2018).

The hydrological situation in the DS coastal aquifer is different from that of the Mediterranean Sea due to the rapid drop of the DS level, estimated to be >1 m/yr, as a result of its negative water balance (Yechieli et al. 1998; Lensky et al. 2005). This negative balance resulted from diversion of the Jordan River and its tributaries that used to flow to the DS until the 1960s, by Israel, Jordan and Syria. It should be noted that lake-level fluctuations were recorded throughout the Late Pleistocene (Torfstein et al. 2013) as well as the last drop of Lake Lisan and in the Holocene period (Bookman et al. 2004; Stein et al. 2010). The DS level is expected to drop in the next 200–300 years until reaching a new steady state at a level lower than the present one by ~150 m (Yechieli et al. 1998; Krumgalz et al. 2000).

The DS and the groundwater systems are hydraulically interconnected (Yechieli et al. 1995). The effect of the drop of the lake level on the adjoining groundwater levels was discussed by Kafri (1982), Yechieli (1993, 2006), Yechieli et al. (1995, 2001, 2009) and Kiro et al. (2008). Contrary to most coastal aquifer around the world, where the FSI move upward and inland mostly due to pumping, the location of the FSI in the DS aquifer is moving downward, following the drop of the DS (Kiro et al. 2008; Yechieli et al. 2010) (see further discussion in Chapter "Dynamic Relationship Between the Sea and the Aquifer"). The drop in groundwater level follows the drop of the DS level. However, the response becomes more moderate inland and westward (Yechieli et al. 2009). This downward movement is profound in the boreholes which are located near the DS and in areas where the permeability is relatively high (in the middle of an alluvial fan), while the response is much less noticeable in distal areas which consist of less permeable sequences. The fresh–saline water interface is responding to the drop of the sea level and the eastward shift of the DS shoreline, resulting in rapid flushing of the DS groundwater brine by brackish water (Kiro et al. 2008).

A unique hydrogeological situation occurs in the northern part of the DS coast, in the area of En Feshcha (Zukim) springs, which act as the main discharge zone of the Mountain JGA into the DS. In this 2–3 km of a coastal nature reserve, the discharge volume is ~70 MCM/yr, with little variation during the last tens of years (Burg et al. 2016). While groundwater levels in some parts of this area are declining, other parts are almost constant with very little response to the rapid drop of the DS, indicating a poor hydraulic connection with the sea (Burg et al. 2016).

Direct groundwater discharge into the DS (submarine groundwater discharge—SGD) was found in several places along the coast by an infrared camera (Mallast et al. 2013a, b). Evidence for relatively fresh groundwater discharge was found also at depth of 10–20 m below the DS level in several locations near the Samar springs (Siebert et al. 2014). In some of these sub-lacustrine springs, extensive biological activity was observed, in comparison with the usual DS conditions (Ionescu et al. 2012).

The groundwater system in the vicinity of the DS is affected by the many faults bordering the Dead Sea Rift. This issue was studied in the vicinity of one of the major marginal faults in the southern part of the DS where the permeability of the rocks at the fault zone was found to be significantly lower than further away from the fault (Gabay et al. 2014). TDEM studies confirmed a sharp decrease in the hydraulic gradient across the fault, as expected by a permeability barrier (Yechieli et al. 2001).

Another coastal aquifer within the Dead Sea Rift is near the Sea of Galilee (see also Chapter "The Eastern Dead Sea Rift Continental Groundwater Base Level"). The local coastal aquifer is in fact the extension of the carbonate JGA of central Galilee which approaches and drains directly to the lake or by emerging springs on the lake shore. Saline water was found in the lower part of the aquifer at a depth of several hundreds of meters, attaining in places a salinity of up to that of normal seawater. The source of these saline waters and its hydraulic connection to other fresh groundwater bodies has been debated for many years. At present, the Sea of Galilee is a freshwater lake (less than 300 mg Cl/l), but in previous times (part of the period of 20–70 kyr), a larger and more saline ancient lake (Lisan Lake) existed in this area, the remnants of which were found in several locations near the current freshwater lake.

## 4 Seawater Intrusion Monitoring

The rate of seawater intrusion has been monitored for several decades by the Hydrological Service of Israel by performing electrical conductivity (EC) depth profiles (Fig. 7) in many observation boreholes, at distance of up to 2 km from the shoreline (Reznik et al. 2019; Melloul and Zeitoun 1999). The Israeli coastal aquifer was divided into strips of ~2 km from south to north (Fig. 8) hosting several boreholes

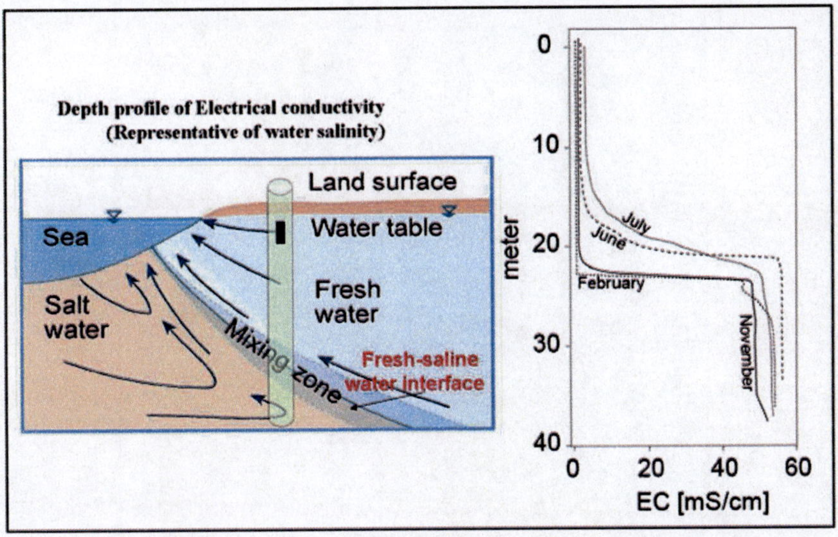


**Fig. 7** EC profile across the fresh–saline water interface showing variations with time (modified after Levanon and Yechieli 2017)

in each strip that penetrate the FSI and enabling its monitoring with time. The basic monitoring concept is that an upward movement of the FSI in a borehole indicates that the seawater intrudes inland and vice versa when the FSI declines. The FSI is usually defined as the 50% seawater salinity contour, but other definitions can also be applied, such as the 10–90% range, the 20–80% range or the depth where a significant change in the EC or salinity occurs (Levanon and Yechieli 2017).

The multi-year monitoring in the Mediterranean coastal aquifer shows that the FSI moved upward in several places within 5 years (Levanon and Yechieli 2017). A broader picture of the situation in the entire coastal aquifer is shown in Fig. 9, where in many parts of the aquifer, the toe of the FSI has moved inland by more than 1 km, while in other areas, no movement was seen (Melloul and Zeitoun 1999). This depends on the specific hydrogeological setting and pumping management in each region. In addition to monitoring of the location of the FSI, there is also monitoring by the Israeli Hydrological Service of water levels in selected representative boreholes and sampling groundwater for analysis of the chemical composition (most frequently Cl and $NO_3$, and less for other parameters). The change in salinity (represented by Cl) combined with chemical composition is usually used to verify the EC values that are measured in the field.

The location of the FSI and its fluctuation is also accurately measured with a set of EC divers, inserted into several depths in an observation borehole, below and above the FSI. Such set of measurements allow for continuous monitoring of the entire fresh–saline water interface or the transition zone (Fig. 10) (see Chapter "Dynamic Relationship Between the Sea and the Aquifer"). Such measurement, as done near

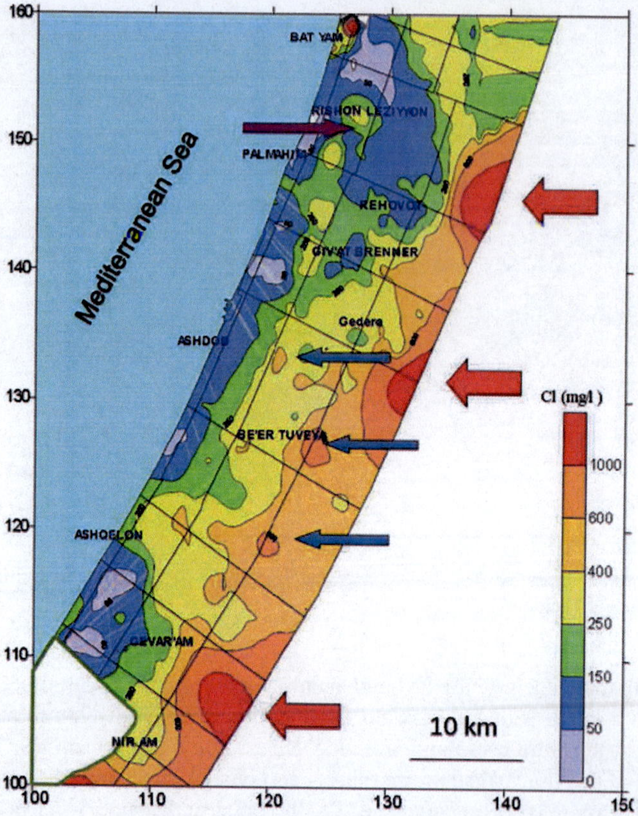


**Fig. 8** Chlorinity map of the monitored strips in the Mediterranean coastal aquifer (data of the Hydrological Service of Israel)

the shoreline, produced an artifact of exaggerated movement of the interface in long screened casing in boreholes.

A new method of monitoring the location of the FSI (SMD—Fig. 11) was recently implemented in the Mount Carmel coastal aquifer (Tal et al. 2019). The advantages of this method are the telemetric system, the high-frequency measurements (daily) and that it measures the EC not only inside the borehole but mainly outside the casing, namely in the aquifer. Thus, the monitoring does not suffer from borehole artifacts that distort the measurement in most long perforated boreholes.

Monitoring of the FSI can be done also with the geoelectric method of TDEM (Time-Domain Electromagnetic, see Chapter "Geoelectric, Geoelectromagnetic and Combined Geophysical Methods in Groundwater Exploration in Israel"), which is less accurate than direct measurements of EC profiles but does not suffer from the specific artifact of monitoring boreholes. Moreover, if the goal is to track changes

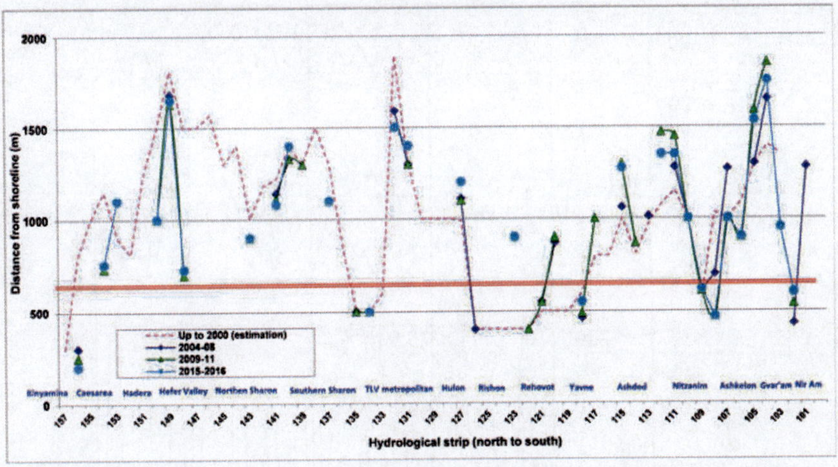


**Fig. 9** Variations with time in the distance of the fresh–saline water interface from the shoreline in the whole Mediterranean coastal aquifer (after the Hydrological-Service-of-Israel 2018)

with time, and not accurate specific values, then monitoring can be done since EC is the only parameter that changes with time.

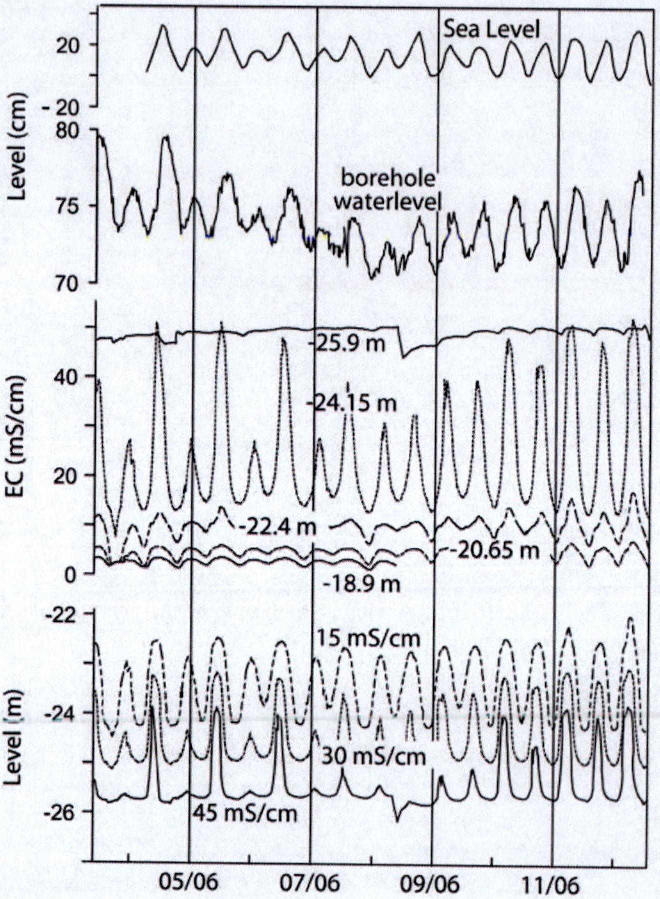


**Fig. 10** Monitoring the interface with a set of EC divers (used by permission of Wiley from Shalev et al. (2009), Groundwater 47:49–56, Fig. 2)

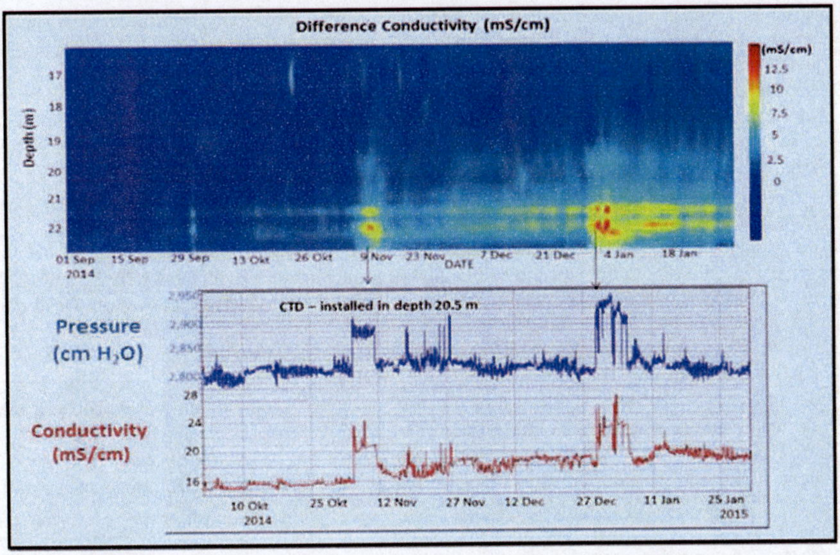


**Fig. 11** Monitoring the interface in Ma'agan Michael with different methods, including the new SMD (after Tal et al. 2019)

# References

Acworth RI, Dasey GR (2003) Mapping of the hyporheic zone around a tidal creek using a combination of borehole logging, borehole electrical tomography and cross-creek electrical imaging, New South Wales, Australia. Hydrogeol J 11(3):368–377. https://doi.org/10.1007/s10040-003-0258-4

Amir N, Kafri U, Herut B, Shalev E (2013) Numerical simulation of submarine groundwater flow in the coastal aquifer at the Palmahim Area, the Mediterranean Coast of Israel, vol 27. https://doi.org/10.1007/s11269-013-0392-2

Arad A, Michaeli A (1967) Hydrogeological investigations in the western catchment of the Dead Sea. Israel J Earth Sci 16:181–196

Baram S, Arnon S, Ronen Z, Kurtzman D, Dahan O (2012) Infiltration mechanism controls nitrification and denitrification processes under dairy waste lagoon. J Environ Qual 41(5):1623–1632. https://doi.org/10.2134/jeq2012.0015

Bear J, Kapuler I (1981) A numerical solution for the movement of an interface in a layered coastal aquifer, vol 50. https://doi.org/10.1016/0022-1694(81)90074-3

Bookman R, Enzel Y, Agnon A, Stein M (2004) Late holocene levels of the Dead Sea, vol 116. https://doi.org/10.1130/b25286.1

Burg A, Yechieli Y, Galili U (2016) Response of a coastal hydrogeological system to a rapid decline in sea level; the case of Zuqim springs—the largest discharge area along the Dead Sea coast, vol 536. https://doi.org/10.1016/j.jhydrol.2016.02.039

Ecker A (1999) Atlas of selected geological cross-sections and subsurface maps of the coastal aquifer of Israel. Geological Survey of Israel Report, Jerusalem

Fink M (1986) Hydrogeological survey of the Pleistocene aquifer of the Gaza Strip: salinity trend analysis. Tahal Report, Tel Aviv

Gabay R, Shalev E, Yechieli Y, Sagy A, Weisbrod N (2014) The permeability of fault zones: a case study of the Dead Sea rift (Middle East), vol 22. https://doi.org/10.1007/s10040-013-1055-3

Gavrieli I, Yechieli Y, Halicz L, Spiro B, Bein A, Efron D (2001) The sulfur system in anoxic subsurface brines and its implication in brine evolutionary pathways: the Ca-chloride brines in the Dead Sea area, vol 186. https://doi.org/10.1016/s0012-821x(01)00247-3

Gvirtzman H (2002) Israel water resources. Yad Ben-Zvi Press, Jerusalem

Hydrological-Service-of-Israel (2018) Status report. Jerusalem

Ionescu D, Siebert C, Polerecky L, Munwes YY, Lott C, Häusler S, Bižić-Ionescu M, Quast C, Peplies J, Glöckner FO, Ramette A, Rödiger T, Dittmar T, Oren A, Geyer S, Stärk H-J, Sauter M, Licha T, Laronne JB, de Beer D (2012) Microbial and chemical characterization of underwater fresh water springs in the Dead Sea. PLoS ONE 7(6):e38319. https://doi.org/10.1371/journal.pone.0038319

Issar A (1968) Geology of the central coastal plain of Israel. Israel J Earth Sci 17:16–29

Kafri U (1982) Relationship between Dead Sea and groundwater levels in the western Dead Sea catchment area. Israel Geol Surv Curr Res 1981:90–94

Kafri U (2000) Survey of drilling for seawater in the southern coastal plain as a source for desalination plants. Isr. Geol. Surv. Rep. GSI/27/2000. Geological Survey of Israel, Jerusalem

Kafri U, Ecker A (1964) Neogene and quaternary subsurface geology and hydrogeology of the Zevulun Plain. Israel Geol Surv Bull 37:13p

Kafri U, Goldman M (2006) Are the lower subaquifers of the Mediterranean coastal aquifer of Israel blocked to seawater intrusion? Results of a TDEM (time domain electromagnetic) study. Israel J Earth Sci 55:55–68

Kass A, Gavrieli I, Yechieli Y, Vengosh A, Starinsky A (2005) The impact of freshwater and wastewater irrigation on the chemistry of shallow groundwater: a case study from the Israeli Coastal Aquifer, vol 300. https://doi.org/10.1016/j.jhydrol.2004.06.013

Kessler A, Kafri U (2007) Application of a cell model for operational management of the Na'aman groundwater basin, Israel. Israel J Earth Sci 56:29–46

Kiro Y, Yechieli Y, Lyakhovsky V, Shalev E, Starinsky A (2008) Time response of the water table and saltwater transition zone to a base level drop. Water Resour Res 44(12). https://doi.org/10.1029/2007wr006752

Kolton Y (1988) Examination of the connection between seawater in the Pleistocene aquifer in the Mediterranean shelf of central Israel. Tahal, Water Planning for Israel. 01/88/31

Krumgalz B, Hecht A, Starinsky A, Katz A, Krumgalz BS, Hecht A, Starinsky A, Katz A (2000) Thermodynamic constraints on Dead Sea evaporation: can the Dead Sea dry up? Chem Geol 165, #1–2, 1–11, vol 165. https://doi.org/10.1016/s0009-2541(99)00156-4

Lensky N, Dvorkin Y, Lyakhovsky V, Gertman I, Gavrieli I (2005) Water, salt, and energy balances of the Dead Sea, vol 411. https://doi.org/10.1029/2005wr004084

Levanon E, Yechieli Y (2017) Monitoring the fresh-salt water interface of the Israeli Coastal Aquifer. Geological Survey of Israel Report. GSI/06/2017. Jerusalem

Levi E, Goldman M, Tibor G, Herut B (2018) Delineation of subsea freshwater extension by marine geoelectromagnetic soundings (SE Mediterranean Sea). Water Resour Manage 32(11):3765–3779. https://doi.org/10.1007/s11269-018-2018-1

Linderfelt WR, Turner JV (2001) Interaction between shallow groundwater, saline surface water and nutrient discharge in a seasonal estuary: the Swan-Canning system. Hydrol Process 15(13):2631–2653. https://doi.org/10.1002/hyp.302

Livshitz Y, Issar A (2010) Applying the conceptual-model of progressive development to the contaminated aquifer underlying Israel's Coastal Plain, vol 2. https://doi.org/10.4236/jwarp.2010.25046

Livshitz Y, Livshitz Y, Ben-Avraham Z (2002) The effect of fault zones on Coastal Aquifer salinization. Final Report. Submitted to Israel Water Commission. Tel-Aviv University

Mallast U, Schwonke F, Gloaguen R, Geyer S, Sauter M, Siebert C (2013a) Airborne thermal data identifies groundwater discharge at the north-western coast of the Dead Sea. Remote Sens 5(12):6361–6381

Mallast U, Siebert C, Wagner B, Sauter M, Gloaguen R, Geyer S, Merz R (2013b) Localisation and temporal variability of groundwater discharge into the Dead Sea using thermal satellite data. Environ Earth Sci 69(2):587–603. https://doi.org/10.1007/s12665-013-2371-6

Melloul AJ, Zeitoun DG (1999) A semi-empirical approach to intrusion monitoring in Israeli Coastal Aquifer. In: Bear J, Cheng AHD, Sorek S, Ouazar D, Herrera I (eds) Seawater intrusion in coastal aquifers—concepts, methods and practices. Springer, Dordrecht, pp 543–558. https://doi.org/10.1007/978-94-017-2969-7_16

Mercado A (1985) The use of hydrogeochemical patterns in carbonate sand and sandstone aquifers to identify intrusion and flushing of Saline Water, vol 23. https://doi.org/10.1111/j.1745-6584.1985.tb01512.x

Mercado A (2000) Hydrological assessment of the alternative of sea water supply to desalination plants through wells in the southern coastal plain. Progress report no. 2. Prepared for the Geological Survey of Israel

Nativ R, Weisbrod N (1994) Hydraulic connections among subaquifers of the coastal plain aquifer, Israel. Groundwater 32(6):997–1007. https://doi.org/10.1111/j.1745-6584.1994.tb00939.x

Reznik IJ, Netzer L, Vanunu O, Yechieli Y, Rabinowitz N, Nof R (2019) Freshwater-saltwater interface monitoring along Israel's Coastal Plain and the eastern pumping line. Israel Geological Survey Report GSI/15/2019, 381 p

Ronen D, Kanfi Y, Magaritz M (1983) Sources of nitrates in groundwater of the coastal plain of Israel Evolution of ideas. Water Res 17(11):1499–1503. https://doi.org/10.1016/0043-1354(83)90004-0

Russak A, Sivan O, Yechieli Y (2016) Trace elements (Li, B, Mn and Ba) as sensitive indicators for salinization and freshening events in coastal aquifers, vol 441. https://doi.org/10.1016/j.chemgeo.2016.08.003

Shalem Y, Weinstein Y, Levi E, Herut B, Goldman M, Yechieli Y (2015) The extent of aquifer salinization next to an estuarine river: an example from the eastern Mediterranean. Hydrogeol J 23(1):69–79. https://doi.org/10.1007/s10040-014-1192-3

Shalem Y, Yechieli Y, Herut B, Weinstein Y (2019) Aquifer response to estuarine stream dynamics. Water 11(8):1678

Shalev E, Yechieli Y (2007) The effect of Dead Sea level fluctuations on the discharge of thermal springs. Israel J Earth Sci 56:19–27

Shalev E, Lazar A, Wollman S, Kington S, Yechieli Y, Gvirtzman H (2009) Biased monitoring of fresh water-salt water mixing zone in coastal aquifers. Groundwater 47(1):49–56. https://doi.org/10.1111/j.1745-6584.2008.00502.x

Shavit U, Furman A (2001) The location of deep salinity sources in the Israeli Coastal aquifer, vol 250. https://doi.org/10.1016/s0022-1694(01)00406-1

Siebert C, Mallast U, Rödiger T, Strey M, Ionescu D, Häusler S, Noriega B, Pohl T, Merkel B (2014) Submarine groundwater discharge at the Dead Sea. In: Proceedings of 23rd salt water intrusion meeting, Husum, Germany

Sivan O, Yechieli Y, Herut B, Lazar B (2005) Geochemical evolution and timescale of seawater intrusion into the coastal aquifer of Israel, vol 69. https://doi.org/10.1016/j.gca.2004.07.023

Smith AJ, Turner JV (2001) Density-dependent surface water–groundwater interaction and nutrient discharge in the Swan-Canning Estuary. Hydrol Processes 15(13):2595–2616. https://doi.org/10.1002/hyp.303

Starinsky A (1974) Relationship between Ca-chloride brines and sedimentary rocks in Israel. The Hebrew University, Jerusalem

Stein M, Torfstein A, Gavrieli I, Yechieli Y (2010) Abrupt aridities and salt deposition in the post-glacial Dead Sea and their North Atlantic connection. Quatern Sci Rev 29(3):567–575. https://doi.org/10.1016/j.quascirev.2009.10.015

Stein S, Russak A, Sivan O, Yechieli Y, Rahav E, Oren Y, Kasher R (2016) Saline groundwater from coastal aquifers as a source for desalination, vol 50. https://doi.org/10.1021/acs.est.5b03634

Stein S, Yechieli Y, Shalev E, Kasher R, Sivan O (2019) The effect of pumping saline groundwater for desalination on the fresh-saline water interface dynamics. Water Res 156:46–57. https://doi.org/10.1016/j.watres.2019.03.003

Tal A, Weinstein Y, Yechieli Y, Borisover M (2017) The influence of fish ponds and salinization on groundwater quality in the multi-layer coastal aquifer system in Israel. J Hydrol 551:768–783. https://doi.org/10.1016/j.jhydrol.2017.04.008

Tal A, Weinstein Y, Wollman S, Goldman M, Yechieli Y (2018) The interrelations between a multi-layered coastal aquifer, a surface reservoir (fish ponds), and the Sea, vol 10. https://doi.org/10.3390/w10101426

Tal A, Weinstein Y, Baïsset M, Golan A, Yechieli Y (2019) High resolution monitoring of seawater intrusion in a multi-aquifer system-implementation of a new downhole geophysical tool. Water 11(9):1877

Torfstein A, Goldstein S, Stein M, Enzel Y (2013) Impacts of abrupt climate changes in the Levant from last glacial Dead Sea levels, vol 69. https://doi.org/10.1016/j.quascirev.2013.02.015

Vengosh A, Ben-Zvi A (1994) Formation of a salt plume in the Coastal Plain aquifer of Israel: The Be'er Toviyya region, vol 160. https://doi.org/10.1016/0022-1694(94)90032-9

Vengosh A, Rosenthal E (1994) Saline groundwater in Israel: its bearing on the water crisis in the country. J Hydrol 156(1):389–430. https://doi.org/10.1016/0022-1694(94)90087-6

Vengosh A, Starinsky A, Melloul A, Fink M, Erlich S (1991) Salinization of the coastal aquifer by Ca-chloride solutions at the interface zone along the Coastal Plain of Israel. Hydrological Service Report 20/1991

Weber N, Yechieli Y, Stein M, Yokochi R, Gavrieli I, Zappala J, Mueller P, Lazar B (2018) The circulation of the Dead Sea brine in the regional aquifer. Earth Planet Sci Lett 493:242–261. https://doi.org/10.1016/j.epsl.2018.04.027

Yechieli Y (1993) The effects of water level changes in closed lakes (Dead Sea) on the surrounding groundwater and country rocks. Weizmann Institute of Science, Rehovot, Israel

Yechieli Y (2000) Fresh-Saline Ground Water Interface in the Western Dead Sea Area, vol 38. https://doi.org/10.1111/j.1745-6584.2000.tb00253.x

Yechieli Y (2006) The response of the groundwater system to changes in the Dead Sea level. In: Enzel AAY, Stein M (eds) New Frontiers in Dead Sea Paleoenvironmental Research, vol Special Paper. Geological Society of America

Yechieli Y, Sivan O (2011) The distribution of saline groundwater and its relation to the hydraulic conditions of aquifers and aquitards: examples from Israel. Hydrogeol J 19(1):71–81. https://doi.org/10.1007/s10040-010-0646-5

Yechieli Y, Ronen D, Berkowitz B, Dershowitz W, Hadad A (1995) Aquifer characteristics derived from the interaction between water levels of a terminal lake (Dead Sea) and an adjacent aquifer, vol 31. https://doi.org/10.1029/94wr03154

Yechieli Y, Ronen D, Vengosh A (1997) Isotopic measurements and groundwater dating at the fresh-saline water interface region of the Mediterranean coastal plain aquifer of Israel. Geological Survey Report GSI/28/96

Yechieli Y, Gavrieli I, Berkowitz B, Ronen D (1998) Will the Dead Sea die? Geology 26(8):755–758. https://doi.org/10.1130/0091-7613(1998)026%3c0755:WTDSD%3e2.3.CO;2

Yechieli Y, Kafri U, Goldman M, Voss CI (2001) Factors controlling the configuration of the fresh-saline water interface in the Dead Sea coastal aquifers: synthesis of TDEM surveys and numerical groundwater modeling. Hydrogeol J 9:367–377

Yechieli Y, Kafri U, Wollman S, Shalev E, Lyakhovsky V (2009) The effect of base level changes and geological structures on the location of the groundwater divide, as exhibited in the hydrological system between the Dead Sea and the Mediterranean Sea. J Hydrol 378(3):218–229. https://doi.org/10.1016/j.jhydrol.2009.09.023

Yechieli Y, Shalev E, Wollman S, Kiro Y, Kafri U (2010) Response of the Mediterranean and Dead Sea coastal aquifers to sea level variations. Water Resour Res 46(12). https://doi.org/10.1029/2009wr008708

Zilberbrand M, Rosenthal E, Shachnai E (2001) Impact of urbanization on hydrochemical evolution of groundwater and on unsaturated-zone gas composition in the coastal city of Tel Aviv, Israel. J Contam Hydrol 50(3):175–208. https://doi.org/10.1016/S0169-7722(01)00118-8

# Dynamic Relationship Between the Sea and the Aquifer

**Elad Levanon, Eyal Shalev, Imri Oz, and Haim Gvirtzman**

The dynamics of sea–aquifer relationship can be studied at different time scales, including short time daily scale of tides, seasonal (winter–summer) scale, trends throughout years or decades (mainly due to pumping or base-level changes—e.g., the Dead Sea (DS) case), and long-term trends, such as sea-level fluctuations due to climate changes. Fresh–saline water interface (hereafter FSI) in coastal aquifers results from the density difference between freshwater originating from rain and saline water intruding from sea. The seawater mixes with freshwater, creating a wide brackish transition zone (Lee and Cheng 1974). The following chapter deals with the dynamic relationship between two coastal aquifers and their adjacent water bodies, namely the Mediterranean Sea and the Dead Sea.

## 1 Mediterranean Coastal Aquifer

The hydraulic connection between the Mediterranean Sea and the coastal aquifer is exhibited in several ways, including the tidal response on groundwater level and on the location of the FSI, as well as the actual tracking of seawater intrusion into the

E. Levanon (✉) · I. Oz (✉) · H. Gvirtzman
Institute of Earth Sciences, The Hebrew University of Jerusalem, Edmond J. Safra Campus, Givat Ram, Jerusalem 91904, Israel
e-mail: eladlevanon@gmail.com

I. Oz
e-mail: imri.oz@mail.huji.ac.il

H. Gvirtzman
e-mail: Haim.Gvirtzman@mail.huji.ac.il

E. Levanon · E. Shalev · I. Oz
Geological Survey of Israel, 32 Leibovitz Str., Jerusalem 9692100, Israel
e-mail: eyal@gsi.gov.il



U. Kafri and Y. Yechieli (eds.), *The Many Facets of Israel's Hydrogeology*,
Springer Hydrogeology, https://doi.org/10.1007/978-3-030-51148-7_5

aquifer near the shoreline. This chapter deals mainly with the physical aspect of this connection, as exhibited by the tidal response (as examined in the field, in the laboratory, and in hydrological simulations) and also with the expected response to future sea-level rise (as examined by simulations). The actual tracking of seawater intrusion into the aquifer and its geochemical aspects is discussed below (this chapter).

The dynamic response to sea tides was examined in the coastal aquifer using field data, laboratory experiments, and hydrological simulations (Levanon et al. 2013, 2016, 2017, 2019). The first stage was to examine the relevance of the observation wells in which the research was done. It was found that long-perforated casings in exploration boreholes are problematic since they cause an artifact in the response of groundwater to sea tide (Shalev et al. 2009; Levanon et al. 2013). Hydrological simulations showed that the fluctuation within a long-perforated section could be several orders of magnitude larger than those within the adjacent aquifer (Fig. 1; Shalev et al. 2009). In order to study the above artifact, new boreholes were drilled

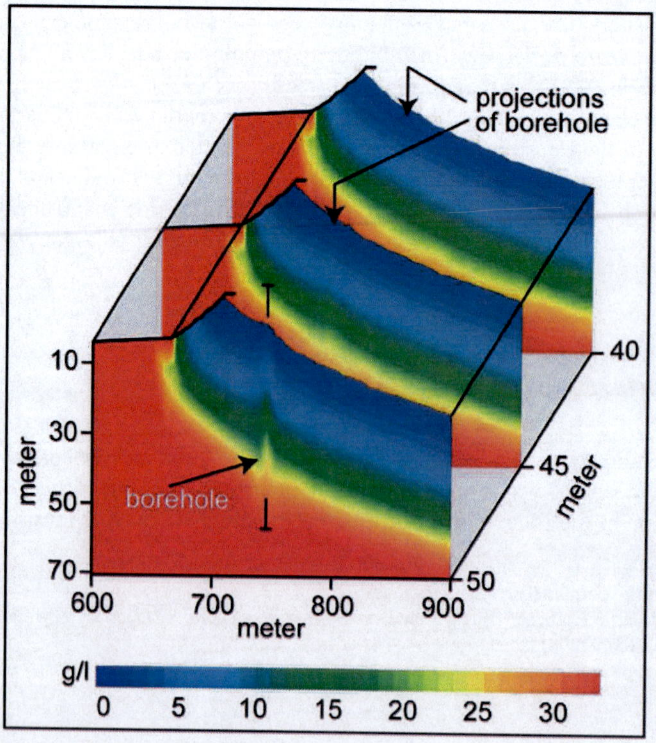


**Fig. 1** Simulation results showing salinity distribution around the borehole for the anisotropic case. The borehole affects the natural flow and elevates the freshwater–salt water mixing zone. Away from the borehole, the natural flow is undisturbed (used by permission of Wiley after Shalev et al. 2009, Groundwater 47:49–56, Fig. 4)

with short-perforated casings (1 m long) at the precise depth of the FSI. These short-perforated wells have shown to exhibit a considerably smaller artifact than in the case of long perforations. Even better results were obtained when no PVC pipes were inserted into the borehole and the EC divers were inserted directly into the drilling hole and covered by the local sand as shown in Fig. 2. The latter case indeed yielded no artifact (Fig. 3; Levanon et al. 2013), proving that the long perforation is the cause of large EC fluctuation that does not represent the actual response of the aquifer to tide.

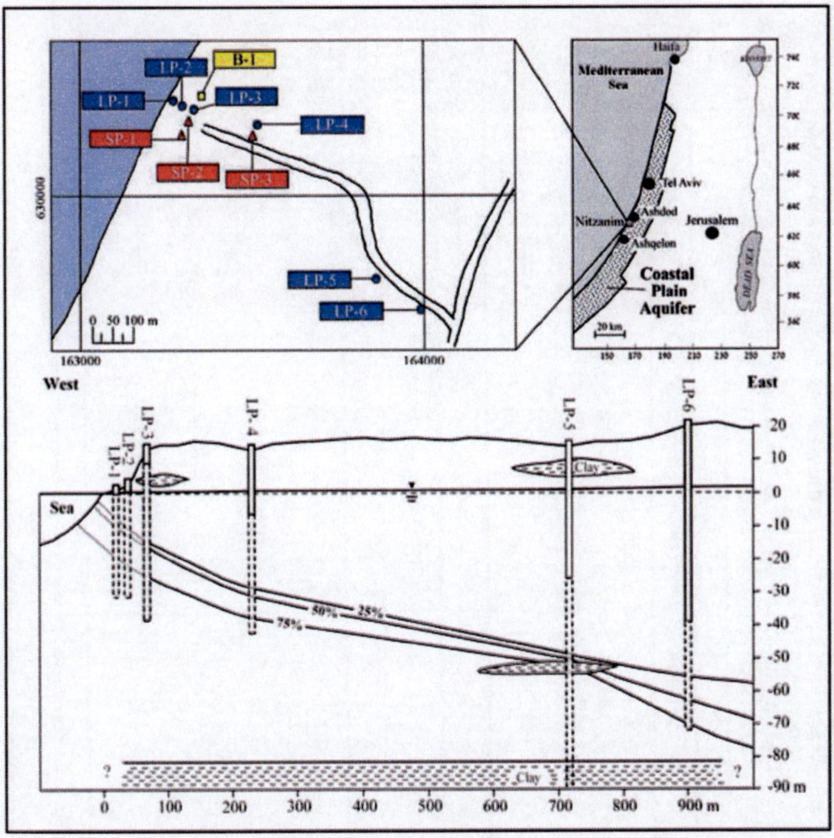


**Fig. 2** Above: Location map and research area. Blue circles: long-perforated (LP-X) boreholes, red triangles: short-perforated (SP-X) boreholes, yellow square: location of the buried (B-X) sensors. Coordinates are in ITM grid. The black lines represent a dirt road. Below: Cross section of the transition zone between fresh and saline waters perpendicular to the shoreline in the research area. Groundwater salinity isopleths are marked by black lines. The broken lines in the boreholes represent the perforated segment. On the vicinity of the shoreline, the location of the transition zone was estimated because the groundwater is consistently affected by sea waves (used by permission of Wiley after Levanon et al. (2013), Groundwater Monitor Remed 33:101–110, Figs. 1 and 3)

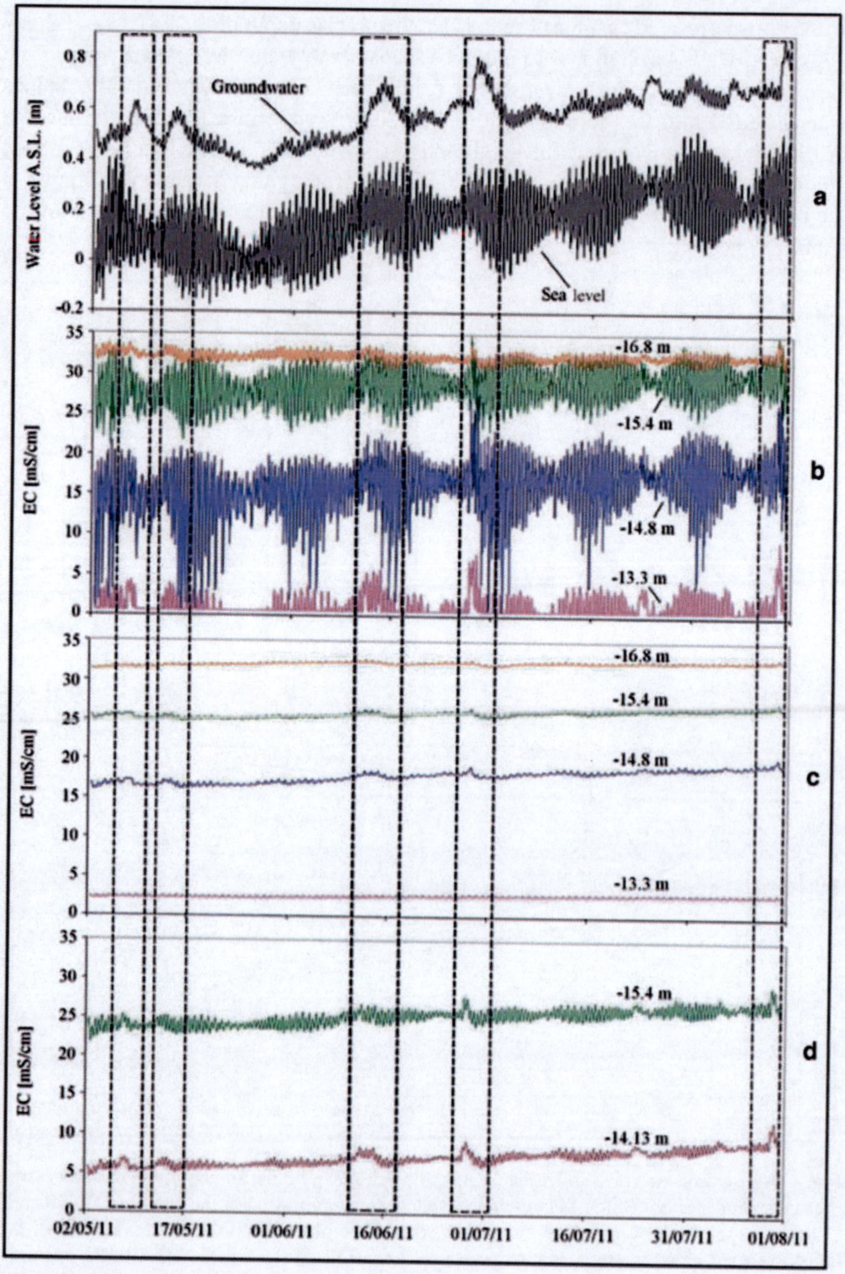
Water Level A.S.L. [m]
Groundwater
Sea level
a
EC [mS/cm]
-16.8 m
-15.4 m
-14.8 m
-13.3 m
b
EC [mS/cm]
-16.8 m
-15.4 m
-14.8 m
-13.3 m
c
EC [mS/cm]
-15.4 m
-14.13 m
d
02/05/11
17/05/11
01/06/11
16/06/11
01/07/11
16/07/11
31/07/11
01/08/11

◂**Fig. 3** Top to bottom: **a** Sea level and groundwater level in LP-3; **b** EC in four sensors located at different depths in long-perforated borehole LP-3; **c** EC in four sensors buried at the same depths in B-1; and **d** EC in two sensors located at different depths in short-perforated boreholes SP-1 and SP-2. The broken frames represent events with a sharp rise in groundwater level due to stormy sea. Numbers on B, C, and D represent the depths of the sensors (used by permission of Wiley after Levanon et al. (2013), Groundwater Monitor Remed 33:101–110, Fig. 4)

The second stage was to examine the detailed response to sea tide of both groundwater level and location of the FSI. This was done by field monitoring and hydrological simulations of the same field conditions (Levanon et al. 2016, 2017). The field results showed an increase in time lag of the response of the water level in the phreatic coastal aquifer as a function of their distance from the sea, up to a distance of 70 m from the shoreline. The simulations showed a head response of the top FSI with a time lag of ~1 h at a distance of 70 m from the shoreline (Fig. 4; Levanon et al. 2016). The time lag to the response in the more saline section of the FSI below, at the same location, was about twice. Somewhat similar results were found in the field, measured by the EC divers (Levanon et al. 2017).

The difference in the time lag was explained in a conceptual model whereby the freshwater level responds faster since it involves pressure propagation, whereas the change in the saline portion involves the actual movement of the entire water body,

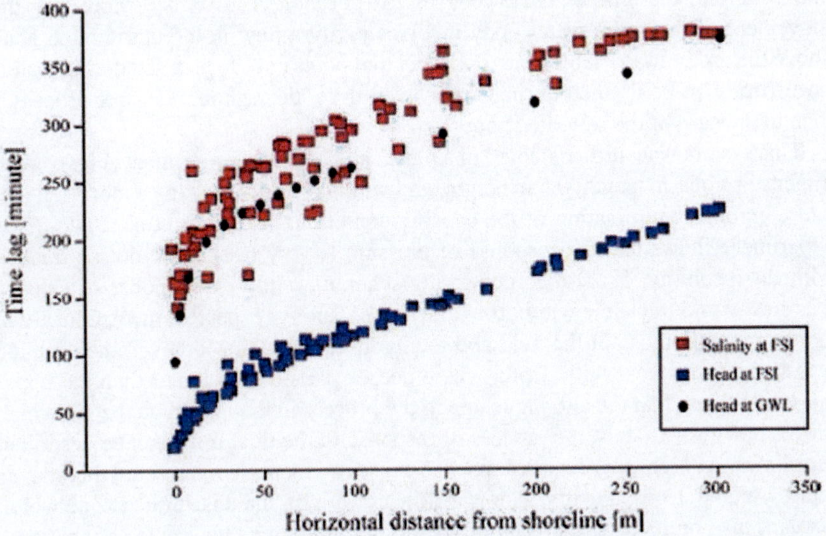


**Fig. 4** Time lags of the head and salinity in 80 observation points which were located along the fresh–saline water interface in the hydrological simulation. The salinity and the hydraulic head were calculated for the same observation points, yet the time lags are completely different. The time lags in the groundwater level (black dots) equal that of the salinity in the fresh–saline water interface (used by permission of Elsevier after Levanon et al. (2016), Ad Water Resources 96:34–42, Fig. 6)

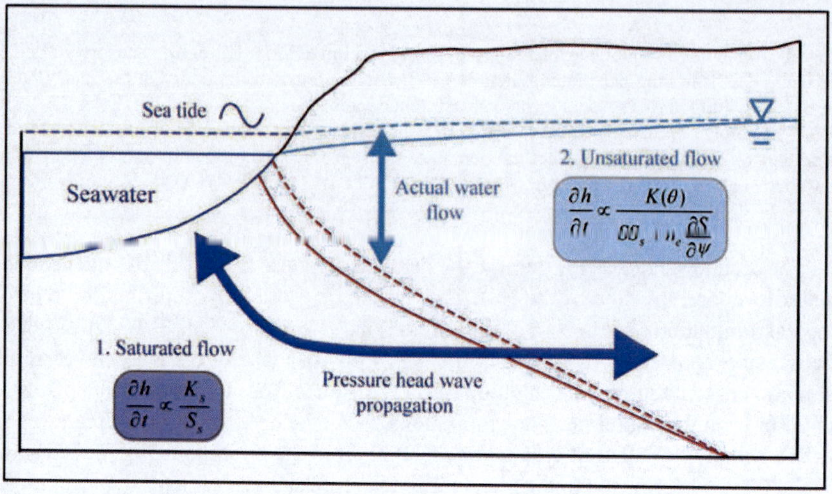


**Fig. 5** Theoretical model explaining schematically the sea tide effect on groundwater fluctuations. The pressure wave propagation is much faster than the fluctuations of the FSI and the GWL, because the different values of the hydraulic conductivity ($K$) and the specific storage ($S_s$) in the saturated and unsaturated zones (used by permission of Elsevier after Levanon et al. (2017), J Hydrol 551:665–675, Fig. 1)

and therefore, the process is slower (Fig. 5; Levanon et al. 2017). Moreover, the movement of the entire water body involves partly a flow in the unsaturated zone above the groundwater table, due to the fact that pores have to be filled during water-level rise or to be drained during water level drop, throughout a slower processes than in the case of the saturated zone.

This system was further studied in a physical laboratory experiment (Fig. 6) with dimensions of 1 m length, 0.5 m height, and width of 3 cm, allowing a more detailed and controlled examination of the specific conditions (Levanon et al. 2019). This experiment shows that the response of pressure is very fast (could not be tracked with the resolution of 20 m/s monitoring scheme), while the response of salinity was slower and trackable within the same time span. As expected, the results are in agreement with those of the field and the hydraulic model. Moreover, the time lag was found to increase with depth, where deeper parts respond later than the upper ones, indicating that the lagging is due to the process in the unsaturated zone which moves downward. Thus, the process of response to the tidal force can be described as follows: At high tide, the sea-level rise causes an increase in pressure that moves rapidly inland. Consequently, a slower flow upward, in the unsaturated zone which depends also on its degree of saturation and not only on its hydraulic conductivity, takes place. The water-level rise in the unsaturated zone causes a delay in the pressure change that effects the salinity profile in the FSI. The opposite process occurs at low tide, where the water level drops. The upward and downward rates are not exactly the same where some hysteresis was found to occur (Levanon et al. 2019).

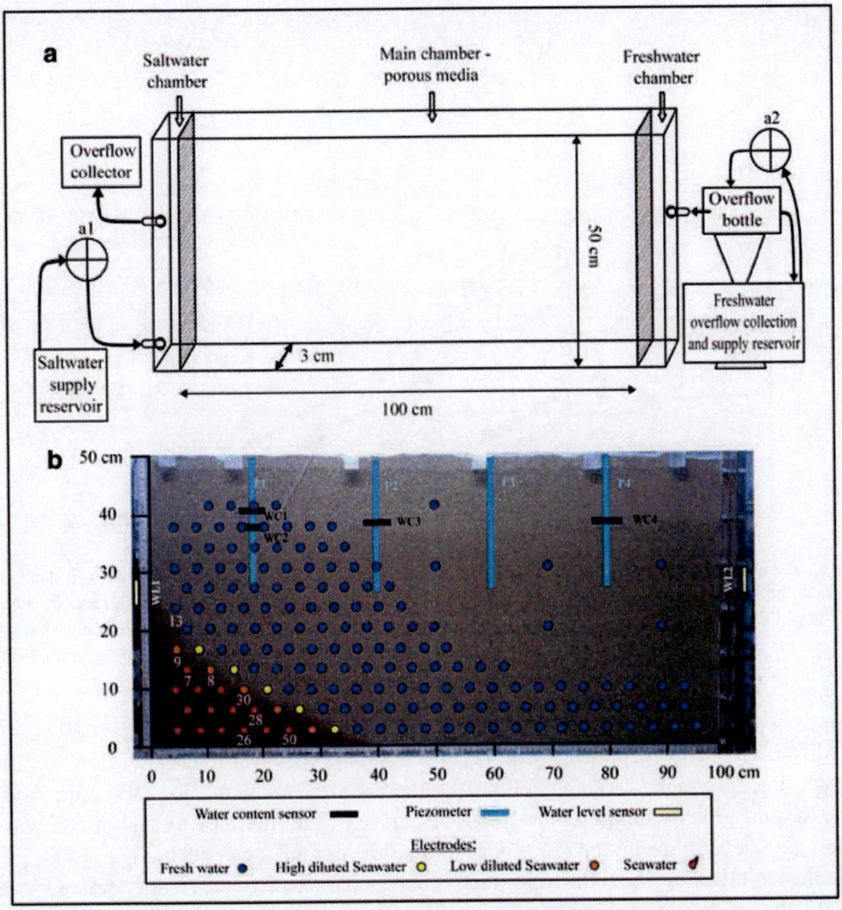


**Fig. 6** **a** Laboratory setup scheme (Levanon et al. 2019). The main flow chamber simulates the aquifer, and the two side chambers define the boundary conditions during the experiments. The left chamber represents the saltwater boundary at the seaside, and the right chamber represents the inland freshwater boundary of the regional aquifer. **b** A picture of the laboratory aquifer. One hundred and sixty-eight electrodes are placed for in situ voltage measurements, which is equivalent to the salinity of the water. Four sensors of water content (WC1–WC4) are placed in the capillary zone above the water level. Four piezometers (P1–P4) located a few centimeters below the water table at different distances from the left boundary. Two divers (WL1–WL2) are placed in the water chambers

The circulation of seawater in the coastal aquifer was also studied using computer simulation of experimental laboratory setups (Oz et al. 2015). The main issue here was to determine as to where does the seawater circulate back to the sea. It was found that it occurs at the lower part of the FSI, mainly below the 90% seawater contour (Fig. 7).

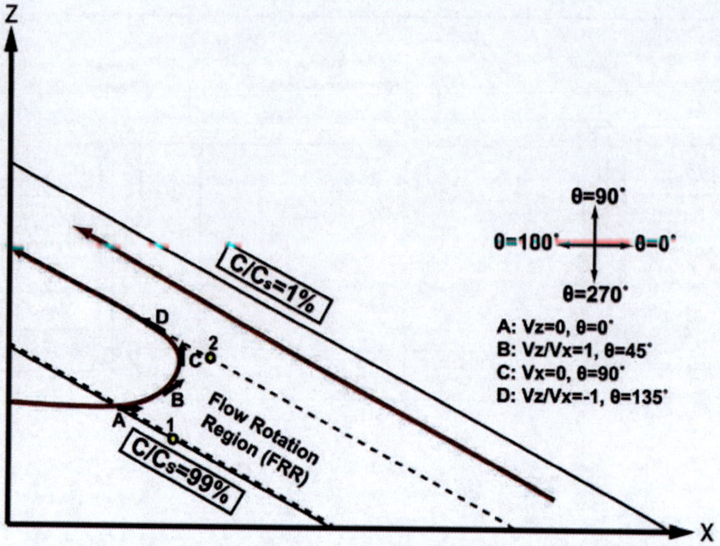


**Fig. 7** Conceptual scheme (not to scale) of the saltwater rotating path lines. The two black dashed lines and the two solid black lines represent the boundaries of the FRR and the FSI, respectively. The red arrows represent the flow lines; black arrows (A–D) represent the flow directions during the rotation. Points 1 and 2 mark the location where sensitivity tests were conducted (used by permission of Elsevier after Oz et al. (2015), J Hydrol 530:734–741, Fig. 6)

The long-term response to the predicted future sea-level rise was examined in hydrological simulation using Feflow, for the case of a 1 m rise in 100 years, suggested by several global climate models. The response of the interface to such a rise was found to depend mostly on the topography of the coastal area (Yechieli et al. 2010). In the Israeli case, where the topography in most parts of the coast is moderate (~1 to 2%), the sea-level rise is not expected to over flood the coastal area to large distance, and therefore, the FSI will not almost move inland (Fig. 8). This is contrary to the situation in coastal low lands and deltas of very mild topography (such as Bangladesh or the Nile delta) where small sea-level rise will assumingly cause flooding of large area and therefore salinization of groundwater to long distance from the shoreline. In cases where sea-level rise will be accompanied by a significant decrease in recharge (due to climate change and decrease in precipitation, as predicted by some of global climate models), the FSI is expected to move significantly inland. In fact, such a decrease in recharge is expected to result in a similar hydrological behavior to that of over-exploitation of fresh groundwater, which was encountered in many coastal aquifers around the world.

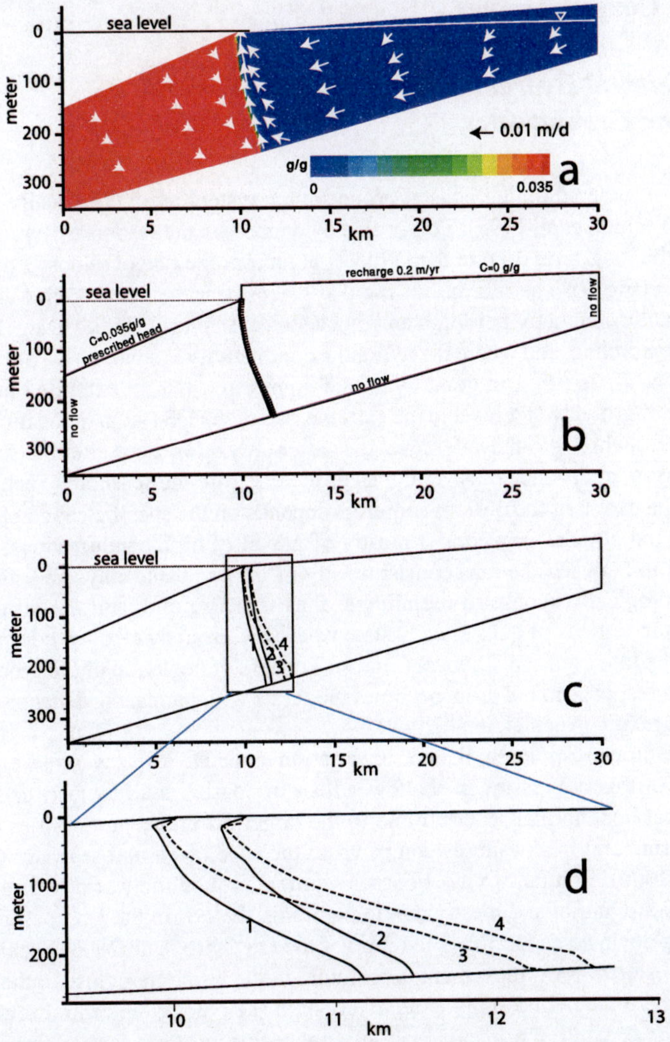


**Fig. 8** Hydrological steady-state simulations of the expected changes in the case of Mediterranean Sea-level rise. The value that represents the fresh–saline water interface is 50% seawater, which is ~11 g/L Cl in the Mediterranean coastal aquifer. **a** Basic simulation of the Mediterranean coastal aquifer showing the fresh and saline water bodies and the interface in between. Also shown are the flow velocity arrows. **b** Simulation of the case of steep (cliff) topography. The effect of sea-level rise is not observed. **c** Simulation of the case of a mild topography with a slope of 2.5‰. Sea-level rise of 1 m exhibits the shift of the fresh–saline water interface inland after 100 yr by 400 m (line 1 is at $T = 0$ years and line 2 is at $T = 100$ years). Also exhibited is the shift of the interface in the case of decrease in recharge by 50%, before and after sea-level rise (lines 3–4, respectively). **d** Blowup of Fig. 4c, showing the changes in the location of the fresh–saline water interface (after Yechieli et al. 2010)

## 2 DS Coastal Aquifer

### 2.1 Current Hydraulic Connection Between the DS and Groundwater

The DS and the adjoining coastal groundwater system are hydraulically interconnected (Yechieli et al. 1995). This is also evidenced by the response of groundwater level to the long-term drop in the DS level at an average rate of almost 1 m/yr since the 1970s (Fig. 9). The response is rapid at places near the DS shoreline and where the hydraulic conductivity is high and much slower in places farther away (few kms) from the shoreline, and where the hydraulic conductivity is smaller (Kiro et al. 2008; Yechieli et al. 2010). The good hydraulic connection is also exhibited in the fast response of groundwater level to the extreme rise of the DS (~2 m in the rainy winter of 1991/92) where groundwater level rose after only a few days after the start of the rainy season, at distance of several tens of meters from the seashore (Yechieli et al. 1995). The extent of the different aquifers depends on the specific geographic areas, whereby the alluvial fans consist mostly of gravel of high conductivities, whereas the areas in between the fans consist mostly of impermeable clays and silts. Thus, the hydrological response in the alluvial fans is very rapid, and the groundwater system is in a status of quasi-steady state where the groundwater level drops almost at the same rate as of the DS level (Fig. 9). The rate of depletion of the groundwater level, as a response to DS drop, becomes moderate with increasing distance from the shoreline inland (Yechieli et al. 2009).

The resultant response of the FSI to variation in the DS levels was also examined. The FSI in this area is much shallower than in coastal aquifers next to seas and oceans that host normal seawater due to the larger difference in densities between the DS brine and freshwater as compared to the case of normal seawater (Fig. 10; Yechieli 2000). Contrary to most coastal aquifers around the world, where the FSI move upward and inland mostly due to pumping, the FSI in the DS coastal aquifer is moving downward, following the drop of the DS (Kiro et al. 2008; Yechieli et al. 2010). This downward movement is profound in the boreholes, close to the DS and in area where the permeability is relatively high (i.e., along the central route of the alluvial fans), while the response is less noticeable in areas farther from the DS shoreline. The drop of the FSI is accompanied by a process of fast flushing of the residual DS brines that are replaced by fresher groundwater. The time response of the FSI to the above is longer than that of the groundwater levels (Kiro et al. 2008). As expected, the drop of the FSI is clearly noticed near the shore and up to a distance of 70 m inland, whereas no drop was observed at a distance of 700 meters from the shoreline (Fig. 10).

Another relevant issue of the hydrological system near the DS is the behavior of the circulation during the drop of the DS. This process, which occurs in any coastal aquifer connected to a saline water body, involves the penetration of the latter into the aquifer and its consequent back flow to the sea, driven by the flow of the fresh groundwater. This same process occurred also at the coastal aquifer of the DS at the

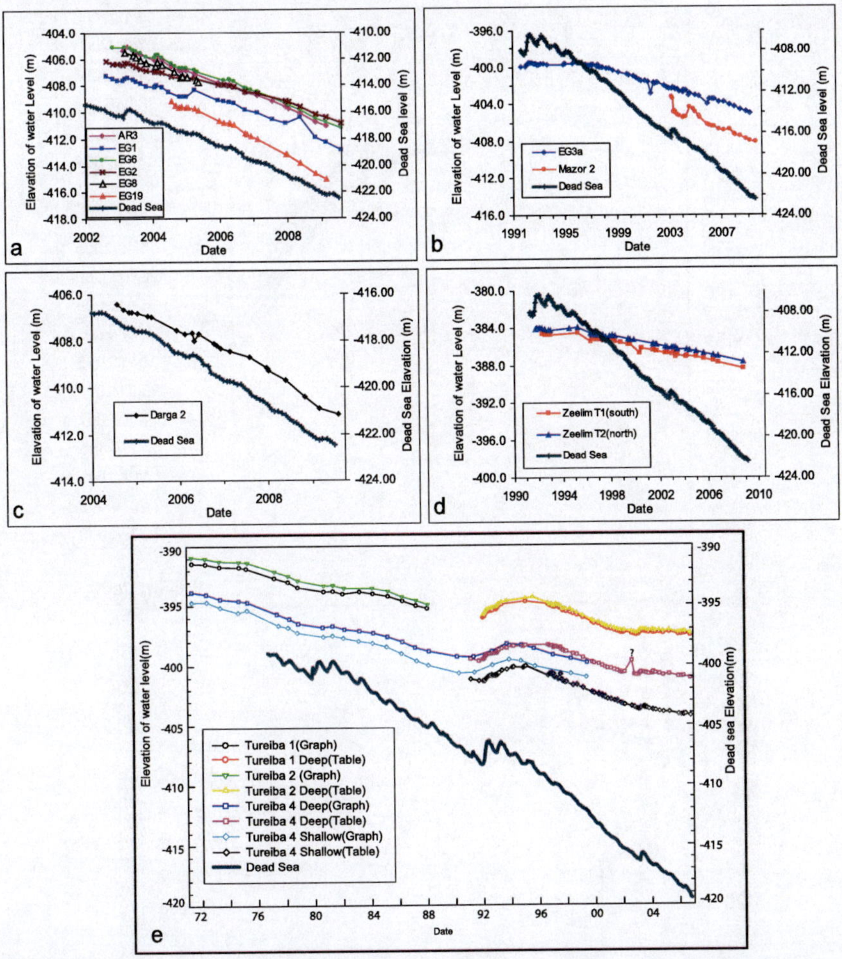


**Fig. 9** Groundwater levels in monitoring boreholes in the Dead Sea coastal aquifer at different locations and distances from the shoreline (in ~2005, given in parentheses) along with the Dead Sea level (data from the Hydrological Service of Israel). **a** AR3, EG1, EG2, EG6, EG8 and EG19, located at Wadi Arugot ~500 m from the Dead Sea shoreline. **b** EG-3a (~900 m), located 3 km south from Wadi Arugot. **c** Darga 2 (700 m), located in Wadi Darga. **d** Zeelim T1 and T2 (~4 km), located in Wadi Zeelim. **e** Turiebe (~500 m), located 10 km north of Wadi Darga. Graph, data from Hydrological Service of Israel (after Yechieli et al. 2010)

steady-state situation prior to the continuous recent drop of the DS levels. When the drop started, the inland brine flow increased accompanied by only very little back circulation to the DS. However, after a while, even though the drop of the DS level continued, the back flow circulation to the sea has resumed at a similar rate to the initial situation (Kiro et al. 2008). All the above indicate that the DS hydrological system is reversed to be in a quasi-steady-state condition.

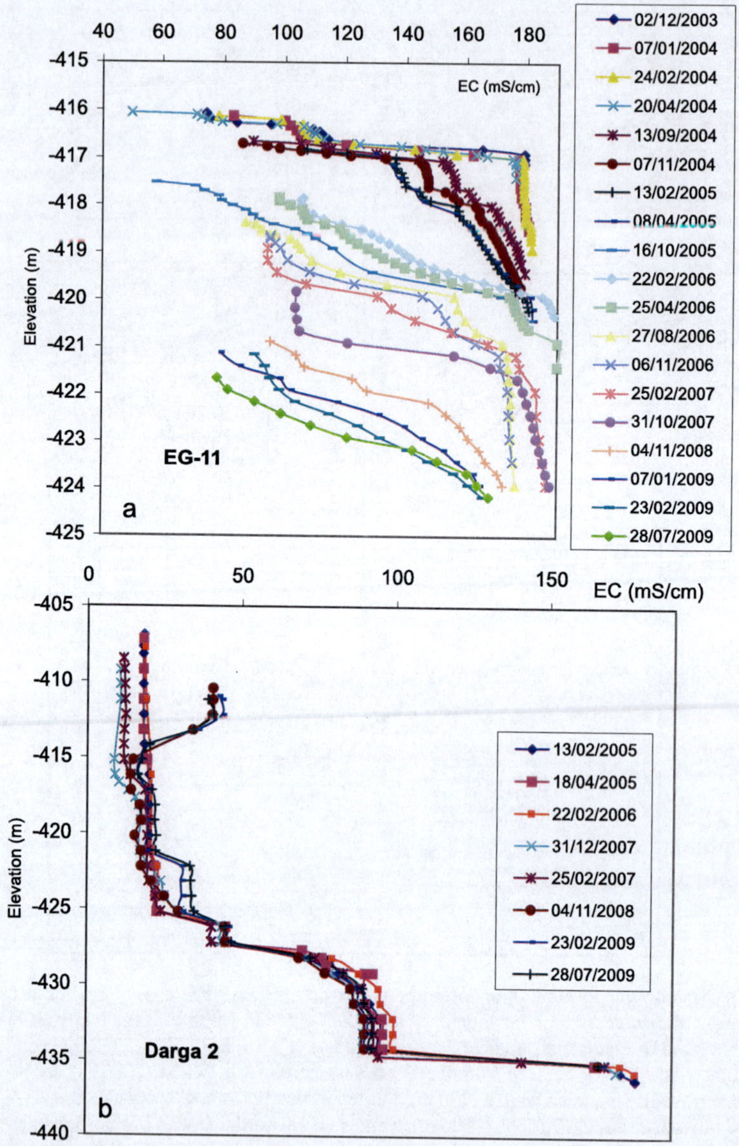


**Fig. 10** Profiles of EC in two boreholes in the Dead Sea coastal aquifer. **a** EG 11 borehole at a distance of 70 m from the shoreline. Significant changes in the location of the fresh–saline water interface are observed. **b** Darga 2 borehole, at a distance of 700 m from the shoreline, showing a small change in the interface depth (after Yechieli et al. 2010)

## 3 Past and Future

The relationship between the DS and groundwater system in its vicinity can be studied also on a larger time scale, either at the time of its Lisan Lake precursor (15–70 kyr B.P.) or in the future, when the DS is expected to reach a new steady state at a level of ~150 m below the current one. Past and future simulations of the hydrogeological setups show that the response to DS level changes is significant close to the DS but is less profound several kms away from the shoreline. Indeed, no significant effect was found on the location of the regional water divide in between the DS and Mediterranean Sea base levels, except for a drop in its elevation due to the future drop of the future DS (Yechieli et al. 2009). This is due to the fact that the location of the divide here is structurally controlled (see further discussion in Chapter "Paleohydrogeology of Israel").

The connection between the discharging hot brines and the DS water body is revealed mostly in the cluster of the Qedem springs. These emerging hot brines seem to be hydraulically connected to the DS, as their outlets follow the movement of the shoreline eastward (Weber et al. 2018). Hydrological simulation showed that when the DS level rises, its brine penetrates more into the surrounding rocks, whereas at times of level drops, the DS brine inland intrusion is smaller, and the thermal brine flow toward the DS increases (Fig. 11; Shalev and Yechieli 2007; Shalev et al. 2007). Preliminary dating of the Qedem brine also implied that its intrusion took place at the Lisan Lake high stand, some 30 kyrs ago (Weber et al. 2018).

## 4 The Groundwater System Next to a Stratified Lake

The DS has been stratified in several time spans of its history (e.g., until the 1980s and during parts of the Lisan Lake existence). A similar setup of a stratified lake may occur in the case that the planned Red Sea–Dead Sea canal will be materialized in the future. During such a period, when stratification occurs and the lower more concentrated water mass is significantly thicker than the upper water mass, the resultant simulated hydrological configuration is also expected to be more complex (Fig. 12; Oz et al. 2011, 2014). In such a case, instead of having the usual interface between fresh groundwater above and saline groundwater below, the expected configuration is of three water bodies with different salinities, and as a result, having three interfaces and circulation cells in between (Fig. 13).

Such stratification, whereby the upper layer is less saline than the deeper DS brine, can cause enhancement of the sinkhole formation, as shown in laboratory experiments (Oz et al. 2016). The general relation between the groundwater system and sinkhole formation is discussed in detail separately (Chapter "Hydrological and Geological Controls on the Evolution of the Dead Sea Sinkholes").

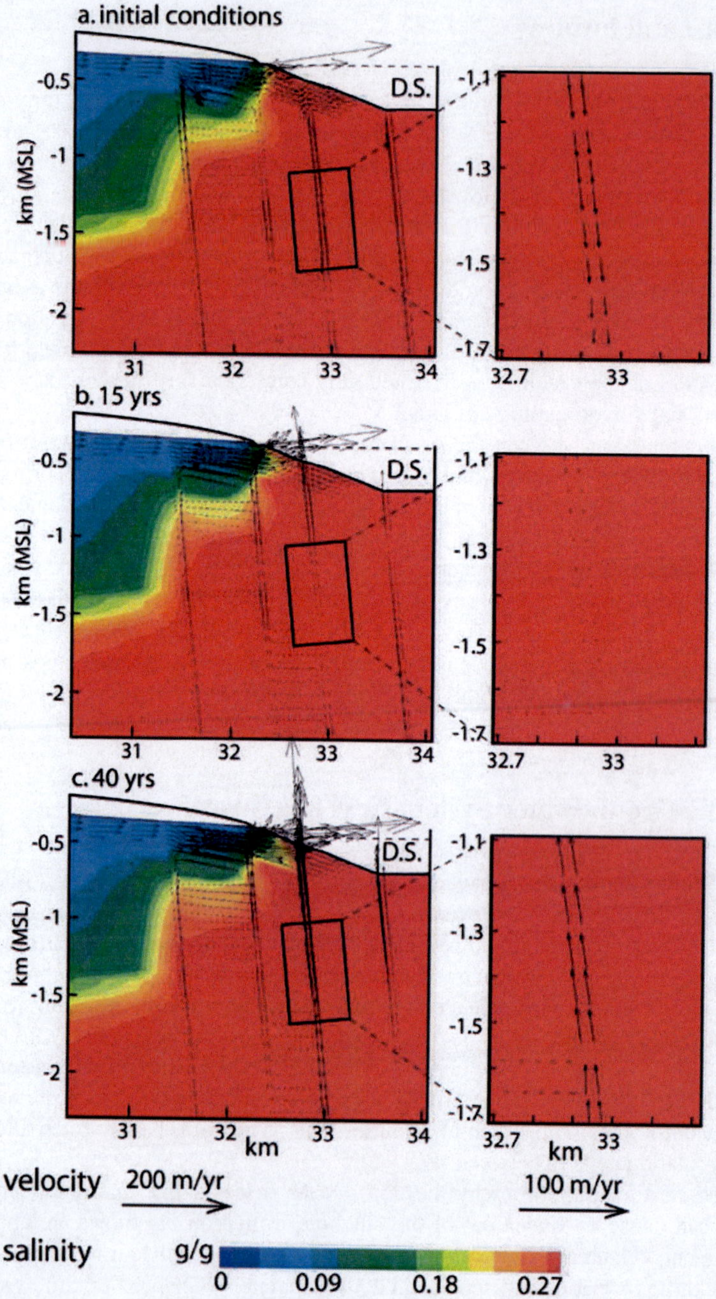


**Fig. 11** Salinity and groundwater velocity results, in the area close to the Dead Sea. **a** Steady-state flow. **b** 15 years after recession has begun (time of change of flow direction, no flow). **c** 40 years after recession has begun (present situation) (after Shalev and Yechieli 2007)

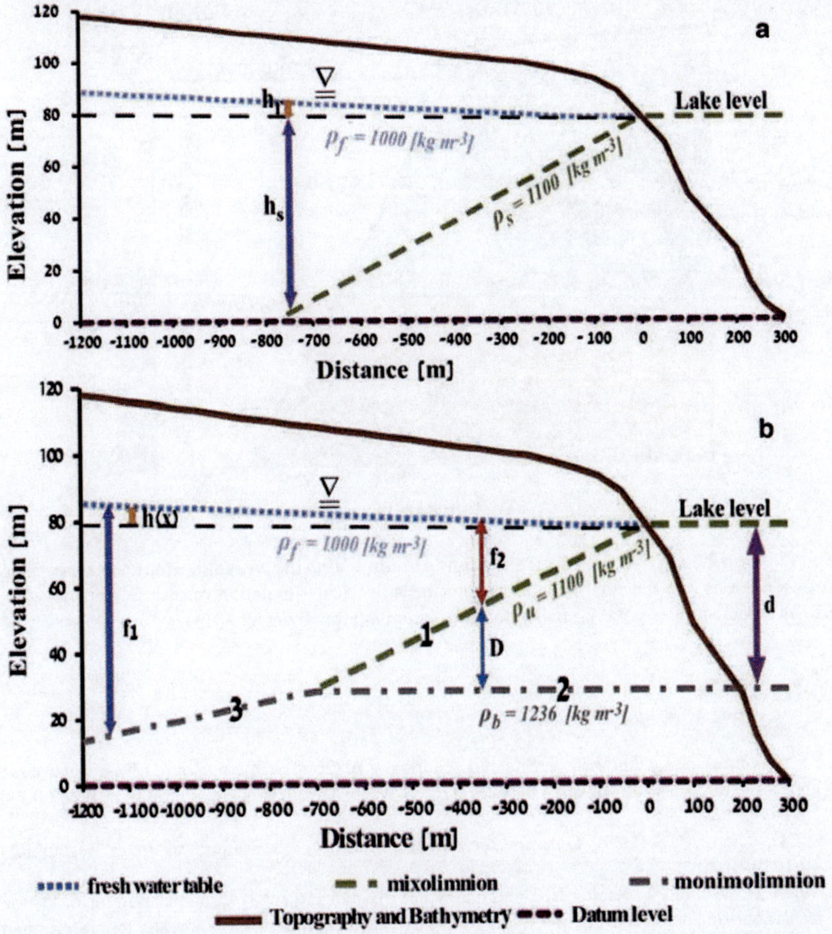


**Fig. 12** Effect of density stratification in the DS on groundwater system—theoretical consideration. Groundwater interface configuration adjacent to **a** a holomictic lake and **b** a meromictic lake. Numbers represent the interfaces between the three water types (after Oz et al. 2011)

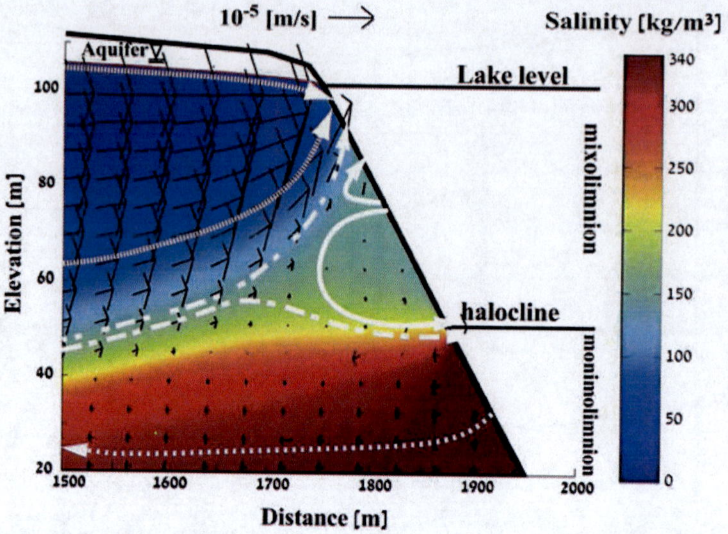


**Fig. 13** Distribution of concentration and flow patterns within the wedge-like intruded zone. Black arrows demonstrate the flow velocities from the numerical simulation results. Schematic white arrows demonstrate general groundwater flow patterns (after Oz et al. 2011)

## References


Kiro Y, Yechieli Y, Lyakhovsky V, Shalev E, Starinsky A (2008) Time response of the water table and saltwater transition zone to a base level drop. Water Resour Res. https://doi.org/10.1029/2007WR006752

Lee CH, Cheng RTS (1974) On seawater encroachment in coastal aquifers. Water Resour Res 10:1039–1043

Levanon E, Yechieli Y, Shalev E, Friedman V, Gvirtzman H (2013) Reliable monitoring of the transition zone between fresh and saline waters in Coastal Aquifers. Groundwater monitor Remed 33:101–110

Levanon E, Shalev E, Yechieli Y, Gvirtzman H (2016) Fluctuations of fresh-saline water interface and of water table induced by sea tides in unconfined aquifers. Ad Water Resour 96:34–42

Levanon E, Yechieli Y, Gvirtzman H, Shalev E (2017) Tide-induced fluctuations of salinity and groundwater level in unconfined aquifers—field measurements and numerical model. J Hydrol 551:665–675

Levanon E, Gvirtzman H, Yechieli Y, Oz I, Ben-Tzur E, Shalev E (2019) The dynamics of sea-tide-induced fluctuations of groundwater level and freshwater-saltwater interface in coastal aquifers—laboratory experiments and numerical modeling. Geofluids. https://doi.org/10.1155/2019/6193134

Oz I, Shalev E, Gvirtzman H, Yechieli Y, Gavrieli I (2011( Groundwater flow patterns adjacent to a long-term stratified (meromictic) lake. Water Resour Res 46. https://doi.org/10.1029/2010wr010146

Oz I, Shalev E, Yechieli Y, Gavrieli I, Gvirtzman H (2014) Flow dynamics and salt transport in a coastal aquifer driven by a stratified saltwater body: lab experiment and numerical modeling. J Hydrol 511:665–674

Oz I, Shalev E, Yechieli Y, Gvirtzman H (2015) Saltwater circulation patterns within the freshwater–saltwater interface in coastal aquifers: laboratory experiments and numerical modeling. J Hydrol 530:734–741

Oz I, Shalev E, Yechieli Y, Gavrieli I, Levanon E, Gvirtzman H (2016) Salt dissolution and sinkhole formation: results of laboratory experiments. J Geophys Res Earth Surf 121:1746–1762. https://doi.org/10.1002/2016JF003902

Shalev E, Yechieli Y (2007) The effect of Dead Sea level fluctuations on the discharge of thermal springs. Isr J Earth Sci 56:19–27

Shalev E, Lyakhovsky V, Yechieli Y (2007) Is convective heat transport significant at the Dead Sea Basin? Geofluids 7:292–300

Shalev E, Lazar A, Wollman S, Kington S, Yechieli Y, Gvirtzman H (2009) Biased monitoring of fresh-salt water mixing zone in coastal aquifers. Groundwater 47:49–56

Weber N, Yechieli Y, Stein M, Yokochi R, Gavrieli I, Zappala J, Mueller P, Lazar B (2018) The circulation of the Dead Sea brine in the regional aquifer. Earth Planet Sci Lett 493:242–261

Yechieli Y (2000) Fresh-saline water interface in the western Dead Sea area. Groundwater 38:615–623

Yechieli Y, Ronen D, Berkovitz B, Dershovitz WS, Hadad A (1995) Aquifer characteristics derived from the interaction between water levels of a terminal lake (Dead Sea) and an adjacent aquifer. Water Resour Res 31:893–902

Yechieli Y, Kafri U, Wollman S, Shalev E, Lyakhovsky V (2009) The effect of base level changes and geological structures on the location of the groundwater divide, as exhibited in the hydrological system between the Dead Sea and the Mediterranean Sea. J Hydrol 378:218–229

Yechieli Y, Shalev E, Wollman S, Kiro Y, Kafri U (2010) Response of the Mediterranean and Dead Sea coastal aquifers to sea level variations. Water Resour Res 46:W12550. https://doi.org/10.1029/2009WR00870

# Geochemical Aspects of Seawater Intrusion into the Mediterranean Coastal Aquifer

**Amos Russak, Boaz Lazar, and Orit Sivan**

## 1 Introduction

Seawater intrusion into coastal aquifers has been observed and discussed for many years. In most cases, seawater intrusion rates were inferred from an increase in salinity (e.g., Acworth and Dasey 2003; Linderfelt and Turner 2001), or movement of the fresh–saline water interface (FSI) (e.g. Melloul and Zeitoun 1999). The change in the chemical composition was also suggested to be a tool for identifying the process of seawater intrusion (e.g., Mercado 1985; Stuyfzand 1993). Direct determination of the seawater intrusion rate was previously done with the use of radiocarbon and tritium tracers (Sivan et al. 2005; Yechieli et al. 2001, 2009). The general topic of dating of groundwater, including the determination of the rate of seawater intrusion, is dealt with in detail in Chapter "General Information and Hydrogeology of the Mediterranean and Dead Sea Coastal Aquifers and Their Relation with Their Base Level" of this book. The following sub-chapter describes the use of geochemical and stable isotopic parameters as a way to monitor seawater intrusion, its rate and the biogeochemical processes that are involved with the intrusion, mostly in the fresh–saline water interface (FSI) or transition zone.

A. Russak (✉)
Zuckerberg Institute for Water Research, Blaustein Institute for Desert Research Studies, Ben Gurion University of the Negev, Sde Boqer, Israel
e-mail: amos.russak@gmail.com

B. Lazar
The Institute of Earth Sciences, The Hebrew University of Jerusalem, Edmond J. Safra Campus, Givat Ram, 91904 Jerusalem, Israel
e-mail: boaz.lazar@huji.ac.il

O. Sivan
Department of Geological and Environment Sciences, Ben Gurion University of the Negev, Beer Sheva, Israel
e-mail: oritsi@bgu.ac.il



U. Kafri and Y. Yechieli (eds.), *The Many Facets of Israel's Hydrogeology*, Springer Hydrogeology, https://doi.org/10.1007/978-3-030-51148-7_6

## 2 Methodology

The chemical evolution of groundwater during seawater intrusion was examined by field sampling, laboratory experiments and complementary geochemical computer modeling. Field sampling was conducted in several locations along the coastal aquifer of Israel and at different seasons to examine the seasonal effect of salinization (potentially in the summer) and freshening (potentially in the winter) of the aquifer, on the water composition.

The first step in the field was to conduct a profile of the FSI as an indicator for the salinization or freshening stages of the aquifer. This was obtained by several methods: the first sampling was performed with hand sampler at an interval resolution of 1–2 m to identify the FSI zone. At a later stage, a much higher-resolution profile was obtained using the multilayered sampler (MLS) (Ronen et al. 1986, 1987), which consists of a solid PVC rod with dialysis cells at intervals of ~10 cm. This PVC rod was inserted into the FSI zone and remained there for ~1 month until a steady-state condition was reached with the surrounding groundwater and the obtained sample represented truly the aquifer. This method provided very detailed information and resulted in high-resolution sampling (about 20 samples in 2 m). However, each sample is very small (about 5 mL) which does not allow to sample other depth intervals at the same time. Another method of sampling was conducted with a submersible pump at different depths (Russak et al. 2016), which allowed to sample larger volumes of water, still with relatively high depth interval resolution. The effect of this pumping was assumed to be about two meters. Therefore, the distance between each sampling point needed to be long (minimum two meters) and practically only 3–4 sampling points was conducted in each bore hole: one sampling point of fresh groundwater, 1–2 points across the FSI and one point in the saline groundwater underneath. A higher sampling resolution was obtained using a peristaltic pump (and a thin tube), in cases of a relatively shallow water table (of a few meters), such as those of some boreholes near the shoreline (Russak et al. 2015b, 2016). The effect of pumping using the peristaltic pump was small, enabling to sample, every 1 m, a volume of up to 0.5 L.

Laboratory column experiments were conducted to mimic the field conditions in a more controlled system. Thus, the experiments were done under aerobic or anaerobic conditions, as observed usually in the fresh groundwater and saline groundwater columns, respectively (Russak et al. 2015a). The anaerobic conditions were achieved by injecting nitrogen gas throughout the experiment and monitoring the dissolved oxygen (DO) levels, making sure that no oxygen was left and no seepage of oxygen occurred during the experiments (Russak et al. 2015a). The column was first filled with sand brought from the study area from a depth of 1–2 m, representing the aquifer sediment and then it was saturated with fresh groundwater taken from the study site. The seawater intrusion was simulated by salinization experiment, which was conducted by injecting seawater (taken from the beach alongside the study site) into the column until the composition of the water in the column reached seawater composition. The flushing was simulated by the freshening experiment, namely by injecting fresh groundwater until the composition in the column returned to be that of the fresh groundwater. Samples for major ion chemistry were taken in all experiments. In some of the experiments, other parameters were examined, including nutrients as phosphate ($PO_4^{3-}$), dissolved silica (DSi), ammonium ($NH_4^+$), nitrite ($NO_2^-$) and nitrate ($NO_3^-$) (including stable isotope of nitrogen in nitrate—$\delta^{15}N_{NO_3}$) and metal ions as boron (B), manganese ($Mn^{2+}$), barium ($Ba^{2+}$) and lithium ($Li^+$).

The experimental results were modeled using PHREEQC (Parkhurst and Appelo 1999) where the only geochemical process taking into consideration is cation exchange. The data input from the experiment into the model was the composition of the end member water (fresh groundwater and seawater), the cation exchange capacity (CEC) of the sediment in the column in mol units as well as the dispersivity coefficient (estimated by the breakthrough curve of the $Cl^-$) as the only transport parameter, whereby diffusion was assumed to be negligible. The CEC of the sediment (in units of meq/100 g) was estimated from the content of the percentage of organic carbon and clay in the sediment. In order to get CEC in mol unit, the mass of the sediment in the column was calculated by multiplying the volume of the column with the porosity of the sediment and with the density of sand.

## 3 Behavior of Major Ions During Salinization and Freshening Processes

The dynamics of major ions at the FSI was quantified by field sampling, laboratory experiments and modeling (Sivan et al. 2005; Russak and Sivan 2010). The good correlation between the seasonal field sampling, the salinization/freshening experiments and the PHREEQC model clearly indicates that there is significant salinization of the aquifer in summer and freshening in the winter. Furthermore, the results of the model simulated well the experiment results (Fig. 1), indicating that cation exchange is the dominant process throughout the experiments.

The study showed that during seawater intrusion (and salinization experiment), there were deviations of the water composition from mixing between fresh water and seawater due to cation exchange processes. These modifications include the increase in $Ca^{2+}$ (Fig. 2) and the decrease in $K^+$, similarly to previous observations in the laboratory (e.g., Appelo et al. 1990) and in the field (e.g., Stuyfzand 1993, 2008). The opposite phenomenon was observed during the flushing of the aquifer by fresh groundwater (and freshening experiment), where there was an increase in $K^+$ and a decrease in $Ca^{2+}$. It should be noted that the change of these ions is much more pronounced during the salinization than during freshening, and the latter seems to be almost a product of a simple mixing. Nevertheless, during the laboratory experiments, which were controlled and therefore could provide better exhibition and understanding of these processes, the effect of freshening was indeed clearly observed as an increase in $K^+$ and decrease in $Ca^{2+}$ (Russak and Sivan 2010). As expected,

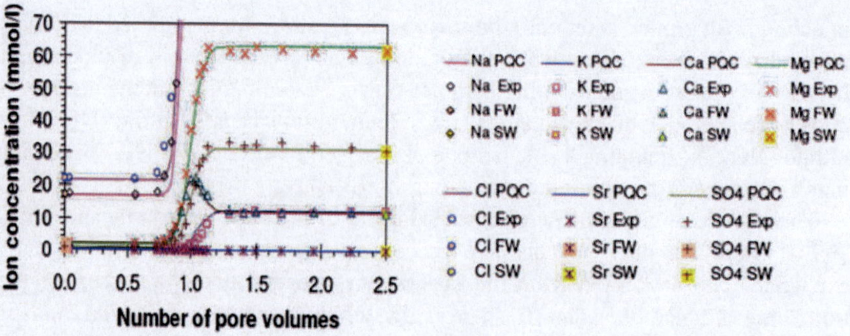


**Fig. 1** Breakthrough curves from the salinization experiment—results (symbols) and model (lines). The results from the experiments and the model are in close agreement with each other, suggesting that the model represents the experiment well. This suggests that cation exchange is the dominant process affecting the major cations during salinization and freshening of the aquifer

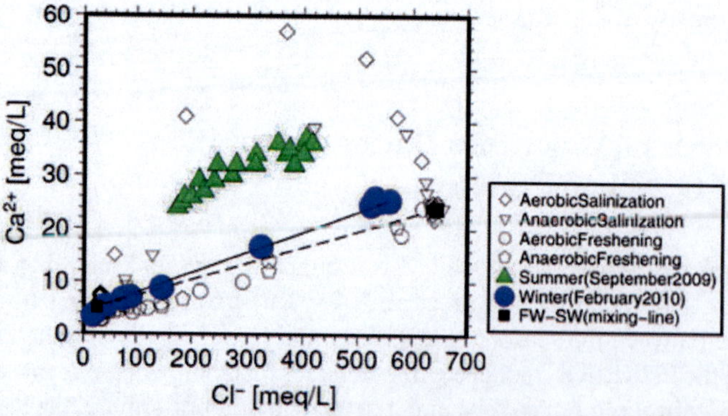


**Fig. 2** $Ca^{2+}$ versus $Cl^-$ from field measurements in summer (September 2009, green triangle) and winter (February 2010, blue circle) from Poleg area and from aerobic and anaerobic salinization and freshening experiments (open rhombus, inverted triangle, octagon and pentagon, respectively). The black dashed line represents mixing between fresh groundwater and seawater (black square), and the black line represents mixing between fresh groundwater and saline groundwater (used by permission of Elsevier after Russak and Sivan 2010; Russak et al. 2015a, Fig. 2)

there is no real effect of the oxygen levels on the behavior of these major ions, which seem to have similar concentrations in both aerobic and anaerobic conditions.

Salinization and freshening processes were found to be reversible for the case of major ions, meaning that a similar amount of $Ca^{2+}$ is desorbed during seawater intrusion as absorbed during freshening (Russak et al. 2016). The opposite trend is observed for $K^+$, where a similar amount of $K^+$ is desorbed during freshening and absorbed during salinization.

These changes in the major ions chemistry were used to develop an index for identifying the status of the aquifer, either at a process of seawater intrusion or at

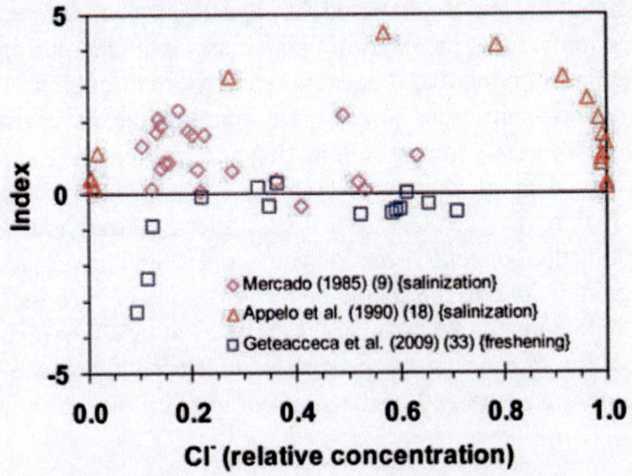


**Fig. 3** Relation between salinization index versus Cl—from data of previous studies. The first study described salinization of the Israeli coastal aquifer (Mercado 1985), the second one described a salinization laboratory experiment (Appelo et al. 1990), and the last one described a freshening of the aquifer near Venice, Italy (Gattacceca et al. 2009). The index results fit quite well the conclusion of these studies (used by permission of the American Chemical Society, after Russak and Sivan 2010, Fig. 5)

flushing (Fig. 3). Index values higher than zero denote status of seawater intrusion while values equal to, below zero denote a freshening process. Other chemical indicators (e.g., the ratio of $Na^+/Cl^-$ vs. $Cl^-$ concentration) for seawater intrusion were previously suggested by Mercado (1985).

The above processes also influence the carbonate system in groundwater in the vicinity of the FSI. The dynamics of seawater intrusion was studied by Sivan et al. (2005) using carbon isotopes ($^{13}C$ and $^{14}C$). The negative $\delta^{13}C$ values of dissolved inorganic carbon (DIC) in saline groundwater (>80% seawater) of −8‰ versus 0‰ of seawater suggest that organic carbon is oxidized to DIC during the penetration of seawater into the aquifer. Mass balance calculations also indicate that this process is also responsible for lowering the $^{14}C$ activities from ~100% of modern carbon (pmc) at the sea to ~60 pmc in saline groundwater below the FSI of the aquifer at a distance of ~70 m from the shoreline, implying that the oxidized carbon is old with $^{14}C$ activity close to zero.

## 4 Behavior of Minor and Trace Elements During Salinization and Freshening Processes

The second stage included the examination of the behavior of minor and trace elements in the FSI using field data and experiments (Russak et al. 2016). While $Br^-$ concentrations fell on a mixing line between fresh water and seawater during both salinization and freshening processes, as expected from this conservative ion, the concentrations of $Li^+$ and B decreased during salinization, similar to the behavior of $K^+$ (Fig. 4). Moreover, $Li^+$ concentration decreased with increasing distance from the shoreline, again implying that it is exchanging more throughout the intrusion.

$Ba^{2+}$ and $Mn^{2+}$ concentrations were also affected by the above processes. These elements behaved differently during salinization and freshening (Russak et al. 2016). Both cations showed along the experiments a breakthrough curve considerably different from that of the conservative ions. Indeed, throughout the process of seawater intrusion, the concentrations of $Ba^{2+}$ and $Mn^{2+}$ increased by more than an order of magnitude, due to the strong effect of desorption of $Mn^{2+}$ and $Ba^{2+}$.

The studied Nitzanim site hosts several boreholes at different distances from the shoreline, thus allow examining the processes with time and space of penetration (Fig. 5). The sampling at the end of the summer showed that the concentrations of

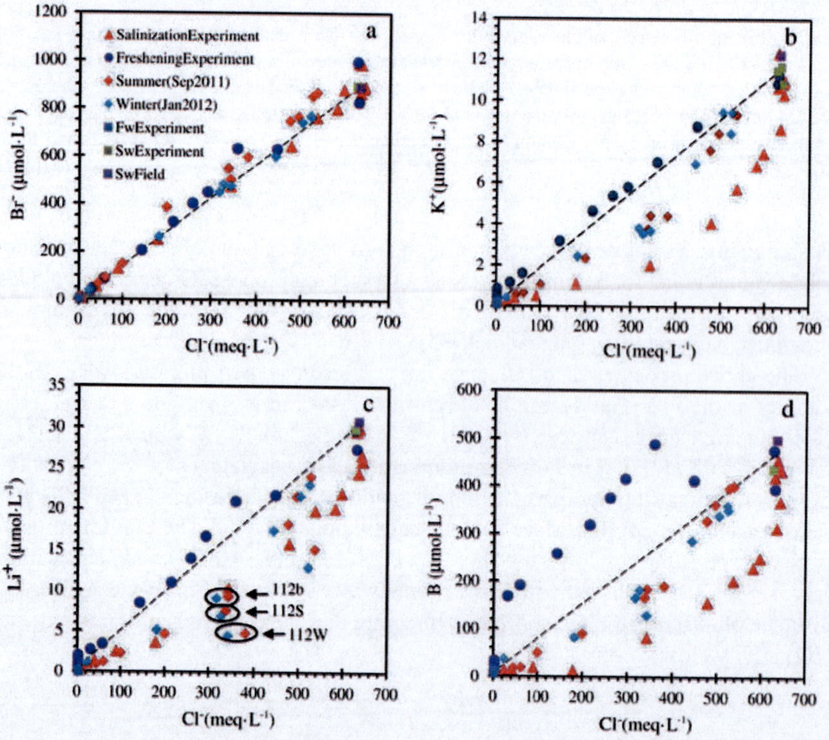


**Fig. 4** $Br^-$ (**a**), $K^+$ (**b**), $Li^+$ (**c**) and B (**d**) versus $Cl^-$ of data obtained from the salinization (red triangle) and freshening (blue circle) experiments and field data from summer (red rhombus) and winter (blue rhombus) from Nitzanim area. Black dashed line represents the mixing between fresh water (FW, blue square) and seawater (SW) from the experiments (green square) and the field (purple square). $Br^-$ is a conservative species, while B and $Li^+$ are non-conservative. Black circles mark the different observation wells. Note that $Li^+$ concentration is depleted in the same order as the distance of the wells from the shore. The legend appears on one graph but refers to all the graphs in the figure (used by permission of Elsevier after Russak et al. 2016, Fig. 6)

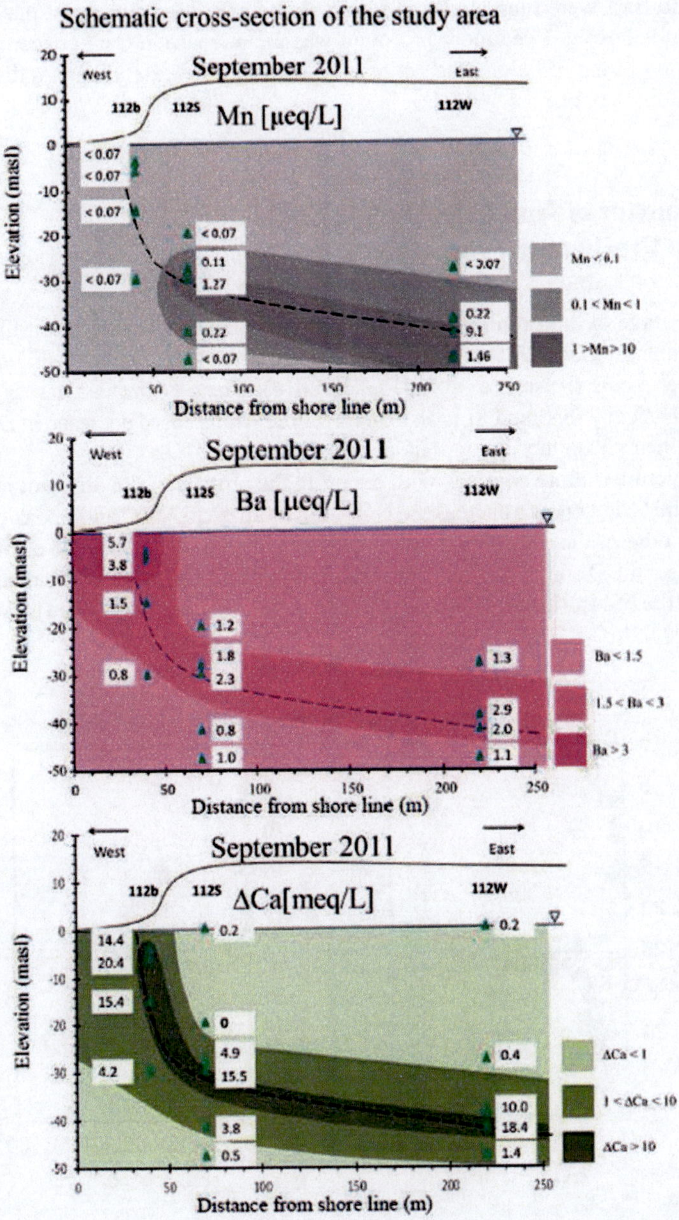


**Fig. 5** A cross section showing variations in $Mn^{2+}$ and $Ba^{2+}$ concentration and ΔCa (difference between measured and expected from mixing) in 2D view from the field in summer in Nitzanim area. Green triangles represent the sampling points. The different areas (in different shade of green) represent range of values to emphasize the area of the enrichment of $Ca^{2+}$. The dashed line represents salinity of about half of the salinity in seawater. The $Ca^{2+}$ enriched mostly in the FSI area (used by permission of Elsevier after Russak et al. 2016, Figs. 8, 9 and 10)

$Mn^{2+}$ and $Ba^{2+}$ were found to be considerably higher in the vicinity of the FSI than below and above it. Even more profound was the increase in $Ca^{2+}$ concentration in the vicinity of the FSI along the entire examined traverse extending to a distance of 230 m from the seashore.

## 5 Behavior of Nutrients During Salinization and Freshening Processes

The last stage of the examination dealt with the behavior of nutrients at the fresh–saline water interface (FSI) during seawater intrusion and freshening using field data and experiments (Russak et al. 2015a, b, 2016). The DO profiles across the FSI showed basically that the FSI is also an oxycline, with a sharp decrease in DO when crossing from fresh to saline groundwater (Fig. 6).

The picture is more complex with regard to the profiles of the different nutrients. Across the FSI, there is a decrease in $NO_3^-$, an increase in $NH_4^+$ and a peak of $NO_2^-$ in the middle of the transition zone of the FSI (Fig. 7a). The above is attributed to the change in DO concentration from a saturation value above the FSI to much lower values at the FSI and below it. In oxic water, the oxidized form of nitrogen is favorable,

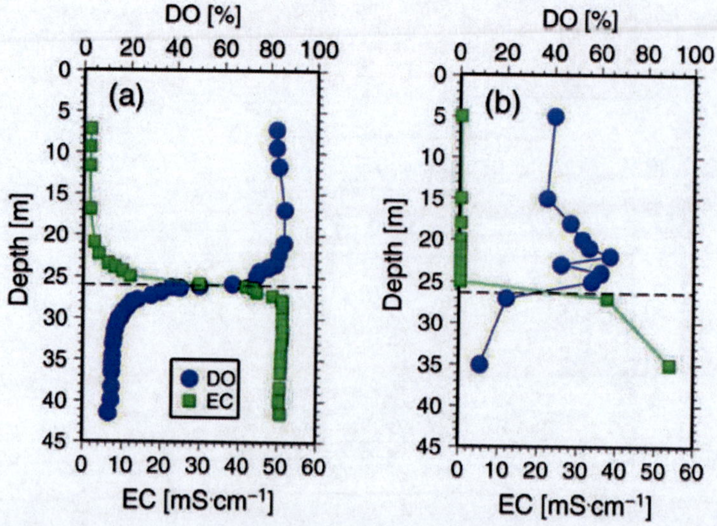


**Fig. 6** EC and DO profiles in Med. aquifer across the FSI. Profiles from well in the Poleg area during **a** 2001 (adapted from Sivan et al. 2005) and **b** 2010 for dissolved oxygen (DO, blue) and electrical conductivity (EC, green). The black dashed line represents the middle of the FSI zone. It can be seen that the FSI, which is the transition zone between fresh and saline groundwater, is also an oxycline, with a transition from aerobic through suboxic to anoxic conditions, respectively (used by permission of Elsevier after Russak et al. 2015a, Fig. 3, and Sivan et al. 2005)

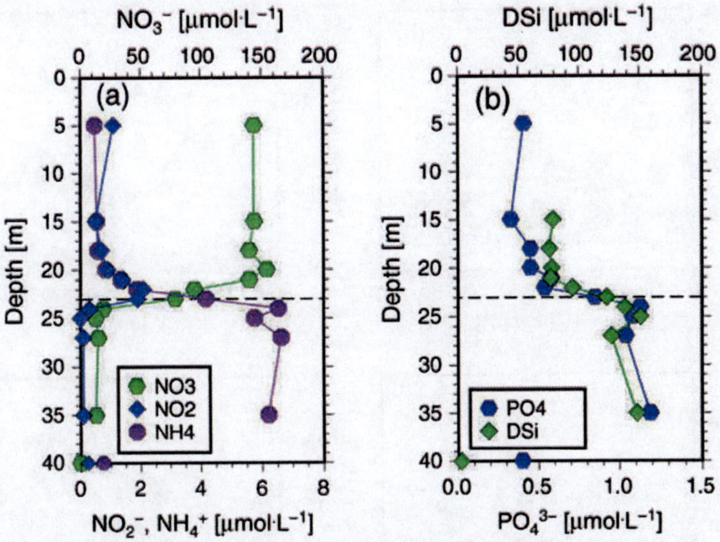


**Fig. 7** Profiles of nutrients from the Poleg well during winter (February 2010) for **a** DIN ($NO_3^-$, green hexagon; $NO_2^-$, blue rhombus; and $NH_4^-$, purple octagon), **b** $PO_4^-$ (blue hexagon) and DSi (green rhombus). Symbols on the 40 m depth line represent the seawater values of the nutrients. The black dashed line represents the middle of the FSI zone. Note that the concentration of $NO_2$ is low in the fresh and saline groundwater and is enriched in the FSI zone (used by permission of Elsevier after Russak et al. 2015a, Fig. 4)

whereas in anoxic conditions the reduced form is favorable. The peak of nitrite is attributed mainly to the microbial processes of denitrification or nitrification, where in both cases the nitrite is the intermediate species. An increase in $PO_4^{3-}$ and DSi was also found at the saline water even though their values in seawater are quite similar to those of fresh groundwater and much lower than those found in the saline water near the FSI (Fig. 7b). This phenomenon of higher concentration of $PO_4^{3-}$ in the saline groundwater was mentioned in other studies (e.g., Price et al. 2006), explaining this increase due to re-mineralization of organic carbon and/or desorption from calcium carbonate. Later experiments, simulating seawater intrusion, supported the latter, whereby the seawater released $PO_4^{3-}$ which was adsorbed when the aquifer was saturated with fresh groundwater (Price et al. 2010). Similar results were obtained in the laboratory experiments, showing the effect of salinization/freshening and thus the DO on the behavior of nutrients (Fig. 8). During salinization, the extension of the saline anoxic zone occurs and thus denitrification takes place, as shown by the decrease of $NO_3^-$ concentration and the increase in $NO_2^-$ concentrations and $\delta^{15}N_{NO_3}$ values (Fig. 9).

Besides the microbial processes, during the anaerobic experiment, $NH_4^+$ was clearly enriched. This enrichment is explained by cation exchange process, since the peak of $NH_4^+$ is almost identical in shape and timing as the peak of $Ca^{2+}$

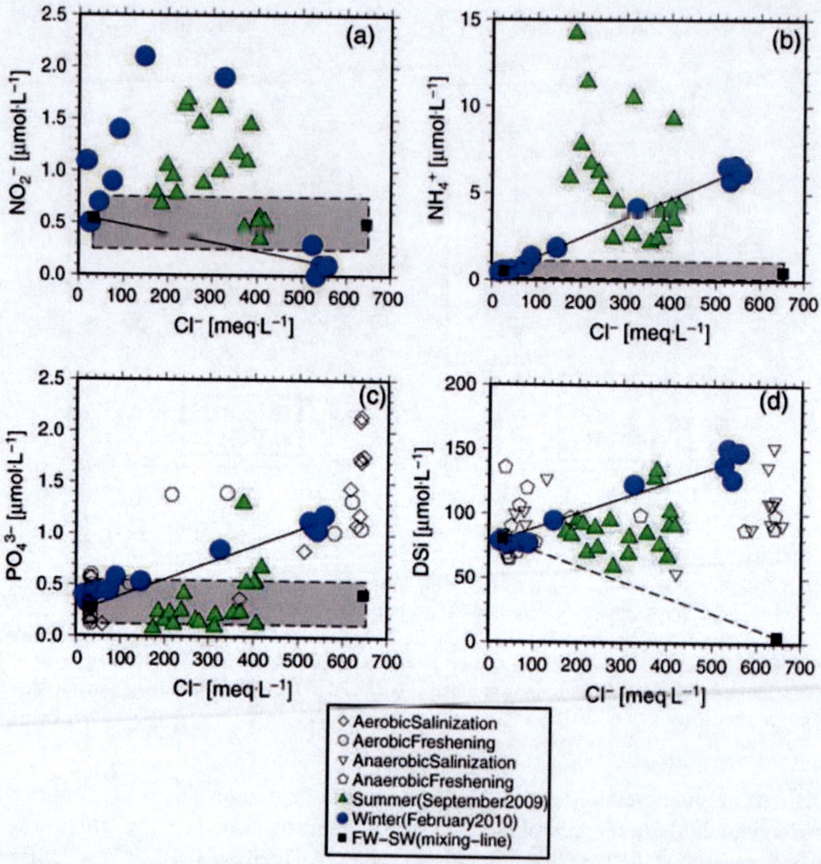


**Fig. 8** $NO_2^-$ (**a**), $NH_4^+$ (**b**), $PO_4^{3-}$ (**c**) and DSi (**d**) versus $Cl^-$ from field measurements in summer (September 2009, green triangle) and winter (February 2010, blue circle) in Poleg area and from aerobic and anaerobic salinization and freshening experiments (open rhombus, inverted triangle, octagon and pentagon, respectively). Mixing line between fresh groundwater and seawater (black square) is represented by a dashed line (for DSi) or gray rectangle (for $NO_2^-$, $NH_4^+$ and $PO_4^{3-}$). The black line represents the mixing line between fresh groundwater and saline groundwater. Note that the nutrients are enriched in the FSI zone (used by permission of Elsevier after Russak et al. 2015a, Fig. 7)

(Fig. 10a). The $PO_4^{3-}$ showed also an enrichment during the salinization experiment, and under anaerobic conditions the enrichment was more significant than under aerobic conditions (Fig. 10b).

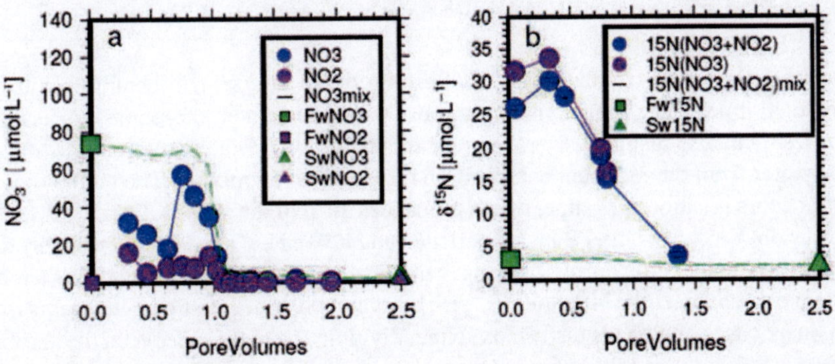


**Fig. 9** Breakthrough curves from anaerobic salinization experiments of the column outlet of $NO_3^-$ (blue circle) and $NO_2^-$ (purple octagon) (**a**) and δ15N (blue circle) (**b**). The green dashed line represents the expected concentration of $NO_3^-$ and $\delta^{15}N$ of $NO_3^-$ from mixing between the fresh water (FW, green square) and seawater (SW, green triangle). The concentrations of $NO_2^-$ in the sources (water tanks) of fresh water and seawater are plotted as purple pentagons and inverse triangles, respectively. It is shown that $NO_3^-$ is depleted, while $NO_2^-$ and $\delta^{15}N$ are enriched, suggesting that denitrification is occurred under anaerobic conditions

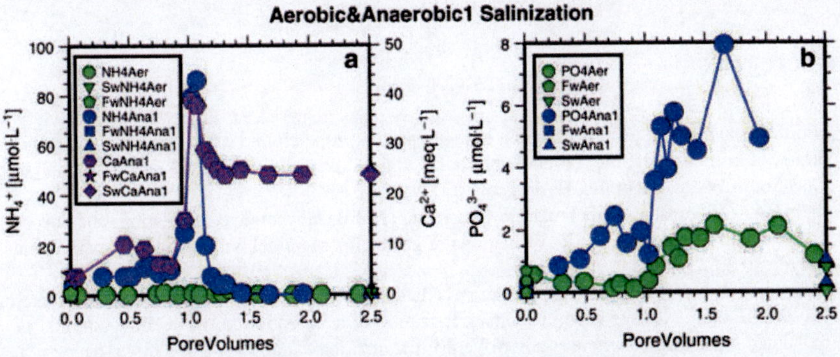


**Fig. 10** Breakthrough curve of $NH_4^+$ from the aerobic (green pentagon) and of $NH_4^+$ (blue circle) and $Ca^{2+}$ (purple hexagon) from the anaerobic 1 salinization experiments (**a**) and breakthrough curve of $PO_4^{3-}$ from the aerobic (green octagon) and anaerobic 1 (blue circle) salinization experiments (**b**) from the column outlet. The fresh water (FW) and seawater (SW) concentrations of $Ca^{2+}$ (purple star and diamond, respectively) and $NH_4^+$ and $PO_4^{3-}$ in aerobic (green pentagon and inverse triangle, respectively) and anaerobic experiment (square and triangle, respectively). $NH_4^+$ and $PO_4^{3-}$ are enriched during the salinization experiment. The similarity between $NH_4^+$ and $Ca^{2+}$ implying that the process affecting the $NH_4^+$ is the same as the one affecting the $Ca^{2+}$—cation exchange. Interesting to note is that the $NH_4^+$ and $PO_4^{3-}$ are enriched more under the anaerobic conditions

## 6 Summary and Conclusions

Seawater intrusion (salinization) and flushing of the aquifer (freshening) occur on seasonal time scale. They do not only cause the groundwater to become more saline or fresh, but also result in several significant processes, deviating the composition of the water from the expected composition resulted just by mixing. The main process that occurs is cation exchange, which influences most of the cations. During seawater intrusion, $Ca^{2+}$, $Mn^{2+}$ and $Ba^{2+}$ are enriched and $K^+$ and $Li^+$ are depleted whereby the opposite trend occurs during flushing. Furthermore, the salinization and freshening changes also affect the $NH_4^+$, $PO_4^{3-}$, B and even DSi concentrations. It appears that in many cases, the FSI is also an oxycline, dividing the aquifer between the aerobic fresh water body and the suboxic to anoxic saline water one. The $NO_3^-$ is depleted due to denitrification in the anoxic zone, while enrichment of $NO_2^-$ is shown in the FSI/oxycline.

**Acknowledgements** We would like first to thank and honor Mr. Haim Hemo from the Geological Survey of Israel (GSI) who passed away on October 2018 and was one of the cornerstones in all the field work of the coastal aquifer in Israel.

## References

Acworth RI, Dasey GR (2003) Mapping of the hyporheic zone around a tidal creek using a combination of borehole logging, borehole electrical tomography and cross-creek electrical imaging, New South Wales, Australia. Hydrogeol J 11(3):368–377

Appelo CAJ, Willemsen A, Beekman HE Griffioen J (1990) Geochemical calculations and observations on salt water intrusions II. Validations of geochemical model with laboratory experiments. J Hydrol 120:225–250

Gattacceca JC, Vallet-Coulomb C, Mayer A, Claude C, Radakovitch O, Conchetto E, Hamelin B (2009) Isotopic and geochemical characterization of salinization in the shallow aquifers of a reclaimed subsiding zone: the southern Venice Lagoon coastland. J Hydrol 378:46–61

Linderfelt WR, Turner JV (2001) Interaction between shallow groundwater, saline surface water and nutrient discharge in a seasonal estuary: the Swan-Canning system. Special issue: integrating research and management for an urban estuarine system: the Swan-Canning Estuary, Western Australia. Hydrol Process 15(13):2631–2653

Melloul AJ, Zeitoun DG (1999) A semi-empirical approach to intrusion monitoring in Israeli coastal aquifer. In: Bear J, Cheng AHD, Sorek S, Ouazar D, Herrera I (eds) Seawater intrusion in coastal aquifers—concepts, methods and practices. Kluwer Academic Publications, Dordrecht, pp 543–558

Mercado A (1985) The use of hydrogeochemical patterns in carbonate sand and sandstone aquifers to identify intrusion and flushing of saline water. Ground Water 23:635–645

Parkhurst DL, Appelo CAJ (1999) User's guide to PHREEQC (version 2): a computer program for speciation, batch-reaction, one-dimensional transport, and inverse geochemical calculation. U.S. Geological Survey, Denver, CO

Price RM, Swart PK, Fourqurean JW (2006) Coastal groundwater discharge—an additional source of phosphorus for the oligotrophic wetlands of the Everglades. Hydrobiologia 569:23–36

Price RM, Savabi MR, Jolicoeur JL, Roy S (2010) Adsorption and desorption of phosphate on limestone in experiments simulating seawater intrusion. Appl Geochem 25(7):1085–1091

Ronen D, Magaritz M, Levy I (1986) A multi-layer sampler for the study of detailed hydrochemical profiles in groundwater. Water Res 20:311–315
Ronen D, Magaritz M, Levy I (1987) An in situ multilevel sampler for preventive monitoring and study of hydrochemical profiles in aquifers. GWMR 7:69–74
Russak A, Sivan O (2010) Hydrogeochemical tool to identify salinization or freshening of coastal aquifers determined from combined field work, experiments, and modeling. ES&T 44:4096–4102
Russak A, Yechieli Y, Herut B, Lazar B, Sivan O (2015a) The effect of salinization and freshening events in coastal aquifers on nutrient characteristics as deduced from as deduced from column experiments under aerobic and anaerobic conditions. J Hydrol 529:1282–1292
Russak A, Yechieli Y, Herut B, Lazar B, Sivan O (2015b) The effect of salinization and freshening events in coastal aquifers on nutrient characteristics as deduced from field data. J Hydrol 529:1293–1301
Russak A, Sivan O, Yechieli Y (2016) Trace elements (Li, B, Mn and Ba) as sensitive indicators for salinization and freshening events in coastal aquifers. Chem Geol 441:35–46
Sivan O, Yechieli Y, Herut B, Lazar B (2005) Geochemical evolution and timescale of seawater intruding into the coastal aquifer of Israel. Geochim Cosmochim Acta 69:579–592
Stuyfzand PJ (1993) Behaviour of major constituents in fresh and salt intrusion waters, in the Western Netherlands. In: Custodio E (ed) Study and modeling of salt water intrusion into aquifers. Proceedings of 12th salt water intrusion meeting, CIHS-CINME, Barcelona, pp 143–160
Stuyfzand PJ (2008) Base exchange indices as indicators of salinization or freshening of (Coastal) aquifers. In: Proceedings of the 20th salt water intrusion meeting, Naples, FL, 23–27 June, Program and proceedings book, pp 262–265
Yechieli Y, Sivan O, Lazar B, Vengosh A, Ronen D, Herut B (2001) Radiocarbon in seawarer intruding into the Israeli Mediterrnean coastal aquifer. Radiocarbon 43:773–781
Yechieli Y, Kafri U, Sivan O (2009) The inter-relationship between coastal sub-aquifers and the Mediterranean Sea, deduced from radioactive isotopes analysis. Hydrogeol J 17(2):265–274

# Geochemical Aspects of Groundwater in the Dead Sea Coastal Aquifer

**Yael Kiro, Naama Avrahamov, Nurit Weber, and Ittai Gavrieli**

## 1 General Introduction About the Geochemistry of the Dead Sea

The Dead Sea is a hypersaline (TDS ~340 g/L) terminal lake located in the Dead Sea Basin, a tectonic depression that is associated with the movements along the Dead Sea Transform (e.g., Garfunkel 1981). The lake's brine composition evolved during the late Neogene-Quaternary through various lakes that filled the basin after the retreat of the late Miocene-early Pliocene "Sedom Lagoon" (Gavrieli and Stein 2006; Neev and Emery 1967; Stein Stein 2001, 2014 and references there). The present lake's Ca-chloride composition ($Na/Cl < 1$ and $Ca/(SO_4 + HCO_3) > 1$) evolved from seawater evaporation, precipitation of minerals (aragonite, gypsum, halite), water–rock interaction (mostly dolomitization) and mixing with terrestrial water (Starinsky 1974). The dissolved salts originated mainly from the Sedom Lagoon brine (Zak 1967; Starinsky 1974; Stein et al. 2000, 2002; Katz and Starinsky 2009), while the freshwaters were flushed to the lake from its large watershed reflecting the regional hydrological conditions (Stein et al. 1997; Haliva-Cohen et al. 2012). These contributed the additional ions, mostly, Ca, $SO_4$ and $HCO_3$, required for the precipitation of the aragonite and gypsum from the terminal lakes that followed the

Y. Kiro (✉) · N. Avrahamov
Weizmann Institute of Science, Rehovot, Israel
e-mail: yael.kiro@weizmann.ac.il

N. Avrahamov
e-mail: Naamakatz@gmail.com

N. Weber
The Hebrew University, Jerusalem, Israel
e-mail: nurit.weber@mail.huji.ac.il

I. Gavrieli
Geological Survey of Israel, Jerusalem, Israel
e-mail: ittai.gavrieli@gsi.gov



U. Kafri and Y. Yechieli (eds.), *The Many Facets of Israel's Hydrogeology*, Springer Hydrogeology, https://doi.org/10.1007/978-3-030-51148-7_7

Sedom Lagoon once it was disconnected from the Mediterranean. Pore waters and halite fluid inclusions from the Dead Sea Deep Drilling Core show fluctuations in the composition of the Dead Sea over the past 220k year. These compositions are similar to the range of compositions of saline groundwater around the lake (Kiro et al. 2017; Levy et al. 2017; Weber et al. 2018).

## 2 Groundwater Geochemistry and Flow Mechanisms

The Quaternary alluvial aquifer in the vicinity of the Dead Sea is comprised of alternating alluvial (gravel, silt and clay) and lacustrine (salt, gypsum and aragonite) sediments. The aquifer is bounded by the marginal faults of the Dead Sea basin. The alluvial aquifer is divided by clay-silt layers to several subunits. Due to the large heterogeneity in the hydraulic parameters of these sediments, there are several types of water bodies within the aquifer differing from each other by their salinity and chemical and isotopic composition (Yechieli and Sivan 2011). These groundwaters are found in many springs and boreholes along the coast of the Dead Sea (see map in Fig. 1). Because of the extremely high density of the Dead Sea, the depth of the fresh–saline water interface (FSI) plunges relatively moderately, and is thus very shallow, allowing drilling and sampling of saline groundwater below the FSI at relatively large distances from the shoreline.

The saline groundwater in the alluvial aquifer share the characteristics of Ca-chloride brines and are generally similar in composition to the Dead Sea brine (low Na/Cl ratio and high $Ca/SO_4 + HCO_3$). The saline springs on the western shores of the Dead Sea emerge along a narrow stretch of coastline, from Enot Zuqim in the north to Hamme Mazor (Ein-Gedi spa area) in the south. The differences in the salinity and the chemical composition between the various brines in the subsurface reflect the different pathways in their evolution and their mixing with freshwater in the subsurface in various proportions. They can be divided into two main types: The Qedem-Shalem and the DSIF-Tapuach groups, differing from each other by several parameters, such as Mg/Ca ratios, $\delta^{34}S$ values and temperature (Fig. 2; Gavrieli et al. 2001). The DSIF-Tappuah brines are encountered in shallow boreholes drilled into the alluvial sediments in the vicinity of the shore. They are characterized by $\delta^{34}S_{SO_4}$ > 30‰, ammonium concentrations of 80–100 mg/L, and salinities and Mg/Ca ratios higher than those of Qedem-Shalem and closer to those of the Dead Sea brine (Fig. 2; Gavrieli et al. 2001). The brines temperature is the average ambient temperature of 25–30°. The Qedem-Shalem springs are thermal brines with temperature of 41–45 °C. The thermal brines flow upward from a depth of ~1 km along the faults of the Dead Sea basin and fill the alluvial aquifer between Mineral Spa and Ein Gedi Spa while traces of these brines are found further north at Enot Zuqim (Fig. 1a). This group includes saline water with salinity of up to 129 g/L Cl, Mg/Ca molar ratios of 2.5–3.5, $\delta^{34}S_{SO_4} = 21–25‰$ (Fig. 2; Gavrieli et al. 2001), and ammonium concentrations of 10–30 mg/L.

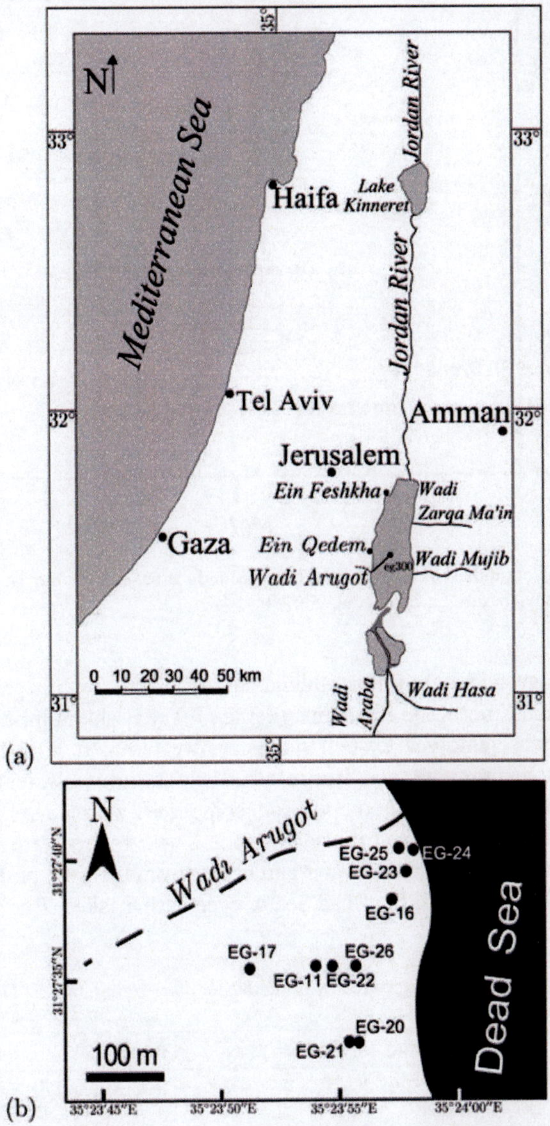


**Fig. 1** **a** Location map of the Dead Sea and the main water sources of the lake. **b** Location map of the boreholes of Wadi Arugot, located on the western side of the Dead Sea (used by permission of Elsevier from Kiro et al. 2013, Fig. 1)

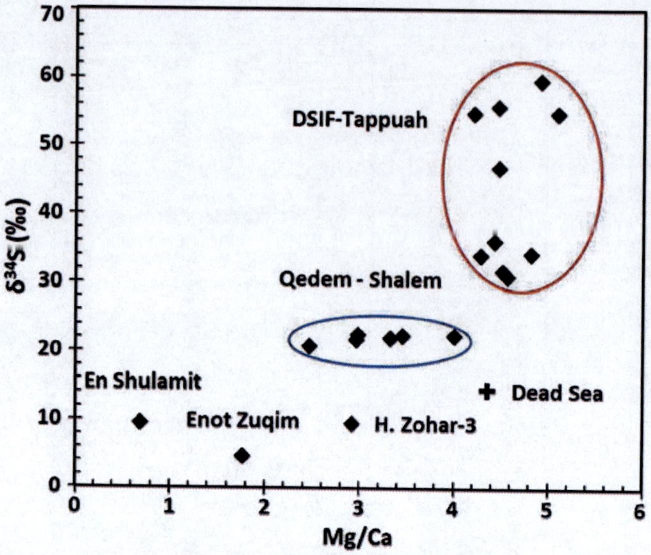


**Fig. 2** Chemical composition of representative groundwaters around the Dead Sea (used by permission of Elsevier from Gavrieli et al. 2001, Fig. 4)

Most of the groundwater in the alluvial aquifer discharges in specific locations: Ein-Feshkha, and Einot Kane and Samar (Table 1). The Feshkha springs are a unique and complex hydrogeological system in the northwestern part of the Dead Sea, with an extremely large discharge zone (Burg et al. 2016). The groundwater has a very large range of salinity and chemical composition, sometimes within a very short vertical (within the same borehole) or horizontal distance. These waters were suggested to be mainly a mixture of fresh groundwater and brines from the time of the Lisan Lake, the Last glacial precursor of the Dead Sea or even earlier lakes (Burg et al. 2016).

**Table 1** Historic main water sources fed by groundwater discharging into the Dead Sea (average values)

| Source | Flux (million $m^3$/year) | References |
|---|---|---|
| Jordan River | 1207 | Salameh and El-Naser (1999) |
| Wadi Hasa | 34 | Salameh (1996) |
| Wadi Mujib | 83 | Salameh (1996) |
| Wadi Zark-Main | 23 | Salameh (1996) |
| Wadi Karak | 18 | Salameh (1996) |
| Ein-Feshkha | 80 | Hydrological Service of Israel |
| Kane and Samar Springs | 20 | Hydrological Service of Israel |
| Wadi Arugot | 5 | Hydrological Service of Israel |

Note that present water inflow is limited to 300 MCM/year (Lensky et al. 2005)

## 3 Circulation of the Dead Sea Brine in the Aquifer

Groundwater in the west shore of the Dead Sea are composed mainly of a mixture between fresh groundwater flowing in from the Judea Group aquifer and the Dead Sea brine penetrating from the east, forming an interface zone in between (Fig. 3a). Despite the high salinity of the Dead Sea, the freshwater component is identified within a few meters from the shoreline (Fig. 3a). However, major elements (Kiro 2006) and $^{14}C$ values do not show a simple mixing process between fresh and saline water bodies (Avrahamov et al. 2010; Fig. 4). Both fresh and saline groundwater in the

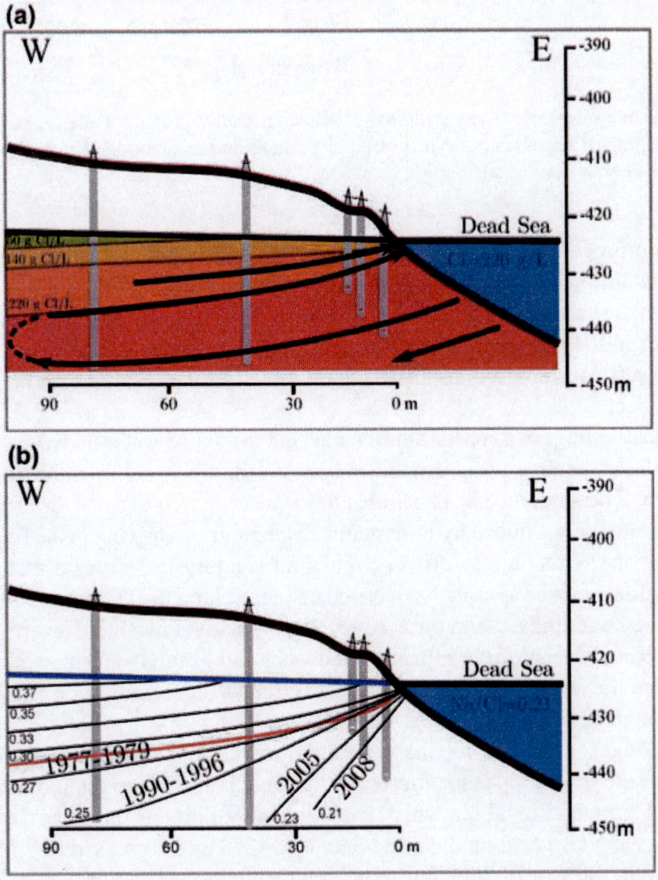


**Fig. 3** **a** The fresh–saline transition zone in Wadi Arugot alluvial fan. **b** Na/Cl contours in the Dead Sea water within the aquifer and the fresh–saline transition zone. Ages are based on the Na/Cl values and the year at which the Dead Sea has similar ratios and represent the timing of Dead Sea water penetration to the aquifer (used by permission of Elsevier from Kiro et al. 2013, Fig. 2)

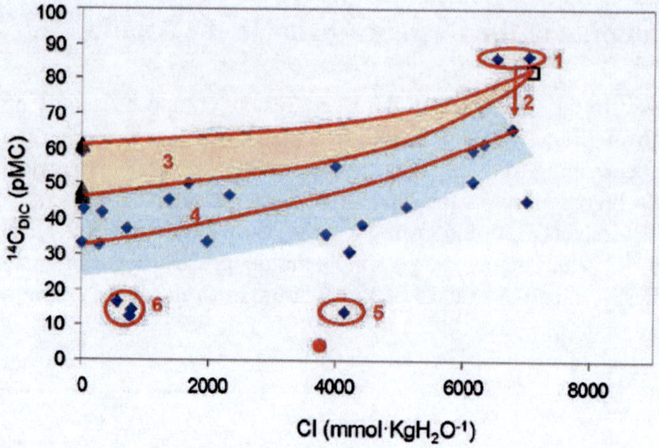


**Fig. 4** Possible stages of $^{14}C_{DIC}$ evolution of saline groundwater in the Dead Sea area. The five possible stages are discussed in the text (used by permission of Cambridge University Press after Avrahamov et al. 2010, Fig. 5)

coastal aquifers have lower $^{14}C$ values than both end-members, indicating the existence of additional geochemical processes (calcium carbonate dissolution, organic matter/methane oxidation, and/or mixing with older water bodies that penetrated into the aquifer in the past). Simplified age calculation indicates that groundwater from the lower sub-aquifer are older than several thousand years (area 5 and 6 in Fig. 4; Avrahamov et al. 2010).

In typical settings of a coastal aquifer adjacent to a saline water body, under steady-state conditions, saline water circulates beneath the fresh–saline transition zone in the aquifer. The saline water penetrates the aquifer due to density differences, and flow back into the sea due to hydrodynamic dispersion and mixing in the fresh–saline transition zone. This density-driven circulation is a physical process that occurs in coastal aquifers can be assumed to have taken place along the Dead Sea shores under steady-state conditions, when the lake level was relatively stable. Nevertheless, the chemical composition of the saline groundwater and groundwater flow simulations indicate that the Dead Sea water continues to circulate at present, even while the lake level is dropping (Kiro et al. 2008; Figs. 3b, 5 and 6).

The strongest evidence for the penetration of Dead Sea brine into the aquifer during the lake level drop is manifested in the Na/Cl ratio in the groundwater within the Arugot stream alluvial fan which has the same salinity as the Dead Sea. During the 1960s, the Na/Cl ratio in the Dead Sea was ~0.28 and decreased to 0.26 in 1985 and to 0.23 in the year of 2000 due to halite precipitation (Gavrieli 1997). Thus, the Na/Cl ratio in the Dead Sea water in the aquifer can be correlated to the age of the water. Figure 3b shows the Na/Cl ratios and the year in which the Dead Sea had this ratio and composition. The decrease in the Na/Cl ratio with depth and toward the lake suggests a continuous inflow of Dead Sea water into the aquifer (Fig. 5). Simulation results for Wadi Arugot alluvial aquifer show that during the Dead Sea

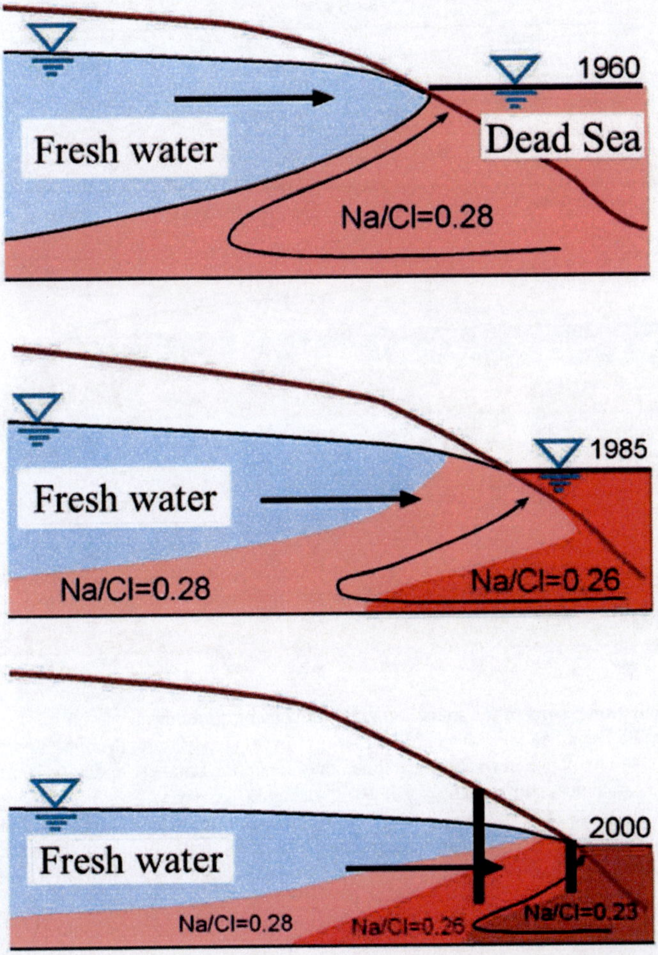


**Fig. 5** Effect of Dead Sea circulation in the aquifer on the Na/Cl ratio in the groundwaters (from Kiro 2006)

lake level drop, the circulation is maintained, but is weaker than under steady-state conditions (Fig. 6). Based on the ages obtained from the Na/Cl ratio, buildup of $^{228}Ra$ ($t_{1/2} = 5.75$ year) in the water by the decay of $^{232}Th$ in the sediments, and tritium and $^{14}C$ values, the rate of the Dead Sea intrusion into the aquifer is 0.5–10 m/year (Avrahamov et al. 2010; Kiro et al. 2013).

During the circulation process of the Dead Sea water in the aquifer, the behavior of many elements is not conservative and therefore their chemical composition is modified. The Arugot stream alluvial fan study area, with its many observation wells (Fig. 1b), is an ideal site for tracking the Dead Sea water and its evolution along its flow path from the lake into the aquifer. Furthermore, it allows to study of the

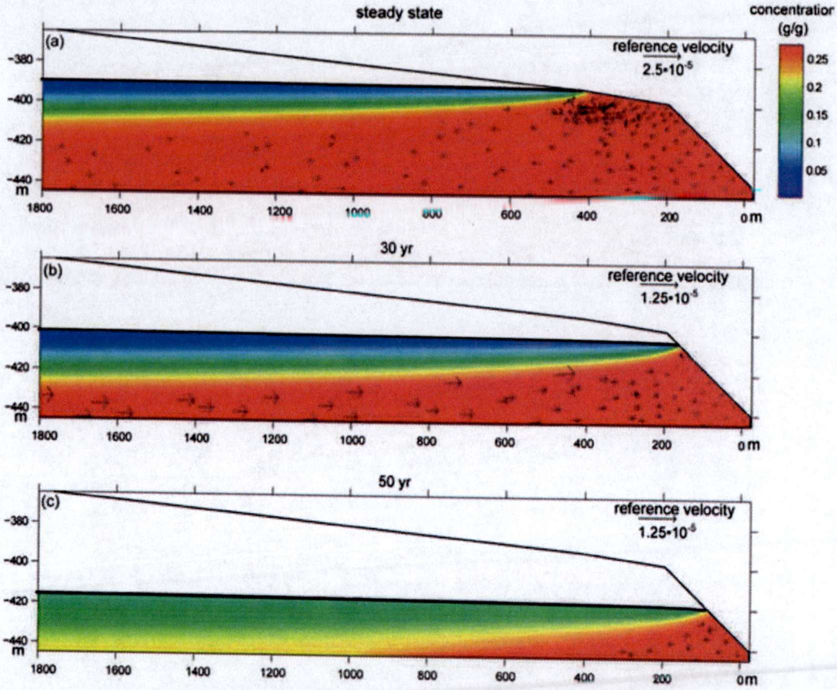


**Fig. 6** Simulation results of the transition zone movement, groundwater levels, and saline water flow during the Dead Sea level drop at Wadi Arugot in (a) steady-state, (b) after 30 years and (c) after 60 years. The parameter values are those that create the smallest hydraulic gradient in the range of the parameter estimation ($k = 4 \times 10^{-10}$ m$^2$ and $Q_0 = 0.03$ kg s$^{-1}$). The topography and bathymetry are those of Wadi Arugot. Only the saline water velocities are shown. The transition zone widens during the Dead Sea level drop (**b**, **c**). The saline water flow is in the direction of the aquifer most of the time during the Dead Sea level drop (This figure is taken from Kiro et al. 2008, WRR)

geochemical processes and rates of water–rock interaction during the brine circulation in the subsurface. As the Dead Sea brine flows into the aquifer and interacts with the sediments, before mixing with freshwater, gypsum and then barite precipitate, thereby removing $SO_4$, Ba, Sr, and $^{226}$Ra from the water. Reducing conditions result in the removal of U, probably due to adsorption that is followed by reduction of Mn and Fe oxides (Kiro et al. 2013; Fig. 7). The changes in the trace elements and isotope concentrations could be very large, sometimes by an order of magnitude within a distance of less than 100 m.

In addition to the redox-driven processes and mineral precipitation, the decay along the U- and Th-series contributes radium and radon isotopes to the Dead Sea brine. Typically, radium is somewhat particle reactive, and in freshwater tends to adsorb to the sediments. However, in the hypersaline Dead Sea brine, it remains soluble with significant less adsorption to the sediments (Kiro et al. 2012). For this reason, the enrichment in radium isotopes and the duration for obtaining steady-state activities is proportional to each of the isotopes' half-life. Thus, the short-lived radium isotopes ($^{223}$Ra and $^{224}$Ra) activities increase within a few days, $^{228}$Ra reaches steady-state activities after ~30 year, whereas $^{226}$Ra ($t_{1/2}$ = 1600 year) does not change in these time-scales.

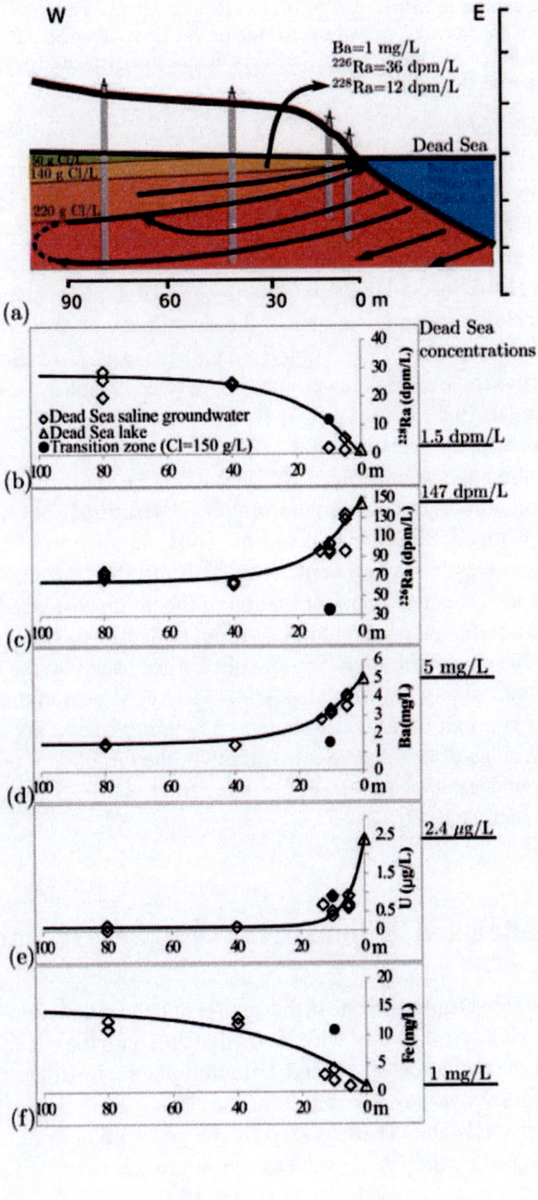

◀**Fig. 7** **a** A cross section showing the fresh–saline transition zone in the Dead Sea aquifer and the estimated direction of groundwater flow in this system. This figure shows typical activities and concentrations found in the mixing zone. **b–f** Concentrations of $^{226}Ra$, $^{228}Ra$, Ba, U and Fe in the circulated Dead Sea water (~220 g Cl/L) from the wells in Fig. 1b along a flow path from the Dead Sea into the aquifer. Multiple samples are from the same well or from wells that are at the same distance from the lake shore. The concentrations of each element in the fresh–saline transition zone are also shown. The solid curves are calculated according to the removal or contribution rate of each element as described in Kiro et al. (2013). Note that the saline water flows back into the Dead Sea after dilution with terrestrial groundwater at the fresh–saline transition zone and therefore the concentrations are lower than the actual end-member of the circulated Dead Sea water (This figure is taken from Kiro et al. 2014, EPSL)

The unique behavior of radium isotopes in the Dead Sea hydrological system provides a method for calculating the volumes of brine involved in the long-term circulation in of the Dead Sea brine in the aquifer. Because of the differences in $^{226}Ra$ ($t_{1/2}$ = 1600 year) and $^{228}Ra$ ($t_{1/2}$ = 5.75 year) half-lives, their activities are very different in the Dead Sea and they are sensitive to different geochemical processes during water–rock interaction. Because of the slow decay of $^{226}Ra$, it is enriched in the Dead Sea with activities of ~145 dpm/L (e.g., Kiro et al. 2012; Stiller and Chung 1984). $^{228}Ra$ activities are much lower, ranging between 1 and 1.5 dpm/L (Somayajulu and Rengarajan 1987; Kiro et al. 2014). When the Dead Sea brine penetrates the aquifer, barite precipitates relatively fast, before $^{228}Ra$ activity increases, and therefore its precipitation affects only the $^{226}Ra$, which decreases to ~60 dpm/L. When the Dead Sea brine continues to flow into the aquifer, as described above, $^{228}Ra$ is added into the water, while $^{226}Ra$ remains constant. Thus, circulation of Dead Sea brine in the aquifer removes $^{226}Ra$ and contributes $^{228}Ra$ to the lake water. This unique behavior allows a well-constrained estimation of the volume of Dead Sea brine that circulates in the aquifer. According to the radium budget, ~400 MCM/year of brine infiltrates from all sides of the Dead Sea and discharges back into the Dead Sea. This annual volume is similar to the estimated ~300–350 MCM/year of the present inflow to the Dead Sea from all other water sources (including rivers, floods and groundwater from both sides of the Dead Sea). Although, the circulation process does not affect the water budget, it is significant for many trace elements in the lake, such as Ba, U, Fe, Mn (Kiro et al. 2014).

## 4 $^{14}C$ Evolution and Tritium in the Dead Sea Groundwater

The circulation of the Dead Sea brine in the aquifer and the interaction of the different brines along the geological history with each other and with the rift valley sediments and rocks affect the evolution of $^{14}C$ and Tritium in the regional groundwaters.

Figure 4 suggests five possible stages in the $^{14}C_{DIC}$ evolution of the Dead Sea coastal aquifer groundwater. The first stage is the penetration of the Dead Sea into the aquifer, indicated by the $^{14}C_{DIC}$ values of groundwater near the Dead Sea shore, which are very similar to the recent Dead Sea value (86 pMC as compared to 82 pMC in the Dead Sea; Lewenberg 2005). This saline groundwater has the same composition of major ions as the Dead Sea and $^{14}C$ values of the 1970s and early 1980s in the Dead Sea. The tritium value (2.9 TU) is also characteristic to the Dead Sea and yields a value of ~12 TU for the early 1980s, which is similar to the measured level at the same time −11.2 TU in 1979 (Carmi et al. 1984).

In the second stage, the gap between the $^{14}C_{DIC}$ values of the Dead Sea and the Dead Sea-like groundwater samples is explained by low $^{14}C_{DIC}$ contribution from the oxidation of old organic matter in the sediment or methane oxidation (Barker et al. 1979). In the third stage, mixing between the hypersaline water and freshwater affects the $^{14}C_{DIC}$ values. Most of the freshwater values spread over a wide range (46–60 pMC) as a result of two different freshwater components; relatively low $^{14}C_{DIC}$ fresh groundwater from Judea Group aquifer of Cretaceous age, to, and recent flood water that is characterized by high $^{14}C_{DIC}$ values.

Low $^{14}C_{DIC}$ (~14 pMC) and tritium (<1 TU) values may indicate older groundwater (areas 5 and 6 in Fig. 4), and are typical to the lower sub-aquifer (Avrahamov et al. 2010). This may indicate slower flow rate of the salt and fresh water bodies and is consistent with the stagnant behavior of the fresh–saline transition zone at the same location (Kiro 2006). The typical mixture between Dead Sea water, fresh water, and old brine in the Wadi Arugot alluvial aquifer, that is indicated by the major elements (Kiro 2006), is reflected also in the $^{14}C_{DIC}$ values. The samples all lie below the mixing line between the present (and past) Dead Sea brine and fresh water. Moreover, the absence of samples lying on a mixing line between the present Dead Sea and fresh water, indicates that most of the geochemical processes and mixing occur with Dead Sea water that are a few decades old.

## 5 Oxidation of Organic Matter in the Dead Sea Coastal Aquifer

Based on methane concentration and DIC-carbon isotope ($^{13}C_{DIC}$) mass balance, it was suggested that methane oxidation occurs in the Dead Sea alluvial aquifer (Avrahamov et al. 2010). This mass balance showed that the negative $\delta^{13}C$ values (−12 to −24‰) can only be explained by oxidation of organic matter in the aquifer. Since there is no significant increase in the DIC concentration in the coastal aquifer, the source of the low $^{13}C$ organic compound was suggested to be methane whose $\delta^{13}C$ values are usually much lower than most other organic compounds. A small fraction of oxidized methane, whose values can reach −60‰, can yield a very low $\delta^{13}C$ value in DIC without changing its concentration significantly. The anaerobic conditions, and the coincident of sulfate depletion and isotope enrichment of sulfur and oxygen isotopes in the sulfate ($\delta^{34}S_{SO_4}$ and $\delta^{18}S_{SO_4}$) points that this anaerobic oxidation of methane (AOM) is likely coupled with bacterial sulfate reduction (BSR) in the anoxic part of the fresh-saline interface zone. The microbial community is also indicative of BSR and association with anaerobic methanotrophic archaea (ANME) (Avrahamov et al. 2014).

Figure 8a, b show the $\delta^{34}S_{SO_4}$ and $\delta^{18}O_{SO_4}$ compositions as a function of the apparent BSR progress. This rough estimation of apparent net sulfate reduction was

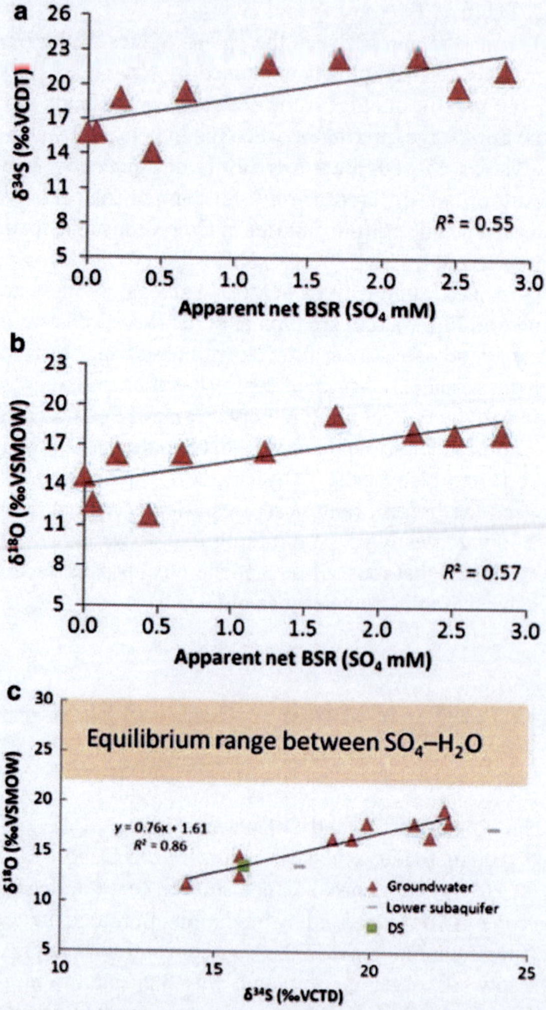


**Fig. 8** **a** $^{34}S_{SO4}$ and **b** $^{18}O_{SO4}$ versus apparent net BSR. The pre-reaction sulfate concentration was determined from the mixing fraction and chemical equilibrium to gypsum as calculated from the Phreeqc model. **c** $\delta^{18}O_{SO_4}$ versus $\delta^{34}S_{SO_4}$ sulfate data in the groundwater samples display a positive relation. The predicted range of equilibrium isotope fractionations between $SO_4$ and $H_2O$ are indicated by the pink area (estimated from Fritz et al. 1989; Brunner et al. 2005; Turchyn et al. 2006; Wortmann et al. 2007; Aller et al. 2010) (used by permission of Wiley after Avrahamov et al. 2014, Figs. 6 and 9)

achieved by subtracting measured sulfate values from the initial sulfate concentration (calculated from the Phreeqc model). Both $\delta^{34}S_{SO_4}$ and $\delta^{18}O_{SO_4}$ show an increase with the progressive sulfate reduction and fit a Rayleigh distillation of $^{32}S$ during sulfate reduction, resulting in isotopic enrichment in the residual sulfate. Based on measurements of the isotope ratios and sulfate concentrations, the enrichment factor of sulfur ($^{34}\varepsilon$) was found to be 17 ± 3.5‰. This $^{34}\varepsilon$ value is in the lower end of the range found for sulfur isotope fractionations (0–72‰) in natural environment (Canfield 2001; Wortmann et al. 2001; Brunner et al. 2005; Sim et al. 2011). Small sulfur isotope fractionation may occur under conditions of limited sulfate supply and/or high sulfate reduction rates (e.g., Aharon and Fu 2000; Canfield et al. 2006; Stam et al. 2011). Given that the sulfate concentrations are high, it was suggested that the small sulfur isotope fractionation here may result from a high rate of BSR, which involves less re-oxidation of sulfur intermediates. In addition, it was previously shown that BSR coupled with AOM generate lower $^{34}\varepsilon$ compared with BSR that is not associated with AOM. As the molecule of methane is smaller and considerably simpler than that of refractory crude oil, some micro-organisms prefer this molecule over more complex organic matter (Aharon and Fu 2000), even though thermodynamically it is less energetically favorable as an electron donor (Devol and Ahmed 1981; Aharon and Fu 2000). This reason could explain the current low $^{34}\varepsilon$ compared with previous work in the Dead Sea aquifer (Gavrieli et al. 2001; $^{34}\varepsilon$ = 30‰) for BSR that is not associated with AOM.

Avrahamov et al. (2015) suggested that the isotopic composition of the methane in the alluvial aquifer ($\delta^{13}C_{CH_4}$ values between −36 and −56‰ and one sample of $\delta^{2}H_{CH_4}$ has a value of −128‰) and the molecular ratios of C1/(C2 + C3) (less than 100) indicate dominance of a thermogenic methane source. The thermogenic origin is suggested to be associated with the initial stage of thermal degradation of the bituminous chalk of Senonian oil shale unit buried at depth.

## 6 The Thermal Brines (Ein Qedem Brine)

The Ein-Qedem hydrothermal saline springs system constitutes the major brine inflow to the lake today. The main springs of this system, Ein Shalem and Ein Qedem, are located between Mineral Spa and Ein Gedi Spa with some diluted appearances as far north as Einot Zuqim (Gavrieli et al. 2001; Fig. 1a). When plotted together with all other water types in the vicinity of the Qedem shore, the Ein Qedem brine has a distinct composition, forming a clear end-member with rather small scatter on plots of $Na^+$ versus $Cl^-$ (Fig. 9a). The plot shows two straight mixing lines, one between Ein Qedem brine and the regional freshwaters, and the other between Ein Qedem brine and modern Dead Sea. The $SO_4^{2-}$ versus $Cl^-$ plot (Fig. 9b) shows considerable scatter on the mixing line between EQ brine and the regional freshwaters, which probably stems from dissolution of gypsum (the vectors in Fig. 9b). It should be noted that the direction of the scatter of the $SO_4^{2-}$ versus $Cl^-$ values in Ein

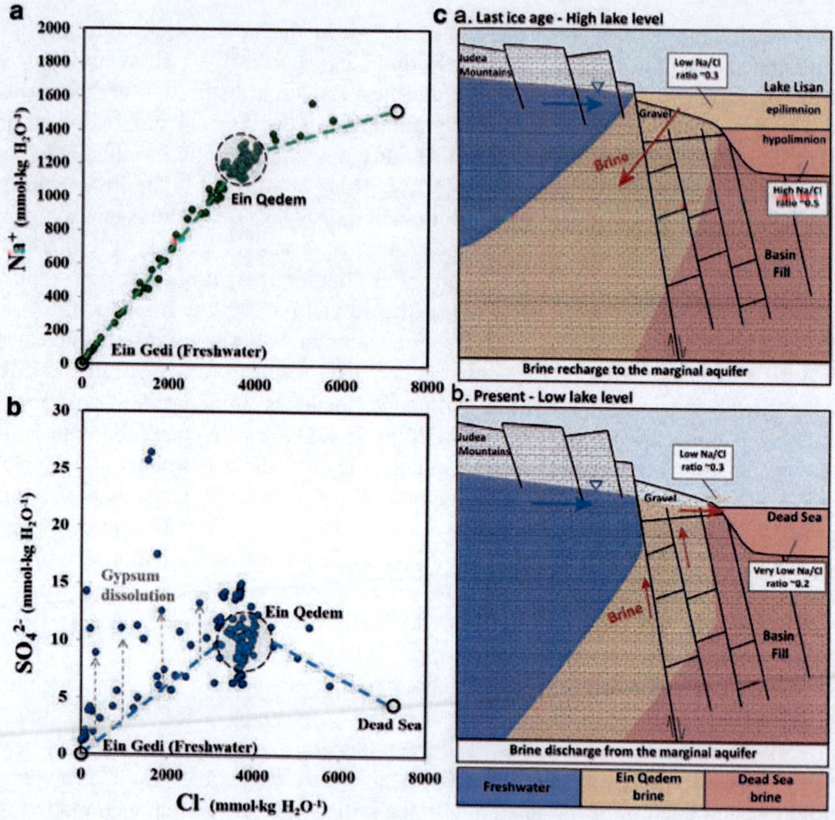


**Fig. 9** **a, b** Major ions ($Na^+$ and $SO_4^{2-}$) versus $Cl^-$ in Ein Qedem springs and vicinity, measured during a time interval of 40 years (1975–2016). The shaded circles represent the scatter of all available analyses of Ein Qedem brine. The gray arrows in $SO_4^{2-}$ are the vectors representing gypsum dissolution by diluted Ein Qedem brine. Right—Schematic hydrological section showing the evolution history of Ein Qedem brine (modified after Shalev and Yechieli 2007): **c-a** During the last glacial period when Lake Lisan filled the basin and it rose to its highest level. The low $Na^+/Cl^-$ ratio in the Ein Qedem brine (0.33) reflects the composition of the upper layer of the lake, which was less affected by salt dissolution. A rise in the hydraulic head drives lake water to the subsurface. **c-b** Current Dead Sea low level. Ein Qedem springs with their higher $Na^+/Cl^-$ value discharge along the shore. A drop in the hydraulic head of the Dead Sea implies flow of the deep groundwater into the lake (modified after Weber et al. 2018)

Qedem brine is parallel to the gypsum dissolution vector, indicating that the brine dissolved small amounts of gypsum during it ascend.

A detailed study of the chemical parameters and direct dating of the groundwater by radiocarbon and radiokrypton suggested that the source of Ein Qedem brines is probably Lisan Lake the precursor of the Dead Sea, at its highest stand some ~40–20 ka (Weber et al. 2018). During that time, the lake water level was much higher than the present Dead Sea (height of ~200 ± 30 m below msl, at ~31–17.4 ka) and the lake was stratified. According to the chemical record that was found in the Dead Sea Deep Drilling Core (Levy et al. 2017), the Qedem brine originated from the upper layer of the Lisan Lake, and penetrated when the lake was at high water level (Fig. 9c; Weber et al. 2018). The conceptual hydrological model for the circulation of this brine suggests a strong relation between the lake's level and the groundwater flow direction. During lake's high stand, the lake's epilimnetic (upper water mass) brine penetrates into the aquifer, while during lake's low stands, such as during the last Interglacial and the Holocene Dead Sea, saline groundwater ascends form a depth of ~1000 m (indicated by the high temperature of ~45 °C), reaching the surface very close to the lake's coastline (Fig. 9c).

## References

Aharon P, Fu B (2000) Microbial sulfate reduction rates and sulfur and oxygen isotope fractionations at oil and gas seeps in deepwater Gulf of Mexico. Geochim Cosmochim Acta 64:233–246. https://doi.org/10.1016/S0016-7037(99)00292-6

Aller RC, Madrid V, Chistoserdov A et al (2010) Unsteady diagenetic processes and sulfur biogeochemistry in tropical deltaic muds: implications for oceanic isotope cycles and the sedimentary record. Geochim Cosmochim Acta 74:4671–4692. https://doi.org/10.1016/J.GCA.2010.05.008

Avrahamov N, Yechieli Y, Lazar B et al (2010) Characterization and dating of saline groundwater in the Dead Sea area. Radiocarbon 52:1123–1140

Avrahamov N, Antler G, Yechieli Y et al (2014) Anaerobic oxidation of methane by sulfate in hypersaline groundwater of the Dead Sea aquifer. Geobiology 12(6):511–528. https://doi.org/10.1111/gbi.12095

Avrahamov N, Gelman F, Yechieli Y et al (2015) Proposed sources of methane along the Dead Sea Transform. Chem Geol 395:165–175. https://doi.org/10.1016/j.chemgeo.2014.11.026

Barker JF, Fritz P, Brown RM (1979) C-14 measurements in aquifers with $CH_4$. In: Isotope hydrology 1978. Proceedings symposium, 19–23 June 1978, Neuherberg, vol II. IAEA, Vienna, pp 661–678

Brunner B, Bernasconi SM, Kleikemper J, Schroth MH (2005) A model for oxygen and sulfur isotope fractionation in sulfate during bacterial sulfate reduction processes. Geochim Cosmochim Acta 69:4773–4785. https://doi.org/10.1016/j.gca.2005.04.017

Burg A, Yechieli Y, Galili U (2016) Response of a coastal hydrogeological system to a rapid decline in sea level; the case of Zuqim springs—the largest discharge area along the Dead Sea coast. J Hydrol 536:222–235. https://doi.org/10.1016/j.jhydrol.2016.02.039

Canfield DE (2001) Isotope fractionation by natural populations of sulfate-reducing bacteria. Geochim Cosmochim Acta 65:1117–1124. https://doi.org/10.1016/S0016-7037(00)00584-6

Canfield DE, Olesen CA, Cox RP (2006) Temperature and its control of isotope fractionation by a sulfate-reducing bacterium. Geochim Cosmochim Acta 70:548–561. https://doi.org/10.1016/j.gca.2005.10.028

Carmi I, Gat JR, Stiller M (1984) Tritium in the Dead Sea. Earth Planet Sci Lett 71:377–389. https://doi.org/10.1016/0012-821X(84)90104-3

Devol AH, Ahmed SI (1981) Are high rates of sulphate reduction associated with anaerobic oxidation of methane? Nature 291:407–408. https://doi.org/10.1038/291407a0

Fritz P, Basharmal GM, Drimmie RJ, Ibsen J, Qureshi RM (1989) Oxygen isotope exchange between sulfate and water during bacterial reduction of sulfate. Chem Geol 79:99–105

Garfunkel Z (1981) Internal structure of the Dead Sea leaky transform (rift) in relation to plate kinematics. Tectonophysics 80:81–108. https://doi.org/10.1016/0040-1951(81)90143-8

Gavrieli I (1997) Halite deposition in the Dead Sea: 1960–1993. In: Niemi T, Ben-Avraham Z, Gat JR (eds) The Dead Sea—the lake and its setting. Oxford University Press, Oxford, pp 161–170
Gavrieli I, Stein M (2006) On the origin and fate of the brines in the Dead Sea basin. Geol Soc Am 2401:1–22. https://doi.org/10.1130/2006.2401(12)
Gavrieli I, Yechieli Y, Halicz L et al (2001) The sulfur system in anoxic subsurface brines and its implication in brine evolutionary pathways: the Ca-chloride brines in the Dead Sea area. Earth Planet Sci Lett 186:199–213. https://doi.org/10.1016/S0012-821X(01)00247-3
Haliva-Cohen A, Stein M, Goldstein SL et al (2012) Sources and transport routes of fine detritus material to the Late Quaternary Dead Sea basin. Quat Sci Rev 50:55–70. https://doi.org/10.1016/j.quascirev.2012.06.014
Katz A, Starinsky A (2009) Geochemical history of the Dead Sea. Aquat Geochem 15:159–194. https://doi.org/10.1007/s10498-008-9045-0
Kiro Y (2006) The effect of the Dead Sea level drop in the past 50 years on the groundwater system in the alluvial aquifer in its vicinity. M.Sc., The Hebrew University, Jerusalem
Kiro Y, Yechieli Y, Lyakhovsky V et al (2008) Time response of the water table and saltwater transition zone to a base level drop. Water Resour Res 44:1–15. https://doi.org/10.1029/2007WR006752
Kiro Y, Yechieli Y, Voss CI et al (2012) Modeling radium distribution in coastal aquifers during sea level changes: the Dead Sea case. Geochim Cosmochim Acta 88:237–254. https://doi.org/10.1016/j.gca.2012.03.022
Kiro Y, Weinstein Y, Starinsky A, Yechieli Y (2013) Groundwater ages and reaction rates during seawater circulation in the Dead Sea aquifer. Geochim Cosmochim Acta 122:17–35. https://doi.org/10.1016/j.gca.2013.08.005
Kiro Y, Weinstein Y, Starinsky A, Yechieli Y (2014) The extent of seawater circulation in the aquifer and its role in elemental mass balances: a lesson from the Dead Sea. Earth Planet Sci Lett 394:146–158. https://doi.org/10.1016/j.epsl.2014.03.010
Kiro Y, Goldstein SL, Garcia-Veigas J et al (2017) Relationships between lake-level changes and water and salt budgets in the Dead Sea during extreme aridities in the Eastern Mediterranean. Earth Planet Sci Lett 464:211–226. https://doi.org/10.1016/j.epsl.2017.01.043
Lensky NG, Dvorkin Y, Lyakhovsky V, Gertman I, Gavrieli I (2005) Water, salt, and energy balances of the Dead Sea. Water Resour Res 41:W12418. https://doi.org/10.1029/2005WR004084
Levy EJ, Stein M, Lazar B et al (2017) Pore fluids in Dead Sea sediment core reveal linear response of lake chemistry to global climate changes. Geology 45(4):315–318. https://doi.org/10.1130/G38685.1
Lewenberg O (2005) The hydrogeology and geochemistry of groundwater in the alluvial fan of Wadi Arugot, En Gedi reservation. The Hebrew University of Jerusalem
Neev D, Emery KO (1967) The Dead Sea—depositional processes and environments of evaporites. Jerusalem
Salameh E (1996) Water quality degradation in Jordan (impacts on environment, economy and future generations resources base). Royal Society for the Conservation of Nature and Friedrich-Ebert-Stiftung, Amman, Jordan
Salameh E, Wl-Naser H (1999) Does the actual drop in Dead Sea level reflect the development of water sources within its drainage basin? Acta Hydrochim Hydrobiol 27(1):5–11
Shalev E, Yechieli Y (2007) The effect of Dead Sea level fluctuations on the discharge of thermal springs. Isr J Earth Sci 56:19–27
Sim MS, Bosak T, Ono S (2011) Large sulfur isotope fractionation does not require disproportionation. Science 333(80):74–77. https://doi.org/10.1089/jmf.2016.3871
Somayajulu BLK, Rengarajan R (1987) $^{228}$Ra in the Dead Sea. Earth Planet Sci Lett 85:54–58
Stam MC, Mason PRD, Laverman AM et al (2011) $^{34}S/^{32}S$ fractionation by sulfate-reducing microbial communities in estuarine sediments. Geochim Cosmochim Acta 75:3903–3914. https://doi.org/10.1016/j.gca.2011.04.022
Starinsky A (1974) Relationship between Ca-chloride brines and sedimentary rocks in Israel. Ph.D. thesis. The Hebrew University, Jerusalem (in Hebrew, English summary)

Stein M (2001) The sedimentary and geochemical record of neogene-quaternary water bodies in the Dead Sea basin—inferences for the regional paleoclimatic history. J Paleolimnol 271–282

Stein M (2014) The evolution of neogene-quaternary water-bodies in the Dead Sea Rift Valley. Springer, Dordrecht, pp 279–316

Stein M, Starinsky A, Katz A et al (1997) Strontium isotopic, chemical, and sedimentological evidence for the evolution of Lake Lisan and the Dead Sea. Geochim Cosmochim Acta 61:3975–3992. https://doi.org/10.1016/S0016-7037(97)00191-9

Stein M, Starinsky A, Agnon A et al (2000) The impact of brine-rock interaction during marine evaporite formation on the isotopic Sr record in the oceans: evidence from Mt. Sedom, Israel. Geochim Cosmochim Acta 64:2039–2053. https://doi.org/10.1016/S0016-7037(00)00370-7

Stein M, Agnon A, Katz A, Starinsky A (2002) Strontium isotopes in discordant dolomite bodies of the Judea Group, Dead Sea basin. Isr J Earth Sci 51:219–224. https://doi.org/10.1560/61UM-UMQF-YU86-JNU0

Stiller M, Chung YC (1984) Radium in the Dead Sea: a possible tracer for the duration of meromixis 1. Limnol Oceanogr 29:574–586

Turchyn A, Sivan O, Schrag D (2006) Oxygen isotopiccomposition of sulfate in deep sea pore fluid: evidence for rapid sulfur cycling. Geobiology 4:191–201

Weber N, Yechieli Y, Stein M et al (2018) The circulation of the Dead Sea brine in the regional aquifer. Earth Planet Sci Lett 493:242–261. https://doi.org/10.1016/j.epsl.2018.04.027

Wortmann UG, Bernasconi SM, Böttcher ME (2001) Hypersulfidic deep biosphere indicates extreme sulfur isotope fractionation during single-step microbial sulfate reduction. Geology 29:647–650. https://doi.org/10.1130/0091-7613(2001)029%3c0647:HDBIES%3e2.0.CO;2

Wortmann UG, Chernyavsky B, Bernasconi SM, Brunner B, Böttcher ME, Swart PK (2007) Oxygen isotope biogeochemistry of pore water sulfate in the deep biosphere: dominance of isotope exchange reactions with ambient waterduring microbial sulfate reduction (ODP Site 1130). Geochim Cosmochim Acta 71:4221–4232

Yechieli Y, Sivan O (2011) The distribution of saline groundwater and its relation to the hydraulic conditions of aquifers and aquitards: examples from Israel. Hydrogeol J 19:71–81. https://doi.org/10.1007/s10040-010-0646-5

Zak I (1967) The geology of Mt. Sedom. Ph.D. thesis, The Hebrew University of Jerusalem (in Hebrew, English abstract)

# Submarine Groundwater Discharge Along the Israeli Eastern Mediterranean Coast and in Inland Basins

**Yishai Weinstein**

**Abstract** Submarine groundwater discharge (SGD) includes all water that discharges directly from the aquifer and coastal sediments to the sea. Fresh water SGD occurs mainly at the coastline, while saline SGD (circulated seawater) occurs up to the outer shelf or even upper slope. SGD affects both coastal aquifer water budgets and coastal seawater chemistry and quality, since it often carries with it solutes and contaminants. In Israel, it mainly occurs from the Mt. Carmel coast northward, where the Cretaceous carbonate aquifer is either in direct contact with the sea or exposed nearby. It was shown that at the southern Carmel coast (Dor Bay), fresh water discharge is on the order of >1 x $10^6$ $m^3$ $yr^{-1}$, which if applied to the whole Carmel coast is more than is withdrawn from the local aquifer. Fresh groundwater discharge to the whole Mediterranean is about 20% of the riverine discharge. However, when total SGD is considered (by Ra isotope mass balance), it sums up to 200–4, 300 x 109 $m^3$ $yr^{-1}$, which is much higher than riverine discharge and is probably mainly due to offshore discharge across shelves and shallow seas. It was shown both at the local scale (Dor Bay) and for the whole Mediterranean that SGD is a major conveyor of nutrients to the Mediterranean, although some of the nitrates are denitrified in the sediments ('subterranean estuary') en route to the sea.

## 1 Submarine Groundwater Discharge (SGD)—General Background

SGD is defined as any direct discharge of groundwater from coastal aquifer or shelf sediments to the sea. As such, SGD includes both fresh (SFGD), meteoric water that is being driven by hydraulic gradient between land and the sea, and seawater that is recirculated through coastal aquifers and sediments (RSGD, Taniguchi et al. 2002; Fig. 1). The latter is not clearly defined. It includes both seawater, circulating for years and decades along the fresh–saline water interface in the aquifer (Cooper

Y. Weinstein (✉)
Department of Geography and Environment, Bar-Ilan University, Ramat Gan, Israel
e-mail: weinsty@biu.ac.il


U. Kafri and Y. Yechieli (eds.), *The Many Facets of Israel's Hydrogeology*, Springer Hydrogeology, https://doi.org/10.1007/978-3-030-51148-7_8

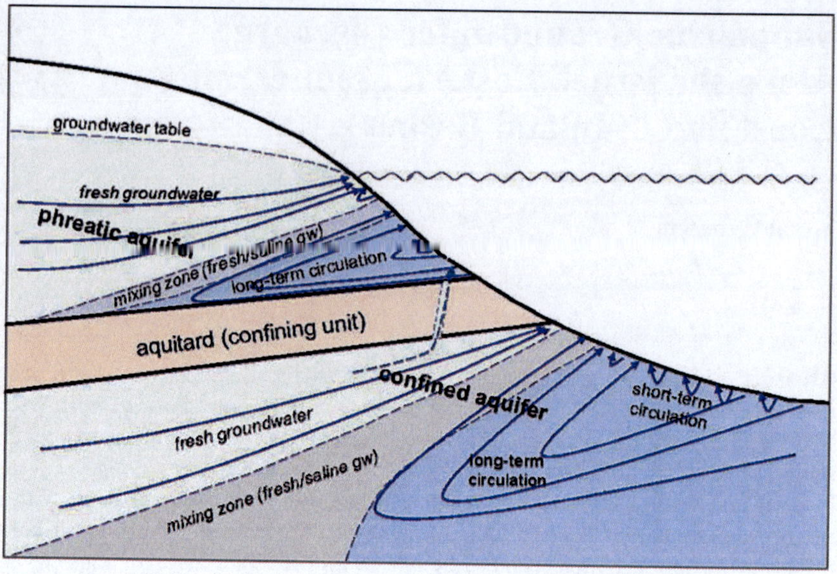


**Fig. 1** Conceptual sketch of SGD scenarios, including terrestrial fresh SGD and seawater circulation, both short-lived and the long-lived, along the fresh–saline water interface

1959; Michael et al. 2005; Kiro et al. 2015), as well as shallow and short-term (hours) circulating seawater, driven by tides and waves (Li et al. 1999). The spatial extent of SGD is also not accurately defined. While most studies focused on close-to-shore SGD (e.g., Boehm et al. 2004; Weinstein et al. 2007a), Moore et al. (2008) showed that RSGD probably occurs across the entire continental shelf. Other studies documented large fresh-to-brackish water reserves that underlie continental shelves up to >100 km offshore (e.g., Johnston 1983; Arad 1983; Post et al. 2013), although most of these were shown to be paleowater from glacial times (e.g., Groen et al. 2000; Person et al. 2003; Cohen et al. 2010), therefore cannot be considered SGD per se.

SGD was shown to convey large volumes of water to the global ocean. While fresh SGD is no more than 10% of the global riverine annual discharge ($0.2–5 \times 10^{12}$ compared with ~$40 \times 10^{12}$ $m^3$ $year^{-1}$, Garrels and MacKenzie 1967; Nace 1970; Cathles et al. 1987; Zektser and Dzhamalov 2007), total SGD is much higher ($90–150 \times 10^{12}$ $m^3$ $year^{-1}$, Kwon et al. 2014). Also, the Atlantic and Mediterranean total SGD is $20–40 \times 10^{12}$ and $0.2–4.3 \times 10^{12}$ $m^3$ $year^{-1}$ (Moore et al. 2008; Rodellas et al. 2015), compared with $25 \times 10^{12}$ and $0.31 \times 10^{12}$ $m^3$ $year^{-1}$ of riverine discharge, respectively (Zektser and Dzhamalov 2007).

## 2 Methodology

SGD is known since ancient times (see review in Taniguchi et al. 2002), but there are hardly any quantitative studies of this process before the 1990s. Quantitative methods may be divided into three categories, including: (1) modeling, which is mainly about FSGD (e.g., Smith and Turner 2001; Oberdorfer 2003; Wilson and Gardner 2006; Zektser and Dzhamalov 2007), (2) direct measurements, usually of total SGD, and mainly by manual or automated seepage meters (e.g., Boyle 1994; Taniguchi et al. 2006; Weinstein et al. 2007b), (3) the use of proxies such as radon and Ra isotopes (e.g., Moore 1996; Cable et al. 1996; Burnett and Dulaiova 2003), nutrients (e.g., Johannes 1980; Valiela and Costa 1988), methane (Dulaiova et al. 2010) and SF6 (Cable and Martin 2008), as well as physical proxies such as temperature variations, traced by remote sensing (e.g., Shaban et al. 2005; Johnson et al. 2008). While (2) is mainly about local SGD, categories (1) and (3) yield areal mass balances and can go all the way from a small beach to the whole-basin scale (e.g., Moore et al. 2008; Kwon et al. 2014; Rodellas et al. 2015).

## 3 The Role of SGD in Oceanic Mass Balances and Water Quality

While fresh SGD is important to coastal aquifer water balances, both fresh discharge and RSGD deliver various solutes and contaminants to the sea. This is mainly known for its coastal water quality effects (e.g., Slomp and Van Cappellen 2004; Charette and Buesseler 2004; Shelenbarger et al. 2006; Burnett et al. 2007), which could also lead to algae blooms (e.g., Smith and Swarzenski 2012; Amato et al. 2016; Kwon et al. 2017). Also, oceanic mass balances of Ra, U, Sr, and Sr isotopes are argued to be strongly affected by SGD (e.g., Basu et al. 2001; Beck et al. 2013; Gonneea et al. 2014).

## 4 SGD Along the Israeli Coast

In Israel, there are several aquifers that are in direct contact with the sea. These are mainly the Quaternary coastal aquifer, the Carmel aquifer and the Western Galilee aquifer. In this paper, we will use the terms (1) Pleistocene aquifer for all the Quaternary sandy aquifers along the coast and (2) Cretaceous aquifer for the karstic, carbonate rocks exposed next to shore, mainly in the northern parts of Mount Carmel area and the western Galilee.

Weinstein et al. (2006) first studied SGD along the Israeli Mediterranean coast, using radon ($^{222}$Rn) as a tracer. The basic concept is that radon activity (concentration) is several orders of magnitude higher in groundwater than in seawater. This is because water is enriched with radon on contact with aquifer rocks (e.g., Burnett and Dulaiova 2003), while in seawater it decays within a relatively short time (half-life of 3.84 days). High radon activities were mainly found along the northern coast, west of Mount Carmel and western Galilee (Fig. 2), while there is hardly any evidence for SGD along the central and southern coast. The confirmed SGD area includes both coastal stretches, where carbonate rocks of the Late Cretaceous aquifer are exposed to the sea, and coastal regions dominated by the sandy Quaternary aquifer.

The discharge was also reflected in reduced salinities, although seawater salinity was hardly lower than 37‰ (compared with 39–40‰ in typical eastern Mediterranean water). Nevertheless, in certain partly enclosed embayments at the northern Mount Carmel and western Galilee, where karstic carbonates of the Cretaceous

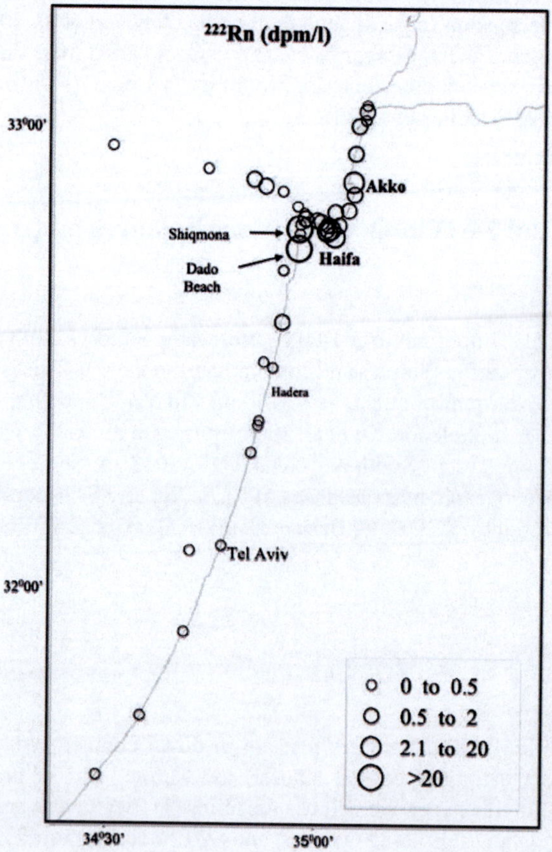


**Fig. 2** Rn survey (in disintegrations per minute per liter) in coastal water (a few to 50 m from shore) along the Israeli Mediterranean coast. Note the higher radon along the northern coast, which suggests enhanced SGD at this area (used by permission of Elsevier from Weinstein et al. 2006, Fig. 4a)

aquifer are exposed to the sea (e.g., Akhziv area, unpublished data), coastal seawater salinity sometimes dropped to less than 20‰.

It should be noted that the above investigation was limited to next-to-shore water (tens of meters), namely the inter-tidal and sub-tidal areas, where most SGD occurs assuming phreatic conditions (e.g., Li et al. 1999). However, as will be noted in the Dor Bay section below, fresh groundwater discharge may also derive from confined units, which are opened to the sea farther offshore. Moreover, considering the long-going debate in Israel coastal hydrogeology about the connection of deeper aquifer units with the sea (see Chapter "Geoelectric, Geoelectromagnetic and Combined Geophysical Methods in Groundwater Exploration in Israel" and reference therein), the possibility of significant offshore fresh groundwater discharge should not be disregarded (e.g., Moore and Wilson 2005; Post et al. 2013). Electrical resistivity studies of Kafri and Goldman (2006) suggest that in some areas, the deep sub-aquifers are physically connected to the sea (Fig. 3). Based on tidal signals in groundwater

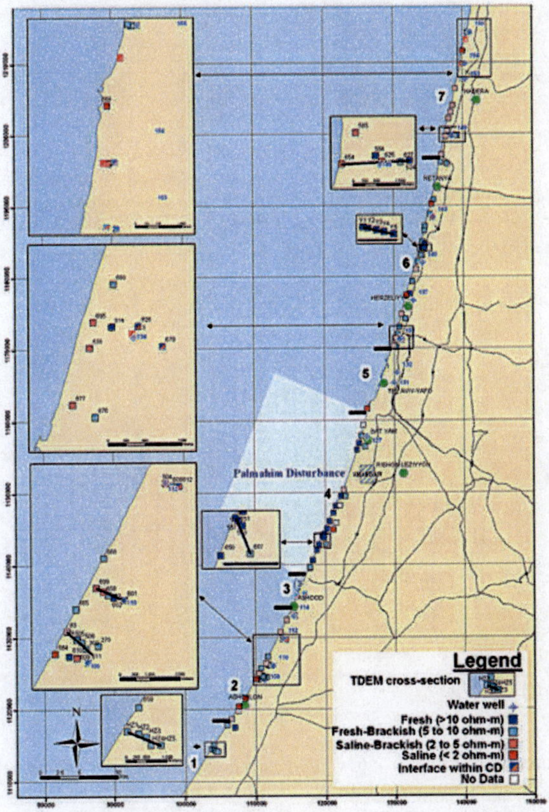


**Fig. 3** Resistivity (TDEM) distribution map within deep sub-aquifers along the Israeli coast. Freshwater (blue dots) is interpreted as no connection with the sea (after Kafri and Goldman 2006)

(Fig. 4a, b), Narovlansky (2018) also suggested that in some areas, the deep units are connected to the sea. Considering the depth of these units (usually >100 m), this connection could occur at the outer shelf, although the alternative possibility of breaching through confining layers closer to shore could not be negated, as is shown by simulation in Amir et al. (2013). Physical connection calls for a fresh–saline water interface, which could include the discharge of fresh or mixed fresh–saline groundwater to the sea. One evidence for an offshore discharge is shown in Fig. 2,

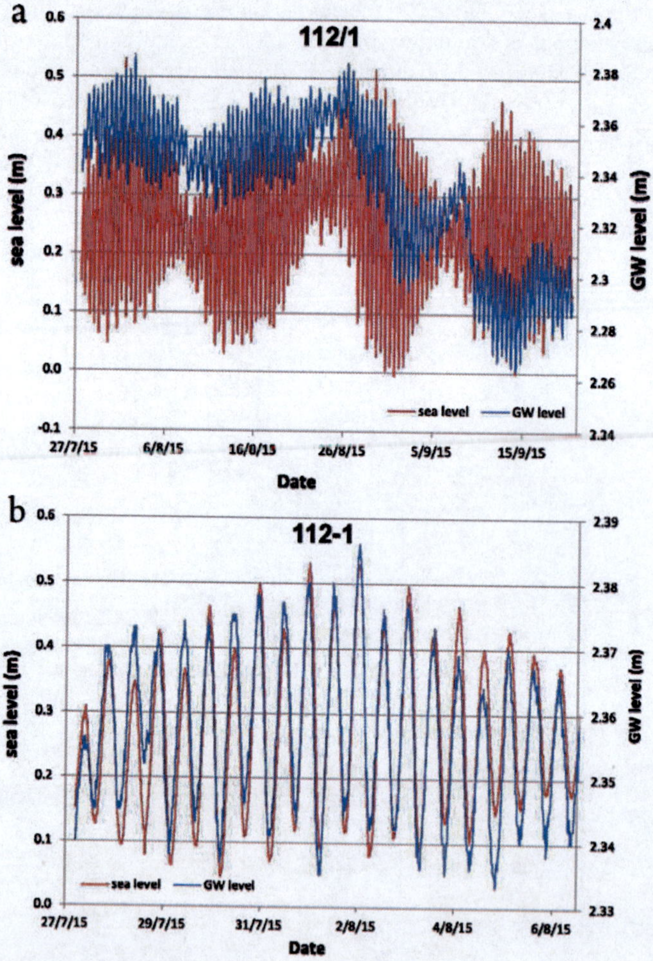


**Fig. 4** Groundwater level from Nitzanim 112-1 borehole, which is opened to deep sub-aquifer *D* (depth of 177–191 m) of the coastal aquifer, ca. 715 m from the sea. Groundwater data is compared with seawater level (Hadera station, data from IOLR). **a** Two months time series; **b** zoom-in for ten days. Note the very good positive correlation both of the bi-tidal and of the two-week spring tide cycles (after Narovlansky 2018)

where high activities (0.6 dpm/L compared to <0.2 in typical seawater) were observed 10–15 km offshore, northwest of Haifa (Mount Carmel), which could be related in this case to the submarine extension of the Carmel Fault. Recently, Paldor et al. (2019, 2020) used numerical simulations to show that deep water SGD could occur even at 15 km offshore northern Israel, where the Cretaceous aquifer is exposed by a submarine canyon. The possibility and extent of offshore discharge should be further studied.

## 5 SGD in Dor Bay

SGD was mainly studied at Dor Bay, in the southern Mount Carmel coast (Fig. 5a). Swarzenski et al. (2006) showed by resistivity profiles that a freshwater 'tongue' is undelaying the bay, and that the freshwater approaches seabed during low tide, which was interpreted as a clear indication for tidal-induced discharge (Fig. 5b). Weinstein et al. (2007a) showed by radon measurements in coastal boreholes and in the bay that the freshwater has been discharging from the confined Kurkar (calcareous sandstone) aquifer, which is irregularly exposed at the bay.

Weinstein et al. (2007a) used radon time series taken during three field campaigns in 2006 (e.g., Fig. 5c) to assess the average SFGD (specific) discharge at 5 cm/d, or a flux to the bay of 5 $m^3\ d^{-1}$ per 1 m length of shoreline, and a total SGD discharge of 8 $m^3\ d^{-1}$ per 1 m shoreline. This was quite similar to the SFGD estimate of Swarzenski et al. (2006) of 7 $m^3\ d^{-1}$ per 1 m shoreline, which was based on the differences between low- and high-tide resistivity profiles in the bay sediments (Fig. 5b). The above was later slightly corrected to a multi-annual estimate of 3.5 cm/d (SFGD; Weinstein et al. 2011).

Using the above SFGD estimate for a coastal stretch of 100 m width (whereby Dor Bay is 120 m wide), it results in ~$1.3 \times 10^6\ m^3\ year^{-1}$ (or total SGD of $2.1 \times 10^6\ m^3\ year^{-1}$) per 1 km coast. Weinstein et al. (2007a) noted that this figure is significantly larger than the natural recharge across the 3.5 km-wide southern Mount Carmel coast, even if one assumes it is entirely drained to the sea, which suggests that part of this discharge should come from the Cretaceous aquifer. Tal et al. (2018) also showed, by mass balances, that the Pleistocene coastal aquifer at Ma'agan Michael, 6 km south of Dor Bay, is being fed by water from the Cretaceous aquifer.

Extending the above estimate to the entire length of the Mount Carmel coast (33 km), one gets an annual freshwater discharge of ca. $42 \times 10^6\ m^3$, which exceeds the groundwater volume utilized annually from the entire Mount Carmel aquifer.

Recently, Ratner-Narovlansky et al. (2019) showed that the partly-confined sub-aquifer (B) is underlain by a deeper confined unit (C), which also discharges to the sea at about 150 m offshore. This extends the width of the discharge belt seaward and may imply that even larger volumes of freshwater are discharging to the sea at this area. Similarly, Tal et al. (2018), using both radon measurements and seismics (CHIRP), identified an SGD 'hotspot' at 500 m from offshore 6 km south of Dor

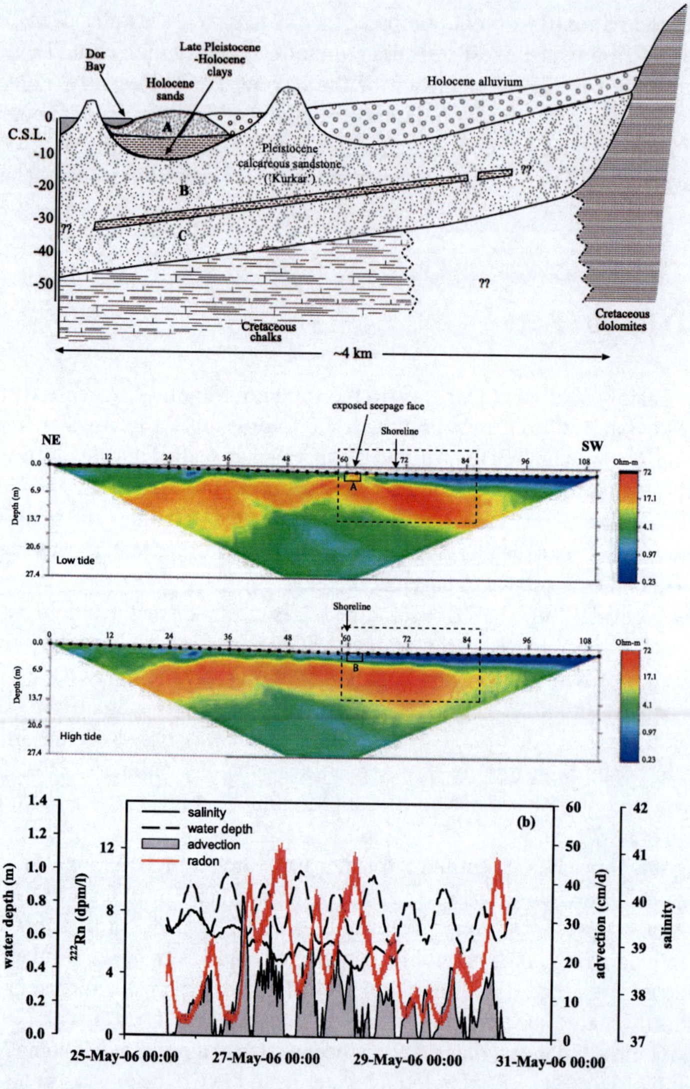


**Fig. 5** **a** Hydrogeological section in the Quaternary coastal aquifer of the southern Carmel plain at Dor (modified from Weinstein et al. 2007a). **b** Shore-diagonal transect of inverted resistivity during low (top) and high (bottom) tide at Dor Bay. Note the seepage close to shoreline during low tide (modified from Swarzenski et al. 2006). **c** Radon time series at Dor Bay (ca. 30 m from shore) during May 2006 compared with sea level and salinity; discharge (advection) of freshwater is deduced from changes in radon inventory, after corrections were made for mixing with the open sea and evasion to the atmosphere (modified from Weinstein et al. 2007a)

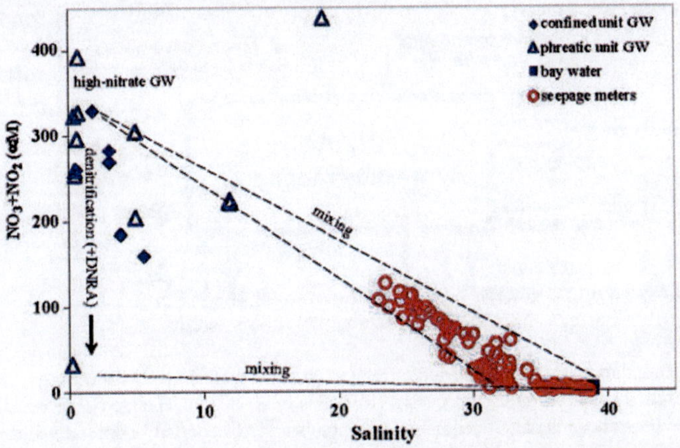


**Fig. 6** Nitrate concentrations in water discharging from bay floor at Dor Bay. The water define a mixing line between low-nitrate saline end member and high-nitrate fresh groundwater. Note that the saline end member does not have a unique seawater salinity, but rather defines a range of salinities between 40 and 32 (PSU units). This is interpreted as another (prior) mixing between seawater and low-nitrate ground water, which underwent denitrification and DNRA (dissimilatory nitrate reduction to ammonia) (modified from Weinstein et al. 2011)

Bay, which suggests that at least at certain places, discharge can extend farther away from shoreline.

Weinstein et al. (2011) showed that FSGD carries large loads of nutrients, mainly nitrate and silica to the Dor Bay (506 and 560 mol year$^{-1}$ per 1 m shoreline) (Fig. 6). They also showed that during the seaward travel through the 'subterranean estuary' (Moore 1999), some of the nitrate is being denitrified (103 mol year$^{-1}$ per 1 m shoreline), while another part undergoes DNRA (dissimilatory nitrate reduction to ammonia, 24 mol year$^{-1}$ per 1 m shoreline). Finally, phosphate was shown to be mainly conveyed by the saline RSGD.

## 6 SGD in the Mediterranean

SGD was reported in the Mediterranean since ancient times, where the Phoenicians and Romans were using this source of water to fuel their boats with drinking water (e.g., Strabo and Plini the Elder in Taniguchi et al. 2002). Recently, it was studied both at karstic/carbonate coastal areas, where discharge is of the focused, submarine spring type, which is mainly FSGD (e.g., Spain, Gulf of Lyon, Croatia, Greece, Lebanon, Libya), (Fleury et al. 2007; Garcia-Solsona et al. 2010; Tsabaris et al. 2011; Tovar-Sánchez et al. 2014), as well as at sandy coasts, such as the Venice Lagoon, where discharge is diffuse and is mainly of the RSGD type (e.g., Rapagliaa

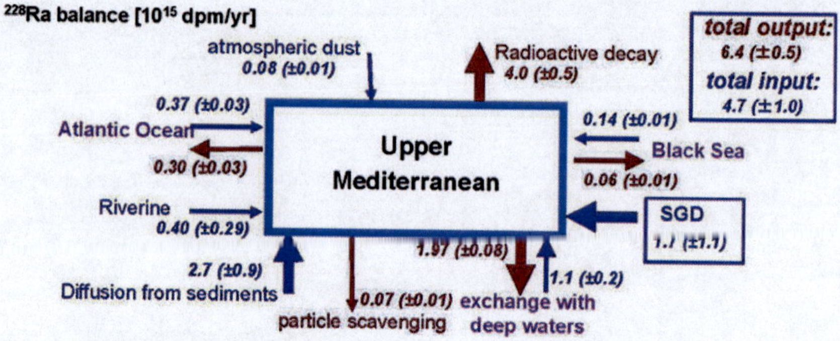


**Fig. 7** $^{228}$Ra balance of the whole Mediterranean (surface water). SGD contribution is calculated as the difference between input and output (Rodellas et al. 2015). Then, SGD water discharge to the total Mediterranean is calculated using end member $^{228}$Ra content in coastal water

et al. 2010; Gattacceca et al. 2011). The discussion herein focuses on the studies dealing with the entire basin SGD.

Zektser and Dzhamalov (2007) used hydrological considerations to assess the fresh, terrestrial SGD to the Mediterranean at 68 × $10^9$ $m^3$ $year^{-1}$, which is 22% of the riverine discharge of 310 × $10^9$ $m^3$ $year^{-1}$. Rodellas et al. (2015) used $^{228}$Ra (half-life of 5.75 year) mass balances (Fig. 7) to assess the total SGD at 200–4300 × $10^9$ $m^3$ $year^{-1}$. This is much larger than the riverine input to the Mediterranean, but quite similar to SGD estimates derived by $^{228}$Ra for the Atlantic and for the entire global oceans (Moore et al. 2008; Kwon et al. 2014). Using a coastal length of 65,000 km, this yields a discharge of 4–93 × $10^6$ $m^3$ $year^{-1}$ per 1 km coast, which is significantly higher than the one derived by radon for the Dor Bay area. However, the value derived for the entire Mediterranean definitely includes water discharging farther away from shore, across continental shelves and shallow seas, such as the Adriatic Sea, while at the Dor Bay, the estimate was limited to the 120 m-wide embayment.

Rodellas et al. (2015) further used these numbers to calculate the load of nutrients carried by SGD to the whole Mediterranean. Using concentrations in coastal groundwater from around the Mediterranean, they estimated the SGD input at 20–1500 × $10^9$ mol $year^{-1}$ DIN, 0.1–4.6 × $10^9$ mol $year^{-1}$ DIP, and 20–650 × $10^9$ mol $year^{-1}$ DSi, which are similar to the loads conveyed by rivers and dust to the sea. It is noteworthy that these numbers should be treated cautiously, since while the total SGD was calculated for the total sea, and may include discharge (of RSGD) from all shelves and shallow seas, nutrient concentrations were mainly taken from coastal (fresh and saline) groundwater, which could be significantly higher and with different composition (e.g., nitrate/ammonia ratios) than those in water discharging farther offshore.

## 7 SGD in Inland Basins

SGD is not limited to marine coastal areas, and it occurs as well in lakes, estuaries and rivers (e.g., Lee 1977; Shaw and Prepas 1990; Boyle 1994; Schwartz 2003). In Israel, this is mainly relevant to the freshwater Lake Kinneret (Sea of Galilee) and to the hypersaline Dead Sea, both located along the Dead Sea Transform. The lakes are being fed by both freshwater and old brines, which have been highly affected by recent human intervention and over-exploitation (e.g., potash industry, water diversion).

A geophysical study of Lake Kinneret showed that SGD is limited to its littoral, while pore water exchange at its deeper part is dominated by diffusion (Hurwitz et al. 1999) (see also Chapter "The Eastern Dead Sea Rift Continental Groundwater Base Level"). SGD specific discharge at the Kinneret littoral was assessed by seepage meters and a gel profiler at 0.2–1.6 m $year^{-1}$ (Dror et al. 1999; Mortimer et al. 1999; Nishri et al. 2002), one order of magnitude lower than at the Mediterranean coast. The total groundwater discharge to the lake was estimated to be about 60–110 × $10^6$ $m^3$ $year^{-1}$ (Mero 1978; Nishri et al. 1999), which delivers to the lake ca. 72,000–90,000 ton of chlorides per year (Smith et al. 1989; Nishri et al. 1999). The Lake Kinneret SGD is composed of both brackish diluted Ca-chloride brines, (e.g., Starinsky 1974; Mazor 1978) and freshwater, and its discharge (as studied in onshore springs) is driven by the hydraulic gradient between land and the lake, as well as by lake level (e.g., Rimmer et al. 1999; Abbo et al. 2003). Chemical mass balances of lake water imply that most of this SGD is of the Fuliya springs type, issuing in the Fuliya area, at the northwestern part of the lake (Nishri et al. 1999; Kolodny et al. 1999). Nevertheless, sub-lacustrine study of discharging water suggests that the lake is probably underlain by several pockets of brines (Weinstein et al. 2003). This should be further studied, by more detailed pore water chemistry (e.g., Br/Cl ratios, Klein-Ben David et al. 2004) or possibly by means of isotopic tracers, such as radium isotopes (e.g., Lazar et al. 2008).

SGD in the Dead Sea was studied by Mallast et al. (2013) at the northwestern part of the lake, next to the onshore springs at Ein Feshkha. Using airborne thermal imaging, they located large number of both focused, submarine springs and seepages through the alluvium. They concluded that although SGD is certainly significant, the fluxes are much lower (~10%) than those at the onshore springs. Other studies detected submarine springs and diffusive discharge of both fresh or brackish water and of brines along both western and the eastern sides of the lake (e.g., Akawwi et al. 2008; Ionescu et al. 2012; Mallast et al. 2013), mostly next to onshore springs. Mallast and Siebert (2018) used unmanned aerial vehicle to study the large spatio-temporal variability in Dead Sea SGD, probably related to the highly dynamic and non-equilibrium conditions in the Dead Sea coastal aquifer.

The extent of recirculation of Dead Sea water in the adjacent alluvium aquifer (RSGD) was studied by Kiro et al. (2014), using the unique antithetic behavior of $^{226}Ra$ and $^{228}Ra$, since the aquifer is a sink for the high $^{226}Ra$ of Dead Sea water and a source for $^{228}Ra$. They also applied mass balance calculations to both. Accordingly, Kiro et al. (2014) were able to assess the extent of lake water circulation below the fresh–saline water interface at 400 × $10^6$ $m^3$ $year^{-1}$ (or 2.5 × $10^6$ $m^3$ $year^{-1}$ per 1 km coast), similar to estimates of the total terrestrial discharge to the Dead Sea of 300 ± 300 × $10^6$ $m^3$ $year^{-1}$ (Lensky et al. 2005).

## 8 Remaining Questions

1. The Israeli coastal hydrogeology is highly variable, both naturally and in terms of human intervention. SGD was studied in detail in the Mount Carmel coastal plain, where water discharges from relatively thin (40–50 m) sandy aquifer. It has not been investigated, and SGD extent was not assessed or characterized neither in the northern coast, where Cretaceous carbonates are exposed to the sea, nor in the southern coast, where the nearshore coastal aquifer is composed of multiple units with debated connection to the sea.
2. Similarly, the loads of nutrients and other contaminants carried by SGD to the sea were characterized only at Dor, which could be very different in other coastal settings. Moreover, contaminant contribution has to be studied in relation with land–sea gradients and the specific hydrogeology involved.
3. Deep, offshore SGD should also be studied, mainly albeit not solely, along the northern coast, where the Cretaceous carbonate aquifer is exposed to the sea.

## References

Abbo H, Shavit U, Markel D, Rimmer A (2003) A numerical study on the influence of fractured regions on lake/groundwater interaction; the Lake Kinneret (Sea of Galilee) case. J Hydrol 283:225–243

Akawwi E, Al-Zouab A, Kakish M, Koehn F, Sauter M (2008) Using thermal infrared imagery (TIR) for illustrating the submarine groundwater discharge into the eastern shoreline of the Dead Sea-Jordan. Am J Environ Sci 4(6):693–700. https://doi.org/10.3844/ajessp.2008.693.700

Amato DW, Bishop JM, Glenn CR, Dulai H, Smith CM (2016) Impact of submarine groundwater discharge on marine water quality and Reef Biota of Maui. PLoS ONE 11(11):e0165825. https://doi.org/10.1371/journal.pone.0165825

Amir N, Kafri U, Herut B, Shalev E (2013) Numerical simulation of submarine groundwater flow in the coastal aquifer at the Palmahim Area, the Mediterranean Coast of Israel. Water Resour Manage 27(11):4005–4020

Arad A (1983) A summary of the artesian coastal basin of Guyana. J Hydrol 63(3–4):299–313

Basu AR, Jacobsen SB, Poreda RJ, Dowling CB, Aggar PK (2001) Large groundwater strontium flux to the oceans from the Bengal Basin and the marine strontium isotope record. Science 293:1470–1473

Beck AJ, Charette MA, Cochran JK, Gonneea ME, Peucker-Ehrenbrink B (2013) Dissolved strontium in the subterranean estuary—implications for the marine strontium isotope budget. Geochim Cosmochim Acta 117:33–52

Boehm AB, Paytan A, Shellenbarger GG, Davis KA (2004) Composition and flux of groundwater from a California beach aquifer: implications for nutrient supply to the surf zone. Cont Shelf Res 26:269–282

Boyle DR (1994) Design of a seepage meter for measuring groundwater fluxes in the nonlittoral zones of lakes—evaluation in a boreal forest lake. Limnol Oceanogr 39(3):670–681

Burnett WC, Dulaiova H (2003) Estimating the dynamics of groundwater input into the coastal zone via continuous radon-222 measurements. J Environ Radioact 69:21–35

Burnett WC, Wattayakorn G, Taniguchi M, Dulaiova H, Sojisuporn P, Rungsupa S, Ishitobi T (2007) Groundwater-derived nutrient inputs to the Upper Gulf of Thailand. Cont Shelf Res 27:176–190

Cable JE, Martin JB (2008) In situ evaluation of nearshore marine and fresh pore water transport into Flamengo Bay, Brazil. Estuar Coast Shelf Sci 76:473–483

Cable JE, Burnett WC, Chanton JP, Weatherly G (1996) Modeling groundwater flow into the northeastern Gulf of Mexico based on $^{222}$Rn. Earth Planet Sci Lett 144:591–604

Cathles LM, Members of Working Group 3 (1987) Fluid circulation in the crust and the global geochemical budget. In: Report of the second conference on scientific ocean drilling, COSOD II, Strasbourg, France, pp 67–86

Charette MA, Buesseler KO (2004) Submarine groundwater discharge of nutrients and copper to an urban subestuary of Chesapeake Bay (Elizabeth River). Limnol Ocean 49(2):376–385

Cohen D, Person M, Wang P (2010) Origin and extent of fresh paleowaters on the Atlantic continental shelf, USA. Ground Water 48:143–158. https://doi.org/10.1111/j.1745-6584.2009.00627.x

Cooper HH (1959) A hypothesis concerning the dynamic balance of fresh water and salt water in a coastal aquifer. J Geophys Res 64:461–467

Dulaiova H, Camilli R, Henderson PB, Charette MA (2010) Coupled radon, methane and nitrate sensors for large-scale assessment of groundwater discharge and non-point source pollution to coastal waters. J Environ Radioact 101(7):553–563

Dror G, Ronen D, Stiller M, Nishri A (1999) Cl/Br ratios of Lake Kinneret, pore water and associated springs. J Hydrol 225(3–4):130–139

Fleury P, Bakalowicz M, de Marsily G (2007) Submarine springs and coastal karst aquifers: a review. J Hydrol 339:79–92

Garcia-Solsona E, Garcia-Orellana J, Masque P, Rodellas V, Mejıas M, Ballesteros B, Domınguez JA (2010) Groundwater and nutrient discharge through karstic coastal springs (Castelló, Spain). Biogeosciences 7:2625–2638

Garrels RM, MacKenzie FT (1967) Evolution of sedimentary rocks. Norton & Co.

Gattacceca JC, Mayer A, Cucco A, Claude C, Radakovitch O, Vallet-Coulomb C, Hamelin B (2011) Submarine groundwater discharge in a subsiding coastal lowland: a $^{226}$Ra and $^{222}$Rn investigation in the Southern Venice lagoon. Appl Geochem 26(5):907–920

Gonneea ME, Charette MA, Liu Q, Herrera-Silveira JA, Morales-Ojeda SM (2014) Trace element geochemistry of groundwater in a karst subterranean estuary (Yucatan Peninsula, Mexico). Geochim Cosmochim Acta 132:31–49. https://doi.org/10.1016/j.gca.2014.01.037

Groen J, Velstra J, Meesters A (2000) Salinization processes in paleowaters in coastal sediments of Suriname: evidence from Δ 7Cl analysis and diffusion modelling. J Hydrol 234:1–20

Hurwitz S, Goldman M, Ezersky M, Gvirtzman H (1999) Geophysical (time domain electromagnetic model) delineation of a shallow brine beneath a freshwater lake, the Sea of Galilee, Israel. Water Resour Res 35(12):3631–3638

Ionescu D, Siebert C, Polerecky L, Munwes YY, Lott C, Haüsler S, Bižić-Ionescu M, Quast C, Peplies J, Glöckner FO, Ramette A, Rödiger T, Dittmar T, Oren A, Geyer S, Stark H-J, Sauter M, Licha T, Laronne JB, de Beer D (2012) Microbial and chemical characterization of underwater fresh water springs in the Dead Sea. PLoS ONE 7(6):e38319. https://doi.org/10.1371/journal.pone.0038319

Johannes RE (1980) The ecological significance of the submarine discharge of groundwater. Mar Ecol Prog Ser 3:365–373

Johnston RH (1983) The salt-water–fresh-water interface in the tertiary limestone aquifer, southeast Atlantic outer continental-shelf of the USA. J Hydrol 61:239–249

Johnson AG, Glenn CR, Burnett WC, Peterson RN, Lucey PG (2008) Aerial infrared imaging reveals large nutrient-rich groundwater inputs to the ocean. Geophys Res Lett 35:L15606. https://doi.org/10.1029/2008GL034574

Kafri U, Goldman M (2006) Are the lower subaquifers of the Mediterranean coastal aquifer blocked to seawater intrusion? Results of a TDEM (time domain electromagnetic) study. Isr J Earth Sci 55:55–68

Kiro Y, Weinstein Y, Yechieli Y, Starinsky A (2014) The role of long-term aquifer seawater circulation in elemental mass balances: a lesson from the Dead Sea. Earth Planet Sci Lett 394:146–158

Kiro Y, Weinstein Y, Starinsky A, Yechieli Y (2015) Application of radon and radium isotopes to groundwater flow dynamics: an example from the Dead Sea. Chem Geol 411:155–171

Klein-Ben David O, Sass E, Katz A (2004) The evolution of marine evaporitic brines in inland basins: the Jordan-Dead Sea Rift valley. Geochim Cosmochim Acta 68(8):1763–1775

Kolodny Y, Katz A, Starinsky A, Moise T, Simon E (1999) Chemical tracing of salinity sources in Lake Kinneret (Sea of Galilee), Israel. Limnol Oceanogr 44(4):1035–1044

Kwon EY, Kim G, Primeau F, Moore WS, Cho H-M, DeVries T, Sarmiento JL, Charette MA, Cho Y-K (2014) Global estimate of submarine groundwater discharge based on an observationally constrained radium isotope model. Geophys Res Lett 41. https://doi.org/10.1002/2014GL061574

Kwon HK, Kang H, Oh YH, Park SR, Kim G (2017) Green tide development associated with submarine groundwater discharge in a coastal harbor, Jeju, Korea. Sci Rep 7: 6325. https://doi.org/10.1038/s41598-017-06711-0

Lazar B, Weinstein Y, Paytan A, Magal E, Bruce D, Kolodny Y (2008) Ra and Th adsorption coefficients in lakes—Lake Kinneret (Sea of Galilee) "natural experiment". Geochim Cosmochim Acta 72(14):3446–3459

Lee DR (1977) A device for measuring seepage flux in lakes and estuaries. Limnol Oceanogr 22(1):140–147

Lensky NG, Dvorkin Y, Lyakhovsky V, Gertman I, Gavrieli I (2005) Water, salt, and energy balances of the Dead Sea. Water Resour Res 41:W12418

Li L, Barry DA, Stagnitti F, Parlange JY (1999) Submarine groundwater discharge and associated chemical input to a coastal sea. Water Resour Res 35:3253–3259

Mallast U, Siebert C (2018) Combining continuous spatial and temporal scales for SGD investigations using UAV-based thermal infrared measurements. Hydrol Earth Syst Sci Discuss. https://doi.org/10.5194/hess-2018-361

Mallast U, Schwonke F, Gloaguen R, Geyer S, Sauter M, Siebert C (2013) Airborne thermal data identifies groundwater discharge at the North-Western Coast of the Dead Sea. Remote Sens 5:6361–6381. https://doi.org/10.3390/rs5126361

Mazor E (1978) Mineral waters of the Kinneret basin and possible origin. In: Serruya C (ed) Lake Kinneret. Monographiae Biologicae 32, pp 103–120

Mero F (1978) The water balance of Lake Kinneret. In: Serruya C (ed) Lake Kinneret. Monographiae Biologicae 32, pp 99–102

Michael HA, Mulligan AE, Harvey CF (2005) Seasonal oscillations in water exchange between aquifers and the coastal ocean. Nature 436:1145–1148

Moore WS (1996) Large groundwater inputs to coastal waters revealed by $^{226}Ra$ enrichments. Nature 380:612–661

Moore WS (1999) The subterranean estuary: a reaction zone of groundwater and sea water. Mar Chem 65:111–126

Moore WS, Wilson AM (2005) Advective flow through the upper continental shelf driven by storms, buoyancy, and submarine groundwater discharge. Earth Planet Sci Lett 235:564–576

Moore WS, Sarmiento JL, Key RM (2008) Submarine groundwater discharge revealed by $^{228}Ra$ distribution in the upper Atlantic Ocean. Nat Geosci 1:309–311

Mortimer RJG, Krom MD, Boyle DR, Nishri A (1999) Use of a high-resolution pore-water gel profiler to measure groundwater fluxes at an underwater saline seepage site in Lake Kinneret, Israel. Limnol Oceanogr 44:1802–1809

Nace RL (1970) World Hydrology: status and prospects. In: World water balance. In: I. Symposium for the Association of Internationale D'Hydrologie Scientifique Publication No. 92, pp 1–10

Narovlansky Y (2018) Examining the connectivity of coastal confined aquifers to the Mediterranean Sea by tidal fluctuation. M.Sc. thesis, Bar-Ilan University

Nishri A, Stiller M, Rimmer A, Geifman Y, Krom M (1999) Lake Kinneret (the Sea of Galilee): the effects of diversion of external salinity sources and the probable chemical composition of the internal salinity sources. Chem Geol 158:37–52

Nishri A, Boyle DR, Koren N, Stiller M (2002) The contribution of water and chloride to Lake Kinneret through unfocused seepage, based on in situ seepage measurements. Isr J Earth Sci 51:269–279

Oberdorfer JA (2003) Hydrogeologic modeling of submarine groundwater discharge: comparison to other quantitative methods. Biogeochemistry 66:159–169

Paldor A, Aharonov E, Katz O (2019) Dynamics of saltwater intrusion and submarine groundwater discharge in confined coastal aquifers: a case study in northern Israel. Hydrogeol J. https://doi.org/10.1007/s10040-019-01958-5

Paldor A, Katz O, Aharonov E, Weinstein Y, Roditi-Elasar M, Lazar A, Lazar B (2020) Deep submarine groundwater discharge—evidence from Achziv submarine canyon at the exposure of the Judea group confined aquifer, Eastern Mediterranean. J Geophys Res Oceans 125:e2019JC015435. https://doi.org/10.1029/2019JC015435

Person M, Dugan B, Swenson JB, Urbano L, Stott C, Taylor J, Willett M (2003) Pleistocene hydrogeology of the Atlantic continental shelf, New England. Bul Geol Soc Am 115(11):1324–1343. https://doi.org/10.1130/B25285.1

Post VEA, Groen J, Kooi H, Person M, Ge S, Edmunds WM (2013) Offshore fresh groundwater reserves as a global phenomenon. Nature 504:71–78

Rapagliaa J, Ferrarin C, Zaggia L, Moore WS, Umgiesser G, Garcia-Solsonac E, Garcia-Orellanac J, Masqué P (2010) Investigation of residence time and groundwater flux in Venice Lagoon: comparing radium isotope and hydrodynamical models. J Environ Radioact 101(7):571–581

Ratner-Narovlansky Y, Weinstein Y, Yechieli Y (2019) Tidal fluctuations in a multi-unit coastal aquifer. J Hydrol 580:124222

Rimmer A, Hurwitz S, Gvirtzman H (1999) Spatial and temporal characteristics of saline springs: Sea of Galilee, Israel. Ground Water 37:663–673

Rodellas V, Garcia-Orellana1 J, Masqué P, Feldman M, Weinstein Y (2015) Submarine groundwater discharge: a major source of nutrients to the Mediterranean Sea. Proc Natl Acad Sci 112(13):3926–3930

Schwartz MC (2003) Significant groundwater input to a coastal plain estuary: assessment from excess radon. Estuar Coast Shelf Sci 56:31–42.dd

Shaban A, Khawalie M, Abdallah C, Faour G (2005) Geologic controls of submarine groundwater discharge: application of remote sensing to north Lebanon. Environ Geol 47. https://doi.org/10.1007/s00254-004-1172-3

Shaw RD, Prepas EE (1990) Groundwater-lakes interactions: II. Nearshore seepage patterns and the contribution of groundwater to lakes in central Alberta. J Hydrol 119:121–136

Shelenbarger GG, Monismith SG, Genin A, Paytan A (2006) The importance of submarine groundwater discharge to the nearshore nutrient supply in the Gulf of Aqaba (Israel). Limnol Oceanogr 51(4):1876–1886

Slomp CP, Van Cappellen P (2004) Nutrient inputs to the coastal ocean through submarine groundwater discharge: controls and potential impact. J Hydrol 295:64–86

Smith AJ, Turner JV (2001) Density-dependent surface water–groundwater interaction and nutrient discharge in the Swan-Canning Estuary. Hydrol Proc 15:2595–2616

Smith CG, Swarzenski PW (2012) An investigation of submarine groundwater–borne nutrient fluxes to the west Florida shelf and recurrent harmful algal blooms. Limnol Oceanogr 57(2):471–485

Smith SV, Serruya S, Geifman Y, Berman T (1989) Internal sources and sinks of water, P, N, Ca, and Cl in Lake Kinneret, Israel. Limnol Oceanogr 34(7):1202–1213

Starinsky A (1974) Ca-chloride brines and sedimentary rocks relations in the northern rift valley, Israel. PhD thesis, The Hebrew University, Jerusalem, 176 p

Swarzenski PW, Burnett WC, Greenwood WJ, Herut B, Peterson R, Dimova N, Shalem Y, Yechieli Y, Weinstein Y (2006) Combined time-series resistivity and geochemical tracer techniques to examine submarine groundwater discharge at Dor Beach Israel. Geophys Res Lett 33:L24405. https://doi.org/10.1029/2006GL028282

Tal A, Weinstein Y, Walman S, Goldman M, Yechieli Y (2018) The interrelations between a multi-layered coastal aquifer, a surface reservoir (fish ponds) and the sea. Water 10:1426. https://doi.org/10.3390/w10101426

Taniguchi M, Burnett WC, Cable JE, Turner JF (2002) Investigation of submarine groundwater discharge. Hydrol Proc 16:2115–2129

Taniguchi M, Ishitobi T, Shimada J (2006) Dynamics of submarine groundwater discharge and freshwater-seawater interface. J Geophys Res 111:C01008. https://doi.org/10.1029/2005jc002924

Tovar-Sánchez A, Basterretxea G, Rodellas V, Sánchez-Quiles D, García-Orellana J, Masqué P, Jordi A, López JM, Garcia-Solsona E (2014) Contribution of groundwater discharge to the coastal dissolved nutrients and trace metal concentrations in Majorca Island: Karstic vs detrital systems. Environ Sci Technol 48(20):11819–11827

Tsabaris C, Anagnostou MN, Patiris DL, Nystuen JA, Eleftheriou G, Dakladas Th, Papadopoulos V, Prospathopoulos A, Papadopoulos A, Anagnostou EN (2011) A marine groundwater spring in Stoupa, Greece: shallow water instrumentation comparing radon and ambient sound with discharge rate. Proc Earth Planet Sci 4:3–9

Valiela I, Costa JE (1988) Eutrophication of Buttermilk Bay, a Cape Cod coastal embayment: concentrations of nutrients and watershed nutrient budgets. Environ Manage 12:539–553

Weinstein Y, Katz A, Kastner M, Nishri A, Jannasch H (2003) Brine heterogeneity and dispersed interstitial advective flow underneath the Sea of Galilee, Israel. European Geophysical Society Meeting, Geophysical Research Abstracts 5, p 7706

Weinstein Y, Less G, Kafri U, Herut B (2006) Submarine groundwater discharge in the southeastern Mediterranean (Israel), preliminary results. Radioact Environ 8:360–372

Weinstein Y, Burnett WC, Swarzenski PW, Shalem Y, Yechieli Y, Herut B (2007a) The role of coastal aquifer heterogeneity in determining fresh groundwater discharge and seawater recycling: an example from the Carmel coast, Israel. J Geophys Res 112:C12016. https://doi.org/10.1029/2007JC004112

Weinstein Y, Shalem Y, Burnett WC, Swarzenski PW Herut B (2007b) Temporal variability of submarine groundwater discharge: assessments via radon and seep meters, the southern Carmel Coast, Israel. In: Sanford W (ed) A new focus on groundwater–seawater interactions. IAHS Publ. 312. IAHS Press, Wallingford, UK, pp 125–133

Weinstein Y, Yechieli Y, Shalem Y, Burnett WC, Swarzenski PW, Herut B (2011) What is the role of fresh groundwater and recirculated seawater in conveying nutrients to the coastal ocean? Environ Sci Technol 45(12):5195–5200. https://doi.org/10.1021/es104394r

Wilson AM, Gardner LR (2006) Tidally driven groundwater flow and solute exchange in a marsh: numerical simulations. Water Resour Res 42:W01405. https://doi.org/10.1029/2005wr004302

Zektser IS, Dzhamalov RG (2007) Submarine groundwater. CRC Press, Boca Raton, FL, 466 pp

# The Nubian Sandstone Aquifer in the Sinai Peninsula and the Negev Desert

Roi Ram, Avihu Burg, and Eilon M. Adar

## 1 Preface

A huge amount of groundwater, estimated to be hundreds of billions of cubic meters of freshwater to brackish water, is found in the Mesozoic Nubian sandstones of the Sinai Peninsula and the Negev Desert. The first hypotheses for the existence of a regional aquifer were raised back in the 1940s, with observations from steady-flow springs, not affected by annual changes in precipitation, located at the margins of the Gulf of Suez and south of the Dead Sea (Issar 1985; Shiftan 1958). This finding led researchers to postulate that these springs are fed by a regional aquifer that stretches underneath the Sinai Peninsula and the Negev Desert (defined hereafter as the "Sinai–Negev Basin"). The first practical evidence for the existence of this groundwater reservoir was found during oil drillings in central Sinai and in the northeastern Negev (Aharoni 1966; Issar et al. 1972).

Although separated today by an international boundary, the Sinai–Negev Basin (Fig. 1) constitutes one tectonic unit, which has been partly separated from the

---

This chapter is dedicated to Prof. Arie S. Issar, a researcher, hydrogeologist and among the pioneers who discovered and assessed the enormous hidden groundwater potential in the Nubian Sandstone Aquifer under the Sinai–Negev deserts. His numerous and thorough hydrogeological researches revealed, tier upon tier, the secrets of the groundwater in the aquifer. These paved the road for the development of agriculture and industries in the Negev and the Arava Valley to this very day.

---

R. Ram (✉) · E. M. Adar
Zuckerberg Institute for Water Research, Blaustein Institutes for Desert Research, Ben-Gurion University of the Negev, Sede Boqer Campus, 8499000 Midreshet Ben Gurion, Israel
e-mail: roira@post.bgu.ac.il

E. M. Adar
e-mail: Eilon@bgu.ac.il

R. Ram · A. Burg
Geological Survey of Israel, 32 Yeshayahu Leibowitz St., 9692100 Jerusalem, Israel
e-mail: burg@gsi.gov.il



U. Kafri and Y. Yechieli (eds.), *The Many Facets of Israel's Hydrogeology*, Springer Hydrogeology, https://doi.org/10.1007/978-3-030-51148-7_9

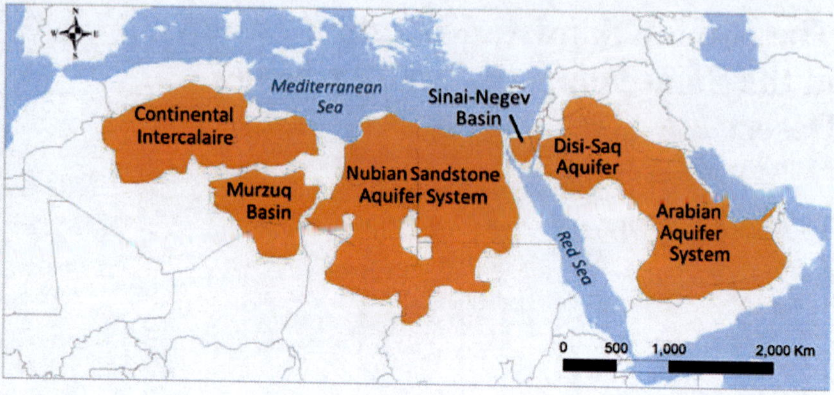


**Fig. 1** Major groundwater basins across North Africa, the Middle East and the Arabian Peninsula. Modified after Schulz et al. (2017)

African plate by the Gulf of Suez rift since the Late Oligocene period (Ben-Avraham 1978). Located at the northern margins of the global desert belt, this area is currently subjected to hyper-arid conditions. However, geological and geochemical evidence, such as speleothems (Vaks et al. 2006, 2010), paleo-lakes (Ginat et al. 2018), travertine deposits (Livnat and Kronfeld 1985; Enmar 1999; Waldmann et al. 2010) and the depleted isotopic composition of the groundwater (Issar et al. 1972), indicates that different and more rainy climatic conditions prevailed in the region during earlier epochs. It was suggested that the main recharge to the Nubian Sandstone Aquifer (NSA) took place during these more pluvial periods.

Nubian sandstones are well known in North Africa and the Arabian plate, where this clastic sequence reaches thicknesses of up to several kilometers (McKee 1963; Thorweihe 1990). In the Sinai–Negev Basin, the Nubian sandstones are buried at considerable depths under most parts of the basin, with the exception of several localities where the sandstones are uplifted and exposed, and where recharge could have taken place (Aharoni 1966; Kroitoru 1980). The burial under a thick sequence of carbonates and marls creates confined conditions and disconnects the system from the troposphere (Issar et al. 1972). However, disagreements have revolved around the question of whether most of the aquifer is indeed completely isolated, or whether it is perhaps linked to the troposphere through other shallower units and, hence, has also been recharged through additional areas along the flow path. Furthermore, the existence of a regional aquifer in such an arid region raises several additional questions: Where and when has the recharge to the aquifer taken place? What were the climatic conditions under which this area, now characterized by very low precipitation amounts and high evaporation rates, received its significant recharge? Was it a singular massive recharge period or a result of several distinct pluvial epochs? And is the system being recharged even under the present hyper-arid climate? Many studies tried to answer these and other fundamental questions using various techniques. The current summary reviews the questions, the dilemmas and the discoveries of the relevant studies, and tries to examine whether the combination of a wide range of geochemical and geophysical tools has led to a complete understanding of the flow system in the Nubian Sandstone Aquifer of the Sinai–Negev Basin.

The discovery of a huge groundwater reservoir within the ancient Nubian sandstones is not unique to the Sinai–Negev Basin. Transboundary deep sandstone aquifers are also well known in different groundwater basins over the Sahara and Arabian Deserts (Fig. 1), from the Continental Intercalaire in Algeria and Tunisia, through the great Nubian Sandstone Aquifer System (NSAS) under the Western Desert of Egypt, eastern Libya, northwestern Sudan and northeastern Chad, and further east to the regional aquifers in southern Jordan and Saudi Arabia (Richts et al. 2011; Macdonald et al. 2012; UN-ESCWA and BGR 2013; Sultan et al. 2019). In these arid basins, currently characterized by low precipitation rates (mostly <100 mm/year) (Nicholson 2000), groundwater reservoirs constitute an indispensable resource for life. During the past century, these reservoirs have been under constant development and increasing exploitation in order to meet human needs and to enable economic growth (Richey et al. 2015). Hence, although this summary is dedicated to the Nubian Sandstone Aquifer of the Sinai–Negev Basin, a short review of the hydrogeological findings in North Africa will contribute to understanding the magnitude and volume of this ancient water body in context.

## 2 The Study of the Nubian Sandstone Aquifers of North Africa

The existence of freshwater springs in the middle of the Sahara Desert, one of the most arid regions on earth, led researchers in the late nineteenth century to study the nearby groundwater resources, first in Algeria (Rolland 1890) and later on in the northeastern Sahara. Pioneering studies in the NSAS (Fig. 1) provided water-level maps indicating the groundwater flow pattern from the elevated area in the southwest, where the sandstone units are exposed and form the intake areas for natural recharge (Fig. 2a), toward the north and northeast (Ball 1927; Sandford 1935). These studies hypothesized that the aquifer is currently being recharged through these southern intake areas, which are located at lower latitudes characterized by higher precipitation. However, consequent studies included $^{14}C$ measurements and, as a result, hypothesized that the groundwater is, in fact, fossil[1] and was recharged during a more pluvial period that prevailed some 25,000–35,000 years ago (Knetsch 1962; Knetsch et al. 1963). Thorweihe (1990) continued this approach and re-examined the role of local recharge to the aquifer over the currently hyper-arid central and northern parts of the basin. In contrast to the hypothesis of Ball (1927) and Sandford (1935), who regarded the southern outcrops as the main intake area of the basin, Thorweihe

[1] Fossil groundwater, according to UNESCO: "Water that infiltrated usually millennia ago and often under climatic conditions different from the present, and that has been stored underground since that time" (Foster and Loucks 2006).

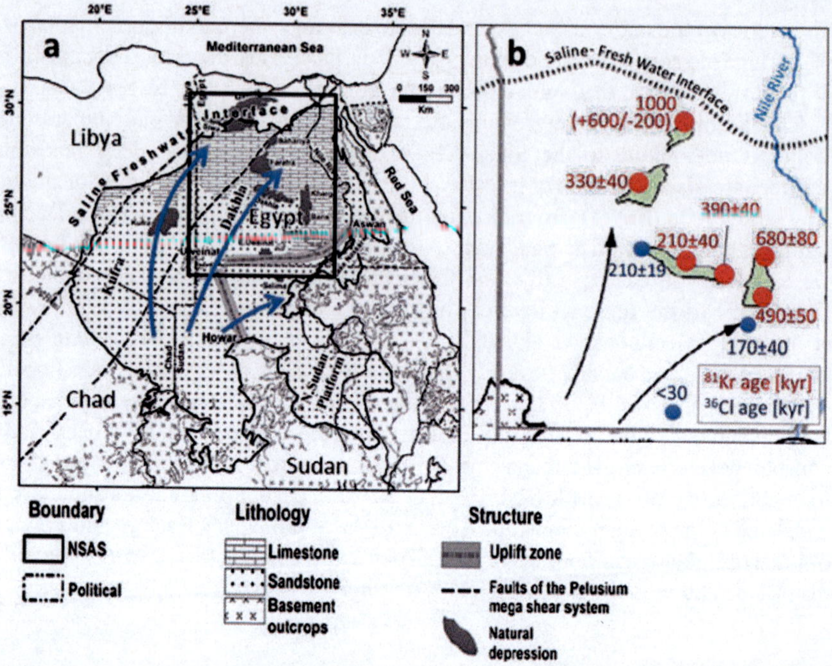


**Fig. 2** **a** The Nubian Sandstone Aquifer System of the northeastern Sahara Desert, modified after Mohamed et al. (2017). General flow directions in the aquifer (blue arrows) are shown as suggested in the pioneering studies by Ball (1927) and Sandford (1935), modified after Thorweihe and Heinl (2002), and **b** obtained $^{81}$Kr and $^{36}$Cl ages in the aquifer, highlighting the antiquity of water in the Nubian system. Green shades represent oases. Modified after Sturchio et al. (2004) and Patterson et al. (2005)

concluded that while groundwater flow from the southern exposures toward the north does exist, this regional flow "seems to be of minor importance" in relation to the impact of the local recharge over the central and northern parts of the basin. It was also suggested in this study that the limitations of the $^{14}$C dating method (up to ~40,000 years) make it an inappropriate tool for the study of groundwater flow in such a regional aquifer, where ages are expected to be much older downstream and away from the intake areas.

The study of stable isotopes in the fossil Nubian aquifers of North Africa highlighted the continental effect pattern of the paleo-rains. Paleo-waters across North Africa revealed a trend of depletion toward the east from −20‰ to −80‰ in the $\delta^2$H, similar to the pattern of present-day rains in Europe (Fig. 3) (Sonntag et al. 1979). This process of depletion in the stable isotopic composition of rains, known as *rainout*, is triggered by precipitation of the "heavier" $^{18}$O and $^2$H isotopes, whereas the remaining vapor becomes more and more depleted in relation to these isotopes as the air mass moves away from the moisture source and proceeds over the continent (Clark and Fritz 1997). Interestingly, this eastward isotopic depletion within

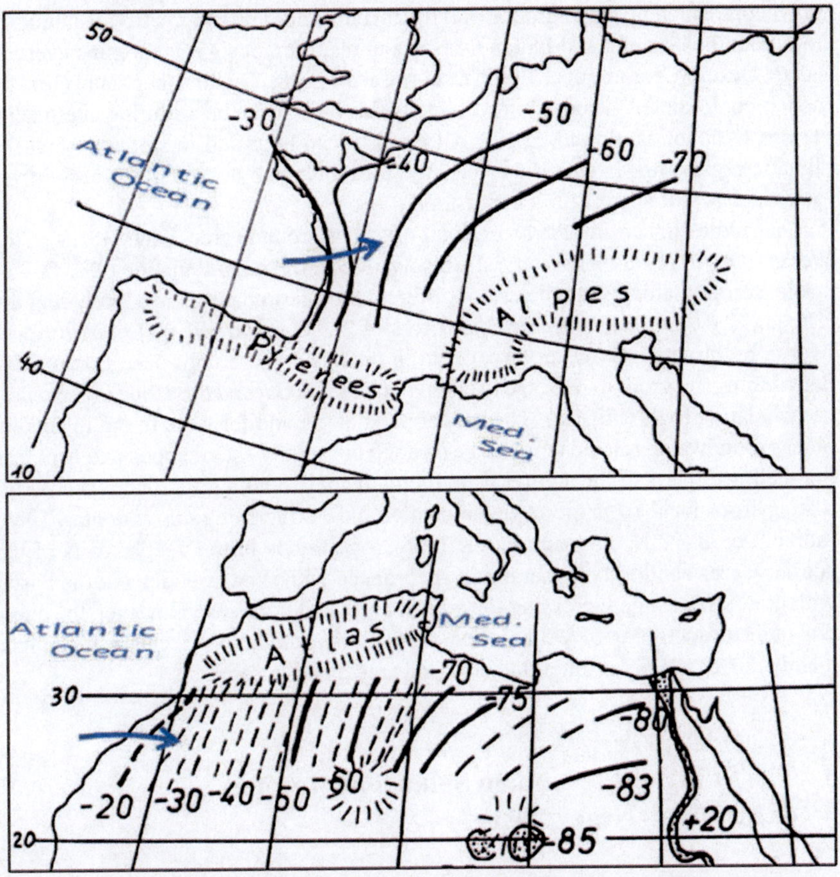


**Fig. 3** Isotopic pattern as reflected in the $\delta^2$H values of paleo-groundwater in North Africa, with significant depletion from west to east, with its equivalent pattern reflected in present rains across Europe. Modified after Sonntag et al. (1979)

the North African paleo-waters is opposed to the modern meteoric waters in this region, in which enriched (and even positive) $\delta^2$H values have been measured in the easternmost area, and more negative values of ~−25‰ have been measured in the western parts of the region (Abouelmagd et al. 2012).

The unique isotopic trend of paleo-waters was explained by paleo-westerlies that transported Atlantic moisture inland (Sultan et al. 1997; Thorweihe and Heinl 2002; Edmunds 2005) during the Last Glacial Period (Sonntag et al. 1979) or during several different glacial periods (Abouelmagd et al. 2012). Other studies argued for interglacial, rather than glacial, wet periods in the region, based on the dating of carbonate precipitation in paleo-lakes (Szabo et al. 1995), travertines (Crombie et al. 1997) and speleothems (El-Shenawy et al. 2018). It was also suggested that the northward migration of the Intertropical Convergence Zone (ITCZ) during interglacial periods was accompanied by the migration of African monsoons northward, driving humidity from both the Atlantic and Indian Oceans and resulting in significant rains over the Sahara Desert (Crombie et al. 1997; Osborne et al. 2008). On the other hand, climatic models contradicted the possibility of intensified rains in the basin during interglacial periods (Abouelmagd et al. 2012). All in all, it can be stated in summary that the climatic regime under which the recharge to the Nubian aquifers of North Africa took place is still not yet fully understood.

The riddle concerning the source and origin of groundwater in the NSAS of the Western Desert of Egypt was partially resolved recently. A combination of $\delta^{18}O$–$\delta^{2}H$ stable isotopes, along with the use of different long-term dating tools, including the $^{81}Kr$ and $^{36}Cl$ radioactive tracers (half-lives of 229,000 and 301,000 years, respectively), highlighted the water circulation in the basin and emphasized that indeed, determining the groundwater age in certain places was beyond the scope of the radiocarbon dating tools (Fig. 2b). The deeper parts of the aquifer were found to contain older groundwater, related to recharge through the southwestern exposures, whereas the shallower parts of the aquifer were found to contain younger groundwater, originating from local recharge over the central and northern regions (Sonntag 1984; Sturchio et al. 2004; Patterson et al. 2005). Age trends in the deeper parts of the aquifer were found to be generally in accordance with flow directions in the basin, with groundwater age increasing toward the north and northeast. However, the question of the exact timing of recharge and, particularly, whether the recharge took place mainly during glacial or interglacial periods remained unresolved.

## 3 The Study of the Nubian Sandstone Aquifer of the Sinai–Negev Basin

Several similarities can be found between the NSAS of the Western Desert and the NSA of the Sinai–Negev Basin, not only in their physical attributes, but also in the challenges, in the scientific conflicts and in the evolution of studies regarding the sources of the groundwater hosted in these aquifers. Nubian sandstones were deposited in North Africa and the Arabian plate, which were later tectonically separated (Hempton 1987), over extended periods in the Paleozoic to Mesozoic eras (McKee 1963). Unlike in the NSAS, where the whole Nubian section is referred as a single hydrogeological unit, in the Sinai–Negev Basin, the term NSA refers mainly to the Lower Cretaceous Kurnub Group clastic section. Other deeper Nubian sections were also found to be saturated in different wells (Nativ et al. 1987), but it is only the Lower Cretaceous section that was found to be hydrologically active throughout the whole basin. The hydrogeological settings, as well as the interrelations between the NSA and the other deeper Nubian sandstone hydrological units, are further discussed in Sects. 3.1 and 3.2.

Initial studies of the Lower Cretaceous Kurnub Group Aquifer in the Sinai–Negev Basin were conducted at the southwestern margins of the Dead Sea (Picard 1953;

Shiftan 1958, 1961; Arad and Michaeli 1967). Artesian water was discovered in this aquifer in 1953 during oil drilling, followed by additional drillings for water exploitation (Picard 1959). While the understanding of this regional aquifer's potential as a water resource developed, the origin of the artesian brackish water remained unclear. Shiftan (1961) hypothesized that it is fossil groundwater, originating from infiltration along the Arava Rift Valley during high stands of the proto-Dead Sea Lake. Others claimed that the water is not of fossil origin, but stemmed from contemporaneous recharge over the northeastern Negev and the Hebron Mountains, located west of the Dead Sea (Arad and Michaeli 1967). Since the Nubian sandstones are barely exposed over these mountains, these authors postulated that water infiltrates through the overlying carbonates of the Judea Group.

Another revolutionary idea for the origin of the NSA's groundwater was proposed in the early 1970s. In a pioneering study, Issar et al. (1972) used the unique $\delta^{18}O$ and $\delta^{2}H$ compositions in wells and springs across the Sinai Peninsula and the Negev Desert, coupled with obtained old $^{14}C$ ages, to conclude that "the various Nubian sandstone aquifers found in the investigated area are parts of one aquifer system." Based on the isotopic composition, Issar et al. (1972) proposed that recharge to the aquifer took place during the Last Glacial Period, when cooler and wetter climatic conditions prevailed within the region. This and later studies paved the way for further development and abstraction of groundwater from the NSA, especially along the Arava Rift Valley, where nowadays more than 20 million $m^3$ of brackish water is abstracted annually for local agricultural and industrial needs (Water Authority of Israel 2014).

### 3.1 Hydrogeological Background

The Lower Cretaceous Kurnub Group clastic sequence marks a non-marine stage that prevailed prior to a prolonged sea transgression over most of the Sinai–Negev Basin (Weissbrod 1969a, b; Bartov 1974). The group consists mainly of sandstone formations, combined with intercalations of clays and silts. The percentage of clays within the sequence increases northward and northwestward, where it passes through facies change into marine carbonates and marls (Aharoni 1966; Kroitoru 1980; Jenkins 1990; Rosenthal et al. 1998). However, sands constitute more than 80% of the total thickness of the group under vast parts of the basin (Kroitoru 1980; Nativ 1984). The group thickens from 180 m in the south, to up to more than 500 m in northern Sinai (Abd El Samie and Sadek 2001).

The NSA extends under vast parts of Sinai and the Negev, and covers an estimated area of 35,000 $km^2$ (Kroitoru 1980). A rough calculation of the groundwater storage, with an average saturated thickness of 300 m and effective porosity of 10%, yields a water volume of about $10^{12}$ $m^3$. This emphasizes the undeniable importance of this reservoir, especially in such an arid basin. The limited sandstone exposures that could have served as direct intake areas (Fig. 4) include those at the foothills of the Arabian–Nubian Massif in southern Sinai (Issar et al. 1972; Kroitoru 1980) and the

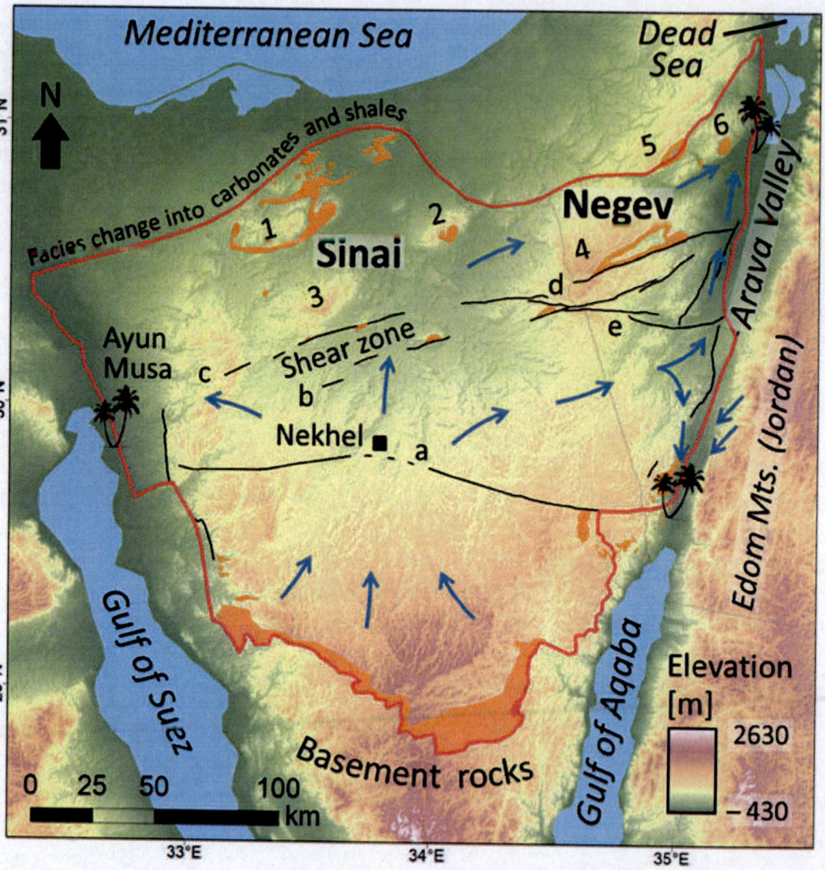


**Fig. 4** The Nubian Sandstone Aquifer of the Sinai–Negev Basin. Orange polygons mark intake areas, and arrows indicate main flow directions in the aquifer. Main outlets of the aquifer are marked by black trees. Major anticlines and faults are annotated by numbers and letters, accordingly, as follows: 1: Maghara; 2: Halal; 3: Yelek; 4: Ramon; 5: Hatira; and 6: Hazera; a: Ragabet El-Naam-Themed; b: Abu Kandu–Arif Bator; c: Minshera; d: Ramon; and e: Paran. Red line represents the estimated boundaries of the NSA, after WPRP (1998) and JICA (1999) in Sinai, and after Fink (1994), Guttman et al. (1999) and Dafny (2012) in the Negev region. Note the small flow arrows from the margins of the Edom Mountains toward the southern Arava, as discussed in Sect. 3.2

smaller exposures over the eroded anticlines of northern Sinai and the Negev (Arad and Kafri 1980; Guttman et al. 1999; Vengosh et al. 2007; Shalaby et al. 2012).

Recharge rates have been estimated to be relatively low under the current arid climate. For example, Kroitoru (1980) calculated an average active recharge of 3.5 $\times 10^6$ $m^3$/year for the whole basin, with 0.5 $\times 10^6$ $m^3$/year recharges through the northern Negev outcrops and the rest through the exposures in Sinai. A slightly higher recharge rate of 3.4 $\times 10^6$ $m^3$/year over the intake areas of the northern Negev was proposed by Guttman et al. (1999), and an order of magnitude higher recharge rate of

$32.8 \times 10^6$ $m^3$/year was proposed for the intake areas in the Sinai Peninsula (Dames and Moore 1985; WPRP 1998). A recent study used the Soil Water Assessment Tool model to calculate a recharge rate of $13.0 \times 10^6$ $m^3$/year for the southern Sinai intake areas and suggested that the current recharge is not negligible even under the present arid climate (Sultan et al. 2011). Active recharge to the aquifer in the last several thousands of years is reflected in the relatively young $^{14}C$ ages close to the intake areas in southern Sinai and the northeastern Negev (Abouelmagd et al. 2014; Vengosh et al. 2007). Some studies discussed the recharge–discharge balance in the system (before the disruption of equilibrium by production) and hypothesized that the natural discharge rate is dictated by the current recharge rate (Issar et al. 1972; Kroitoru 1980; Nour and Yehia 1993; WPRP 1998; Guttman et al. 1999).

The boundaries of the Sinai–Negev Basin (Fig. 4) are defined by geological structures and faults as follows: the uplifted basement rocks of the Arabian–Nubian Massif in the south, the Dead Sea Transform in the east, the Suez Rift in the west and the facies change into less permeable shales and carbonates, accompanied by increased salinities, in the north (Issar 1985; Mills and Shata 1989; WPRP 1998). The Kurnub Group is overlain by the thick carbonate sequence of the Upper Cretaceous Judea Group (Fig. 5). A regional angular disconformity appears at its base, thus bringing the Kurnub Group into direct contact with various rock formations (Fig. 6), from the carbonates and clastics of the Jurassic Arad Group in the northern parts of the basin, through the Triassic Ramon Group in the central Negev, and to the Cambrian Nubian sandstones of the Yam Suf Group in the southern parts of the basin (Kroitoru 1980). Such geological settings presumably create complicated interconnections between the different hydrogeological units in the region (Guttman et al. 1999; Bein et al. 2001; Rosenthal et al. 2007; Vengosh et al. 2007).

The flow pattern in the Sinai–Negev Basin (Fig. 4) generally exhibits a fanlike structure from the elevated area in southern Sinai toward the natural outlets along the eastern and western rifts (Issar et al. 1972; Rosenthal et al. 2007). The western discharge zone was identified near the Gulf of Suez (Ayun Musa area), and eastern outlets were identified near the Gulf of Aqaba and south of the Dead Sea (Issar et al. 1972). The discharge through the eastern outlets was estimated to be 2.5 times higher than that of the western outlets (Nour and Yehia 1993; WPRP 1998). For the eastern outlets, the precise outlet locations are dictated by the fault system of the Dead Sea Transform, and the groundwater finally discharges into local alluvial aquifers and saline sabkhas (the northeastern outlet is demonstrated in Fig. 7). Dominant anticlines dissect the entire basin (Fig. 4) and play a major role in directing the in-basin groundwater flow. Major E-W faults (Fig. 4) abruptly cut the lateral continuity of the rock units and may serve either as a barrier for the groundwater flow or, alternately, as preferential flow paths (discussed later in Sect. 3.2). In the northern parts of the basin, groundwater flow paths are governed by the Syrian Arc folds, where the sandstone units are uplifted high above the regional water level to become hydrologically unsaturated or even eroded in places (Arad and Kafri 1980; Yechieli et al. 1992; Rosenthal et al. 2007). Except for these local uplifted structures, under most of the basin, the Kurnub Group is buried at considerable depths, and the confined aquifer is characterized by high water heads and even artesian flow in places.

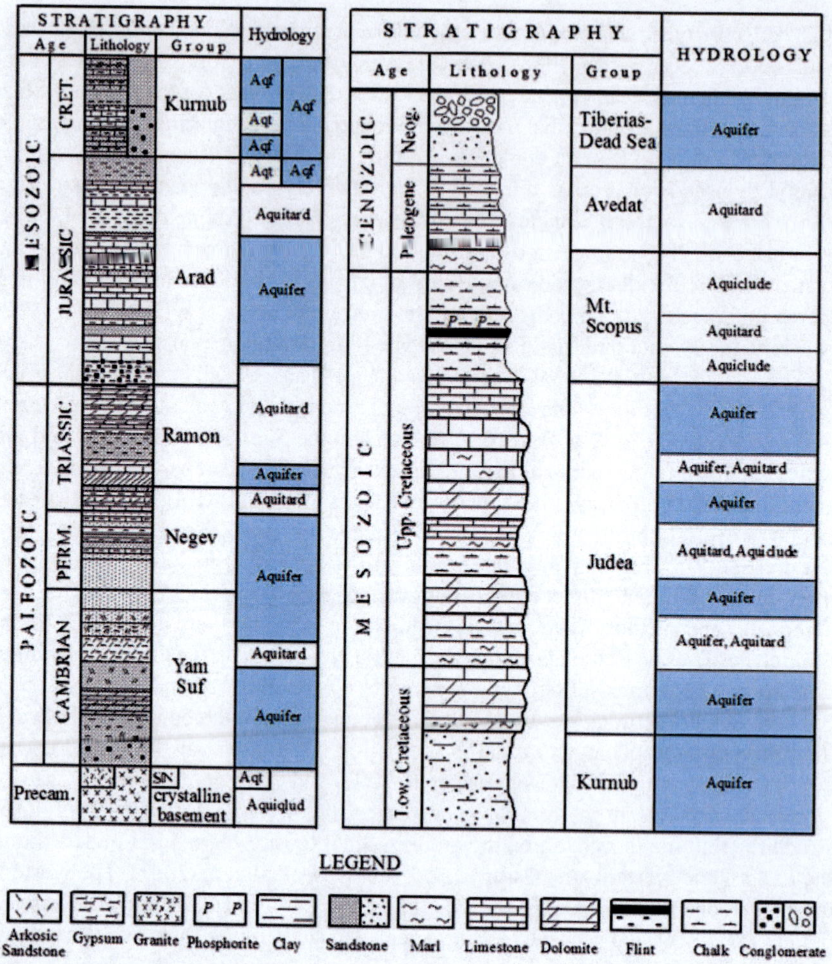


**Fig. 5** Generalized hydro-stratigraphic sections for the Negev region. The left section is a modification after Nativ et al. (1987) and Calvo and Gvirtzman (2013), and illustrates the deeper groups in the Negev Desert based mainly on deep oil drillings. The right section is modified after Burg et al. (2017) and presents the litho-stratigraphy of shallower units in the northern Negev

## 3.2 *Evolution of Studies of the NSA*

### 3.2.1 Initial Studies on the Source and Origin of Groundwater

Numerous studies focused on the water levels, the geological structure, and the chemical and isotopic compositions of the groundwater for interpretation of the flow patterns within the NSA. Dozens of wells have been drilled over the years in order to increase the utilization of the NSA reservoir. This, in turn, provided further access

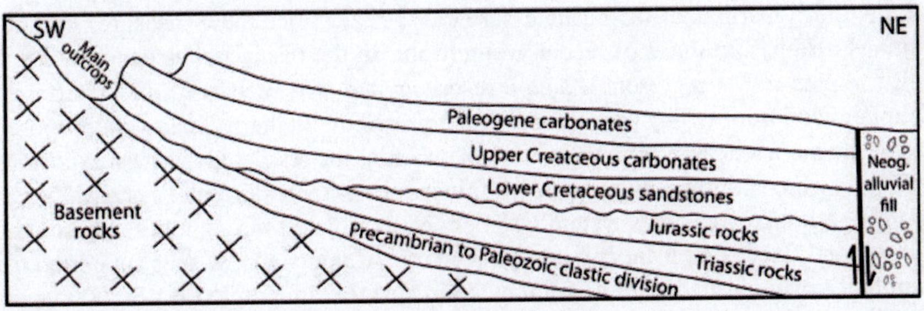


**Fig. 6** A schematized cross section from the sandstone outcrops of southern Sinai, through the southern Negev and toward the northern Arava Valley, illustrating the angular disconformity at the base of the Lower Cretaceous rocks. The disconformity deepens to the south, thus creating contacts with rocks of increasing ages. The geological settings may create different hydrological relations between the regional aquifers in the basin

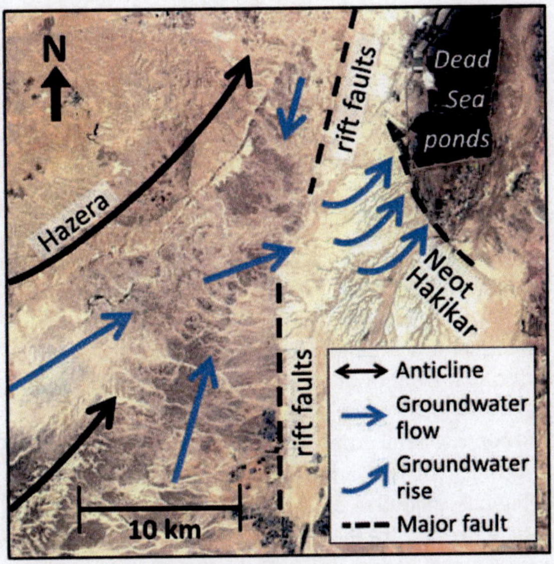


**Fig. 7** Northeastern outlets of the system, where the regional aquifers of the Negev discharge into the alluvial shallow aquifer through the listric Neot Hakikar Fault. Background image is taken from Google Earth, Image ©2019 Maxar Technologies, CNES/Airbus

to groundwater samples and basic hydrological parameters, thereby allowing the knowledge regarding this resource to expand.

As mentioned above, it was mainly the unique stable isotope composition of oxygen and hydrogen in the groundwater that led to understanding the NSA's continuity throughout the basin: Low $\delta^{18}O$ and $\delta^{2}H$ values were measured in central Sinai, along the Arava Rift margins in the east, and in the Ayun Musa Springs in the western margins of the basin (Issar et al. 1972). Since these measured compositions differ from those of recent groundwater in the basin, it was concluded that all samples rather represent a single reservoir that was recharged under different climatic conditions. The vast sandstone exposures of southern Sinai were proposed to be the main recharge zone to the system, whereas the restricted northern exposures (Fig. 4) were estimated to play only a minor role in recharging the system. Except for these intake areas, the system was defined as isolated throughout the basin and thus mostly detached from any present recharge. Calculated $^{14}C$ ages supported the antiquity of the water, with a groundwater residence time of 13,000–30,000 years, indicating that the aquifer contains primarily fossil water (Issar et al. 1972).

Other studies tried to re-evaluate the flow pattern. A slightly different approach was suggested by Arad and Kafri (1980), who claimed that the northern local Nubian sandstone exposures, as well as the larger Judea Group exposures scattered in the Sinai–Negev Basin, could not be excluded as old and current intake areas to the aquifer. Furthermore, according to their approach, the dominant shear zone in central Sinai (Fig. 4) breaks the continuity and the homogeneity of the flow system by diverting groundwater flow parallel to the faults. Therefore, these authors claimed that north of the shear zone groundwater is mainly formed by recharge through the northern exposures, whereas the southern part is dominated by recharge through the southern intake areas. In contrast, Rosenthal et al. (1981) claimed that the major faults of the shear zone are not continuous lineaments, and therefore, groundwater flow is indeed continuous in places where the vertical displacement is small.

### 3.2.2 Interrelations Between the NSA and Other Regional Aquifers

Although the NSA system was defined as isolated in earlier studies, some complex interrelations with other groundwater bodies were suggested for different parts of the basin.

By far, the most controversial proposed configuration is the connection of the NSA with the overlying Judea Group carbonate aquifer. Arad and Kafri (1980) claimed that no hydrogeological separation exists between the Judea Group Aquifer and the underlying NSA in many parts of the Sinai–Negev Basin; thus, recharge to the latter occurs through exposures of both groups simultaneously. Furthermore, by examining the geological evidence, water levels and ionic ratios of groundwater within the two aquifers, these authors hypothesized that the two hydrogeological units function as a single hydrological unit, in which groundwater flows from one to the other and vice versa. Rosenthal et al. (1981) responded to this, claiming that the water levels, the chemical composition and the $\delta^{18}O$–$\delta^{2}H$ values in these two aquifers differ significantly, and hence, there is no basis for the hypothesis of a combined hydrological system. The discussion about whether or not the two aquifers are connected has continued to preoccupy researchers in recent years, as further discussed below.

Interrelations and groundwater connectivity were also proposed for the NSA and the underlying deeper Arad Group (Jurassic) Aquifer (Fig. 5) in the northern parts of the Sinai–Negev Basin (Nativ et al. 1987). Based on geological and hydrological evidence from deep oil drillings, it was suggested that no hydrogeological separation exists between the two groups, dictating hydrological continuity for the two aquifers. Ascending flow from the lower (Jurassic) to the upper (NSA) aquifer was proposed to be the source of higher sulfate concentrations in the NSA in central and northern Sinai (Rosenthal et al. 2007) and in the northeastern Negev (Vengosh et al. 2007). Based on a calibrated dynamic hydrological model, the contribution of water from the Jurassic aquifer into the NSA in the northeastern Negev amounted to only ~3 millions of cubic meters per year (Guttman et al. 1999). Since the Jurassic aquifer is still almost entirely unexploited, with only a single active well tapping this aquifer in the entire Negev, no regional conclusions can be made regarding the existence of such connectivity. However, water in the above-mentioned well was found to be isotopically depleted (−7.8‰ in $\delta^{18}O$), which raised the possibility that the depleted water in the Jurassic aquifer and that in the NSA originated from a common source (Vengosh et al. 2007).

Complicated hydrological interrelations were identified within the southern Arava Rift Valley. Major faults divide this region into several faulted blocks, thus creating complex lateral contacts and hydrological interrelations between different units in this sub-basin (Kroitoru 1980). Besides the regional flow component from central Sinai, through the southern Negev and toward the southern Arava, an additional contribution of groundwater to the NSA was identified from the east. This eastern component was assumed to originate from recharge through the graben fill units at the foothills of the high Edom Mountains of Jordan (Fig. 4), which are characterized by significantly higher amounts of precipitation (Kroitoru et al. 1981; Bein et al. 2001). Based on uranium isotopes (234,238), hydrogeological interrelations were proposed between the NSA and other regional aquifers in the southern Arava, including the Upper Cretaceous Judea Group and the Paleozoic Yam Suf Group aquifers (Fig. 5), as well as with the shallower alluvial aquifer (Kroitoru et al. 1981). Later studies tried to make a quantitative deconvolution of all mixed groundwater sources in this sub-basin by using the $\delta^{18}O$–$\delta^{2}H$ isotopes and the chemical concentrations (Gat and Galai 1982; Rosenthal et al. 1990; Adar et al. 1992; Bein et al. 2001). In all of these studies, $\delta^{18}O$–$\delta^{2}H$ values were found to serve as good indicators for the differentiation between the isotopically depleted NSA water flowing regionally from the west and the local, isotopically more enriched components that are mixed in the southern Arava sub-basin.

A depleted $\delta^{18}O$–$\delta^{2}H$ isotopic composition was also observed in the deeper regional aquifer in the southern Arava—the Yam Suf Group (Bein et al. 2001). This group also consists of Nubian sandstones and outcrops exclusively in the southern parts of the basin, with broad exposures tangential to those of the NSA. In the southern Arava, the angular disconformity at the base of the Lower Cretaceous (Fig. 6) creates a direct contact between the Kurnub and the Yam Suf Groups (Kroitoru 1980). Hence, it was proposed that these two aquifers are hydrologically undifferentiated in this region (Nativ et al. 1987). Both aquifers were found to be affected by mixture with the local brine (the "Timna Brine," further discussed below). In contrast, under most of the Negev, including its central and northern parts, the Yam Suf Group is located much deeper and was found to contain brines, evidencing the lack of flushing in these very deep parts of the aquifer (Nativ et al. 1987).

### 3.2.3 Tracking the Source of Salts

The source of salts in the NSA is an additional controversial issue. Salinity varies widely across the NSA (Fig. 8); the chloride ($Cl^-$) concentrations increase from ~300 mg/L in the southern part of the basin in Sinai up to ~2000 mg/L in the vicinity of the Dead Sea (Rosenthal et al. 2007; Vengosh et al. 2007). Relatively high salinities of 900–2000 mgCl/L were measured within and north of the shear zone (Fig. 4) in Sinai (Shalaby et al. 2012). Most groundwater in the NSA of the Negev and the Arava regions has typical $Cl^-$ concentrations of 600–800 mg/L, with the exceptions of higher salinities found locally along the Arava margins, and of low salinities in wells

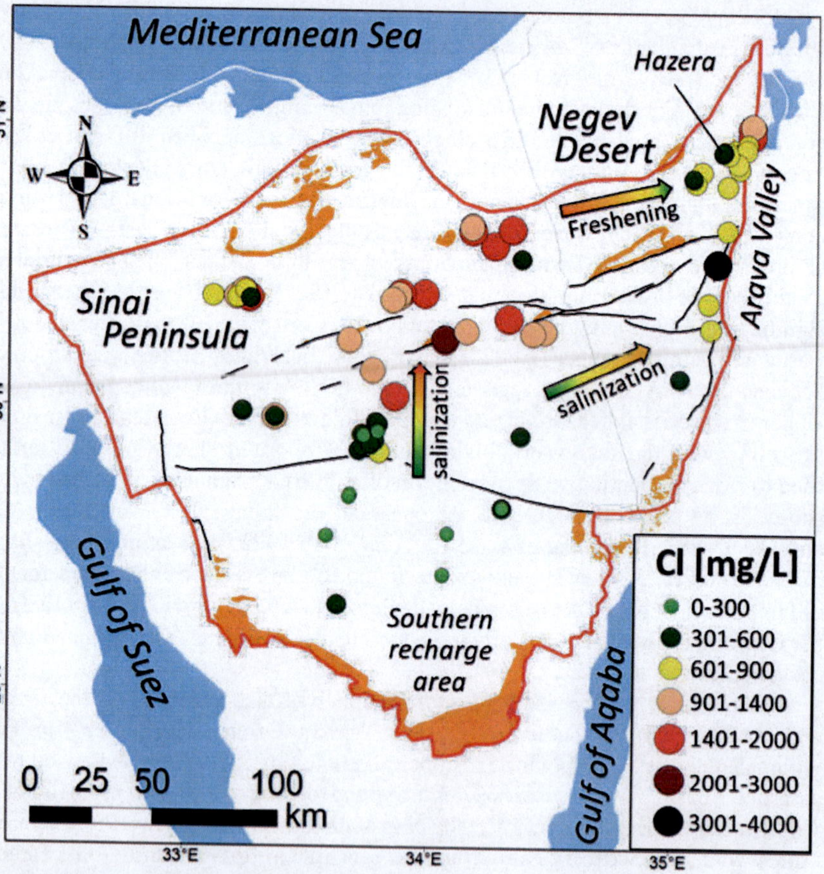


**Fig. 8** $Cl^-$ concentrations in the aquifer. Data from the Japan International Cooperation Agency (JICA) (1992, 1999), Abd El Samie and Sadek (2001), Rosenthal et al. (2007), Shalaby et al. (2012) and Ram et al. (2020). The intake areas of the NSA are marked in orange shades and the assessed boundaries of the basin are in red line. Black lines denote major faults. Modified after Ram et al. (2020)

within the recharge zone of the Hazera cirque (Vengosh et al. 2007). The spatial fluctuations in salinity, accompanied by changes in ionic ratios, were explained by several mechanisms: (1) flushing of entrapped paleo-seawater by paleo-meteoric water, as evidenced by similarities in isopotential and iso-salinity lines (Issar 1979, 1981); (2) water–rock interactions, including the dissolution of evaporates and carbonates (Rosenthal et al. 1998; Vengosh et al. 2007); and (3) mixing with different types of brines, mainly, but not only, along the Arava Rift Valley (Starinsky 1974; Yechieli et al. 1992; Rosenthal et al. 2007). The freshwaters in the Hazera cirque, as well as the trend of freshening along the flow path from northern Sinai toward the northeastern Negev (Fig. 8), indicate the importance of the northern Negev exposures as intake areas.

Relatively higher salinities were measured in the more carbonatic parts of the Kurnub Group within northern Sinai (Fig. 8) and in the northwestern Negev (Rosenthal et al. 1998, 2007; Shalaby et al. 2012). It was suggested that the facies change in this part of the basin results in low permeability, thereby resulting in a poor flushing of residual salts (Issar 1979; Mills and Shata 1989; Kronfeld et al. 1993). Rosenthal et al. (2007) used PHREEQC modeling to relate the salinization to two different processes: (1) mixing with seawater remnants, which could have penetrated into the NSA through the incised erosive Neogene tunnels found in northern Sinai and the northwestern Negev during the Early Pliocene (Rosenthal et al. 2007), and (2) the intrusion of sulfate-rich groundwater from the underlying Jurassic aquifer, enabled by the vertical displacement that occurs in the shear zone (Mills and Shata 1989; Rosenthal et al. 2007). The intrusion of sulfate-rich groundwater from the Jurassic rocks into the NSA was also demonstrated for the northeastern Negev (Vengosh et al. 2007), as is further discussed below.

Far from the Mediterranean, in the eastern parts of the aquifer, mixing with local brines was suggested as the main cause for salinization in the NSA. In the northeastern margins of the aquifer, closer to the Dead Sea, a correlation was observed between pumping rates and increased salinities, accompanied by a decrease in the Na/Cl and Cl/Br ratios. This was explained by the abstraction of brines flowing from the east, through the adjacent faults of the rift, and into the NSA. The geochemistry of these brines differs from that of present-day Dead Sea water, and hence, the source of the brine was estimated to be an ancient pre-Dead Sea Lake (Yechieli et al. 1992; Rosenthal et al. 1998, 2007; Burg et al. 2000). In the southern Arava, in the area of the ancient Timna copper mines, another type of brine was found as the source of salinization (Adar et al. 1992; Bein et al. 2001). This brine (known as Timna Brine) is also characterized by a low Na/Cl ratio (Starinsky 1974) and is distinct due to its extremely low $^{87}Sr/^{86}Sr$ ratio of 0.706, which probably reflects the interaction of the local brine with magmatic rocks (Beyth et al. 1984). Various explanations were given as to the origin of the Timna Brine. One hypothesis related it to the intrusion of evaporated water during a high stand of a Pliocenic proto-Dead Sea Lake (Starinsky 1974), whereas the second hypothesis related the formation of the brine to a transgression and evaporation of seawater from the Gulf of Aqaba during the Late Pleistocene to Early Holocene period (Bein et al. 2001). However, a time-domain electromagnetic (TDEM) mapping of the different saline water bodies along the

Arava Valley (Kafri et al. 2008) exhibits a somewhat different salinity configuration in this region. It was suggested in this study that Pliocene brines intruded into the Arava Valley from the north down to the structural and hydrological divide in the center of the Arava Valley. From the Gulf of Aqaba northward, only the Pliocene marine transgression was detected and the existence of brines in the Timna area is related to lateral density-driven surficial brines known from the nearby Yotvata and Timna Sabkhas.

Advanced isotope techniques, including the use of stable isotopes in sulfate, were recently used to illuminate the sources of water and solutes, as well as the geochemical processes, in the aquifer (Vengosh et al. 2007). Shallow groundwater in the phreatic part of the aquifer in the northeastern Negev (Hazera anticline, Fig. 4) was found to have low $\delta^{34}S$ and low $\delta^{18}O_{SO_4}$, indicative of the oxidation of pyrite. In the confined part of the northeastern Negev, two distinct groups were observed. The first group, with relatively enriched $\delta^{18}O_{H_2O}$ and $\delta^{2}H$ values, was related to recharge through the local exposures and, hence, was assumed to go through processes during infiltration that were similar to those presented for the unconfined zone. However, this group was found to contain lower $SO_4$ concentrations in comparison with the unconfined zone, accompanied by higher $\delta^{34}S$ and $\delta^{18}O_{SO_4}$ values, believed to derive from the bacterial reduction of $SO_4$ in the confined zone. In contrast, the second group in the confined part of the northeastern Negev has lower $\delta^{18}O$ and $\delta^{2}H$ values, a higher $SO_4$ content, and relatively high $\delta^{34}S$, high $\delta^{18}O_{SO_4}$ and low $^{87}Sr/^{86}Sr$ ratios. This was interpreted as an evidence for the intrusion of water from the underlying Jurassic aquifer, in which the dissolution of gypsum of marine origin from the underlying Triassic rocks occurred (Rosenthal et al. 2007; Vengosh et al. 2007). In the southern Arava, sulfate isotopes were used to highlight the importance of the mixing with the underlying Timna Brine, which is characterized, in addition to the low $^{87}Sr/^{86}Sr$ ratio, by a high $SO_4$ content and a high $\delta^{34}S$ value (Vengosh et al. 2007).

### 3.2.4 The Role of Faults

The accumulation of geological data in various studies, together with findings from tens of drillings, enabled a very detailed study on the role of the faults that cut across the basin. For instance, the Paran Fault (Fig. 4), with its dominant vertical displacement (Sakal 1967; Bartov 1974; Frieslander 2000), was considered for more than 20 years of research as the groundwater divide between the water drainage southward to the Gulf of Aqaba base level and the northward drainage to the Dead Sea (Kroitoru 1980). Later studies found this obstacle to be a hydrogeological barrier only in its central and western parts, whereas continuous flow across this fault toward the north was identified at its easternmost part, where the vertical throw along the fault is reduced (Guttman et al. 1999). Hence, the location of the subsurface water divide between the two drainage bases was redefined to be several tens of kilometers southward, at the same place, more or less, as the surface water divide (Guttman et al. 1999).

The role of major faults across the Sinai–Negev Basin as either major barriers or, in contrast, as preferential paths for groundwater flow, was widely discussed in different studies. Mills and Shata (1989) depicted a water-level map of the NSA in which the groundwater flow from southern to northern Sinai is interrupted by the Themed Fault (also known as the Ragabet El-Naam Fault, Fig. 4). These authors hypothesized that the fault zone serves as a preferential flow path, thus diverting some of the water to the east and to the west. Abd El Samie and Sadek (2001) studied the vertical movement along this fault and found that significant displacement occurs in both the eastern and the western parts of the fault, whereas in its central part, in the Nekhel area (Fig. 4), the displacement is minimal. Hence, it was argued that groundwater flow from southern Sinai is directed toward this area of minimal displacement and from there spreads east, north and west toward the natural outlets. Further north, the major faults within the shear zone of central Sinai (Fig. 4) were related to the occurrence of an intrusion from the deeper Jurassic aquifer into the NSA, as well as to an intrusion of the NSA water into the Upper Cretaceous Judea Aquifer (Rosenthal et al. 2007). Shalaby et al. (2012) reviewed the hydrochemistry of the NSA in Sinai in relation to the shear zone and concluded, similarly to what was proposed by Arad and Kafri (1980), that the shear zone serves as a barrier for northward groundwater flow in the NSA.

### 3.2.5 Additional Outlets in the Northwestern Negev

While there is a general agreement regarding the natural outlets of the NSA at its eastern and western margins, there has been a long debate regarding its northern outlets. Those advocating for the existence of such outlets claim that the groundwater with increased salinity and high temperatures, which is abstracted from the Judea Group Aquifer in the northwestern Negev (north of the northern border of the Sinai–Negev Basin; Fig. 4), is actually fed by the underlying NSA (Kronfeld et al. 1993). According to their approach, this flow is the result of the dynamic depletion of water levels within the Judea Group Aquifer following massive groundwater abstraction and is enabled by dissection of the hydrological units by faults and/or by the lack of impervious rock formations between the groups (Rosenthal et al. 1992; Weinberger et al. 1992; Weinberger and Rosenthal 1994). A contradictory explanation relates the increased salinity in the Judea Group Aquifer in this region to the prolonged percolation of brackish water from the overlying Mt. Scopus Group. According to this approach, the extra heat of the Judea Group Aquifer represents exothermic reactions that occur mainly due to the oxidation of organic matter derived from the bituminous deposits found in the Mt. Scopus Group rocks (Burg et al. 2017).

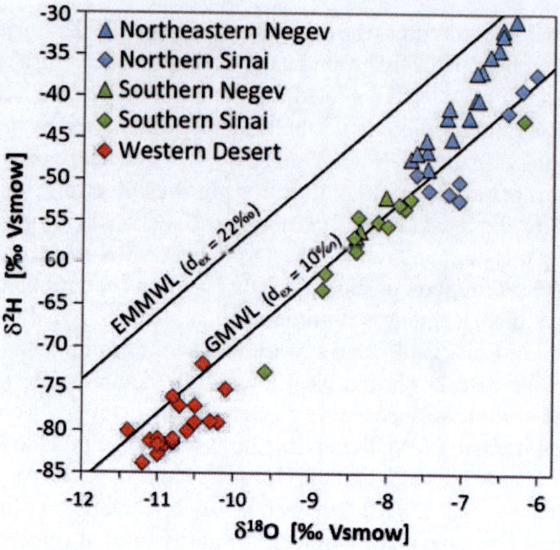


**Fig. 9** Isotopic compositions in the different parts of the NSA in the Sinai–Negev Basin. Note the relative enrichment in the isotopic compositions of the groundwater in the northern parts in comparison with those in the southern parts of the aquifer. Also shown are the isotopic compositions of the NSAS in the Western Desert of Egypt. Data gathered from JICA (1992, 1999), Abd El Samie and Sadek (2001), Patterson et al. (2005), Abouelmagd et al. (2014), Ram et al. (2020)

### 3.2.6 Recharge During Different Epochs and Under Different Climates

The timing of the main recharge to the NSA, as well as the climate that prevailed during more pluvial epochs in the basin, was addressed in many studies. The isotopically depleted groundwater of the NSA was found to have low *d*-excess values ($d_{ex} = \delta^2H - 8\delta^{18}O$) of ~10‰ (Fig. 9) (Issar et al. 1972; Gat and Issar 1974). Such low $d_{ex}$ values differ from those of present rains in the Mediterranean coastal regions and fit to the Global Meteoric Water Line (GMWL) (Gat et al. 1969). The low $d_{ex}$ values were used as an indication for precipitation under more temperate climates than that of the present day (Issar et al. 1972; Gat and Issar 1974; Levin et al. 1980). Studies on the $\delta^{18}O$ and $\delta^2H$ of modern rainstorms in the Sinai–Negev Basin were conducted in order to identify possible moisture sources that could have produced such depleted isotopic compositions. High variability was observed between different storms and even within a single event (Levin et al. 1980; Dody 1995; Adar et al. 1998); however, the most depleted rains in the region, also characterized by very low $d_{ex}$ values, were found to be correlated to storms in which the southwestern moisture source was dominant, whereas storms that originated from the Mediterranean had higher $\delta^{18}O$–$\delta^2H$ and $d_{ex}$ values (Leguy et al. 1983). Hence, it was suggested that the different isotopic compositions of the NSA paleo-waters reflect a reduced role of the Mediterranean Sea as the moisture source in comparison with its dominance in present-day rains in the basin (Gat and Carmi 1987).

The depleted isotopic composition of ~−8‰ in $\delta^{18}O$, accompanied by the low $d_{ex}$ values of ~10‰, is preserved between southern Sinai and the southern Negev. However, toward the northeastern Negev, the isotopic composition changes into higher $\delta^{18}O$ values of up to −6‰, accompanied by increased $d_{ex}$ values of up to 20‰ (Fig. 9). This spatial distribution of the isotopic compositions has led to the hypothesis that the groundwater in the northeastern parts of the basin is highly affected by recharge that took place through the local intake areas (Gat and Galai 1982). The timing of the local recharge through the northern intake areas was assumed to be later than that of the $^{18}O$-depleted groundwater encountered in the southern Negev (Guttman et al. 1999). With the use of $^{14}C$ activities, and by taking into consideration additional processes that may affect $^{14}C$ activities such as $CaCO_3$ dissolution and bacterial $SO_4$ reduction (a reaction that liberates $HCO_3^-$ into the water), the groundwater ages of 21,000–38,000 years were estimated for the northeastern Negev. Hence, the isotopically heavier waters were associated, once again, with recharge during the Last Glacial Period (Vengosh et al. 2007).

A similar interpretation of the local recharge signature was given to the isotopic enrichment encountered between the southern parts of the basin in Sinai ($\delta^{18}O$ of about −8‰) and the central and northern Sinai (heavier values of up to −6‰, Fig. 9). Here again, the change in the isotopic composition was related to the occurrence of local recharge to the NSA, which in this case could have taken place over the northern Sinai intake areas (Fig. 4). It was hypothesized that the gap between the isotopic signatures in southern and northern Sinai derives from a *rainout* process, which takes place as the mass of vapors is transported from the Mediterranean, through northern Sinai, and south to the uplifted recharge zone in the southern parts of the basin (Abd El Samie and Sadek 2001).

The $^{14}C$ ages of the groundwater in the NSA were combined recently with noble gas-derived temperatures in Sinai (Abouelmagd et al. 2014). Noble gases are used in hydrology as paleo-climate markers: Their solubility is highly dependent on the temperature of the percolating water, especially for the heavier ones (Ar, Kr and Xe). Furthermore, due to the inert properties of the noble gases, they do not tend to participate in any chemical reactions in the subsurface, and hence, the original signature of the percolating water is preserved (Mazor 1972; Aeschbach-Hertig and Solomon 2013). Noble gas temperatures of 17.5–22.0 °C were calculated for five NSA wells in the Sinai–Negev Basin, 3 °C lower on average than the present-day mean annual air temperature of 23 °C at the recharge area (Abouelmagd et al. 2014). Combined with modeled $^{14}C$ ages of 24,000–31,000 years for these waters, it was stated once again that the recharge to the aquifer took place during the cooler and wetter climate that prevailed during the Last Glacial Period (Abouelmagd et al. 2014).

The $^{14}C$ dating method was repeatedly used in numerous studies for identifying the age of the water in the NSA since the early 1970s (e.g., Gat and Issar 1974; Issar et al. 1972) and up to recent works (Vengosh et al. 2007; Burg et al. 2013; Abouelmagd et al. 2014). The measured $^{14}C$ activities were always very low and even below the detection limit. The sensitivity of the method, especially in the case of low $^{14}C$ activities, raises questions regarding the reliability of using the low $^{14}C$ activities for age calculations (Aggarwal et al. 2014). Furthermore, the lack of age gradients over hundreds of kilometers across the Sinai–Negev Basin and along the head gradient contradicts the hydrological logic of groundwater aging downstream of the trajectories. An ongoing study, implemented by an international group of researchers, including the authors of this chapter, combines $^{14}C$ with the new $^{81}Kr$ long-term dating tool in order to illuminate the precise groundwater ages in the Negev region (Ram et al. 2020; Yokochi et al. 2019). The advantage of the long half-life, coupled with its inert properties, places $^{81}Kr$ as a powerful tool with a wide dating range of 20,000–1,300,000 years (Purtschert et al. 2013; Lu et al. 2014), far beyond the $^{14}C$ groundwater dating capacity.

The $^{81}Kr$ ages of the groundwater in the confined parts of the NSA in the Negev Desert and the Arava Valley were found to range between 46,000 and 627,000 years (Fig. 10), far beyond the limits of the $^{14}C$ dating method (Ram et al. 2020). The wide age range serves as a proof that replenishment to the aquifer did not occur during a single epoch, as suggested previously by the $^{14}C$ ages (e.g., Issar et al. 1972), but was the result of several consecutive events, with an Early Mid-Pleistocene $^{81}Kr$

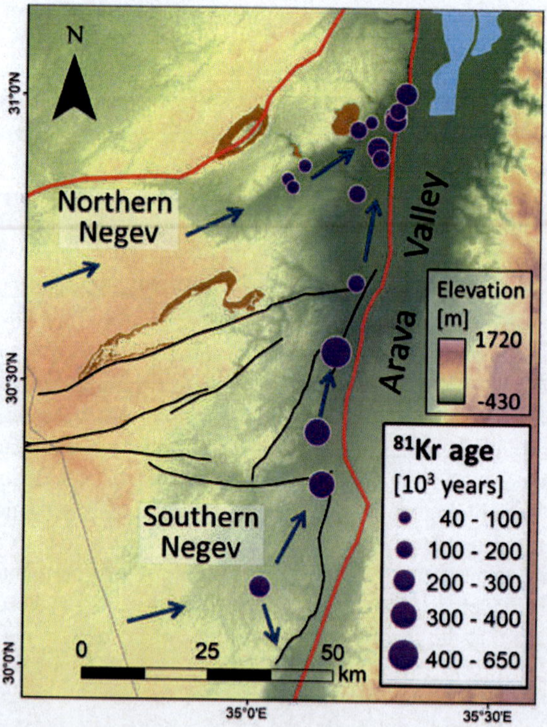


**Fig. 10** $^{81}Kr$ ages in the NSA of the Negev Desert and the Arava Valley, showing wider age ranges in comparison with the former $^{14}C$ age estimations. Brown polygons denote Kurnub Group outcrops, red and black lines are assessed NSA boundaries and major E-W faults, respectively. Blue arrows mark main flow directions. Modified after Ram et al. (2020)

groundwater age as the oldest event found thus far in the basin, and the young $^{14}C$-derived ages close to the intake areas (Abouelmagd et al. 2014; Vengosh et al. 2007), evidence for very recent recharge to the aquifer. The effect of local recharge through the northern exposures is emphasized by the rejuvenation in the $^{81}Kr$ ages in the northeastern parts of the basin (Fig. 10). A correlation between the obtained $^{81}Kr$ groundwater ages and $\delta^{18}O$ compositions in the northeastern Negev supports the existence of mixing between two main components: (1) older, $^{18}O$-depleted water, which probably originated from recharge through the southern Sinai outcrops, and (2) younger, relatively $\delta^{18}O$-enriched water, which originated from local recharge through the northern sandstone exposures (Yokochi et al. 2019).

The $^{14}C$ activities in the NSA were re-estimated recently with the implementation of a novel sampling method, involving the extraction of dissolved $CO_2$ gas out of the groundwater, instead of using the traditional water sample collection (Yokochi et al. 2018). This advanced sampling method possesses several advantages, including simultaneous collection for the $^{14}C$ and $^{81}Kr$ methods, as well as eliminating the risk for undetectable atmospheric contamination during the sampling. The measured $^{14}C$ activities in gas samples from the Negev Desert were found to be lower than those obtained in former studies, where the "traditional" sampling method for DIC-$^{14}C$ analysis had been used, and, in most cases, were lower than the reported detection limit (<0.195 pmc). Very low $^{14}C$ activities were also encountered in the confined parts of the aquifer in the northeastern Negev, in places where the rejuvenation of groundwater ages due to local recharge was highlighted by the $^{81}Kr$ dating (Yokochi et al. 2019). As discussed above, the $^{81}Kr$–$\delta^{18}O$ correlation in this region indicates a mixture between younger and older groundwater, possessing heavier and more depleted isotopic compositions, respectively. Since active recharge to the aquifer is evidenced in the unconfined parts of the aquifer in this region (Vengosh et al. 2007), the possibility of a very young groundwater component, which could have been diluted by much older groundwater, must be taken into consideration. However, the finding of very low $^{14}C$ activities refutes the possibility of significant impact from very recent recharge in the confined part of the aquifer in the northeastern Negev and demonstrates that the mixing processes in this area include different types of mainly pre-Holocene waters.

## 4 Epilogue

Quite incidentally, a huge aquifer was discovered in the Nubian sandstones under the arid regions of the Sinai Peninsula and the Negev Desert. The combination of detailed geological studies and different geochemical and isotopic tools helped to uncover the hydrological characteristics of the groundwater system, including its recharge areas, the timing of recharge, its flow trajectories, its salinization processes, the hydrological relations between the NSA and other regional aquifers in the basin, etc. The isotopic compositions of the water were used in order to learn about the climatic conditions that prevailed during the main recharge to the aquifer, and were also used for delineating between the NSA of the Sinai–Negev Basin and the transboundary NSAS of the Western Desert in northeastern Africa.

The study of this regional aquifer has been ongoing for more than half a century, and yet, a few questions have not yet been fully answered, such as the climate under which the recharge took place, the interrelations between the NSA and other regional aquifers, the reason for such a depleted isotopic composition of the groundwater, the groundwater flow regime in ancient times and the paleo-discharge points in the ancient, more rainy climatic periods. However, the long list of different tools used for studying the flow pattern has given rise to an updated hydrogeological scheme, which can be summarized as follows: (1) Groundwater flows from the southern Sinai outcrops toward the main outlets close to the Gulf of Suez, the Gulf of Aqaba and south of the Dead Sea; (2) groundwater ages indicate that recharge to the aquifer took place during several rainy periods over the Pleistocene; (3) active recharge is identified only within and close to the recharge areas of the system; (4) high salinities are encountered locally in the aquifer, particularly in the northern parts of Sinai and very close to the eastern outlets in the Negev; (5) mixing of waters of different types (in terms of $^{18}O$ and $^{2}H$ stable isotopes) and of different ages takes place mainly in the northern parts of the aquifer, where the regional flow component of $^{18}O$-depleted water is mixed with a local $^{18}O$-enriched component. The latter is a result of recharge through the northern restricted outcrops. Since groundwater ages in this region are close to (and beyond) $^{14}C$ dating capacity, it is assumed that this mixing is between different types of mainly "**fossil**" (pre-Holocene) waters.

Finally, the NSA of the Sinai–Negev Basin is located in between other immense regional basins composed of thick fossil Nubian aquifers, including the Sahara Desert in the west and the Arabian Plate in the east. Implementation of the varied methods and tools described throughout this chapter will enable researchers not only to resolve some local questions, but also to gain a broad perspective on the climatic conditions under which past recharge took place in these present-day vast hyper-arid regions, as well as on the various recharge events.

## References

Abd El Samie SG, Sadek MA (2001) Groundwater recharge and flow in the Lower Cretaceous Nubian Sandstone aquifer in the Sinai Peninsula, using isotopic techniques and hydrochemistry. Hydrogeol J 9:378–389

Abouelmagd A, Sultan M, Milewski A et al (2012) Toward a better understanding of palaeoclimatic regimes that recharged the fossil aquifers in North Africa: inferences from stable isotope and remote sensing data. Palaeogeogr Palaeoclimatol Palaeoecol 329–330:137–149

Abouelmagd A, Sultan M, Sturchio NC et al (2014) Paleoclimate record in the Nubian Sandstone Aquifer, Sinai Peninsula, Egypt. Quat Res 81:158–167

Adar EM, Rosenthal E, Issar AS, Batelaan O (1992) Quantitative assessment of the flow pattern in the southern Arava Valley (Israel) by environmental tracers and a mixing cell model. J Hydrol 136:333–352

Adar EM, Dody A, Geyh M et al (1998) Distribution of stable isotopes in arid storms I. Relation between the distribution of isotopic composition in rainfall and in the consequent runoff. Hydrogeol J 6:50–65

Aeschbach-Hertig W, Solomon DK (2013) Noble gas thermometry in groundwater hydrology. In: Burnard P (ed) The noble gases as geochemical tracers. Advances in isotope geochemistry. Springer, Berlin, Heidelberg, pp 81–122

Aggarwal PK, Araguas-Araguas L, Choudhry M et al (2014) Lower groundwater $^{14}C$ age by atmospheric $CO_2$ uptake during sampling and analysis. Groundwater 52:20–24

Aharoni E (1966) Oil and gas prospects of Kurnub Group (Lower Cretaceous) in southern Israel. Bull Am Assoc Pet Geol 50:2388–2403

Arad A, Kafri U (1980) Hydrogeological interrelationship between the Judea Group and the Nubian sandstone aquifers in Sinai and the Negev. Isr J Earth Sci 29:67–72

Arad A, Michaeli A (1967) Hydrogeological investigations in the western catchment of the Dead Sea. Isr J Earth Sci 16:181–196

Ball J (1927) Problems of the Libyan Desert. Geogr J 70:21–38

Bartov J (1974) A structural and paleogeographical study of the central Sinai faults and domes. Dissertation, Hebrew University of Jerusalem (in Hebrew, English abstract)

Bein A, Yechieli Y, Bensabat J (2001) Quantifying the groundwater resources of the southern Arava Rift Valley: a confined desert system recharged laterally by external sources. Isr J Earth Sci 50:217–236

Ben-Avraham Z (1978) The structure and tectonic setting of the levant continental margin, Eastern Mediterranean. Tectonophysics 46:313–331

Beyth M, Starinsky A, Lazar B (1984) Low temperature geothermal waters, Timna, southern Israel. Isr Geol Surv Curr Res 17–20

Burg A, Lomelski S, Shachar H et al (2000) Hydrological survey for examination of the water potential in the southern Dead Sea and northern Arava Valley. Report no. 3. Tahal Consulting Engineers Ltd., Report 00.312, Tel Aviv (in Hebrew)

Burg A, Zilberbrand M, Yechieli Y (2013) Radiocarbon variability in groundwater in an extremely arid zone—the Arava Valley, Israel. Radiocarbon 55:963–978

Burg A, Gavrieli I, Guttman J (2017) Concurrent salinization and development of anoxic conditions in a confined aquifer, Southern Israel. Groundwater 55:183–198

Calvo R, Gvirtzman Z (2013) Assessment of $CO_2$ storage capacity in southern Israel. Int J Greenhouse Gas Control 14:25–38

Clark I, Fritz P (1997) Environmental isotopes in hydrogeology. Lewis Publishers, Boca Raton, New York

Crombie MK, Arvidson RE, Sturchio NC et al (1997) Age and isotopic constraints on Pleistocene pluvial episodes in the Western Desert, Egypt. Palaeogeogr Palaeoclimatol Palaeoecol 130:337–355

Dafny E (2012) Groundwater flow and solute transport within the Yarqon-Taninim aquifer, Israel. Dissertation, Hebrew University of Jerusalem (in Hebrew)

Dames M, Moore TC (1985) Sinai development study, phase 1. Final report 5. Ministry of Development, Cairo, Egypt

Dody A (1995) Isotopic composition of rainfall and runoff in a small Arid Basin with implications for deep percolation. Dissertation, Ben Gurion University of the Negev

Edmunds W (2005) Groundwater as an archive of climatic and environmental change. In: Aggarwal PK, Gat JR, Froehlich KF (eds) Isotopes in the water cycle. Springer, Dordrecht, pp 341–352

El-Shenawy MI, Kim S, Schwarcz HP et al (2018) Speleothem evidence for the greening of the Sahara and its implications for the early human dispersal out of sub-Saharan Africa. Quat Sci Rev 188:67–76

Enmar L (1999) The travertines in the northern and central Arava: stratigraphy, petrology and geochemistry. Dissertation, Hebrew University of Jerusalem (in Hebrew, English abstract)

Fink M (1994) Evaluation of groundwater exploitation possibilities for the Parsa pumped storage project. Tahal Consulting Engineers Ltd., Report R-94/54, Tel Aviv

Foster S, Loucks DP (2006) Non-renewable groundwater resources—a guidebook on socially sustainable management for water policy makers. UNESCO IHP

Frieslander U (2000) The structure of the Dead Sea transform emphasizing the Arava, using new geophysical data. Dissertation, The Hebrew university of Jerusalem (in Hebrew, English abstract)

Gat JR, Carmi I (1987) Effect of climate changes on the precipitation patterns and isotopic composition of water in a climate transition zone: case of the Eastern Mediterranean Sea area. IAHS Spec Publ 168, pp 513–524

Gat JR, Galai A (1982) Groundwaters of the Arava Valley—an isotopic study of their origin and interrelationships. Isr J Earth Sci 31(1):25–38

Gat JR, Issar A (1974) Desert isotope hydrology: water sources of the Sinai Desert. Geochim Cosmochim Acta 38:1117–1131

Gat JR, Mazor E, Tzur Y (1969) The stable isotope composition of mineral waters in the Jordan Rift Valley, Israel. J Hydrol 7:334–352

Ginat H, Opitz S, Ababneh L et al (2018) Pliocene-Pleistocene waterbodies and associated deposits in southern Israel and southern Jordan. J Arid Environ 148:14–33

Guttman Y, Burg A, Lifshitz Y et al (1999) Hydrological model for estimation of water potential in the Arava aquifers—final report. Tahal Consulting Engineers Ltd., Report 6127-d99.201, Tel Aviv (in Hebrew)

Hempton MR (1987) Constraints on Arabian Plate motion and extensional history of the Red Sea. Tectonics 6:687–705

Issar A (1979) The paleohydrology of southern israel and its influence on the flushing of the Kurnub and 'Arad groups (Lower Cretaceous and Jurassic). J Hydrol 44:289–303

Issar A (1981) The rate of flushing as a major factor in determining the chemistry of water in fossil aquifers in southern Israel. J Hydrol 54:285–296

Issar A (1985) Fossil water under the Sinai-Negev peninsula. Sci Am 253:104–110

Issar A, Bein A, Michaeli A (1972) On the ancient water of the Upper Nubian Sandstone aquifer in central Sinai and southern Israel. J Hydrol 17:353–374

Jenkins DA (1990) North and Central Sinai. In: Said R (ed) The geology of Egypt. A.A. Balkema, Rotterdam, pp 361–380

JICA (1992) North Sinai groundwater resources study in the Arab Republic of Egypt. Japan International Cooperation Agency and Water Resources Research Institute, Pacific Consultants International and Sanyu Consultants Inc. Final report, Tokyo

JICA (1999) South Sinai groundwater resources study in the Arab Republic of Egypt. Japan International Cooperation Agency and Water Resources Research Institute, Pacific Consultants International and Sanyu Consultants Inc. Final report, Tokyo

Kafri U, Goldman M, Levi E (2008) The relationship between saline groundwater within the Arava Rift Valley in Israel and the present and ancient base levels as detected by deep geoelectromagnetic soundings. Environ Geol 54:1435–1445

Knetsch G (1962) Geologische Überlegungen zu der Frage des artesischen Wassers in der westlichen ägyptischen Wüste (Geological reflections on the question of artesian water in the western Egyptian desert). Geol Rundsch 52:640–650

Knetsch G, Shata A, Degens E et al (1963) Untersuchungen an grundwässern der Ost-Sahara (Investigations on groundwaters of the eastern Sahara). Geol Rundsch 52:587–610

Kroitoru L (1980) The hydrogeology of the Nubian Sandstone in southern Israel. Dissertation, Tel Aviv University (in Hebrew, English abstract)

Kroitoru L, Kronfeld J, Ginzburg A (1981) The geohydrology of the Gerofit-Ya'alon area. Isr J Earth Sci 30:24–30

Kronfeld J, Rosenthal E, Weinberger G et al (1993) The interaction of two major old water bodies and its implication for the exploitation of groundwater in the multiple aquifer system of the central and northern Negev, Israel. J Hydrol 143:169–190

Leguy C, Rindsberger M, Zangwil A et al (1983) The relation between the $^{18}O$ and deuterium contents of rain water in the Negev Desert and air-mass trajectories. Chem Geol 41:205–218

Levin M, Gat J, Issar A (1980) Precipitation, flood-and groundwaters of the Negev highlands: an isotopic study of desert hydrology. In: Arid zone hydrology: investigations with isotope techniques. IAEA, Vienna, pp 3–22

Livnat A, Kronfeld J (1985) Paleoclimatic implications of U-Series dates for lake sediments and travertines in the Arava Rift Valley, Israel. Quat Res 24:164–172

Lu ZT, Schlosser P, Smethie WM et al (2014) Tracer applications of noble gas radionuclides in the geosciences. Earth-Sci Rev 138:196–214

Macdonald AM, Bonsor HC, Dochartaigh ÉO, Taylor RG (2012) Quantitative maps of groundwater resources in Africa. Environ Res Lett 7:024009 (7 pp)

Mazor E (1972) Paleotemperatures and other hydrological parameters deduced from noble gases dissolved in groundwaters; Jordan Rift Valley, Israel. Geochim Cosmochim Acta 36:1321–1336

McKee ED (1963) Origin of the Nubian and similar sandstones. Geol Rundsch 52:551–587

Mills AC, Shata A (1989) Ground-water assessment of Sinai, Egypt. Ground Water 27:793–801

Mohamed A, Sultan M, Ahmed M et al (2017) Aquifer recharge, depletion, and connectivity: inferences from GRACE, land surface models, and geochemical and geophysical data. Geol Soc Am Bull 129:534–546

Nativ R (1984) The water potential of the deep Jurassic to Paleozoic aquifers in the Negev, Israel. Dissertation, Ben-Gurion University of the Negev (in Hebrew, English abstract)

Nativ R, Bachmat Y, Issar A (1987) Potential use of the deep aquifers in the Negev desert, Israel—a conceptual model. J Hydrol 94:237–265

Nicholson SE (2000) The nature of rainfall variability over Africa on time scales of decades to millenia. Glob Planet Change 26:137–158

Nour D, Yehia M (1993) Hydrogeology of Cretaceous aquifer systems in Sinai Peninsula. Water Research Center, Cairo

Osborne AH, Vance D, Rohling EJ et al (2008) A humid corridor across the Sahara for the migration of early modern humans out of Africa 120,000 years ago. Proc Natl Acad Sci USA 105:16444–16447

Patterson LJ, Sturchio NC, Kennedy BM et al (2005) Cosmogenic, radiogenic, and stable isotopic constraints on groundwater residence time in the Nubian Aquifer, Western Desert of Egypt. Geochem Geophys Geosys 6(1):19 pp

Picard L (1953) The history of groundwater exploration in Israel. Proc Int Symp Desert Res Res Counc Isr 2(4):17

Picard L (1959) Geology and oil exploration of Israel. Bull Res Counc Isr G8:1–30

Purtschert R, Yokochi R, Sturchio NC (2013) Krypton-81 dating of old groundwater. In: Suckow A, Aggarwal PK, Araguas-Araguas LJ (eds) Isotope methods for dating old groundwater. IAEA, Vienna, pp 91–124

Ram R, Burg A, Zappala JC et al (2020) Identifying recharge processes into a vast "fossil" aquifer based on dynamic groundwater 81Kr age evolution. J Hydrol 587:124946. https://doi.org/10.1016/j.jhydrol.2020.124946

Richey AS, Thomas BF, Lo M-H et al (2015) Quantifying renewable groundwater stress with GRACE. Water Resour Res 51:5217–5238

Richts A, Struckmeier WF, Zaepke M (2011) WHYMAP and the groundwater resources map of the World 1:25,000,000. In: Sustaining groundwater resources. Springer, Dordrecht, Netherlands, pp 159–173

Rolland G (1890) Géologie et hydrologie du Sahara algérien (Geology and hydrology of the Algerian Sahara). Imprimerie nationale, Paris

Rosenthal E, Issar A, Kroitoru L et al (1981) Hydrogeological inter-relationship between the Judea Group and the Nubian Sandstone aquifers in Sinai and the Negev. Isr J Earth Sci 30:145–153

Rosenthal E, Adar E, Issar AS, Batelaan O (1990) Definition of groundwater flow patterns by environmental tracers in the multiple aquifer system of southern Arava Valley, Israel. J Hydrol 117:339–368

Rosenthal E, Weinberger G, Berkowitz B et al (1992) The Nubian Sandstone aquifer in the central and northern Negev, Israel: delineation of the hydrogeological model under conditions of scarce data. J Hydrol 132:107–135

Rosenthal E, Jones BF, Weinberger G (1998) The chemical evolution of Kurnub Group paleowater in the Sinai-Negev province—a mass balance approach. Appl Geochem 13:553–569

Rosenthal E, Zilberbrand M, Livshitz Y (2007) The hydrochemical evolution of brackish groundwater in central and northern Sinai (Egypt) and in the western Negev (Israel). J Hydrol 337:294–314

Sakal E (1967) The geology of Rekhes Menuha. Dissertation, The Hebrew University of Jerusalem (in Hebrew, English abstract)

Sandford KS (1935) Sources of water in the north-western Sudan. Geogr J 85:412–431

Schulz S, Walther M, Michelsen N et al (2017) Improving large-scale groundwater models by considering fossil gradients. Adv Water Resour 103:32–43

Shalaby A, Embaby A, Seiam A (2012) Structural constraints on the groundwater regime of the Cretaceous aquifers in central Sinai, Egypt. J African Earth Sci 75:37–56

Shiftan ZL (1958) An artesian aquifer of the southern Dead Sea Basin. Bull Res Counc Isr 7:27–52

Shiftan ZL (1961) New data on the artesian aquifers of the southern Dead Sea Basin and their geological evolution. Bull Res Counc Isr 10G:267–291

Sonntag C (1984) Autochthoneous groundwater in the confined Nubian Sandstone Aquifers. In: Klitzsch E, Said R, Schrank E (eds) Berliner Geowissenschaftlich Abhandlungen, Ser. A, vol 50. Von Dietrich Reimer, Berlin, pp 217–220

Sonntag C, Muennich KO, Junghans C et al (1979) Palaeoclimatic information from deuterium and oxygen-18 in carbon-14-dated north Saharian groundwaters. Groundwater formation in the past. In: Int. At. Energy Agency V (ed) Isotope hydrology 1978, 10th edn. Internat. Atomic Energy Agency, pp 569–581

Starinsky A (1974) Relationship between Ca-chloride brines and sedimentary rocks in Israel. Dissertation, The Hebrew university of Jerusalem (in Hebrew, English abstract)

Sturchio NC, Du X, Purtschert R et al (2004) One million year old groundwater in the Sahara revealed by krypton-81 and chlorine-36. Geophys Res Lett 31:2–5

Sultan M, Sturchio N, Hassan F et al (1997) Precipitation source inferred from stable isotopic composition of Pleistocene groundwater and carbonate deposits in the Western Desert of Egypt. Quat Res 48:29–37

Sultan M, Metwally S, Milewski A et al (2011) Modern recharge to fossil aquifers: geochemical, geophysical, and modeling constraints. J Hydrol 403:14–24

Sultan M, Sturchio NC, Alsefry S et al (2019) Assessment of age, origin, and sustainability of fossil aquifers: a geochemical and remote sensing-based approach. J Hydrol 576:325–341

Szabo BJ, Haynes CV, Maxwell TA (1995) Ages of Quaternary pluvial episodes determined by uranium-series and radiocarbon dating of lacustrine deposits of Eastern Sahara. Palaeogeogr Palaeoclimatol Palaeoecol 113:227–242

Thorweihe U (1990) Nubian aquifer system. In: Said R (ed) The geology of Egypt. A.A. Balkema, Rotterdam, pp 601–612

Thorweihe U, Heinl M (2002) Groundwater resources of the Nubian aquifer system NE-Africa, modified synthesis. Observatoire du Sahara et du Sahel, Paris

UN-ESCWA, BGR (2013) Inventory of shared water resources in Western Asia. United Nations economic and social commission for western Asia; Bundesanstalt für Geowissenschaften und Rohstoffe, Beirut

Vaks A, Bar-Matthews M, Ayalon A et al (2006) Paleoclimate and location of the border between Mediterranean climate region and the Saharo-Arabian Desert as revealed by speleothems from the northern Negev Desert, Israel. Earth Planet Sci Lett 249:384–399

Vaks A, Bar-Matthews M, Matthews A et al (2010) Middle-Late Quaternary paleoclimate of northern margins of the Saharan-Arabian Desert: reconstruction from speleothems of Negev Desert, Israel. Quat Sci Rev 29:2647–2662

Vengosh A, Hening S, Ganor J et al (2007) New isotopic evidence for the origin of groundwater from the Nubian Sandstone Aquifer in the Negev, Israel. Appl Geochem 22:1052–1073
Waldmann N, Torfstein A, Stein M (2010) Northward intrusions of low- and mid-latitude storms across the Saharo-Arabian belt during past interglacials. Geology 38:567–570
Water Authority of Israel (2014) State of the water sources, management report. Jerusalem (in Hebrew)
Weinberger G, Rosenthal E (1994) The fault pattern in the northern Negev and southern Coastal Plain of Israel and its hydrogeological implications for groundwater flow in the Judea Group aquifer. J Hydrol 155:103–124
Weinberger G, Rosenthal E, Flexer A (1992) The subsurface geology of Central and Northern Negev with possible implications on the regional ground water flow pattern. J Afr Earth Sci (and Middle East) 14:155–172
Weissbrod T (1969a) The Paleozoic of Israel and adjacent countries, part 1: the subsurface Paleozoic stratigraphy of southern Israel. Geol Surv Isr Bull 47(1008):23
Weissbrod T (1969b) The Paleozoic of Israel and adjacent countries, part 2: the Paleozoic outcrops in southwestern Sinai and their correlation with those of southern Israel. Geol Surv Isr Bull 48(1033):32
WPRP (1998) Hydrogeology of deep aquifers in the Western Desert and Sinai. Ministry of Public Works and Water Resources, Water Policy Reform Program, Cairo
Yechieli Y, Starinsky A, Rosenthal E (1992) Evolution of brackish groundwater in a typical arid region: Northern Arava Rift Valley, southern Israel. Appl Geochem 7:361–374
Yokochi R, Bernier R, Purtschert R et al (2018) Field degassing as a new sampling method for $^{14}C$ analyses in old groundwater. Radiocarbon 60:349–366
Yokochi R, Ram R, Zappala JC et al (2019) Radiokrypton unveils dual moisture sources of a deep desert aquifer. Proc Natl Acad Sci 116(33)

# The Eastern Dead Sea Rift Continental Groundwater Base Level

**Uri Kafri and Yoseph Yechieli**

**Abstract** The Dead Sea Rift in eastern Israel forms a continental endorheic groundwater base level, with the deepest Dead Sea proper terminal basin. The hydrogeological setup and behavior of the different segments of the Dead Sea Rift base level are discussed in this chapter.

## 1 General

The Dead Sea Rift (hereafter DSR) (Fig. 1) forms a continental terminal base level, currently with a surface area of some 43,000 km$^2$. A considerable part of this basin lies below sea level, and its deepest, still declining, point is currently below 433 m bsl (in 2020) in the Dead Sea proper (Bookman (Ken-Tor) et al. 2004; Kafri and Yechieli 2010).

The DSR is part of a 1000 km long Transform which extends from the Red Sea spreading area in the south to the Zagros-Taurus zone of plate convergence in the north. The Arabian Plate has moved sinistrally along the Transform with respect to the Sinai sub-plate (Fig. 1) by some 105 km since the Neogene (Garfunkel 1981). Rhomb-shaped grabens formed along the Transform, as expected the following, morphotectonic depressions: the Dead Sea, the Sea of Galilee and the Hula Basin (Garfunkel and Ben-Avraham 1996), which are currently at the elevations of ca 433 m bsl, 210 m bsl and 70 m asl, respectively. The DSR is structurally connected to the Mediterranean Sea by the NW-SE directed Yizre'el and Bet Shean Valleys of northern Israel, which are bounded by active faults. These valleys enabled the Pliocene marine ingression from the Mediterranean to the DSR which initiated the formation of saline water bodies within the DSR (i.e., Kafri and Ecker 1964: Zak 1967; Kafri and Yechieli 2010).

U. Kafri (✉) · Y. Yechieli
Geological Survey of Israel, 32 Leibovitz Str., 9692100 Jerusalem, Israel
e-mail: uri.kafri@gsi.gov.il

Y. Yechieli
e-mail: yechieli@gsi.gov.il


U. Kafri and Y. Yechieli (eds.), *The Many Facets of Israel's Hydrogeology*, Springer Hydrogeology, https://doi.org/10.1007/978-3-030-51148-7_10

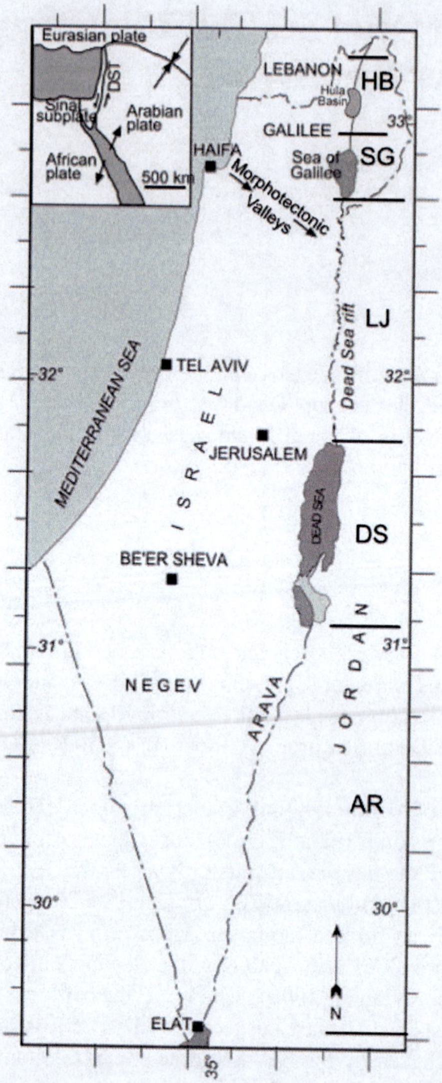


Fig. 1 Map of the DSR Basin, including its different segments (modified after Kafri and Yechieli 2010)

The entire DSR, and its connected valleys, is subjected at present to neotectonic activity starting since the Quaternary (Garfunkel et al. 1981). The history of the DSR since the Neogene, acting as consecutive paleo base levels, is discussed in Chapter "Paleohydrogeology of Israel" and in Kafri and Yechieli (2010). The current Dead Sea is in fact the remnant of its former receding Lake Lisan. Its levels, during the Holocene, were around 400 m bsl, with a maximum fluctuation of several tens of meters (Bookman (Ken-Tor) et al. 2004). The changes of the monitored Dead Sea levels, since ca hundred years ago are described, based on several studies, by Kafri and Yechieli (2010) and below.

Following a long history of subsidence of the DSR since the Neogene, the present-day configuration and segmentation of the eastern DSR base level can be described as follows:

Kafri and Yechieli (2010) have described and typified different known base level types on earth as shown schematically in Fig. 2. Different segments of the DSR base level are exhibited, based on the above scheme, on a north–south cross section (Fig. 3).

The DSR base level is bounded in the north by the western slopes of the uplifted Mount Hermon. Groundwater flows from there southward to the Hula Basin which constitutes an intermediate flow-through base level above sea level. The groundwater flows into it and subsequently flows southward to the Sea of Galilee basin. The latter is also an intermediate flow-through base level, but in this case considerably below

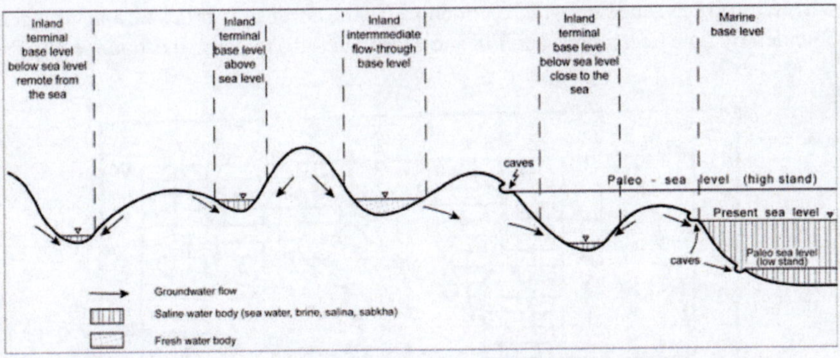


**Fig. 2** Schematic description of the main base level types (after Kafri and Yechieli 2010)

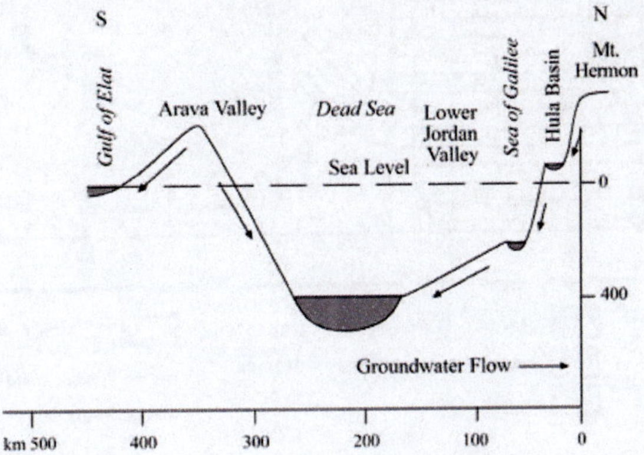


**Fig. 3** A N-S section along different segments of the DSR base level

sea level. The groundwater flows furtherly to the south to the Dead Sea proper basin which constitute the deepest endorheic terminal base level below sea level on earth. The Arava Valley segment, more to the south, is located between the Dead Sea base level and the marine one of the Gulf of Elat (Aqaba). A hydrological divide exists in the middle of the Arava Valley between the two base levels.

# 2 The Hydrogeology of Different Segments of the DSR Base Level

## 2.1 *The Hula Basin Segment (HB on Fig. 1)*

The Hula Basin constitutes the northern segment of the DSR base level at an elevation of some 70 m asl (Figs. 3 and 4). The basin is bounded to the northeast by the uplifted Mount Hermon which consists mainly of highly conductive carbonates, overlain by Early Cretaceous sandstones and carbonates. To the north and northwest and west, it is bounded by Late Cretaceous and Eocene carbonates. To the east, it is bounded by the

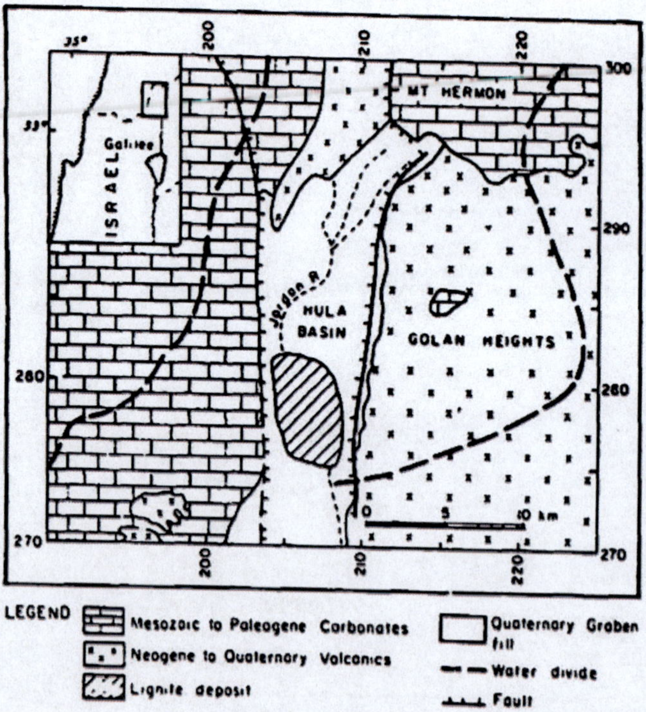


**Fig. 4** Generalized geological map of the Hula Basin (after Kafri and Lang 1987)

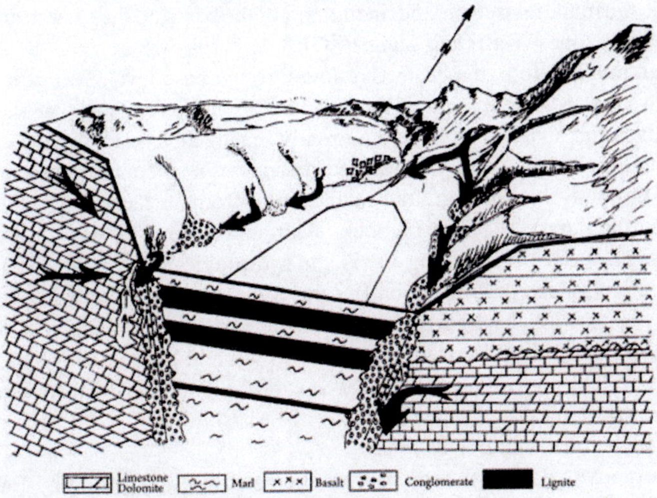


**Fig. 5** Schematic 3D model of the Hula Basin, showing different hydrogeological units and groundwater flow directions (modified after Kafri and Agron 1978)

Golan Heights that consist of Tertiary to Quaternary basalts underlain by Cretaceous and Jurassic carbonates. To the south, the basin is "blocked" by the Korazim structural that consists of basalts underlain by Cretaceous and Jurassic carbonates. It separates between the Hula and the Sea of Galilee Basins. All the above stratigraphic units perform active aquifers that contribute convergent groundwater flow into the basin as shown schematically on Fig. 5.

Part of that convergent flow issues through springs along the margins of the basin and its bordering faults. The Hermon and Dan springs drain the Jurassic aquifer of Mt. Hermon, feeding the tributaries of the Jordan River, on the average, water at the rate of 125 and 250 million $m^3$ accordingly. According to Arad and Bein (1986) and Roded et al. (2013), part of the water that is being recharged on Mount Hermon flows southward along the deep Jurassic aquifer, below the Golan Heights and issue in the springs of the Yarmouk Gorge east of the Sea of Galilee.

The western marginal springs (Einan, Teo and Zahav) also contribute a few tens of million $m^3$ of water annually. The springs that issue from the basalt aquifers on the western slopes of the Golan Heights also yield water at a rate of a few tens of million $m^3$ annually (Dafny et al. 2003).

The subsurface sequence of the basin and its fill was described among others by Kafri and Lang (1987). It consists of at least close to 3 km of basalts, lacustrine shales, lignite beds and sand and gravel. The basin can be divided into two main past environments of deposition. The northern part consists in subsurface mainly of a continuous sequence of gravel beds derived from the mountains in the north and northeast and west. The southern part, however, consists mainly of alternations of shales and lignites which were deposited in lacustrine and swamp environments.

Coarse clastic material is restricted, in this part of the basin, to the eastern and western marginal fault zones (Kafri and Lang 1987).

As mentioned before, the basin is a flow-through base level. Part of the groundwater flows towards the margin of the basin and issues as springs that also feed the Jordan River and some of which are currently managed. The rest of the convergent flow has to find its way out of the basin to keep it fresh, since it is not endorheic (see below). The configuration of groundwater outflow from the basin is somewhat debatable. According to a conceptual model by Kafri and Agron (1978) and further based on pumping tests (Kafri and Lang 1987), the hydrological flow regime is as follows: Whereas the northern part of the basin (gravel beds) is hydrologically conductive, the southern part (shale, lignite) is typified by very low conductivity "blocking" in general a possibility of a southward flow to the Sea of Galilee Basin. The only suggested possibility to such a flow is across the Korazim elevated structure from a head of some 70 m asl to a head of around 210 m bsl in the Sea of Galilee along the marginal fault zones and gravel beds.

An alternative additional outflow from the basin was suggested by Babad et al. (2019). They proposed a groundwater flow path from the deep Jurassic and Cretaceous aquifers in the east, across the conductive sequence of the graben fill in the northern part of the valley to the Cretaceous aquifer west of the valley.

Regarding the salinity of groundwater in the basin, its existence as a flow-through base level and not an endorheic one results in the overall low salinity of its contained groundwater. The Hula Basin was always above sea level and therefore was never intruded by seawater. This is supported by the freshness of its groundwater as well as by the absence of any evaporates all along the deep sequence of the Pliocene to Quaternary basin fill (Kafri et al. 1983; Kafri and Yechieli 2010).

## 2.2 *The Sea of Galilee—Lower Jordan Valley Segment*

**The Sea of Galilee (SG on Fig. 1)**

The Sea of Galilee or Lake Kinneret (Fig. 3), at an elevation of approximately 210 m bsl, is a flow-through base level, down-stream and down-gradient to the Hula Basin. It is being fed by the Jordan River and some smaller streams, as well as by convergent subsurface groundwater flow including nearshore and bottom lake saline springs as well as by seepages from the bottom of the lake (Hurwitz et al. 1999, see also Sect. 4.1 in Chapter "Geoelectric, Geoelectromagnetic and Combined Geophysical Methods in Groundwater Exploration in Israel"). The lake water is fresh, whereas the adjoining groundwater system, which includes also the onshore and bottom lake springs, reveals a wide spectrum of salinities from freshwater through brackish ones and up to a maximum salinity of 35 g/L. The Sea of Galilee used to be in the past the major natural water reservoir and source of the country, whereby some 500 million $m^3$ of water were diverted annually via a major conduit system to the southern, more arid, parts of the country (Gvirtzman et al. 1997). The understanding of the origin and the mechanism of the lake's mild salinization were crucial to the optimal management of the system. Thus, several studies, concerning the above, were carried out in the past and are still taking place as detailed by Kafri and Yechieli (2010) and below. It is accepted by all the researchers that different salinities within the system stem from different proportions of mixtures between cyclic waters, recharged in the surrounding intake areas, and a saline end member from underneath. Three basic approaches tried to explain both the origin of the saline end member and the salinization mechanism. The basic different conceptual models that relate to the salinization mechanism of the system were described by Gvirtzman et al. (1997), Kafri et al. (2007) and Kafri and Yechieli (2010) as shown on Fig. 6. Different approaches are as follows:

According to Goldschmidt et al. (1967) (Fig. 6a), the deep-seated brine within the Rift Valley is not over-pressured and it is being flushed by the circulating convergent fresh groundwater flow. Due to this model, the lowering of the freshwater head through groundwater exploitation might decrease the salinity of the lake and the associated saline springs.

According to Mazor and Mero (1969) (Fig. 6b), the deep-seated brines are over-pressured and are squeezed upward by compaction of the sediments, by tectonic movements and by geothermal activity. Thus, according to this model, a considerable lowering of the head of the groundwater system and the level of the lake by higher water diversion will result in an increased salinization of the system.

According to Kafri and Arad (1979) (Fig. 6c), the current seawater inflow from the Mediterranean Sea also contributes to the salinization of the system. Regarding the origin of the saline end member, most of the researchers claim, based on geochemical parameters, that it represents deep-seated evaporated brines that originated in the ancient saline or hypersaline lakes within the Rift Valley and migrated in the past into the neighboring aquifers. Among several studies that describe the above are: Goldschmidt et al. (1967), Starinsky (1974), Simon and Mero (1992), Gvirtzman et al. (1997), Bergelson et al. (1999), Dror et al. (1999), Kolodny et al. (1999), Rimmer et al. (1999), Hurwitz et al. (2000), Klein-BenDavid et al. (2005), Stiller et al. (2009), Magri et al. (2015). Most of these studies suggested that the driving force to the ascending saline water is the high potential of the fresh groundwater in the adjacent intake area. Mazor and Mero (1969), however, suggested that the saline end member is entrapped, slightly altered Pliocene seawater, which is presently squeezed out from the fill of the DSR as a result of tectonic compression. In this regard, it is important to note that the concentration of all saline mixtures that prevail in the discussed region does not exceed the threshold value of seawater concentration (Kafri and Arad 1979).

On top of the above, an additional saline end member of the current seawater encroachment from the Mediterranean Sea to the northern Rift Valley due to the elevation (head) difference between them was also proposed (Kafri and Arad 1978, 1979; Kafri et al. 2007; Kafri and Yechieli 2010).

Three clusters of saline springs issue from north to west along the western coast of the lake, namely the Tabgha, Fuliya and Tiberias springs. It was found that after a rainy season accompanied by groundwater table rise of the freshwater aquifers and higher spring discharges, those springs behave differently. The solute concentrations

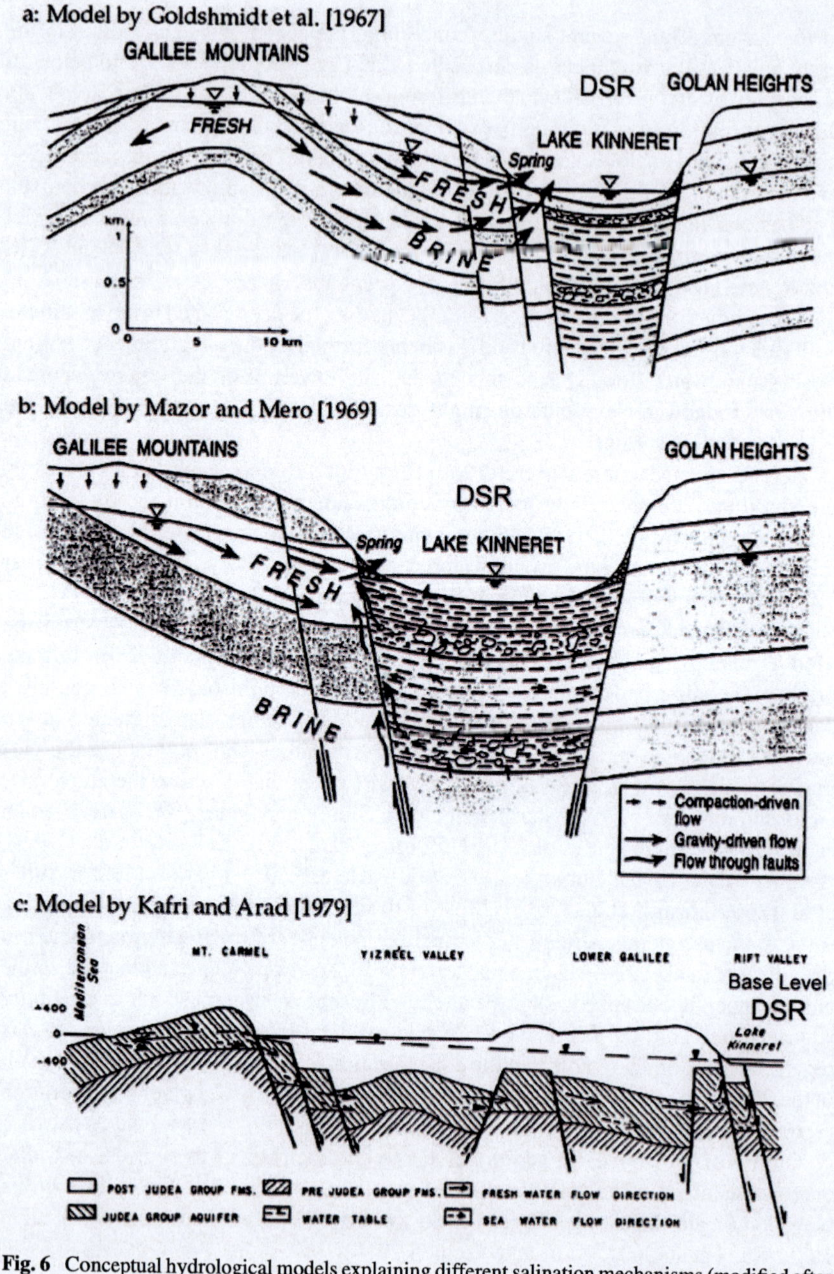


**Fig. 6** Conceptual hydrological models explaining different salination mechanisms (modified after Kafri and Yechieli 2010)

decrease in Tabgha, increase in Fuliya and do not vary in Tiberias, which calls for different salinization mechanisms of those springs (Gvirtzman et al. 1997).

A 3D modeling was conducted for the eastern part of the Kineret Lake (Also named Sea of Galilee; Yechieli et al. 2011, 2016a, b, see Chapter "General Information and Hydrogeology of the Mediterranean and Dead Sea Coastal Aquifers and Their Relation with Their Base Level"). This study indicates that groundwater is affected by thermal saline water body at depth of more than 1000 m, and therefore, most groundwater here is brackish.

**The lower Jordan Valley (LJ on Fig. 1)**

The lower Jordan Valley portion (Fig. 3) that extends between the Sea of Galilee in the north (at around 210 m bsl) and the Dead Sea in the south (at around 430 m bsl) is some 100 km long. The Jordan River flows along it from north to south. In the bottom of the valley, the climate is arid with precipitation between 50 and 150 mm/year, whereas in the mountainous regions that bound the valley to the east and west, precipitation attains values between 600 and 800 mm/year.

The lithostratigraphic units which constitute the main carbonate aquifers that drain toward the valley from west and east were described among others by Salameh (2001) and Moeller et al. (2007, 2009) as follows: In the west (Israeli side), the main carbonate aquifers are those of the upper Cretaceous Judea Group and the Eocene Avedat Group aquifers as well as the Neogene to Quaternary basalt aquifers. On the Jordanian side in the east, the main aquifers are those consisting of upper Cretaceous to Tertiary carbonate formations. The underlying lower Cretaceous sandstone aquifer is mostly recharged through percolating water from the upper Cretaceous aquifers.

The Jordan Rift Valley fill proper consists of thousands of meters of Neogene to Quaternary clastic material that form the Rift Valley fill aquifer, which is topped at the shallow surface by the practically aquicludal Upper Quaternary Lisan Formation. Neogene salt bodies also occur in depth close to the Sea of Galilee.

The lateral groundwater flow from the mountainous aquifers from west and east partly emerges as a series of springs along the marginal fault system of the rift and part of it feeds the Rift Valley fill aquifer.

Several studies were carried out in regard to the salinity and geochemistry of the saline groundwater, hosted in the lower Jordan Rift Valley fill aquifer. Among these are Starinsky (1974), Marie and Vengosh (2001), Salameh (2001), Abu-Jaber and Ismail (2003), Farber et al. (2007) and Moeller et al. (2007, 2009). It is agreed that the groundwater hosted in the Rift Valley aquifer constitutes a mixture of three main end members: (1) Circulating fresh groundwater which feeds the aquifer laterally from the mountainous aquifers from east and west. (2) Thermal and hypersaline Ca-chloride brines which ascend from depth along the fault system. (3) Anthropogenic effluents and agricultural return flows.

### 2.3 *The Arava Valley Segment (AR on Fig. 1)*

The Arava Valley segment of the eastern Rift Valley (Fig. 3) is some 165 km long. It is bounded to the north by the southern tip of the Dead Sea. To the south, it is bounded by the Gulf of Elat (Aqaba) which is a branch of the Red Sea. To the west, it borders the Negev (southern Israel) mountains that rise to elevations of somewhat over 500 m asl, and to the east, it borders the high Edom Mountains in Jordan that rise to elevations over 1000 m asl.

The Arava Valley and its graben fill aquifer are subjected to convergent, both subsurface groundwater and surface flood flows from west and east. The western Negev intake area is practically a desert with annual average rainfall of less than 200 mm and mostly a few tens of mm. The annual average rainfall on the Edom Mountains intake area, on the east, is higher attaining 250 mm.

The main sources of the convergent flows to the Arava Valley are the aquifers of the upper Jurassic to lower Cretaceous, Nubian Sandstone (hereafter NSS) and the upper Cretaceous carbonate Judea Group aquifer (JGA) that feed laterally the graben fill aquifer (hereafter GFA). Also, there is a source of subsurface flows from infiltrating flash floods, typical to the desert regime (i.e., Rosenthal et al. 1990). It was estimated that 75% of the recharge to the GFA originates in the west and only 25% in the east (Adar et al. 1992).

The general hydrogeological setup along the Arava segment based on several available studies as well as of a wide TDEM survey along the segment (Kafri et al. 2008) is as follows:

A topographic divide at the elevation of 230 m asl in the middle of the Arava Valley separates the convergent flow into it to the northern drainage system that flows to the Dead Sea terminal base level and to the southern one that flows to the Gulf of Elat (Aqaba) marine base level. As shown by Kafri et al. (2008), the surface drainage basin of the Arava Valley is divided into the southern and the northern, sub-basins (Fig. 7). In addition to that, it is noticed that rainfall distribution in the Arava drainage basin responsible to runoff and base flow along streams that recharge the GFA is not even. Low rainfall from both eastern and western sides of the valley typifies the southern sub-basin, whereas the northern sub-basin gets low rainfall from the west but rather higher ones from the Edom Mountains in the northeast. The surface area of the southern and the northern sub-basins is 2076 and 8658 km$^2$ accordingly (Fig. 7). As a result, the northern sub-basin gets considerably higher freshwater contribution from rain and floods than the southern one (Kafri et al. 2008).

As mentioned before, the main source that feeds the GFA is the NSS, studied by several authors, only some of which are referred to herein (see also Chapter "The Nubian Sandstone Aquifer in the Sinai Peninsula and the Negev Desert"). The NSS, mostly sandy aquifer of upper Jurassic to lower Cretaceous age, is spread regionally all over Nubia, Egypt, Saudi Arabia, Jordan and Northern Africa. A large part of its contained water is fossil and fresh to brackish. According to their stable isotopes and C-14 composition, they were recharged in more temperate climates than the present one, before 13k year and more than 30k year (Issar et al. 1972). Similarly,

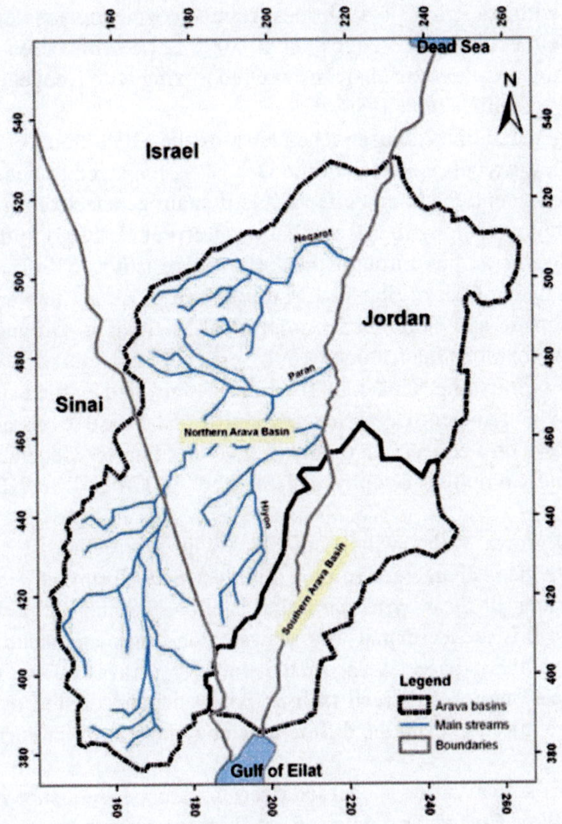


**Fig. 7** Arava Valley drainage basins (after Kafri et al. 2008)

Burg et al. (2016) have shown that the groundwater in the deep NSS has low C-14 values, usually <5 pMC with an age of >26k year. The young shallow groundwater, however, which is recharged via recent flash floods, is characterized as expected by higher C-14 values, up to 70 pMC.

Regarding the age of the NSS groundwater, Vengosh et al. (2007) have shown, based on their geochemistry and isotopic composition that at least in the northeastern Negev they include in addition to the fossil water also a component from the current local recharge, as was also suggested previously by Arad and Kafri (1980).

The salinity and geochemistry of the groundwater within the Arava Valley GFA were discussed in several articles. It is agreed by most of them that the hosted water represents a mixture of NSS water with the current ones recharged by flash floods (i.e., Adar et al. 1992). In addition to the above components, the salinization of the NSS water was attributed to deep-seated Jurassic brines or to mixture with trapped Neogene seawater (Rosenthal et al. 2007; Vengosh et al. 2007). The salinity of the GFA in the southern Arava Valley was also related to a mixture of NSS and flash floods waters with emerging Ca–Cl brines from Paleozoic units in depth (Rosenthal et al, 1990; Bein et al. 2001; Vengosh et al. 2007). The salinization of the GFA in the northern Arava Valley was also attributed to mixing with Neogene brines within the Rift Valley (Yechieli et al. 1992).

Concerning the salinity and geochemistry of the GFA, none of the boreholes within the valley, except one close to the Dead Sea, managed to penetrate in depth the proposed brine end member. A deep time-domain geoelectromagnetic (TDEM) survey was employed along the Arava Valley (Kafri et al. 2008) to detect different saline groundwater bodies through their electric resistivity (Kafri and Goldman 2005). The Dead Sea-Gulf of Elat TDEM traverse (Fig. 76, in Chapter "Geoelectric, Geoelectromagnetic and Combined Geophysical Methods in Groundwater Exploration in Israel") exhibits the following bulk resistivity configuration which enables a groundwater salinity one. This configuration is controlled by the existence of both the present-day and ancient Dead Sea and Gulf of Elat base levels and a divide in between as shown on a conceptual model (Fig. 75 in Chapter "Geoelectric, Geoelectromagnetic and Combined Geophysical Methods in Groundwater Exploration in Israel").

The central Arava Valley which coincides with a structural, topographic and hydrological divide is characterized by a high resistive subsurface sequence, representing freshwater all the way to the rather shallow basement at a depth close to 1000 m. Thus, this area escaped any seawater or brine encroachment from the northern and southern base levels. In the northern Arava Valley, interfaces are detected between interpreted fresh to brackish water and diluted or concentrated brines that were intruded from the different paleo (Bira, Lisan) or current Dead Sea system lakes.

The southern Arava Valley is characterized in its upper sequence by interpreted freshwater to diluted seawater, related to the Pliocene marine ingression, overlying the rather shallow high resistive basement.

It is important to note that no extremely low resistive bodies were detected at depth that could hint to the suggested existence of brines there. The thin low resistivity bodies, encountered up to a depth of 100 m, were interpreted as density-driven brines from the existing (i.e., Yotvata) or ancient sabkhas.

## 2.4 *The Dead Sea Segment*

The Dead Sea segment (Fig. 3) is the current endorheic terminal Dead Sea lake which is the remnant of the pre-existing larger Lake Lisan. It serves as a base level to all convergent surface as well as of groundwater flows from all sides (see also Chapters "Introduction to Studies of Coastal Aquifers of Israel"–"Submarine Groundwater Discharge Along the Israeli Eastern Mediterranean Coast and in Inland Basins"). The hydrogeological setup of this segment based on several studies is summarized below the following Kafri and Yechieli (2010).

The current salinity of the Dead Sea water is around 340 g/L. The input of fresh surface water to the lake is from the north via the Jordan River and from streams that originate in its western, eastern and southern catchment areas. The input of fresh groundwater to the system is from the surrounding intake areas (Arad and Michaeli 1967) in addition to brackish paleo-groundwater that were recharged in more remote intake areas (Issar et al. 1972; Yechieli et al. 1992). The latter also flow in a convergent pattern to the Dead Sea base level. These above described sources mix with the saline (brine) end members ending up as brackish to hypersaline subsurface flows or saline springs that issue close to the lake's shoreline. The origin and geochemistry of the saline end member were discussed in several studies (i.e., Starinsky 1974; Yechieli et al. 1996; Klein-BenDavid et al. 2004; Gvirtzman 2006; Waldmann et al. 2007; Katz and Starinsky 2009). According to the above, the saline end member was initially originated following the Pliocene marine ingression that filled the DSR to form later on the Sedom Lagoon and the consecutive lakes (see also Chapter "Paleohydrogeology of Israel"). The intruded water ended up, through concentration, as Ca-chloride brines, from a marine lagoon to an inland more concentrated saline lake. The brines were subjected later on to periods of inter-mixing processes, water rock interaction, evaporation and deposition of evaporates, such as halite, as well as of "fresher" water aragonite. The brines were and are at present density driven laterally to the adjoining groundwater systems (Stanislavsky and Gvirtzman 1999) where they mix with fresher groundwater. As expected, the interface, or interfaces in the case of a multi aquifer system, between the brines and the overlying fresher groundwater is shallower and of a milder slope, as compared to a case of normal seawater (Yechieli 2000). The interfaces were detected by borehole electrical logging and by time-domain electromagnetic (TDEM) measurements (Fig. 40 in Chapter "Geoelectric, Geoelectromagnetic and Combined Geophysical Methods in Groundwater Exploration in Israel") (Kafri et al. 1997; Yechieli et al. 2001, 2010; Yechieli 2006; Batayneh 2006). The current and paleo interfaces were also detected, and their configuration was also proposed, based on the TDEM method, in the northern Arava Valley, south of the Dead Sea (Fig. 75 in Chapter "Geoelectric, Geoelectromagnetic and Combined Geophysical Methods in Groundwater Exploration in Israel") (Kafri et al. 2008).

Kafri et al. (2014) also proposed, on top of the above, the current subsurface seawater intrusion through the Beer Sheva Valley from the Mediterranean Sea to the DSR, similar to the mechanism suggested in Northern Israel. Moreover, it was suggested that, based on the same mechanism, the higher salinities of the DSR paleo lakes were gained when the head difference between Mediterranean and the DSR base level was maximal (Kafri 2016).

The continuous monitored decline of the Dead Sea levels, since the 1960s, affects the different aspects of the groundwater regime in the last decades. The effect of the lake level drop on the adjoining groundwater levels was discussed by Kafri (1982), Yechieli (1993, 2006), Yechieli et al. (1995, 2001, 2009), Kiro et al. (2008) and Kafri and Yechieli (2010). It is recognized that the continuous drop of the lake level, interrupted by short periods of level rise, resulted in adjacent parallel groundwater level changes (Fig. 9a in Chapter "Dynamic Relationship Between the Sea and the

Aquifer"). Moreover, it was found that the extend of that response is affected by both aquifer properties and the distance of the monitored site from the receding shoreline (Yechieli et al. 1995, 2010; Kiro et al. 2008) as follows: The drop of the groundwater level resembles that of the lake level close to the shoreline and becomes smaller in more distant sites. However, a more considerable inland response of groundwater level drop was noticed along alluvial fans and or stream beds typical of a higher aquifer permeability (Fig. 9b in Chapter "Dynamic Relationship Between the Sea and the Aquifer").

The drop of the lake level and the resultant drop of the groundwater level also dropped the brine-fresher water interface as a result as is schematically shown in Fig. 8 (Kafri and Yechieli 2010). Those were detected by TDEM measurements, by borehole water salinity sampling and by borehole electric conductivity (EC) repeated profiling (Fig. 10a in Chapter "Dynamic Relationship Between the Sea and the Aquifer"), (Kiro et al. 2008; Yechieli et al. 2010). The TDEM detected of 1997 interface and the previous higher one the 1960s by existence the un-flushed saline remnants in the unsaturated zone (Fig. 40 in Chapter "Geoelectric, Geoelectromagnetic and Combined Geophysical Methods in Groundwater Exploration in Israel"), as described by Kafri et al. (1997), Yechieli et al. (2001) Yechieli (2006) and Kafri and Yechieli (2010). The interpretation of the above described setup suggests that throughout the decline of the interface, the portion of the aquifer, occupied previously by saline water, was replaced and flushed by fresher water within a relative

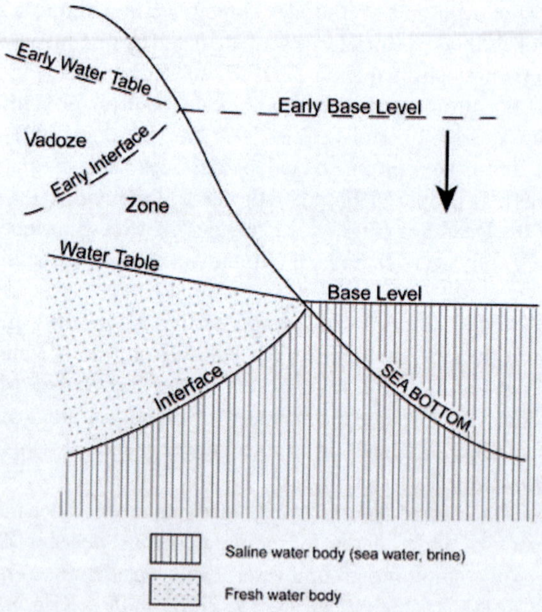


**Fig. 8** A scheme showing the resultant drop of the groundwater table and the fresh–saline groundwater interface due to the decline of the base level (after Kafri and Yechieli 2010)

short time. Thus, due to dewatering process of the upper portion of the aquifer, it was diverted into a vadose zone. The rate of almost complete flushing of the dewatered zone was found to be rapid, within a few decades.

The lake level changes also affected the discharge of the near shore thermal springs. Hydrological simulations exhibited the disturbed equilibrium between the lake level and the springs' discharge, following lake level changes (Shalev and Yechieli 2007). It was shown that lake level decline results in increased spring discharges, whereas a rising lake level reduces the discharge or even results in the complete chocking of the springs.

An additional by-product of the current decline of both the Dead Sea and the groundwater level, is the continuing formation of sinkholes. The hydrogeological setup and the mechanism responsible for this process were discussed by Abelson et al. (2006) and by Yechieli et al. (2006, 2016a, b) as is shown in Fig. 9, (see also Chapter "Hydrological and Geological Controls on the Evolution of the Dead Sea Sinkholes"). The sinkhole formation occurs due to the dissolution of a 10,000 year old halite layer found in the region at a depth of a few tens of meters. This layer was below the brine-fresher water interface, prior to the groundwater level and associated interface decline. Following the decline of the interface, the halite layer underneath was "exposed" this salt layer to dissolution by fresher water, saturation threshold of halite, and sinkholes started to form.

The response of the groundwater system to the expected continuing Dead Sea level decline was obtained by simulations based on reasonable forecast of future lake level elevations. The Dead Sea level is expected to decline at a decreasing rate until gaining an equilibrium between input and evaporation at an elevation of approximately 550 m bsl, approximately within the next 300 years (Yechieli et al. 1998). A FEFLOW simulation of the above scenario for the next 300 years yielded the following results in regard to the response of the groundwater system (Yechieli et al. 2009):

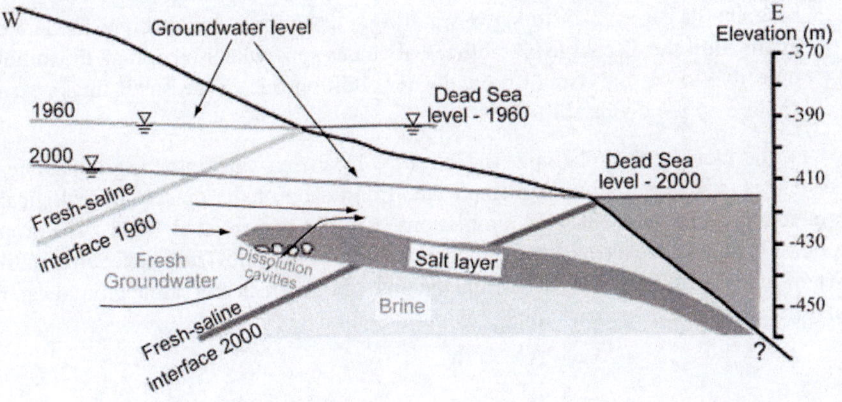


**Fig. 9** A conceptual model showing the decline of the Dead Sea level and the resultant sinkhole formation (after Kafri and Yechieli 2010)

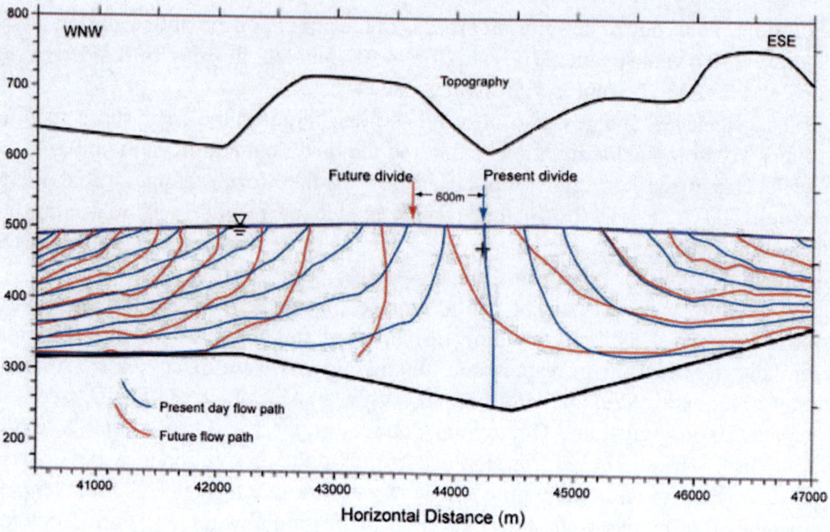


**Fig. 10** Simulated divide shift between the Mediterranean and the DSR base levels as the result of the future Dead Sea level decline (after Kafri and Yechieli 2010)

1. The groundwater level will continue to follow the lake's level drop at a decreasing drop with distance from the lake shoreline (Fig. 9b in Chapter "Dynamic Relationship Between the Sea and the Aquifer").
2. A westward shift of the regional groundwater divide, west of the Dead Sea, by only several hundred meters is expected, since such a shift is limited by the structural axis of the mountain crest in the west (Fig. 10).
3. The fresh–saline interface is expected to shift some 2 km eastward of its present location near the shoreline (Fig. 11).
4. Groundwater discharge to the base level will considerably increase in the beginning, on the expense of storage dewatering. Further on, it will become moderate again, until the discharge is stabilized at values somewhat higher than the initial ones due to the only small groundwater addition due to the small intake area capture by the divide shift.

Furthermore, the circulation of the intruding Dead Sea water into the groundwater system is expected to continue despite the rapid drop of the present hydrological system, as exhibited by SUTRA simulations (Kiro 2006; Kiro et al. 2008). The drop of the Dead Sea level is expected to decrease the intruding flow into the aquifer until a new quasi steady-state is reached and the flow of both fresh and saline groundwater return to be similar to their initial previous situation.

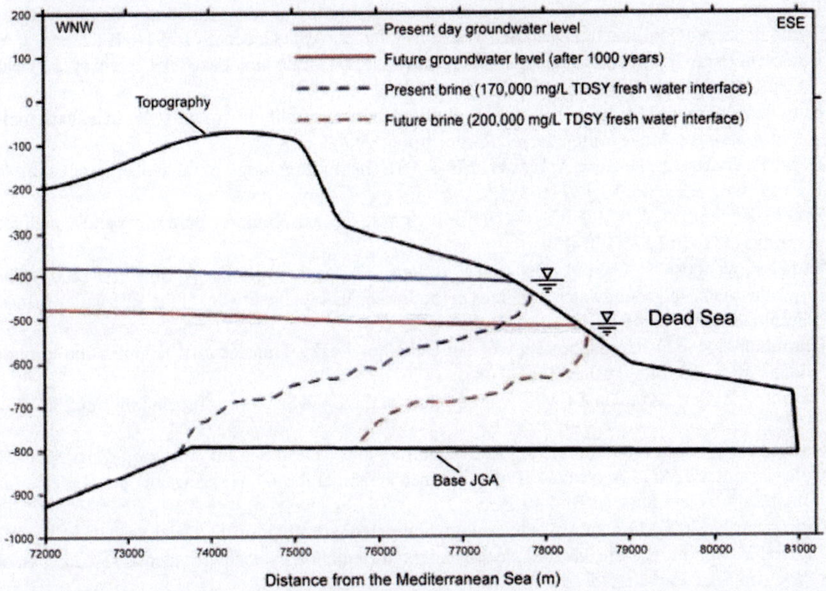


**Fig. 11** Simulation of the present fresh–saline groundwater interface and the future one (after 1000 years) following the predicted decline of the Dead Sea level (after Kafri and Yechieli 2010)

## References

Abelson M, Yechieli Y, Crouvi O, Baer G, Wachs D, Bein A, Shtivelman V (2006) Evolution of the Dead Sea sinkholes. In: Enzel Y, Agnon A, Stein M (eds) New frontiers in the Dead Sea paleoenvironmental research. Geol Soc Am Spec Pap 401:241–253

Abu-Jaber N, Ismail M (2003) Hydrogeochemical modeling of the shallow groundwater in the northern Jordan Valley. Environ Geol 44:391–399

Adar EM, Rosenthal E, Issar AS, Batelaan O (1992) Quantitative assessment of the flow pattern in the southern Arava Valley (Israel) by environmental tracers and a mixing cell model. J Hydrol 136:333–352

Arad A, Bein A (1986) Saline-versus freshwater contribution to the thermal water of the northern Jordan Rift Valley, Israel. J Hydrol 83:49–66

Arad A, Kafri U (1980) Hydrogeological interrelationship between the Judea Group and the Nubian Sandstone aquifers in Sinai and the Negev. Isr J Earth Sci 29:67–72

Arad A, Michaeli A (1967) Hydrogeological investigations in the western catchment of the Dead Sea. Isr J Earth Sci 16:181–196

Babad A, Burg A, Adar EM (2019) Conceptual hydrological approach to a geologically complex basin with scarce data: the Hula Valley, Middle East. Hydrogeol J. https://doi.org/10.1007/s10040-019-02031-x

Batayneh AT (2006) Use of electrical resistivity methods for detecting subsurface fresh and saline water and delineation their interfacial configuration: a case study of the eastern Dead Sea coastal aquifers, Jordan. Hydrogeol J 14:1277–1283

Bein A, Yechieli Y, Bensabat J (2001) Quantifying the groundwater resources of the southern Arava Rift Valley: a confined desert system recharged laterally by external sources. Isr J Earth Sci 50:217

Bergelson G, Nativ R, Bein A (1999) Salinization and dilution history of ground water discharging into the Sea of Galilee, the Dead Sea Transform, Israel. Appl Geochem 14:91–118, 236

Bookman (Ken-Tor) R, Enzel Y, Agnon A, Stein M (2004) Late Holocene lake levels of the Dead Sea. Geol Soc Am Bull 116:555–571

Burg A, Zilberbrand M, Yechieli Y (2016) Radiocarbon variability in groundwater in an extremely arid zone—the Arava Valley, Israel. Radiocarbon 55:963–978

Dafny E, Gvirtzman H, Burg A, Fleischer L (2003) The hydrogeology of the Golan basalt aquifer, Israel. Isr J Earth Sci 52:139–153

Dror G, Ronen D, Stiller M, Nishri A (1999) Cl/Br ratio of Lake Kinneret, pore water and associated springs. J Hydrol 225:130–139

Farber E, Vengosh A, Gavrieli I, Marie A, Bullen TD, Mayer B, Polak A, Shavit U (2007) The geochemistry of groundwater resources in the Jordan Valley: the impact of the Rift Valley brines. Appl Geochem 22:494–514

Garfunkel Z (1981) Internal structure of the Dead Sea Leaky Transform (rift) in relation to plate kinematics. Tectonophysics 80:81–108

Garfunkel Z, Ben-Avraham Z (1996) The structure of the Dead Sea basin. Tectonophysics 266:155–176

Garfunkel Z, Zak Y, Freund R (1981) Active faulting in the Dead Sea rift. Tectonophysics 80:1–26

Goldschmidt MJ, Arad A, Neev D (1967) The mechanism of the saline springs in the Lake Tiberias depression. Geol Surv Isr Bull 45:1–19

Gvirtzman H (2006) Groundwater hydrology and paleohydrology of the Dead Sea rift valley. In: Enzel Y, Agnon A, Stein M (eds) New frontiers in Dead Sea paleoenvironmental research. Geol Soc Am Spec Pap 401:95–111

Gvirtzman H, Garven G, Gvirtzman G (1997) Hydrogeological modeling of the saline hot springs at the Sea of Galilee, Israel. Water Resour Res 33:913–926

Hurwitz S, Goldman M, Ezersky M (1999) Geophysical (time domain electromagnetic model) delineation of a shallow brine beneath a freshwater lake, the Sea of Galilee, Israel. Water Resour Res 35:3631–3638

Hurwitz S, Stanislavsky E, Lyakhovsky V, Gvirtzman H (2000) Transient groundwater-lake interactions in a continental rift: Sea of Galilee, Israel. Geol Soc Am Bull 112:1694–1702

Issar A, Bein A, Michaeli A (1972) On the ancient water of the upper Nubian Sandstone aquifer in central Sinai and southern Israel. J Hydrol 17:353–374

Kafri U (1982) Relationship between Dead Sea and groundwater levels in the western Dead Sea catchment area. Isr Geol Surv Curr Res 1981:90–94

Kafri U (2016) Subsurface seawater intrusion, an additional factor influencing the limnological history of the Dead Sea Basin. Environ Manage Sustain Dev 5:244–254. https://doi.org/10.5296/emsd.v5i.10130

Kafri U, Agron NA (1978) A proposed hydrogeological model for future lignite mining and groundwater drainage in the Hula Basin, northern Israel. In: Water in mining and underground works. SIAMOS symposium, Granada, Spain, pp 173–181

Kafri U, Arad A (1978) Paleohydrology and migration of the ground-water divide in regions of tectonic instability in Israel. Geol Soc Am Bull 89:1723–1732

Kafri U, Arad A (1979) Current subsurface intrusion of Mediterranean seawater—a possible source of groundwater salinity in the Rift Valley System, Israel. J Hydrol 44:267–287

Kafri U, Ecker A (1964) Neogene and Quaternary subsurface geology and hydrogeology of the Zevulun Plain. Isr Geol Surv Bull 37:1–13

Kafri U, Goldman M (2005) The use of the time domain electromagnetic (TDEM) method to delineate saline groundwater in granular and carbonate aquifers and to evaluate their porosity. J Appl Geophys 57:167–178

Kafri U, Lang B (1987) New data on the young Quaternary fill of the Hula Basin, Israel. Isr J Earth Sci 36:73–82

Kafri U, Yechieli Y (2010) Groundwater base level changes and adjoining hydrological systems. Springer, Berlin, 171 pp. https://doi.org/10.1007/978-3-642-13944-4

Kafri U, Kaufman A, Magaritz M (1983) The rate of Pleistocene subsidence and sedimentation in the Hula Basin as compared with those of other time spans in other Israeli tectonic regions. Earth Planet Sci Lett 65:126–132

Kafri U, Goldman M, Lang B (1997) Detection of subsurface brines, freshwater bodies and the interface configuration in-between by the time domain electromagnetic method in the Dead Sea Rift, Israel. Environ Geol 31:42–49

Kafri U, Goldman M, Lyakhovsky V, Scholl C, Helwig S, Tezkan B (2007) The configuration of the fresh-saline groundwater interface within the regional Judea Group carbonate aquifer in northern Israel between the Mediterranean and the Dead Sea base levels as delineated by deep geoelectromagnetic soundings. J Hydrol 344:123–134

Kafri U, Goldman M, Levi E (2008) The relationship between saline groundwater within the Arava Rift Valley in Israel and the present and ancient base levels as detected by deep geoelectromagnetic soundings. Environ Geol 54:1435–1445

Kafri U, Goldman M, Levi E, Wollman S (2014) Detection of saline groundwater bodies between the Dead Sea and the Mediterranean Sea, Israel, using the TDEM method and hydrochemical parameters. Environ Process 1:21–41. https://doi.org/10.1007/s40710-014-0001-2

Katz A, Starinsky A (2009) Geochemical history of the Dead Sea. Aquat Geochem 15:159–194

Kiro Y (2006) The effect of the Dead Sea level drop in the past 50 years on the groundwater system in the alluvial aquifer in its vicinity. MSc thesis, Hebrew Univ Jerusalem, Israel, 116 pp (in Hebrew)

Kiro Y, Yechieli Y, Shalev E, Lyakhovsky V, Starinsky A (2008) Time response of the water table and saltwater transition zone to a base level drop. Water Resour Res. https://doi.org/10.1029/2007WR006752

Klein-BenDavid O, Sass E, Katz A (2004) The evolution of marine evaporitic brines in inland basins: the Jordan-Dead Sea Rift valley. Geochim Cosmochim Acta 68:1763–1775

Klein-BenDavid O, Gvirtzman H, Katz A (2005) Geochemical identification of fresh water sources in brackish groundwater mixtures; the example of Lake Kinneret (Sea of Galilee), Israel. Chem Geol 214:45–59

Kolodny Y, Katz A, Starinsky A, Moise T (1999) Chemical tracing of salinity sources in Lake Kinneret (Sea of Galilee), Israel. Limnol Oceanogr 44:1035–1044

Magri F, Inbar N, Siebert C, Rosenthal E, Guttman J, Moeller P (2015) Transient simulations of large-scale hydrogeological processes causing temperature and salinity anomalies in the Tiberias Basin. J Hydrol 520:342–355

Marie A, Vengosh A (2001) Sources of salinity in groundwater from Jericho area, Jordan Valley. Groundwater 39:240–248

Mazor E, Mero F (1969) The origin of the Tiberias-Noit mineral water association in the Tiberias-Dead Sea Rift Valley, Israel. J Hydrol 7:318–333

Moeller P, Rosenthal E, Geyer S, Guttman J, Dulski P, Rybakov M, Zilberbrand M, Jahnke C, Flexer A (2007) Hydrochemical processes in the lower Jordan valley and in the Dead Sea area. Chem Geol 239:27–49

Moeller P, Rosenthal E, Geyer S (2009) Characterization of aquifer environments by major and minor elements and stable isotopes of sulfate. In: Hoetzl H, Moeller P, Rosenthal E (eds) The water of the Jordan Valley. Scarcity and deterioration of groundwater and its impact on the regional development. Springer, Berlin, pp 83–121

Rimmer A, Hurwitz S, Gvirtzman H (1999) Spatial and temporal characteristics of saline springs: sea of Galilee, Israel. Groundwater 37:663–673

Roded R, Shalev E, Katoshevski D (2013) Basal heat-flow and hydrothermal regime at the Golan—Ajloun hydrological basins. J Hydrol 476:200–211

Rosenthal E, Adar E, Issar AS, Batelaan O (1990) Definition of groundwater flow pattern by environmental tracers in the multiple aquifer system of southern Arava Valley, Israel. J Hydrol 117:339–368

Rosenthal E, Zilberbrand M, Livshitz Y (2007) The hydrochemical evolution of brackish groundwater in central and northern Sinai (Egypt) and in the western Negev (Israel). J Hydrol 337:2
Salameh E (2001) Sources of water salinities in the Jordan Valley Area/Jordan. Acta Hydrochim Hydrobiol 29:329–362
Shalev E, Yechieli Y (2007) The effect of Dead Sea level fluctuations on the discharge of thermal springs. Isr J Earth Sci 56:19–27
Simon E, Mero F (1992) The salinization mechanism of Lake Kinneret. J Hydrol 138:327–343
Stanislavsky E, Gvirtzman H (1999) Basin-scale migration of continental rift brines: paleohydrological modeling of the Dead Sea basin. Geology 27:791–794
Starinsky A (1974) Relationship between Ca-chloride brines and sedimentary rocks in Israel. PhD thesis, Hebrew Univ Jerusalem, 208 pp
Stiller M, Rosenbaum JM, Nishri A (2009) The origin of brines underlying Lake Kinneret. Chem Geol 262:293–309
Vengosh A, Hening S, Ganor J, Mayer B, Weyhenmeyer CE, Bullen TD, Paytan A (2007) New isotopic evidence for the origin of groundwater from the Nubian Sandstone Aquifer in the Negev, Israel. Appl Geochem 22:1052–1073
Waldmann N, Starinsky A, Stein M (2007) Primary carbonates and Ca-chloride brines as monitors of a paleo-hydrological regime in the Dead Sea basin. Quat Sci Rev 26:2219–2228
Yechieli Y (1993) The effects of water level changes in closed lakes (Dead Sea) on the surrounding groundwater and country rock. PhD thesis, Weizmann Institute of Science, Rehovot, Israel
Yechieli Y (2000) Fresh-saline water interface in the western Dead Sea area. Groundwater 38:615–623
Yechieli Y (2006) Response of the groundwater system to changes in the Dead Sea level. In: Enzel Y, Agnon A, Stein M (eds) New frontiers in Dead Sea paleoenvironmental research. Geol Soc Am Spec Pap 401:113–126
Yechieli Y, Starinsky A, Rosenthal E (1992) Evolution of brackish groundwater in a typical arid region: Northern Arava Rift Valley, southern Israel. Appl Geochem 7:361–374
Yechieli Y, Ronen D, Berkovitz B, Dershovitz WS, Hadad A (1995) Aquifer characteristics derived from the interaction between water levels of a terminal lake (Dead Sea) and an adjacent aquifer. Water Resour Res 31:893–902
Yechieli Y, Ronen D, Kaufman A (1996) The source and age of groundwater brines in the Dead Sea area, as deduced from $^{36}$Cl and $^{14}$C. Geochim Cosmochim Acta 60:1909–1916
Yechieli Y, Gavrieli I, Ronen D, Berkovitz B (1998) Will the Dead Sea die? Geology 26(8):755–758
Yechieli Y, Kafri U, Goldman M, Voss CI (2001) Factors controlling the configuration of the fresh-saline water interface in the Dead Sea coastal aquifer: synthesis of TDEM surveys and numerical groundwater modeling. Hydrogeol J 9:367–377
Yechieli Y, Abelson M, Bein A, Crouvi O, Shtivelman V (2006) Sinkhole "swarms" along the Dead Sea coast: reflection of disturbance of lake and adjacent groundwater systems. Geol Soc Am Bull 118:1075–1087
Yechieli Y, Kafri U, Wollman S, Shalev E, Lyakhovsky V (2009) The effect of base level changes and geological structures on the location of the groundwater divide, as exhibited in the hydrological system between the Dead Sea and the Mediterranean Sea. J Hydrol 378:218–229
Yechieli Y, Shalev E, Wollman S, Kafri U (2010) The response of the Mediterranean and Dead Sea coastal aquifers to sea level variations. Geol Surv Isr Int Rep
Yechieli Y, Shaliv G, Wollman S, Kessler A, Rosensaft M, Berger D, Bein A (2011) Three-dimensional modeling of the Kalanit Basin (Fulia-Tabkha Basin). Report GSI/38/2011. Geological Survey of Israel (in Hebrew)
Yechieli Y, Abelson M, Baer G (2016a) Sinkhole formation and subsidence along the Dead Sea coast, Israel. Hydrogeol J 24:601–612. https://doi.org/10.1007/s10040-015-1338-y
Yechieli Y, Shaliv G, Wollman S (2016b) Update of the 3D model of the Fulia Tabkha basin (stage B of the Kalanit project). Report GSI/29/2016. Geological Survey of Israel (in Hebrew)
Zak I (1967) The geology of Mount Sdom. PhD thesis, Hebrew University Jerusalem (in Hebrew)

# Paleohydrogeology of Israel

Uri Kafri

## 1 Introduction

The present chapter deals with different aspects related to the paleohydrogeology of Israel, based on existing local and global data. The discussion refers to the paleo-groundwater systems and not to surface drainage systems. Due to the restrictions of resolution of the data, most of the interpretations are rather qualitative than quantitative.

When dealing with the memory of hydrogeological systems, it should be emphasized that it is limited mostly to young and to current periods, namely since the Neogene. A rough idea of the paleohydrogeology setup that prevailed prior to Neogene times can be gained or speculated through proxies. Among these are: global climate changes, reconstructed paleogeography and environments of deposition, paleo-karst phenomena, delineation of different lithologies, as well as location and delineation of ancient base levels.

The reconstructed paleogeography reveals the on-land potential intake areas as well as the delineation of both marine and continental base levels and suggest the paleo-groundwater flow directions. The paleo-climatic records might give an idea as to the amount of rainfall and evaporation and in turn the natural recharge to the groundwater system. In addition, it might also explain the salinity of the water in the aquifers and the base levels. Tectonism combined with climatic records might explain the base levels elevations, the closure of basins to end up as endorheic base levels, capturing phenomena, and paleo-flow gradients. The lithological sequences might reveal their hydrogeological properties (aquifers, aquicludes) as well as to the type of water, whether fresh, saline or brine, that prevailed in their environment of deposition.

U. Kafri (✉)
Geological Survey of Israel, 32 Leibovitz Str., 9692100 Jerusalem, Israel
e-mail: uri.kafri@gsi.gov.il


U. Kafri and Y. Yechieli (eds.), *The Many Facets of Israel's Hydrogeology*, Springer Hydrogeology, https://doi.org/10.1007/978-3-030-51148-7_11

## 2 The Paleohydrogeology of Israel Prior to, and During Neogene Times

During the geological history, the discussed area was most of the time at the fringe of the Afro-Arabian Plate or the Arabo-Nubian massive in the south and east. Marine conditions prevailed in Israel almost continuously from the Mesozoic to Neogene time when the onset of regional uplift resulted in the establishment of continental conditions over most of Israel (Neev 1960; Gvirtzman 1970, Kafri and Arad 1978). The continental hydrological, both surface and subsurface systems were assumingly drained to the only single marine base level of the Tethys in the west. The retreats of the sea to the west started in the Upper Eocene-Oligocene. In the Miocene, Israel was already continental draining to the Mediterranean Sea in the west all the way from east of the present-day Dead Sea Rift (hereafter DSR) prior to its formation as a base level (Gvirtzman 1970; Roegl 1999; Ziegler 2001). The non-existence of the present-day DSR as an eastern base level were indicated by surface westward paleo-drainage systems that crossed it (Neev 1960; Kolodny 1965; Garfunkel and Horowitz 1966; Kafri and Heimann 1994; Kafri 2002; Kafri and Horowitz 2003).

The Dead Sea Transform was formed in Neogene times (Garfunkel 1981). Only in Middle Miocene the tectonic activity and subsidence along it resulted in the formation of pull-aparts which acted as depressions and base levels along parts of the DSR, as described below.

## 3 The Paleo-Eastern Mediterranean Sea Base Level

The marine Neogene base levels of the Paratethys–Mediterranean Sea were discussed in numerous studies, i.e., Mocochain et al. (2006), Popov et al. (2006), Nunn and Harris (2007), Govers (2009) and Kafri and Yechieli (2010a). According to them, until the end of Tortonian–early Messinian (Late Neogene), the Mediterranean Sea was connected to the Atlantic Ocean. The subsequent partial or complete closure and isolation of the Mediterranean Sea from the Atlantic Ocean resulted in restricted circulation and a subsequent drop of sea. The complete closure and isolation of the Mediterranean Sea, named the Messinian Salinity Crisis (hereafter MSC), between 5.6–5.32 Ma resulted in the choking of, and the partial or complete desiccation of the Mediterranean Sea. The sea level dropped to a depth exceeding 1500 m, or even to over 2800 m (Nunn and Harris 2007). At a later stage, re-flooding of the Mediterranean basin by Atlantic water had started in the Pliocene (Popov et al. 2006) or some 5.33 Ma ago (Nunn and Harris 2007).

The relationship between karst systems and the different Mediterranean base levels since the Neogene was discussed, among others by Audra et al. (2004) and Mocochain et al. (2006). The drop of the paleo-groundwater tables, following the Messinian decline of the base level is indicated, around the Mediterranean Sea by deep paleokarst systems that led to that Messinian deep base level. A shallower karst system level was formed later on as a result of the subsequent rise the Pliocene base level and the adjoining paleo-groundwater levels to a level around 80 m asl.

The Israeli eastern Mediterranean Neogene base levels were discussed among others by Neev (1960), Kafri and Ecker (1964), Gvirtzman (1970) and Druckman et al. (1995). The Messinian sea level drop, as elsewhere in the Mediterranean, was accompanied by subaerial deep incision of channels and canyons along the basin's margins. The desiccation of the sea and its closure resulted in the formation of concentrated brines which deposited evaporates and in addition migrated inland to form the C-type brines, described by Starinsky (1974) and followed by several other scholars (i.e., Vengosh and Rosenthal 1994). The drop of the sea (base) level caused the decline of the paleo-groundwater tables adjacent to it. An upper Miocene to Pliocene paleo-water table is indicated by a series of karstic caves (presently at 400 m asl) on the western slopes of the Judea Mountains, as well as an abrasion plain in western Galilee (presently at 300 m asl) and the adjoining interpreted paleo-water tables Kafri and Arad (1978) (Fig. 1). At a later stage, the re-flooding of the Mediterranean basin by Atlantic water resulted in the sea (base) level and assumingly the adjoining paleo-groundwater levels to rise to elevations above the present one.

The Mediterranean base level was subjected to global climate changes which occurred during the Pleistocene Era, attributed to changes in solar radiation (Milankovitch cycles) and atmospheric circulation, resulting in glacial and inter-glacial periods. The Mediterranean Sea (base) level declined and rose during glacial and interglacial periods respectively. It is reasonable to assume that during the rise of the sea level the adjoining coastal aquifers were encroached by seawater whereas during its subsequent decline the hosted seawater was flushed back to the retreating base level.

The considerable drop of the eustatic sea level in the Last Glacial Maximum (hereafter LGM), some 18 ka ago, to approximately 120 m below sea level (hereafter bsl), and its subsequent rise to the present-day sea level since then, was discussed among others by Fairbanks (1989) and Fleming et al. (1998) (Fig. 2). The Mediterranean Pleistocene coastal aquifer of Israel was re-intruded by seawater during the rise of sea level since the LGM. The aquifer is known to be subdivided into sub-aquifers and aquicludes (Issar 1968). The possibility that the lower sub-aquifers are blocked to the sea in some areas by impervious layers and thus escape seawater encroachment was discussed by Kafri and Goldman (2006), as also shown in Fig. 3. Yechieli et al. (2009a) have shown accordingly that the ages of the saline groundwater consisting of old seawater, become younger upward from the lower to the upper sub-aquifers reflecting the sea level rise since the LGM (Fig. 8, Chapter "Dating of Groundwater in Israeli Aquifers and Determination of Groundwater Flow Velocities").

## 4 The Paleo-Eastern DSR Base Level

As mentioned before, the DSR was formed in Neogene times (Garfunkel 1981). The history and the different stages of its existence as a base level are described herein:

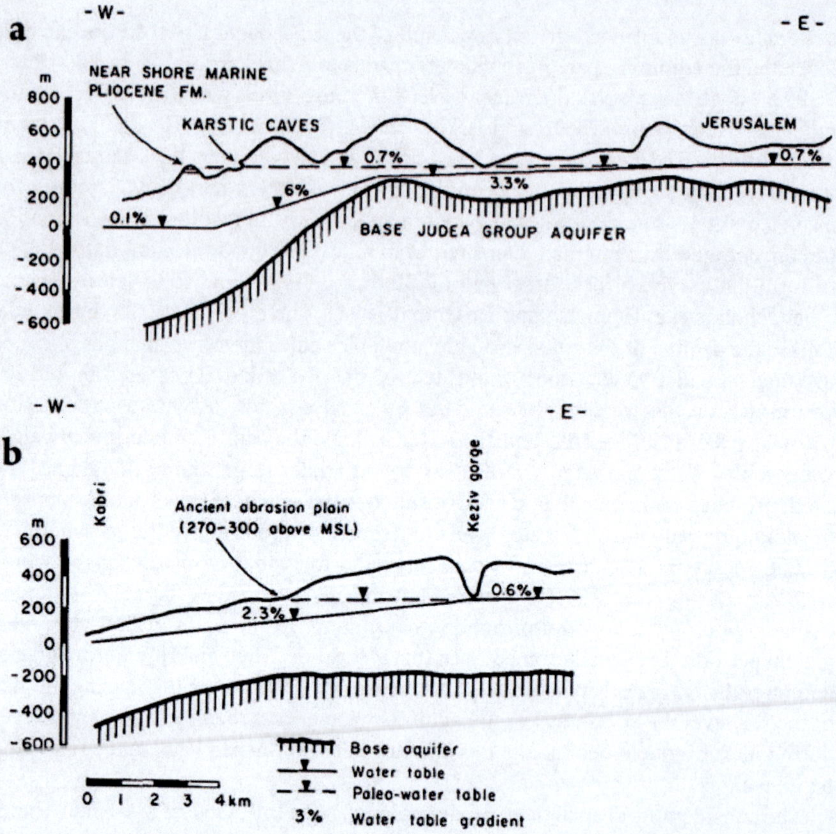


**Fig. 1** Paleo- and present-day groundwater tables in **a** Judea Mountains. **b** Western Galilee (modified after Kafri and Yechieli 2010a)

Until the Middle Neogene, the basin did not exist as a depression, and the paleo-drainage systems originated east of the present-day DSR, flowing all the way to the Mediterranean base level in the west (Horowitz 2001). This seems to be the case also to the groundwater system.

The Pre-Middle Neogene drainage systems were described from southern Israel, based on detrital material (pebbles) which could have been derived and transported along the systems only from areas east of the present-day DSR (Neev 1960; Calvo and Bartov 2001; Zilberman and Calvo 2013). In northern Israel, Kafri (2002) described similar Neogene E-W directed paleo-drainage systems in the Lower Galilee region that drained areas east of the present-day DSR. Similarly, Kafri and Heimann (1994) described the same configuration from west of the Sea of Galilee, indicating that the latter part of the DSR did not exist as a depression and a base level prior to some 2 m.y. Regarding the Hula Basin, at the northernmost part of the DSR in Israel, Kafri

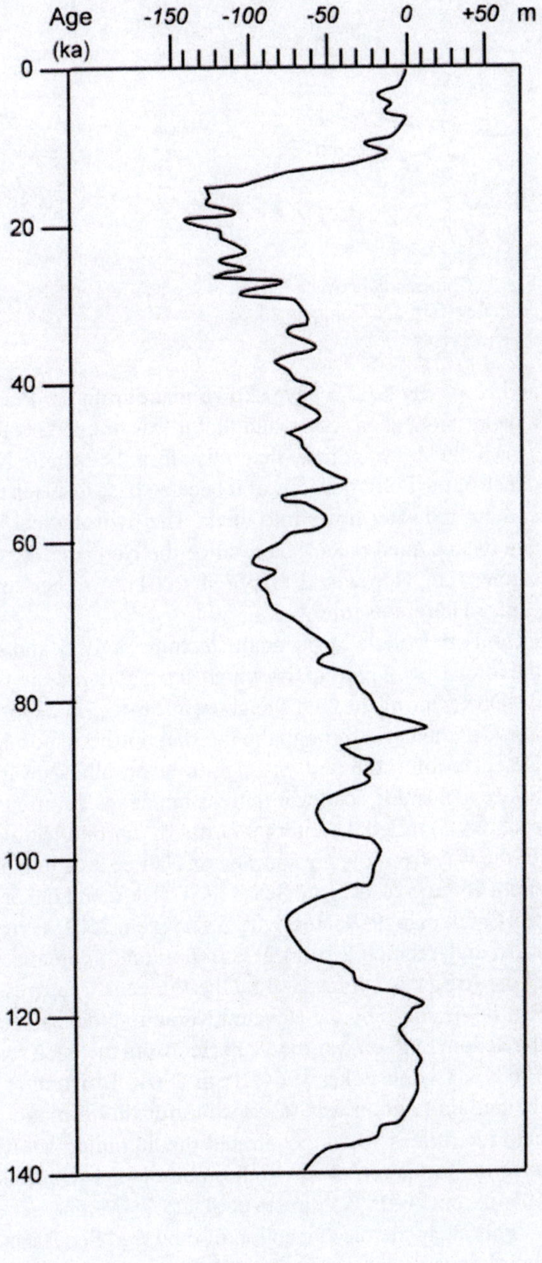


**Fig. 2** Global sea levels since 140 ky (after Kafri and Yechieli 2010a)

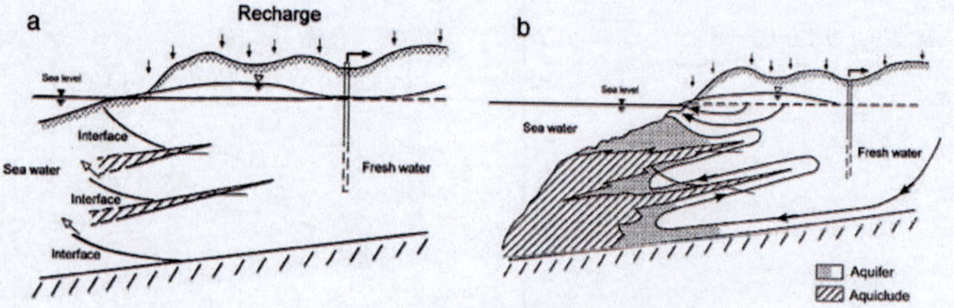


**Fig. 3** Connection between the lower sub-aquifers and the sea. **a** Open to the sea, **b** closed to the sea (after Kafri and Goldman 2006)

and Horowitz (2003) have shown that during the Pliocene this area did not exist as a depression, as of today, and that it was drained northward.

All the above indicate that only since the Middle Neogene, due to tectonic subsidence of the DSR, portions of it became base levels in the east, capturing both surface and groundwater flows into them. The hydrological history of the Dead Sea Basin and its occupied paleo-lakes, since the Neogene, was discussed in detail in several studies (i.e., Horowitz 2001; Stein 2001). A rather concise description of the above is given herein as follows:

Only in Middle Miocene the tectonic activity and subsidence along it resulted in the formation of pull-aparts which acted as depressions and base levels along parts of the DSR. During the Late Neogene, due to tectonic subsidence, the Mediterranean Sea intruded the basin through the internal northern morpho-tectonic valleys of northern Israel, forming the Sedom Lagoon as an inland evaporitic basin (Zak 1967). The Pliocene eustatic sea level ingression, at an elevation around 50–100 m asl (Miller et al. 2005), filled the internal northern morpho-tectonic valleys and the Rift Valley, all the way from the present Sea of Galilee area to a distance of several tens of kms south of the present Dead Sea (Fig. 4) (Kafri and Ecker 1964; Zak 1967; Stanislavsky and Gvirtzman1999; Horowitz 2001; Stein 2001; Gvirtzman 2006; Kafri et al. 2008; Kafri and Yechieli 2010a). It is important to note that the Hula Basin, which is part of the Rift, was not invaded by the Pliocene transgression being separated from the Rift in the south by the elevated Korazim block, as described below. Subsequently, the Sedom Lagoon was disconnected from the open sea and started to desiccate (Zak 1967; Kafri and Ecker 1964; Stein 2001; Torfstein et al. 2009). The desiccation of the Sedom Lagoon was responsible for the formation of concentrated, mostly Ca-chloride, brines which penetrated the adjoining aquifers of the DSR basin as well as to the formation of salt bodies and diapirs within the DSR (Zak 1967; Starinsky 1974; Stein 2001; Waldmann et al. 2007; Weinberger et al. 2007).

Following the disconnection of the Dead Sea Basin from the Mediterranean Sea, three known major lacustrine and hypersaline water bodies (Katz et al. 1977; Torfstein et al. 2009) consecutively occupied the basin throughout the Pleistocene–Holocene and until today. These are Lake Amora (including Lake Samra), Lake Lisan and the present remnant Dead Sea. The water levels of these lakes have risen and declined as

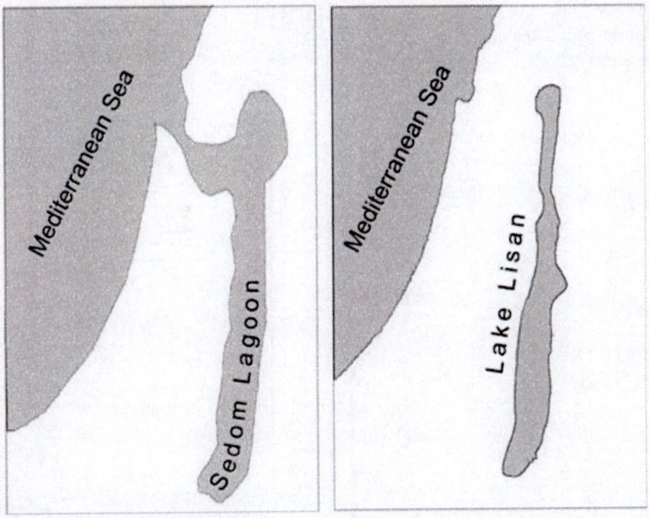


**Fig. 4** Map showing the maximal extend of the Sedom Lagoon and the Lisan Lake (after Kafri and Yechieli 2010a)

a result of both tectonic subsidence and climate changes (Fig. 5) (Stein 2001; Bartov et al. 2006; Waldmann et al. 2009).

Lake Amora existed at least between 740 and 70 ka, whereas its inclusive younger Lake Samra existed between 135 and 70 ka. The limnological history of Lake Amora displays cyclic shifts between dry lowstands and wet highstands, which prevailed during time spans exceeding tens of thousands years. The levels of Lake Samra fluctuated between 320 and 380 m bsl (Stein 2001; Torfstein et al. 2009). The lakes' level curve of the Samra and the Lisan lakes is shown in Fig. 5. Lake Lisan, discussed in numerous studies, existed between 70 and 15 ka and its levels fluctuated between 330 and 150 m bsl (Stein 2001; Torfstein et al. 2009). The maximal extent of the Lisan Lake, within the DSR, is shown on Fig. 4. Additional information about the DS lacustrine and geochemical record along the last 225 k years was obtained from the deep research corehole that was drilled in the deeper part of the DS to a depth of ~450 m (Neugebauer et al. 2014). Best records were obtained from pore water that was extracted from the clayey sediments. Regarding the paleo-levels of the Dead Sea segment proper it should be mentioned that a 455-m-long continuous core was retrieved from a deep core hole drilled in 2010–2011 in the middle of the DS at a water depth of 300 m. The analysis of the core yielded information as to the different sedimentary facies of the penetrated Pleistocene to Holocene sequence and their ages (Neugebauer et al. 2014; Torfstein et al. 2015a). The findings exhibited a 40 m halite dominated core dated to 110–120 ka at a level of 235 m below lake's floor, namely at about 950 m bsl. Torfstein et al. (2015a) related this finding to a catastrophic drawdown of the lake level to the above level. Katz and Starinsky (2015), however, rejected this interpretation, claiming that the proposed water depth was not sufficient

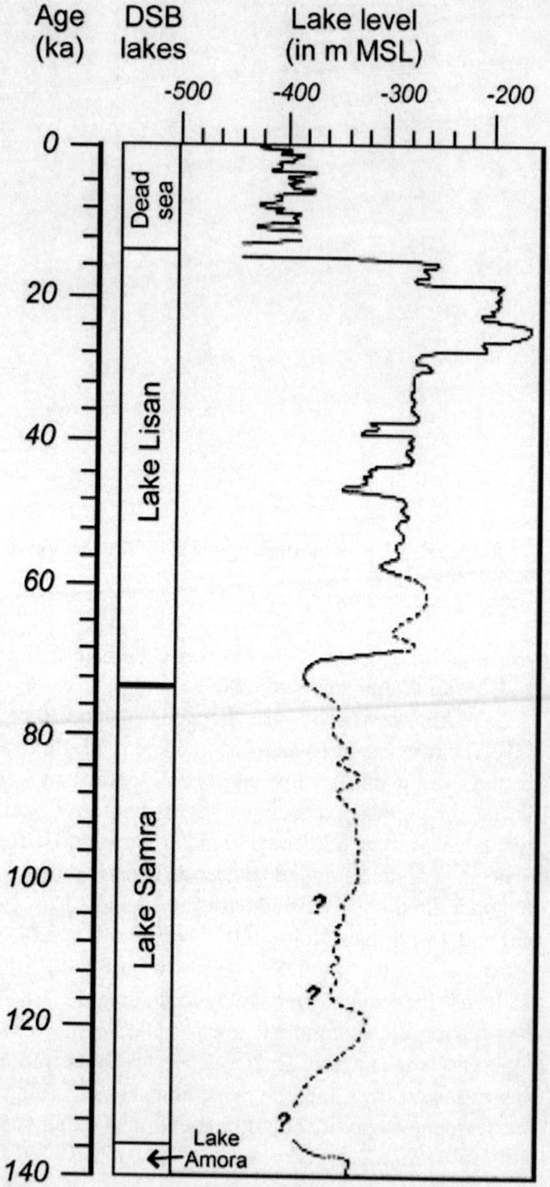


**Fig. 5** Lake level curve of the Samra and Lisan lakes (after Kafri and Yechieli 2010a)

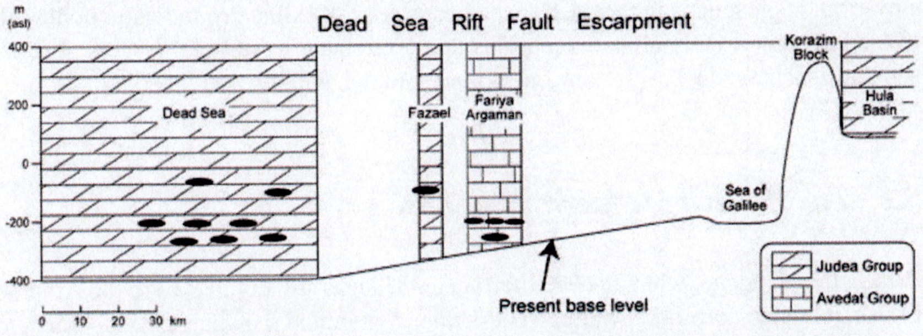


**Fig. 6** A simplified north–south cross-section along the western fault escarpment of the DSR displaying the cave geographical distribution (after Kafri and Yechieli 2010b)

enough to deposit a 40 m thick halite sequence and proposed an external import of salts by leaching of salt bodied in the catchment area. The paleo-lake or base levels, mostly of the Lisan Lake, were also evidenced by aligned series of sub-horizontal caves along the same levels on the western side of the DSR (Fig. 6) (Kafri and Yechieli 2010b). The elevation of the Hula Basin segment of the DSR during all that time was assumingly similar to its present one, namely above the Pliocene eustatic level, and thus was not invaded by the Pliocene ingression. This is evidenced by the fact that neither marine or lacustrine saline sediments, nor trapped seawater or brines, are found within the thick Neogene to Quaternary basin fill (Kafri et al. 1983; Kafri and Yechieli 2010a). The Sea of Galilee segment of the DSR did not constitute a base level until at least 2 my BP (Kafri and Heimann 1994). The lower Jordan Valley and the Dead Sea segment, as far south as to the central Arava Valley segment, were subjected to the above described base level changes. The central Arava Valley segment assumingly, did not constitute a base level. Farther south, the southern Arava Valley was actually part of the Gulf of Elat (Aqaba) marine base level since the Pliocene (Kafri et al. 2008).

## 5 Paleo-Salinity, Salinization, and Interface Configuration

Similar to the present-day situation it is logical to assume that also in the past, the salinization of the paleo-groundwater systems was connected to the salinity of the adjoining base levels in the west and the east. The mechanisms involved were mainly: seawater or brines encroachment into them, flushing of the aquifers and climate changes which effected rainfall, evapotranspiration and in turn the natural recharge.

Since most of the salinization of the aquifers has to do with lateral migration of saline water from the adjoining base levels, the elevation and salinity of the latter serve as the main factor of the process. In fact, the sea level in the west, which is responsible also to intruded salinity to the DSR base level, forms an upper threshold elevation, at present as in the past, to the occurrence of saline groundwater bodies all over the country. Only a few exceptions that are related to surficial or very shallow sabkhas, such as the Yotvata one, are found somewhat above sea level.

## 5.1 The Eastern Mediterranean Base Level

During the Neogene, when the Mediterranean Sea was still connected to the Atlantic Ocean, it hosted normal seawater. It is assumed that inland seawater encroachment to the paleo-coastal hydrogeological systems was controlled by the Ghyben–Herzberg approximation. Thus, due to the assumed rather steep freshwater/seawater interface, the toe of the encroachment was supposed to be only a few km to the east of the paleo-shore lines.

Following the drastic decline of the Mediterranean sea level during the MSC it turned into a hypersaline endorheic huge base level. The inland encroachment of those brines is evident in the deep subsurface of the western coastal plain of Israel. Those brines were studied and defined as C-type brines, with salinities around 100 g/l (Starinsky 1974; Stanislavsky and Gvirtzman 1999). The reopening of the Mediterranean Sea to the Atlantic Ocean and the resultant rise of its level in the Pliocene, resulted in the ingression of normal seawater through the northern valleys into the already existing DSR base level, as described before.

Later on, during the Pleistocene and until to today the Mediterranean Sea level was subjected to considerable changes controlled by global climate changes. During the interglacial periods, sea level rose to levels that prevail at present, whereas, as mentioned before, during the glacial periods it dropped considerably, e.g., in the LGM. However, during that entire period, the Mediterranean seawater salinity assumingly remained the same as at present, namely somewhat higher than global normal seawater salinity. As a result, the paleo-shore lines were during the interglacial periods close to their current location, whereas during glacial periods they retreated considerably westward followed by flushing of the system. As mentioned before a consequent rise of sea level resulted in the opposite process of salinization through seawater encroachment (Yechieli et al. 2009a; Kafri and Yechieli 2010a) (Fig. 8, Chapter "Dating of Groundwater in Israeli Aquifers and Determination of Groundwater Flow Velocities"). Also, it was concluded that a few tens of thousands years are sufficient to a complete flushing of the system during sea level decline, as described from elsewhere in the world (Kafri and Yechieli 2010a).

Regarding the process of seawater encroachment to the coastal aquifer, the possibility of the blockage of its lower sub-aquifers to the sea was studied and delineated (Kafri and Goldman 2006). Based on the above, Amir et al. (2013) simulated the configuration of seawater encroachment since the LGM for the scenario whereby the lower sub-aquifers are blocked to the sea (Fig. 3). They have shown that during the LGM some 15 ky ago, at a level of 100 m bsl, the aquifers were completely flushed from seawater. The consecutive rise of the sea to its present level enabled seawater encroachment to the upper sub-aquifers. The blocked lower ones, however, remained

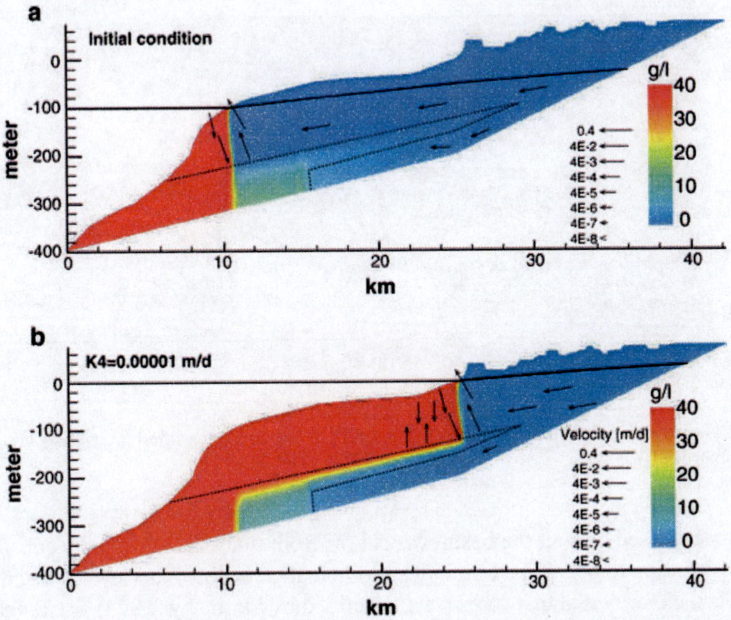


**Fig. 7** Sea level rise simulation result of the "closed" scenario with impermeable separating layer. **a** Initial condition after 125 ky. **b** transient sea level rise after 15 ky (after Amir et al. 2013)

fresh, provided that the lower and the upper sub-aquifers are offshore hydraulically separated (Fig. 7) (Amir et al. 2013).

## *5.2 The Eastern DSR Base Level*

The salinity of the DSR was initially controlled by the Pliocene marine invasion into it through the northern morpho-tectonic valleys to form an elongated sea arm and the later on Sedom Lagoon along the DSR as mentioned above. This water body with normal sea salinity was bounded to the north by the Korazim High and to the south by the central Arava Valley structural and topographic divide. Thus, both sub-basins, namely the Hula Valley in the north and the central Arava divide escaped salinization as evident to a depth of a few km. However, along the rest of the DSR, a normal marine environment assumingly a resultant lateral seawater encroachment to the aquifers prevailed, has known in marine coastal aquifers. The southern part of the DSR, namely the southern Arava Valley, was assumingly encroached from the south by Pliocene seawater up to the central Arava divide (Fig. 75, Chapter "Geoelectric, Geoelectromagnetic and Combined Geophysical Methods in Groundwater Exploration in Israel") as suggested by Kafri et al. (2008).

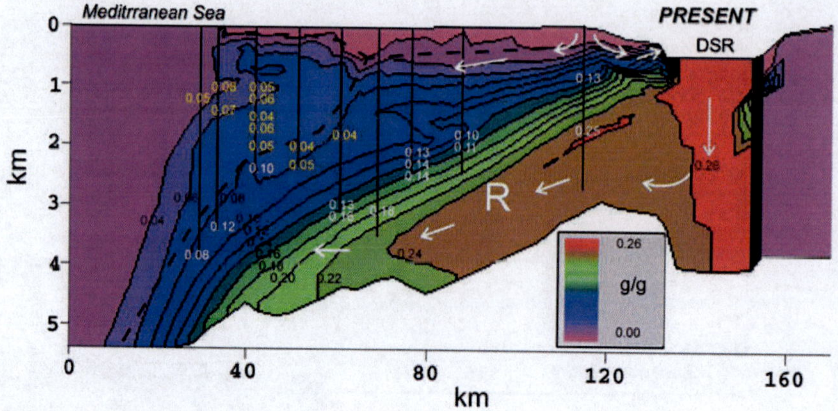


**Fig. 8** Numerical simulations of the present salinity distribution (modified after Stanislavsky and Gvirtzman 1999)

As mentioned before, the basin was cut off from the sea, followed by subsidence and desiccation to become a terminal endorheic hypersaline base level. The concentrated brines formed at that stage were described by Starinsky (1974) as Ca-chloride brines which deposited halite and migrated laterally to salinize the adjoining aquifers. Stanislavsky and Gvirtzman (1999) tested the feasibility of the proposed brine migration through a density driven mechanism and have shown the following (Fig. 8): The R-type brines with a concentration of ten folds of normal seawater, started to be formed some 3 my ago in the DSR and consequently descended to depth and started to migrate westward along the deep Mesozoic aquifers, and underneath the western C-type brines, currently up to a 100 km west of the DSR.

Later on, the base level started to decline from sea level and assumingly the uppermost part of the hydrological system adjacent to the DSR in the west was assumingly subjected to some freshening by flushing of the seawater and the later formed brines back to the DSR base level. The same process of flushing both to the declining Dead Sea basin and to the Gulf of Elat (Aqaba) as a result of the Red Sea level drop was suggested in the Arava Valley by Kafri et al. (2008).

Following the further decline of the base level during the Pleistocene, Lake Lisan was formed, extending also roughly between the Korazim high in the north and central Arava Valley in the south (Fig. 4), namely that it did not encroach those areas (Kafri et al. 1983, 2008; Kafri and Yechieli 2010a).

The salinity of Lake Lisan, the precursor of the present Dead Sea, was about half or less that of the present Dead Sea brine (Katz et al. 1977), exhibiting a north–south salinity gradient along the DSR (Begin 2002; Begin et al. 2004). It is assumed that it has an intermediate density and an intermediate slope of fresh/saline water interface between those of seawater and those of the more concentrated brines. Thus, it can be speculated that the westward encroachment then was farther than in the case of seawater and less than in the case of the more concentrated brine. It can also be

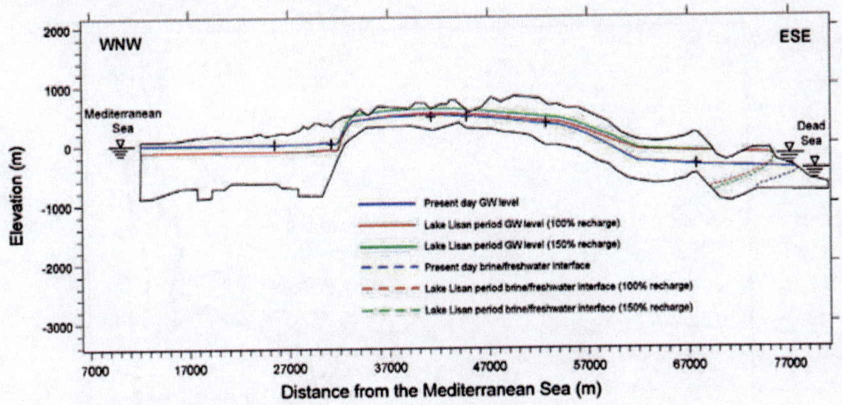


**Fig. 9** Simulation results of Lake Lisan period as compared with the present situation (used by permission of Elsevier, after Yechieli et al. 2009b, Fig. 9)

assumed that during the fluctuations of Lake Lisan levels, declines were accompanied by flushing of the adjoining aquifers and vice versa encroached by brine during rise of levels, as also suggested for the Arava Valley (Kafri et al. 2008).

In this regard, Yechieli et al. (2009b) simulated the hydrogeological setup, for two recharge scenarios, during the LGM, when Lake Lisan level was at 180 m bsl and compared it to present-day configuration (Fig. 9). As expected, the freshwater/brine interface was found to be higher and 2–3 km west of the present-day interface.

## 6 Paleo-Groundwater Divide Locations, and Flow Gradients

The location and configuration of the present-day and paleo-regional groundwater divides between the western and eastern base levels, as well as the related flow gradients was discussed by Kafri and Arad (1978). As shown in Fig. 10, the present-day divide coincides roughly with the mountain crest structural high of the Galilee, Samaria and Judea mountains. Exceptions are the Yizre'el and Be'er Sheva morpho-tectonic depressions which cut across them. There the divide was considerably and asymmetrically shifted toward the Mediterranean base level in the west. The mechanism, responsible to that shift is schematically described as follows (Fig. 11):

Following the formation of the DSR base level and the invasion of Pliocene Sea through the Yizre'el Valley into it, the two, namely the western and eastern base levels had the same eustatic level. The divide at that time was assumingly located symmetrically in between (Fig. 11I). Subsequently, a further decline of the eastern base level, the eastward groundwater flow gradient steepened. Through a re-adjustment process that tends to moderate the gradient to a milder one, the divide not only was shifted toward the Mediterranean base level but was also lowered to

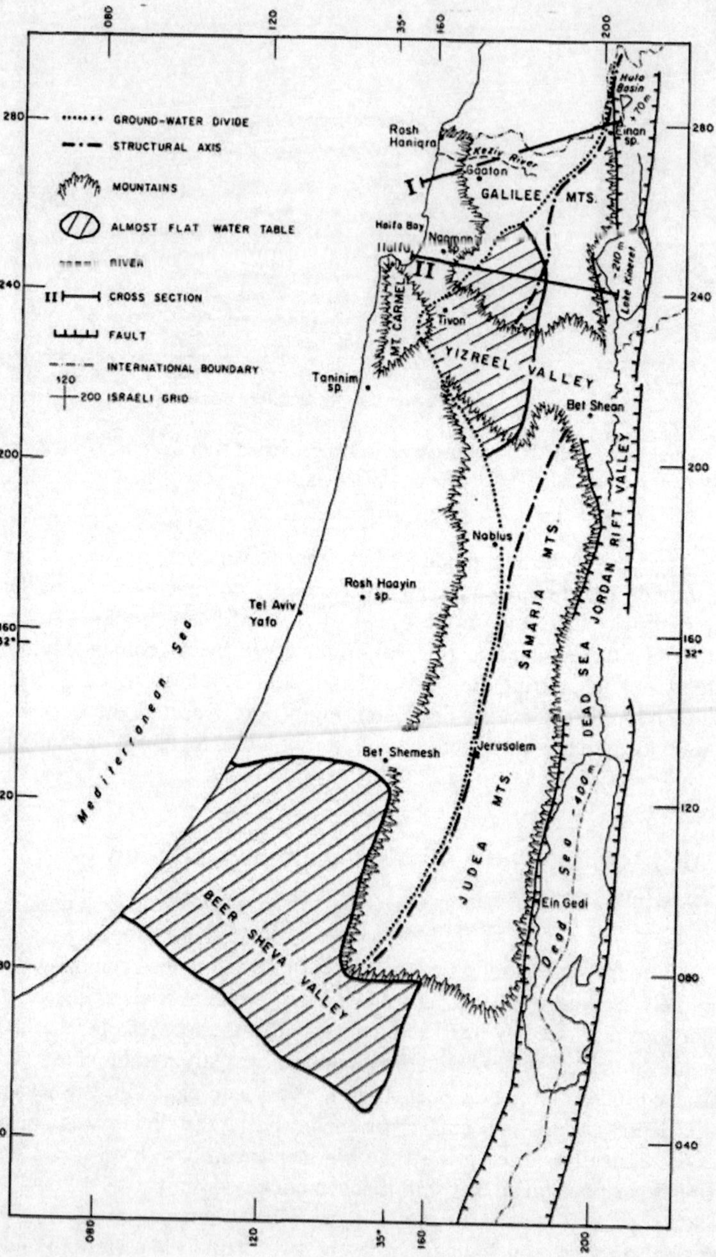


**Fig. 10** Map of central and northern Israel showing the main morpho-tectonic features and groundwater divide (modified after Kafri and Arad 1978)

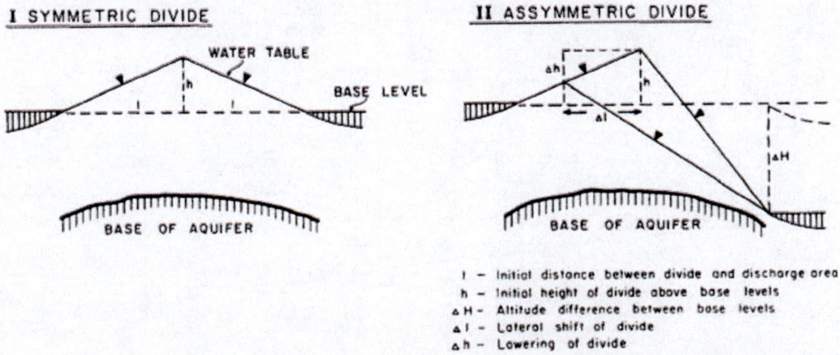


**Fig. 11** Schematic model showing the divide shift due to the subsidence of one base level (after Kafri and Yechieli 2010a)

gain the present-day configuration (Fig. 11II). The ongoing process of the decline of the DSR base level to that of Lake Lisan and further to the still declining Dead Sea level is responsible for the still steep eastward flow gradients.

As mentioned above the location of the divide along the mountain crest is basically controlled by the structure which perches the divide to a level above the base of the hydrological system. Thus, the assumed divide shift there is minimal. This was also deduced through a simulation of the future expected considerable decline of the Dead Sea level. The amount of the westward shift was found to be only some 600 m (Fig. 10 in Chapter "The Eastern Dead Sea Rift Continental Groundwater Base Level") (Yechieli et al. 2009b).

Paleo-water tables and their adjoining flow gradients to the western Mediterranean base level during the late Neogene were described by Kafri and Arad (1978) and Kafri and Yechieli (2010a). Those are evidenced by paleo-shore lines, by fossil hanging karstic caves, which are at present above the regional water table. In some cases, the fossil system serves as an outlet to perched springs that issue at present above the regional water table. The reconstructed paleo-water tables exhibit mild and mature flow gradients of less than 1 per mil whereby the present-day ones, influenced by a subsequent decline of the base level and/or uplift of the watershed, exhibit steeper gradients around 2–3%. Two examples to the above-described configurations were shown from western Galilee and from the Judea Mountains (Fig. 1) by Kafri and Arad (1978).

Similarly, Kafri and Yechieli (2010b) described sub-horizontal hanging fossil karst systems aligned along the western scarp of the DSR indicative of the levels of the eastern paleo-base levels and mostly that of Lake Lisan. The Fariya Valley, approximately perpendicular to the DSR exposes to aligned and inclined to the east fossil cave levels representing paleo-water tables reaching the DSR (Fig. 12).

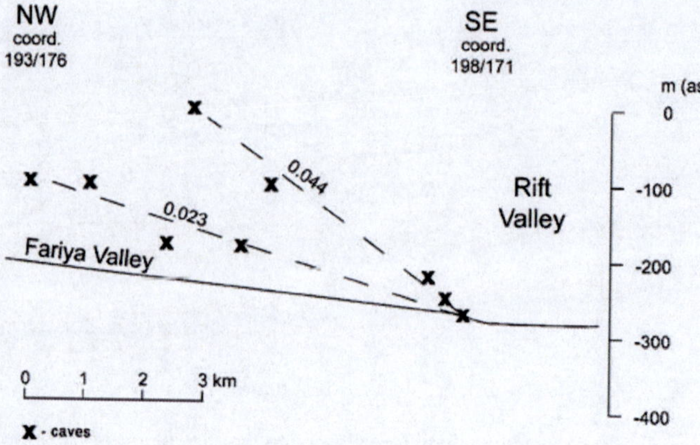


**Fig. 12** Caves alignment in the Fariya Valley, perpendicular to the base level (after Kafri and Yechieli 2010b)

## 7 The Hydrological Connection Between the Western and Eastern Base Levels

As mentioned before, the eastern DSR base level was formed in the Neogene. In the Pliocene, a surficial connection between it and the Mediterranean Sea base level took place whereby the latter intruded into the DSR forming an elongated marine lagoon along it. Consequently, the lagoon was cut off the sea forming a terminal endorheic declining base level hosting hypersaline lakes, which by lateral migration to the west affected the adjoining aquifers.

A subsurface hydraulic connection between both base levels since then was proposed by Kafri and Arad (1978). As schematically shown in Fig. 11II, the continuous decline of the DSR below sea level resulted in, where possible, a lowering and a shift of the regional groundwater divide toward the upper Mediterranean Sea base level. Through this process part of the preexisting western watershed was captured to the eastern DSR base level as shown in Fig. 11II. Due to the elevation difference between both base levels, a subsurface seawater intrusion was proposed from the Mediterranean Sea all the way to the DSR as an additional factor that contributed to the salinity of the deep aquifers in between as well as to the DSR base level (Kafri and Arad 1979; Kafri et al. 2007, 2014). The same phenomenon of subsurface seawater intrusion into continental endorheic base levels considerably below sea level was also described from other places all over the world (Kafri 1982; Kafri and Yechieli 2010a, 2011; Kafri et al. 2013).

The changes of salinity within the DSR base level were analyzed and discussed by Neugebauer et al. (2014) and Torfstein et al. (2015a), based on the 225,000 years sequence, penetrated by the deep drilling core in the middle of the Dead Sea. The different retrieved sedimentary facies was attributed basically to climate changes, lake level and salinity. Accordingly, a clastic facies assumingly represents high lake level and fresher water body whereas an evaporitic one represent low levels and a hypersaline water body. In regard to one segment of the sequence, namely 40 m of halite, Katz and Starinsky (2015) commented that at the interpreted low lake level the thickness of the brine water column was not sufficient enough to deposit that halite thick body. They proposed an external additional salinity source overcome this discrepancy by adding salinity from dissolved adjacent salt bodies, which was agreed by Torfstein et al. (2015b). Following this discussion, Kafri (2016) proposed an additional or alternative salinity source of subsurface seawater intrusion from the Mediterranean base level as described before. It was suggested that the salinity of the DSR base level at that time span correlates roughly not solely with the DSR levels, but also with the elevation difference between both base levels at a certain time span (Fig. 13). Namely, the greater is the elevation difference, the higher is the salinity flux from the west to the DSR, as evident by the sedimentary different facies.

## 8 Paleo-Natural Recharge

Subsequent to the retreat of the sea in the west and the rise of the continent after the Neogene a considerable part of the country and its mountain aquifers were exposed to natural recharge to the aquifers. The latter, in turn, depended on the prevailing climate as well as on climate changes. An insight into the prevailing climates mostly during the Pleistocene was obtained through stable isotopes composition of old groundwater and related speleothems, as well as of sapropels, coupled with their age determination. Most of the data concerning the above was gained through the analyzed speleothems of the Soreq Cave in the Judea Mountains, which yields a record of the last 140 kyr (Bar-Matthews et al. 1997, 1999; Bar-Matthews and Ayalon 2003). Their findings were correlated also with synchronous Sapropel events in the eastern Mediterranean Sea (Bar-Matthews et al. 2000).

According to their findings, the 140 kyr isotopic record of speleothems from the Soreq cave and the marine sapropels show five major low $d^{18}O$ events which corresponds to periods of enhanced rainfall. Out of these, the major enhanced rainfall periods were around 120 and 8 kyr. The time periods of 107–100 and 85–79 kyr, however, corresponds to a climate similar to the present one but with some higher rainfall. At the time of the LGM, some 20 kyr the average temperature was assumed to be 10 degrees cooler (McGarry et al. 2004). At that time, C-4 type temperate grass vegetation prevailed and the rainfall was claimed to be around 250–300 mm/y (Bar-Matthews and Ayalon 2003).

Enzel et al. (2003) showed a relationship between the Dead Sea levels and rainfall and claimed, somewhat different from the above findings that the highest Lisan Lake level is indicative of a higher average rainfall than the present-day one. Miebach et al. (2017), however, have shown, based on pollen assemblages, that the vegetation prevailing in the northern watershed of the Sea of Galilee Sea during its last glacial high stand, calls for dry conditions.

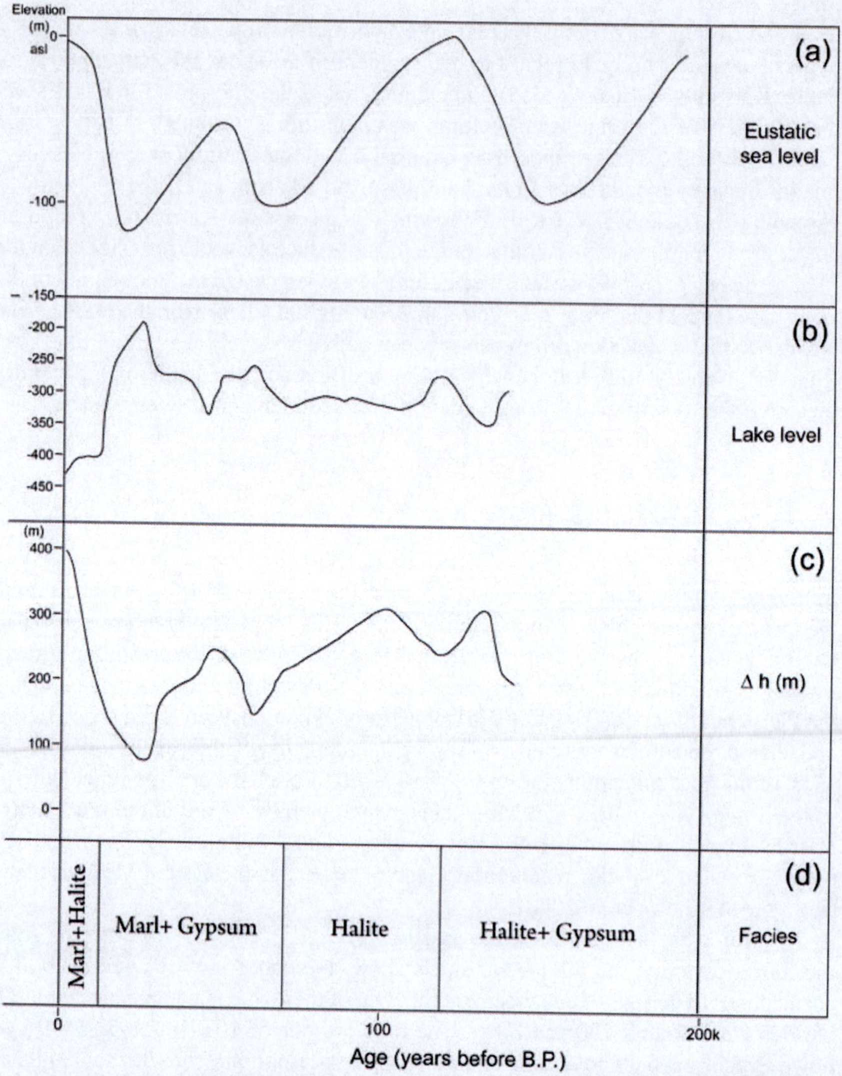


**Fig. 13** Simplified **a** Pleistocene-Holocene global and Mediterranean Sea levels. **b** DSR lakes levels. **c** Elevation difference between the Mediterranean Sea and the DSR lakes levels. **d** Depositional facies in the center of the Dead Sea (modified after Kafri 2016)

When dealing with natural recharge to the aquifer it should be mentioned that the relationship between rainfall and recharge is not necessarily linear. In a hot, humid and rainy (interglacial) climate one should expect the base level to rise. However, when most of the input to the base level is through groundwater flow, things might be different. In the case of the intake areas in Israel, most of the flow to the base levels is groundwater flow which feed them or alternatively emerge as springs that flow directly or through streams into the base levels. Namely, most of the waters discussed, underwent evapotranspiration and recharge. In this case, even cooler (glacial) climates, with less rainfall, C-4 temperate grass vegetation, and thus less evapotranspiration, can also generate a considerable natural recharge (Bar Matthews et al. 1997; Yechieli et al. 2009b).

## 9 The Paleo-Nubian Sandstone Groundwater System

The Nubian Sandstone, mostly sandy aquifer of upper Jurassic to lower Cretaceous age (see also Chapter "The Nubian Sandstone Aquifer in the Sinai Peninsula and the Negev Desert"), is spread regionally all over Nubia, Egypt, Saudi Arabia, Jordan, and northern Africa. A large part of its contained water is fossil and fresh to brackish. According to their stable isotopes and C-14 composition, they were recharged in more temperate climates than the present arid one, before 13 ky and more than 30 ky (Issar et al. 1972). Similarly, Burg et al. (2013) have shown that the groundwater in the deep NSS has low C-14 values, usually <5 pMC with an age of >26 ky. The young shallow groundwater, however, which is recharged via recent flash floods is characterized as expected by higher C-14 values, up to 70 pMC. Regarding the age of the NSS groundwater, Vengosh et al. (2007) have shown, based on their geochemistry and isotopic composition that at least in the northeastern Negev, they include in addition to the fossil water also a component from current local recharge, as was also suggested previously by Arad and Kafri (1980).

The areal distribution of the Nubian Sandstone aquifer intake areas in Sinai and the presumed related groundwater flow directions to the eastern and western base levels were described by Issar et al. (1972). As suggested by Arad and Kafri (1980) the Judea Group aquifer exposures in the southern part of the country (Negev) also serve as current intake areas to the system as is also supported by the Vengosh et al. (2007). The hydrogeological configuration is exhibited in Fig. 4 (Chapter "The Nubian Sandstone Aquifer in the Sinai Peninsula and the Negev Desert").

## References

Amir N, Kafri U, Herut B, Shalev E (2013) Numerical simulation of submarine groundwater flow in the coastal aquifer at the Palmahim area, the Mediterranean coast of Israel. Water Resour Manag. https://doi.org/10.1007/s11269-013-0392

Arad A, Kafri U (1980) Hydrogeological interrelationship between the Judea Group and the Nubian sandstone aquifers in Sinai and the Negev. Isr J Earth Sci 29:67–72

Audra P, Mocochain L, Camus H, Gilli E, Clauzon G, Bigot JY (2004) The effect of the Messinian deep stage on karst development around the Mediterranean Sea. Examples from southern France. Acta Geodin 17:27–38

Bar-Matthews M, Ayalon A (2003) Climatic conditions in the eastern Mediterranean during the last glacial (60–10 ky) and their relations to the upper Palaeolithic in the Levant as inferred from oxygen and carbon isotope systematics of cave deposits. In: Gorring-Morris N, Belfer-Cohen A (eds) More than meets the eye. Studies on upper paleolithic diversity in the Near East, pp 13–18

Bar-Matthews M, Ayalon A, Kaufman A (1997) Late Quaternary paleoclimate in the eastern Mediterranean region from stable isotope analysis of speleothems at Soreq cave, Israel. Quat Res 47:155–168

Bar-Matthews M, Ayalon A, Kaufman A, Wasserburg GJ (1999) The eastern Mediterranean paleoclimate as a reflection of regional events: Soreq cave, Israel. Earth Planet Sci Lett 156:85–95

Bar-Matthews M, Ayalon A, Kaufman A (2000) Timing and hydrological conditions of Sapropel events in the eastern Mediterranean, as evident from speleothems, Soreq cave, Israel. Chem Geol 169:145–156

Bartov Y, Agnon A, Enzel Y, Stein M (2006) Late Quaternary faulting and subsidence in the central Dead Sea basin. Isr J Earth Sci 55:17–31

Begin ZB (2002) Computer simulation of salinity in the Late Quaternary Lake Lisan. Isr J Earth Sci 51:225–232

Begin ZB, Stein M, Katz A, Machlus M, Rosenfeld A, Buchbinder B, Bartov Y (2004) Southward migration of rain tracks during the last glacial, revealed by salinity gradient in Lake Lisan (Dead Sea rift). Quat Sci Rev 23:1627–1636

Burg A, Zilberbrand M, Yechieli Y (2013) Radiocarbon variability in groundwater in an extremely arid zone—the Arava Valley, Israel. Radiocarbon 55:963–978

Calvo R, Bartov Y (2001) Hazeva Group, southern Israel; New observations, and their implications for its stratigraphy, paleogeography, and tectono-sedimentary regime. Isr J Earth Sci 50:71–99

Druckman Y, Buchbinder B, Martinotti GM, Siman Tov R, Ahron P (1995) The burial Afiq Canyon (eastern Mediterranean, Israel): a case study of a tertiary submarine canyon exposed in Late Messinian. Mar Geol 123:167–185

Enzel Y, Bookman (Ken Tor) R, Sharon D, Gvirtzman H, Dayan U, Ziv B, Stein M (2003) Late Holocene climates of the near east deduced from Dead Sea level variations and modern regional winter rainfall. Quat Res 60:263–273

Fairbanks RG (1989) A 17,000-year glacio-eustatic sea level record: influence of glacial melting dates on the Younger Dryas event and deep ocean circulation. Nature 342:637–642

Fleming K, Johnson P, Zwart D, Yokoyama Y, Lambeck K, Chappell J (1998) Refining the eustatic sea-level curve since the last glacial maximum using far- and intermediate-field sites. Earth Planet Sci Lett 163:327–342

Garfunkel Z (1981) Internal structure of the Dead Sea leaky transform (rift) in relation to plate kinematics. Tectonophysics 80:81–108

Garfunkel Z, Horowitz A (1966) The upper Tertiary and Quaternary morphology of the Negev, Israel. Isr J Earth Sci 15:101–117

Govers R (2009) Choking the Mediterranean to dehydration: the Messinian salinity crisis. Geology 37:167–170

Gvirtzman G (1970) The Saqiye Group (late Eocene to early Pleistocene) in the coastal plain of the Hashphela regions, Israel. Ph.D. thesis, The Hebrew University of Jerusalem, 153 p (in Hebrew)

Gvirtzman H (2006) Groundwater hydrology and paleohydrology of the Dead Sea rift valley. In: Enzel Y, Agnon A, Stein M (eds) New frontiers in Dead Sea paleoenvironmental research. Geol Soc Am Spec Pap 401:95–111

Horowitz A (2001) The Jordan Rift Valley. Balkema, Lisse, The Netherlands, 730 p

Issar A (1968) Geology of the central coastal plain of Israel. Isr J Earth Sci 17:16–29

Issar A, Bein A, Michaeli A (1972) On the ancient water of the upper Nubian sandstone aquifer in central Sinai and southern Israel. J Hydrol 17:353–374

Kafri U (1982) Relationship between Dead Sea and groundwater levels in the western Dead Sea catchment area. Isr Geol Surv Curr Res 1981:90–94

Kafri U (2002) Neogene to early Quaternary drainage systems in the lower Galilee, Israel and their relationship to young tectonics. Isr J Earth Sci 51:79–102

Kafri U (2016) Subsurface seawater intrusion, an additional factor influencing the limnological history of the Dead Sea basin. Environ Manag Sustain Dev 5:244–254. https://doi.org/10.5296/emsd.v5i.10130

Kafri U, Arad A (1978) Paleohydrology and migration of the ground-water divide in regions of tectonic instability in Israel. Geol Soc Am Bull 89:1723–1732

Kafri U, Arad A (1979) Current subsurface intrusion of Mediterranean seawater—a possible source of groundwater salinity in the Rift Valley System, Israel. J Hydrol 44:267–287

Kafri U, Ecker A (1964) Neogene and Quaternary subsurface geology and hydrogeology of the Zevulun Plain. Isr Geol Surv Bull 37:1–13

Kafri U, Goldman M (2006) Are the lower subaquifers of the Mediterranean coastal aquifer of Israel blocked to seawater intrusion? Results of a TDEM (time domain electromagnetic) study. Isr J Earth Sci 55:55–68

Kafri U, Heimann A (1994) Reversal of paleodrainage system in the Sea of Galilee area as an indicator of the formation timing of the Dead Sea Rift valley base level in northern Israel. Palaeogeogr Palaeoclimatol Palaeoecol 109:101–109

Kafri U, Horowitz A (2003) Possible northward drainage of the Pliocene Hula Valley. Isr J Earth Sci 52:185

Kafri U, Yechieli Y (2010a) Groundwater base level changes and adjoining hydrological systems. Springer, Berlin, 171 p. https://doi.org/10.1007/978-3-642-13944-4

Kafri U, Yechieli Y (2010b) The role of hydrological base level in the formation of sub-horizontal caves horizons, example from the Dead Sea Basin, Israel. Environ Earth Sci 61:1145–1152. https://doi.org/10.1007/s12665-009-0435-4

Kafri U, Yechieli Y (2011) The relationship between current and paleo groundwater base levels. Quat Int 257:83–96. https://doi.org/10.1016/j.quaint.2011.08.028

Kafri U, Kaufman A, Magaritz M (1983) The rate of Pleistocene subsidence and sedimentation in the Hula Basin as compared with those of other time spans in other Israeli tectonic regions. Earth Planet Sci Lett 65:126–132

Kafri U, Goldman M, Lyakhovsky V, Scholl C, Helwig S, Tezkan B (2007) The configuration of the fresh-saline groundwater interface within the regional Judea Group carbonate aquifer in northern Israel between the Mediterranean and the Dead Sea base levels as delineated by deep geoelectromagnetic soundings. J Hydrol 344:123–134

Kafri U, Goldman M, Levi E (2008) The relationship between saline groundwater within the Arava Rift Valley in Israel and the present and ancient base levels as detected by deep geoelectromagnetic soundings. Environ Geol 54:1435–1445

Kafri U, Shalev E, Lyakhovsky V, Wollman S, Yechieli Y (2013) Numerical modeling of seawater intrusion into endorheic hydrological systems. Hydrogeol J. https://doi.org/10.1007/s10040-013-0972

Kafri U, Goldman M, Levi E, Wollman S (2014) Detection of saline groundwater bodies between the Dead Sea and the Mediterranean Sea, Israel, using the TDEM method and hydrochemical parameters. Environ Process 1:21–41. https://doi.org/10.1007/s40710-014-0001-2.-5

Katz A, Starinsky A (2015) No draw down and no hyperaridity in the ancient Dead Sea (Comments to Torfstein's et al. (2015) paper, EPSL 412:235–244). Earth Planet Sci Lett 427:303–305

Katz A, Kolodny Y, Nissenbaum A (1977) The geochemical evolution of the Pleistocene Lake Lisan-Dead Sea system. Geochim et Cosmochim Acta 41:1609–1626

Kolodny Y (1965) Lithostratigraphy of the Mishash formation in the Nahal Ef'e region and the petrography of the siliceous rocks. M.Sc. thesis, The Hebrew University of Jerusalem, 75 p (in Hebrew)

McGarry S, Bar-Matthews M, Matthews A, Vaks A, Schilman B, Ayalon A (2004) Constrains on hydrological and paleo-temperature variations in the eastern Mediterranean region in the last 140 ka given by the D values of speleothem fluid inclusions. Quat Sci Rev 23:919–934

Miebach A, Chunzhu C, Schwab MJ, Stein M, Litt T (2017) Vegetation and climate during the last glacial high stand (ca.28-22 ka BP) of the Sea of Galilee, northern Israel. Quat Sci Rev 156:47–56

Miller KG, Kominz MA, Browning JV, Wright JD, Mountain GS, Katz ME, Sugarman PJ, Cramer BS, Christe-Blick N, Pekar SF (2005) The Phanerozoic record of global sea-level change. Science 310:1293–1298

Mocochain L, Clauzon G, Bigot J-Y, Brunet P (2006) Geodynamic evolution of the peri-Mediterranean karst during the Messinian and the Pliocene: evidence from the Ardeche and Rhone Valley systems canyons, southern France. Sediment Geol 188–189:219–233

Neev D (1960) A pre-Neogene erosion channel in the southern coastal plain of Israel. Geol Surv Isr Bull 25:20

Neugebauer I, Brauer A, Schwab MJ, Waldmann ND, Enzel Y, Kitagawa H, Torfstein A, Frank U, Dulski P, Agnon A, Aritztegui D, Ben-Avraham Z, Goldstein SL, Stein M, Scientific Party DSDDP (2014) Lithology of the long sediment record recovered by the ICDP Dead Sea Deep Drilling Project (DSDDP). Quat Sci Rev 102:149–165

Nunn JA, Harris NB (2007) Subsurface seepage of seawater across a barrier: a source of water and salt to peripheral salt basins. Geol Soc Am Bull 119:1201–1217

Popov SV, Shcherba IG, Llyina LB, Nevesskaya LA, Paramonova NP, Khondkarian SO, Magyar I (2006) Late Miocene to Pliocene palaeogeography of the Paratethys and its relation to the Mediterranean. Palaeogeogr Palaeoclimatol Palaeoecol 238:91–106

Roegl F (1999) Mediterranean and Paratethys. Facts and hypotheses of an Oligocene to Miocene paleogeography (short overview). Geol Carpath 50:339–349

Stanislavsky E, Gvirtzman H (1999) Basin-scale migration of continental-rift brines: paleohydrological modeling of the Dead Sea basin. Geology 27:791–794

Starinsky A (1974) Relationship between Ca-chloride brines and sedimentary rocks in Israel. Ph.D. thesis, The Hebrew University of Jerusalem, 208 p

Stein M (2001) The sedimentary and geochemical record of Neogene-Quaternary water bodies in the Dead Sea basin-inferences for the regional paleoclimatic history. J Paleolimnol 26:271–282

Torfstein A, Haase-Schramm A, Waldmann M, Kolodny Y, Stein M (2009) U-series and oxygen isotope chronology of the mid-Pleistocene Lake Amora (Dead Sea basin). Geochim Cosmochim Acta 73:2603–2630

Torfstein A, Goldstein SL, Kushnir Y, Enzel Y, Haug G, Stein M (2015a) Dead Sea drawdown and monsoonal impacts in the Levant during the last interglacial. Earth Planet Sci Lett 412:235–244

Torfstein A, Goldstein SL, Kushnir Y, Enzel Y, Haug G, Stein M (2015b) Response to comments on: "Dead Sea drawdown and monsoonal impacts in the Levant during the last interglacial". Earth Planet Sci Lett 427:306–308

Vengosh A, Rosenthal E (1994) Saline groundwater in Israel: its bearing on the water crisis in the country. J Hydrol 156:389–430

Vengosh A, Hening S, Ganor J, Mayer B, Weyhenmeyer CE, Bullen TD, Paytan A (2007) New isotopic evidence for the origin of groundwater from the Nubian Sandstone Aquifer in the Negev, Israel. Appl Geochem 22:1052–1073

Waldmann N, Starinsky A, Stein M (2007) Primary carbonates and Ca-chloride brines as monitors of a paleo-hydrological regime in the Dead Sea basin. Quat Sci Rev 26:2219–2228

Waldmann N, Stein M, Ariztegui D, Starinsky A (2009) Stratigraphy, depositional environments and level reconstruction of the last interglacial Lake Samra in the Dead Sea basin. Quat Res 72:1–15

Weinberger R, Bar-Matthews M, Levi T, Begin ZB (2007) Late-Pleistocene rise of the Sedom diapir on the backdrop of water-level fluctuations of Lake Lisa, Dead Sea basin. Quat Int 175:53–61

Yechieli Y, Kafri U, Sivan O (2009a) The inter-relationship between coastal sub-aquifers and the Mediterranean Sea, deduced from radioactive isotopes analysis. Hydrogeol J 17:265–274

Yechieli Y, Kafri U, Wollman S, Shalev E, Lyakhovsky V (2009b) The effect of base level changes and geological structures on the location of the groundwater divide, as exhibited in the hydrological system between the Dead Sea and the Mediterranean Sea. J Hydrol 378:218–229

Zak I (1967) The geology of Mount Sdom. Ph.D. thesis, The Hebrew University of Jerusalem (in Hebrew)

Ziegler MA (2001) Late Permian to Holocene paleofacies evolution of the Arabian plate and its hydrocarbon occurrences. GeoArabica 6:445–504
Zilberman E, Calvo R (2013) Remnants of Miocene fluvial sediments in the Negev Desert, Israel, and the Jordanian Plateau: evidence for the extensive subsidence basin in the northwestern margins of the Arabian plate. J Afr Earth Sci 82:33–53

# Unsaturated-Zone Hydrology in Israel: A Subjective Review

**Daniel Kurtzman**

**Abstract** The unsaturated zone, overlaying Israeli aquifers, is usually tens of meters thick, and it has significant water and solute storage, as well as reactive capacity. Therefore, being the passage of water recharging the aquifers, the unsaturated zone has great impact on quantity and quality of aquifer water. This review gives a historical perspective of the unsaturated-zone hydrology in Israel, and elaborates some details of studies that focused on the impacts of the unsaturated zone on groundwater. These impacts include aquifer recharge rates under different soils and land uses (crops) and recharge temporal characteristics, aquifer salinization, aquifer contamination with nitrate, contamination with organic compounds and major-ion geochemistry.

## 1 Introduction

The term unsaturated-zone hydrology (hereafter UZH) is being used herein rather than the more popular vadose-zone hydrology (e.g., the Hydrology section in the European Geosciences Union is divided to surface, vadose-zone and groundwater hydrology). The term "vadose" stems from the Latin meaning for "shallow," and the present review includes mostly studies dealing with the deep unsaturated zone, below the root zone. Therefore, the term deep vadose zone, although used in several studies (e.g., Rimon et al. 2007; Turkeltaub et al. 2014, 2015a), is somewhat awkward.

Within the scope of Unsaturated-Zone Hydrology, only works that relate to a water resource, mostly, the aquifer underneath the unsaturated zone, that is being investigated, are included in this review. This excludes studies that focus on agricultural aspects, related to water in the unsaturated porous medium, which are a great deal of soil sciences. "In Israel" means that this review contains studies of the unsaturated zone under Israeli land (or overlaying Israeli aquifers). Theoretical studies, including numerical simulations that are not site specific, or observations-based studies of the unsaturated zone of other countries, published by Israeli scientists, are also out of

D. Kurtzman (✉)
Institute of Soil, Water and Environmental Sciences, Agricultural Research Organization, The Volcani Center, 7505101 Rishon LeZion, Israel
e-mail: daniel@volcani.agri.gov.il


U. Kafri and Y. Yechieli (eds.), *The Many Facets of Israel's Hydrogeology*, Springer Hydrogeology, https://doi.org/10.1007/978-3-030-51148-7_12

scope of this review. Another filter, used for the scientific significance and availability for the international readership, is that ~90% of the studies that are mentioned in the present review were published in peer-reviewed English-language journals. All review articles are critical and subjective, this one in particular has no pretension to be comprehensive, surely there are significant studies that are not mentioned in it, and I apologize for missing them. The main bias of this review is naturally toward studies I was involved in, secondly, research of colleagues that I am more familiar with their work, and third, the more recent research.

## 2 Recent History of the UZH in Israel

Historically, one can divide the UZH in Israel in the last 5 decades to three main periods:

Late 1960s–1986: Significant field studies, many of which aim at the "nitrate bomb" waiting in the deep unsaturated zone to contaminate the coastal aquifer. Unsaturated flow is addressed by both soil scientists and hydrologists. Soil chemists address the faith of contaminants in the deep unsaturated zone.

1987–2003: A relative slowdown in publication of field studies related to the deep unsaturated zone in contrast to significant development in calculation methods of flow and transport in this domain. Studies on unsaturated fractured chalk.

2004–Present: New methods of sampling and especially monitoring of the deep unsaturated zone are developed and applied; unsaturated flow modeling becomes more common in the Israeli groundwater hydrology community. Many studies involving field observations and modeling are published.

From the first period, let me mention the work of Mercado (1976), that summarized an intensive effort, made mainly by Tahal (Water Planning for Israel) to evaluate the nitrate problem in the coastal aquifer. This effort includes the work of Raveh et al. (1972)—the most extensive survey of the deep unsaturated zone above the coastal aquifer that I know which was very briefly summarized by Avnimelech and Raveh (1976). The Mercado (1976) paper is of pioneering quality; namely, it used diverse field data to calibrate a multi-parameter transport model; it addressed uncertainty with the Monte Carlo technique and produced a large set of scenario simulations. In addition, it showed quantitatively that changes need to be made to decrease nitrate fluxes from the unsaturated zone to groundwater (i.e., fertilizer use, source of irrigation water, wastewater discharge, etc.). Nevertheless, I choose to quote a sentence from this paper's abstract that is of more generality for unsaturated-zone hydrology above the Israeli coastal aquifer: "An aggravating factor is the time lag between measures taken to decrease the pollutant load on the surface and the response of the aquifer system."

A book titled "Pollutants in porous media, the unsaturated zone between soil surface and groundwater" edited by Yaron et al. (1984) includes chapters of important international scientists (e.g., Raats, Jury). However, prominent Israeli hydrologists, soil physicists and chemists of that time (e.g., B Yaron, G Dagan, E Bresler, Y

Avnimelech, U Mingelgrin, D Ronen, M Magaritz, A Mercado and others), wrote most of the book. Dagan concluded the introductory chapter of the book in: "… progress has been made in modelling pollutants under laboratory conditions, there is a need to further develop the tools necessary for predicting the motion of pollutants under field conditions." Development of such tools matured after 2004 giving a blossom of works in the UZH in Israel.

As said, the nitrate problem was the prime motive for the first period of relatively intensive research in the unsaturated zone in Israel between 1970 and 1986. However, the study of Ronen et al. (1983) might be one of the reasons for the decline of unsaturated-zone studies in the second period between 1987 and 2003. It was suggested in this study, which is based on groundwater data, that the rate of nitrate fluxes to groundwater decreased dramatically from the 1960s to 1980s. Their theory was that the source of most of the nitrate in groundwater was rich organic nitrogen soils and swamps that were drained and became intensively cultivated between the 1930s and the 1960s, hence mineralized and nitrified, and most of their nitrogen already drained to the aquifer by 1980. This "reliving call" suggested that earlier alarming works' (e.g., Mercado 1976) should be re-thought is a possible cause of the slowdown in the unsaturated-zone analysis of the second period. Nevertheless, significant advance in modeling flow and transport beyond the root zone came from expanding the interest of soil physicists to include environmental concerns, rather than only agricultural yields (Russo and Bouton 1992; Russo 1995). The reduction of the drinking water standard for nitrate from 90 to 70 mg/L in the beginning of the 2000s and the consequent shutdown of many wells were responsible to diversion back to intensive effort in the unsaturated zone in the last 15 years.

The development and application of the Vadose-zone monitoring system (VMS) by Ofer Dahan (Dahan et al. 2003; Rimon et al. 2007) and the increase in the availability of direct-push sampling machines in Israel at the 2000s (Kurtzman and Scanlon 2011) made monitoring and sampling of the deep unsaturated zone relatively simple in comparison to previous decades. Non-intrusive geo-electrical surveys [e.g., electrical resistivity tomography (ERT)] that enable hydrologic characterization (although debatable) of the deep unsaturated zone in 2D were applied in Israel (Haaken et al. 2016). Hydrologist research positions were manned at the Agricultural Research Organization, and hydrology positions at Ben Gurion University and the Geological Survey of Israel were filled with researchers with soil-science background. Unsaturated flow modeling capabilities spread to research groups with groundwater\water resources orientation (e.g., Arnon et al. 2008; Gvirtzman et al. 2008; Kurtzman et al. 2013; Dvory et al. 2016). These together with the global continuous shift of interest in groundwater hydrology from production issues to contamination problems, brought to the second bloom of UZH studies in Israel between 2004 and 2019.

# 3 Organization of the Review

This review is organized according to hydrological issues, the UZH studies deal with. In this categorization, some focus on general characterization of the unsaturated zone is lost. For example, storage and retention time of water and solutes or preferential paths versus slow flow and buffering characteristics of the unsaturated zone. However, these features are discussed in the context of the different themes: i.e., water storage in the recharge theme; nitrogen transport in the nitrate contamination theme; chloride storage and salinization dynamics in the salinization theme, etc.

The following UZH themes compose the review:

Section 4: Groundwater Recharge and Flow in the Unsaturated Zone (the largest section)
Section 5: UZH and Aquifer Salinization
Section 6: UZH and Nitrate Contamination of Aquifers
Section 7: UZH and Organic Contaminants
Section 8: Unsaturated-Zone Geochemistry.

# 4 Groundwater Recharge and Flow in the Unsaturated Zone

## *4.1 Steady-State Annual Recharge Estimates Using Deep Unsaturated-Zone Data*

The first approximation of monthly/annual recharge depth as a fraction of the monthly/annual precipitation, and under irrigated land a fraction of the monthly/annual depth of irrigation, still holds in most groundwater models that are set in Israel today. The conceptual model of the unsaturated zone that stands behind this approximation was called "the train-car model" or "Mercado's train-car model" (both in Hebrew); The only written description of it I found, with no reference, was in the Hebrew textbook on Water Resources of Israel by Gvirtzman (2002). Perhaps this is Israeli hydrology "oral law." The concept is that a fraction of the water surface input, whereby the fraction size depends on soil type, vegetation, land cover, etc., drops to the top of the deep unsaturated zone, just under the root zone and pushes a train-car of the same size from the bottom of the unsaturated zone into the aquifer. This first approximation fits well with the relatively fast response of aquifer head to precipitation and makes the UZH redundant. Hence, UZH was hardly practiced for water-budget issues.

Reading the works of Raveh et al. (1972), and Avnimelech and Raveh (1976) at the 2000s, it was surprising to me that they calculated the nitrate flux to groundwater under a specific fields as a product of the average concentration of nitrate in the deep unsaturated zone (2–10 m depth) and recharge rate used in groundwater models of large scales. This was done, instead of using a local recharge rate estimate from a chloride mass balance (hereafter CMB), using the deep unsaturated chloride concentration, which they obtained.

To the best of my knowledge, the first to suggest using chloride concentration to estimate recharge were Eriksson and Khunakasem (1969) who used chloride concentrations of the Israeli Coastal Aquifer wells' data to estimate recharge rates. This method uses a rough assumption that the water in a nearby well originates from the unsaturated zone above it. One way to avoid this assumption is to use pore-water concentration of chloride in the deep unsaturated zone below the root zone. In the 1970s, Australian researchers first used CMB of the deep unsaturated zone to estimate recharge under different land uses (Allison and Hughes 1978), and later on, it became a common practice in arid and semi-arid environments.

Gvirtzman et al. (1986) used the tritium concentration in sediment samples from the deep unsaturated zone (down to 27 m below surface) for estimating the recharge flux and the flow velocity under a Hamra soil (red Mediterranean sandy loam) crop field, in Glil Yam, 10 km north of Tel Aviv, overlying the Israeli coastal aquifer. They used water budgets to obtain the recharge coefficients (0.15–0.21) and the average annual recharge of 140–170 mm/yr. Tritium, which was still a good environmental tracer in the 1980s, enabled calculating pore-water velocities independently, which averaged to 0.7 m/yr in the clay loam sequence and 2.3 m/yr in the sandy unsaturated zone. In this study, the authors claimed that the anion exclusion is a significant phenomenon at the field scale causing the velocity of chloride in the clay loam to be double that of the water, and the CMB approximation was not used.

Using CMB with data from the deep unsaturated zone became more common in Israel only recently. These estimates are: (1) a fast coarse estimation of the variability of recharge rates within the same plot (soil, land use, agricultural practice), and between different land uses (i.e., different crops, agricultural practice, etc.,); (2) independent steady-state approximation, based on chloride data that is compared to the average of a more detailed transient flow model which is based on water content data (Kurtzman and Scanlon 2011; Kurtzman et al. 2013; Levy et al. 2017). Steady-state recharge rates calculated by CMB with data of the deep unsaturated zone under different soils and land uses overlying the coastal aquifer are presented in Table 1.

The CMB approximations of recharge show variable recharge rates within the same plot. Nevertheless, some distinct characteristics of recharge related to different crops were obtained. Recharge under citrus orchards is generally small, about 5–20% of the surface inputs. On the other hand, under annual winter crops (e.g., potato, strawberry) or deciduous orchards (persimmon) 20–40% of the total surface input is expected to be recharged. Rain-fed crop fields (usually winter serials) on light soils also exhibit relatively high recharge fractions (~35%). Natural, non-cultivated (rangeland) clays exhibit very low recharge through matrix flow (Table 1). In year 2000, 50% of the land overlaying the coastal aquifer was cultivated, and thus, these steady-state approximations of recharge are significant for a better idea of the regional recharge of this aquifer.

**Table 1** Annual average recharge calculated by CMB with data of the deep unsaturated zone

| Nearby settlement | Soil type texture | Land-use crop | # of boreholes | Average surface flux[a] (mm/yr) | Recharge range[b] (mm/yr) | Source |
|---|---|---|---|---|---|---|
| Revadim | Vertisol clay | Irrigated crop field (corn) | 6 | 970 | 90–190 | Kurtzman and Scanlon (2011) |
| Revadim | Vertisol clay | Rangeland | 2 | 480 | 1–2 (only matrix) | Kurtzman and Scanlon (2011) |
| Bnei Dror | Hamra sandy loam | Oranges | 3 | 1300 | 170–230 | Kurtzman et al. (2013) |
| Bnei Dror | Hamra loamy sand | Oranges | 3 | 1300 | 190–330 | Kurtzman et al. (2013) |
| Tel Yitzhak | Hamra sandy loam | Grapefruit | 3 | 1250 | 90–150 | Kurtzman et al. (2013) |
| Kfar Varburg | Vertisol clay | Red grapefruit | 3 | 1450 | 290–440 | Kurtzman et al. (2013) |
| Nir Galim | Sandy clay loam | Grapefruit | 1 (VMS) | 1300 | 135 | Turkeltaub et al. (2014) |
| Nir Galim | Loamy sand | Rain-fed (+slurry) Crop field | 1 (VMS) | 550 | 200 | Turkeltaub et al. (2014) |
| Nir Eliahu | Hamra loamy sand | Potato | 3 | 1050 | 200–460 | Levy et al. (2017) |
| Nordia | Hamra loamy sand | Persimmon | 3 | 1400 | 370–430 | Levy et al (2017) |

(continued)

**Table 1** (continued)

| Nearby settlement | Soil type texture | Land-use crop | # of boreholes | Average surface flux[a] (mm/yr) | Recharge range[b] (mm/yr) | Source |
|---|---|---|---|---|---|---|
| Kadima | Hamra loamy sand | Strawberry | 3 | 1000 | 360–400 | Levy et al. (2017) |

[a]average annual irrigation + precipitation
[b]range of recharge approximations—1 estimate from each borehole within the same plot
Chloride concentration in precipitation in all these coastal plain locations was taken as 15 mg/l. In irrigation water chloride vary according to the specific water source (wastewater treatment plant, well, national carrier, etc.)

## 4.2 *Unsaturated Flow Models, Transient Recharge Estimates and Annual Averages*

Although using CMB of deep unsaturated-zone data pore new light on recharge in recent studies, these still represent steady-state approximations. Most recent Israeli UZH studies that deal with recharge yield also transient recharge rates through unsaturated flow models, calibrated to single-time data (sampling) or transient data (monitoring). These models shed more light on the recharge phenomena than those of rain and irrigation recharge coefficients, mostly used in groundwater models. Kurtzman and Scanlon (2011) calibrated a flow model to a non-cultivated and an irrigated crop field on vertisols (cracking clays) in Revadim and used it to assess the daily recharge fluxes (Fig. 1). The most significant result about annual recharge that issues from Fig. 1 is that annual recharge is controlled both by recent and previous year precipitation. For example, 2001–2002 is a relatively rainy year with small recharge due to low precipitation in 2000–1. On the other hand, 2003–4 is of low precipitation but high recharge due to the previous rainy year of 2002–3 (Fig. 1). Therefore, the use of a constant recharge coefficient with no consideration of previous year precipitation causes errors in groundwater flow models and a delayed response of the system has to be taken into account. Kurtzman et al. (2013) analyzed, similarly, 24 yrs of recharge of two unsaturated zones underlying citrus orchards in the Sharon region

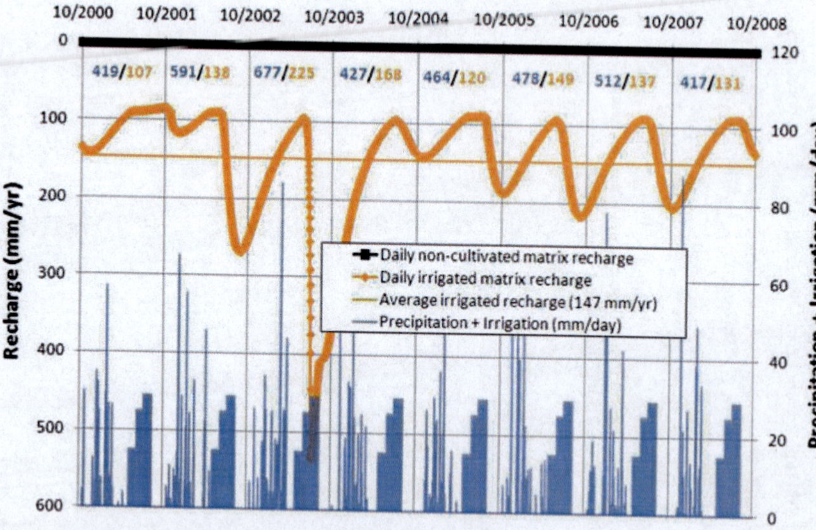


**Fig. 1** Time series of daily recharge fluxes from two calibrated matrix flow models under vertisols in Revadim: (1) under rangeland (black at the top average matrix recharge 2 mm/yr); and (2) under irrigated cropland (orange average 147 mm/yr). At the bottom daily precipitation and irrigation fluxes. Numbers at the top (in mm) are annual precipitation (October–September) (blue) and cumulative annual simulated recharge under irrigated cropland (orange) (modified after Kurtzman and Scanlon 2011)

showing that the calculated recharge under the thicker unsaturated zone (21 m) is not correlated with the same year precipitation but is correlated with the previous one. However, recharge under the thinner unsaturated zone (11 m) correlates well with the concurrent year precipitation and not with the previous year one (Table 4 in Kurtzman et al. 2013).

Turkeltaub et al. (2015a) investigated the dynamics of recharge under an unsaturated zone, 19 m thick, using a model that was calibrated to multiple transient VMS data under a bear sand dune in Ashdod. Monthly recharge correlated significantly and positively with monthly precipitation of the previous 4–10 months, whereas zero lag time between precipitation and recharge usually correlated negatively. After rainy winters, the recharge maximum fluxes are usually in the summer, whereas after less-rainy winters the recharge peaks are in the autumn (Fig. 2). Considering this lag time, Turkeltaub et al (2015a) suggested to those who still prefer to use a simple predictor of annual recharge from the same year precipitation, to correlate the May–April recharge with the annual precipitation of the previous winter (October–April). This recharge calculator is preferable than the simultaneous October–September time span commonly used (Fig. 10 in Turkeltaub et al. 2015a).

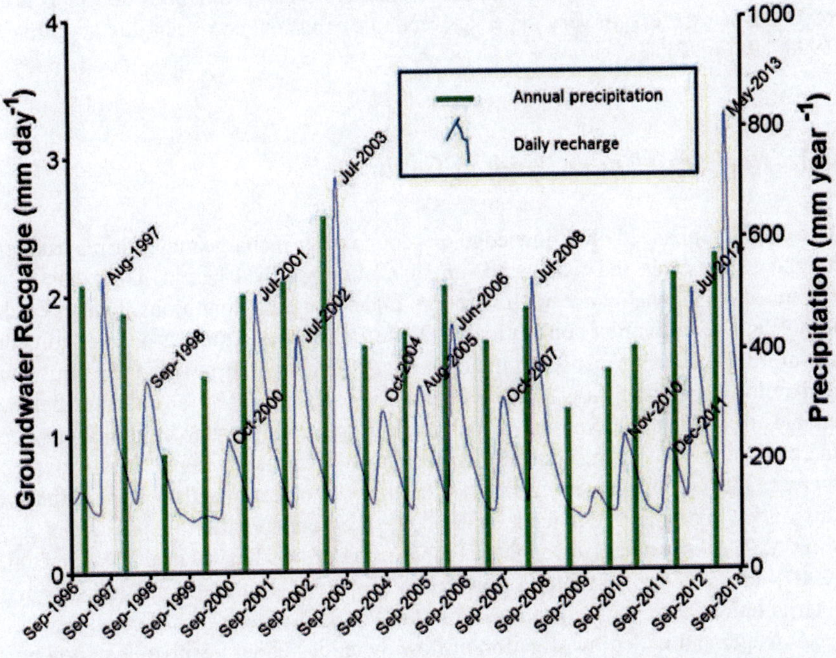


**Fig. 2** Annual precipitation (green bars) and daily simulated groundwater-recharge rates at 19 m depth under the sand dune in Ashdod (purple line). Months in which the recharge rates peaked are noted in a 3-letter abbreviation (modified after Turkeltaub et al. 2015a)

Dafny and Simunek (2016) used hydraulic functions calibrated to an infiltration experiment (Gvirtzman et al. 2008), to estimate recharge rates in bear soils, moderately and densely vegetated rangeland of loess deposits in Ein Habsor in the western Negev. They found that average recharge rates are 63 mm/yr under bear soil and 2–3 mm/yr under dense vegetation. At 22 m depth, recharge in this arid and semi-arid region has no correlation with the concurrent year precipitation whereby high recharge rates due to rainy periods increase recharge only after a few years.

The transient water storage of the deep unsaturated zone, which is high in rainy periods and low in droughts, and the layering with beds of different hydraulic properties (permeability and capillary barrier effects) are the reasons for the time lag between precipitation and recharge. This storage also causes that some recharge occurs under a thick unsaturated zone even if no precipitation occurs since under the root zone; the gravity force continues to affect the water, and it is not totally matched by an upward force until the residual water content is attained. Therefore, some recharge is in fact expected even with no precipitation or irrigation in one year, whereas by using recharge coefficients and the "train-car model" zero recharge is expected (note a minimum-modeled recharge always exists in Figs. 1 and 2). After many years of no surface water input, the entire unsaturated zone will attain a residual water content and recharge will seize; however, this does not happen in reality. Nevertheless, in very dry years recharge can show no rise in flux at all (e.g., 1998–99, Fig. 2).

## 4.3 Recharge Under Ponding Conditions

Israeli UZH enhanced our knowledge on groundwater recharge under drip/sprinkler irrigated fields rain-fed fields and rangeland as described herein. UZH has also advanced our acquaintance with recharge under ponding conditions. Dahan et al. (2007) in one of the first applications of VMS (at this stage only as a water content monitoring system) investigated the dynamics of recharge through coarse alluvium under flooding in the Arava valley. Monitoring water content at different depths enables to estimate the vertical velocity of the infiltrating water by the time since the beginning of flooding that the water content at a certain depth shows a sharp increase (arrival of the wetting front). The flux can be estimated by the product of this velocity and the difference in water content before and after the passage of this front ($\Delta\theta$). Nevertheless, through this monitoring of flooding by itself, Dahan et al. (2007) acknowledge that lateral flow components and preferential paths made the interpretation of recharge dynamics from the VMS data alone, insufficient.

A deeper and wider investigation of flow dynamics under positive-head ponding condition is described by Gvirtzman et al. (2008). This study investigated the flow dynamics in the vadose zone down to 20 m under an infiltration pond (trench) in loess deposits of the Western Negev. The experiment used vertical and slanted VMS to follow the advance of the waterfront in 2D (Z, X) under and on the sides of the pond. A 2D variably saturated flow model was fitted to the measured water contents.

Infiltration fluxes into the dry loess were 6–8 m/day in the beginning of infiltration and stabilized after 6 h on 2 m/day throughout 21 days. The water content data and the model showed a clear onion shape, propagated in the unsaturated zone. The 3 m wide trench formed underneath it an onion shape body of percolating water with 18 m maximum depth and 21 m maximum width after 21 days of ponding.

Different from the small-scale ponding, described above, the Israeli Managed Aquifer Recharge (MAR) facilities (e.g., Shafdan, Menashe Streams System) are based on wide infiltration ponds where the width of the infiltrating surface (the smaller lateral dimension) is significantly larger than the depth to the water table; hence, approximation of 1D flow in the unsaturated zone is more rational. Haaken et al. (2016) explored four cycles of flooding and drying in the 70 m wide Shafdan infiltration pond and depth to the water table around 39 m, with ERT lines. They showed that some parameters acquired from the interpretation of the electrical survey are correlated to some extent with hydraulic properties (e.g., drainage curve) and hence can reflect lateral and vertical heterogeneity of the vadose zone under the infiltration pond. The authors claimed that the ERT surveys may help in optimizing the infiltration-drying cycles, yet it seems that the work did not mature to such optimization. From this study, one can learn that infiltration rates in the ponding cycles can get as high as 2 m/day (annual average is 100 m/yr) and the vadose zone does not get fully saturated during the 12–24 h of ponding.

In the MAR system of Menashe streams, ponding in the infiltration ponds is continuous for days and weeks during the winter as compared to cycles of ponding and drying at the Shafdan. Since 2014, the southern infiltration pond (10.7 ha in area) serves for MAR of surplus desalinated seawater from the Hadera desalination plant ($0.5–3 \times 10^6$ $m^3$/yr). Extensive research on the hydrology and geochemistry of this MAR operation is being conducted since 2014. Ganot et al. (2017) measured, both from surface and the vadose zone, infiltration rates of 10 m/day in the beginning of infiltration to a "dry" unsaturated zone, intermediate infiltration rates of 0.7–3 m/day after 6 h to few days from the onset of ponding, and down to 0.4 m/day after weeks of ponding. A 1D flow model from ground surface down to 30 m depth was calibrated and validated to pond-scale infiltration rates (Fig. 3). The pre-ponding water table in Dec 2014 (before the infiltration began) was 24.4 m below surface. This model is unique since both top and bottom boundary conditions are transient-head boundaries that are constrained to pressure data from the pond surface and an observation well installed in the pond, respectively.

A more general conclusion was deduced from the findings of Ganot et al. (2017). The reduction of the infiltration rate throughout this operation is not due to surface clogging, as reported in many MAR studies, but rather it occurs as the wetting front reaches the clayey layers in the deep unsaturated zone. The final infiltration (and recharge) rate can be approximated quite well prior to any infiltration event if samples of sediments of the entire (pre-ponding) unsaturated zone are taken during drilling (cores are preferable). Based on the particle-size distribution and bulk density data of these samples, one can estimate the saturated hydraulic conductivity of the layers using the Pedo-transfer Functions, widely used in soil science. Alternatively, if good cores are obtained saturated hydraulic conductivity can be measured by Darcy

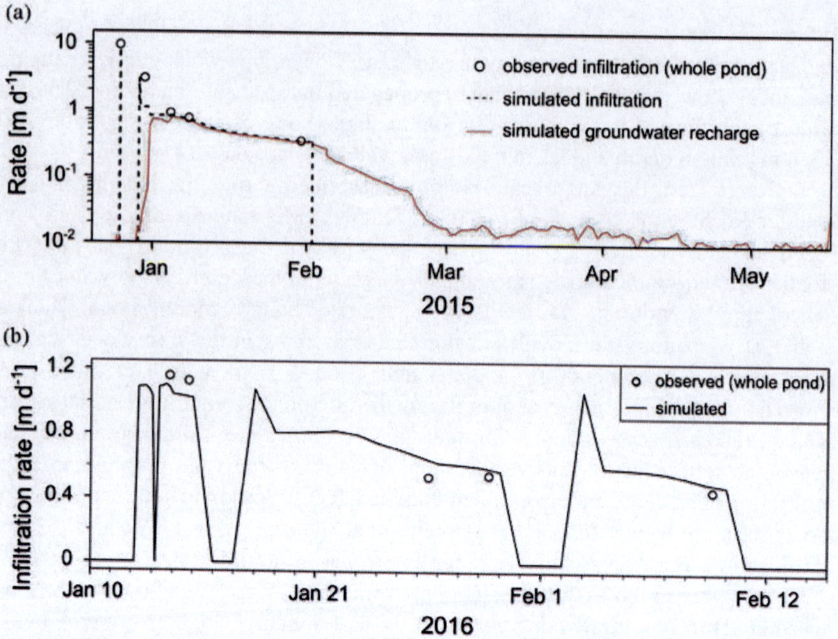


**Fig. 3** **a** Infiltration and recharge rates of the variably saturated model of the Menashe MAR operation with desalinated seawater in winter 2015. The ponding started with 4 h during Dec 25, 2014, stopped for 4 days and started again with almost continuous ponding from Dec 29 to February 2, 2015. Note the recharge rates rise (in red) in the first few days of ponding and after about a week they reach the infiltration rate, whereby in this stage the entire vadose zone is saturated according to the model. Note also that recharge rates of this MAR operation are 2–3 orders of magnitude larger than the natural recharge rates under sands as compared to the rates in Fig. 2. **b** The model that was calibrated to the 2015 large MAR event was validated with infiltration rates in a smaller ($1.3 \times 10^6$ $m^3$) and less continuous event of 2016 (modified after Ganot et al. 2017)

experiments. The effective saturated hydraulic conductivity of the entire unsaturated zone can be estimated with the harmonic mean of the layer's saturated hydraulic conductivity. As shown, after about a week of ponding infiltration, the assumption of recharge occurring under saturated conditions is reasonable (Fig. 3). Therefore, a unit gradient can also be assumed and the infiltration/recharge rate is equal to the unsaturated-zone effective saturated hydraulic conductivity (Ganot et al. 2017).

## 4.4 *Matrix Flow Versus Flow Along Preferential Paths in the Unsaturated Zone of the Coastal Aquifer*

Goldenberg et al. (1996) suggested (based on observations) to adopt the "short cut" perspective of the transport of contaminants in the unsaturated zone rather than the "buffer-protective" one that is being usually used. These popular expressions reflect the two end members of flow characteristics in sub-surface hydrology, namely preferential pathways versus uniform matrix flow. The coastal aquifer and its overlying unsaturated zone consist mostly of granular sediment and sandstone; therefore, naturally, a uniform flow through the porous media matrix was the common conceptual model for flow in the unsaturated zone.

Observations and monitoring of the deep unsaturated zone under vertisols (cracking clays) that cover about half of recharge areas of the coastal aquifer and 2/3 of the southern part of it (Kurtzman 2011) brought new insight as to the matrix vs preferential flow perspective in the unsaturated zone of the coastal aquifer. Very low water contents and extremely high chloride concentrations (>3000 mg/l) exist in the deep unsaturated zone, under non-cultivated and rain-fed fields (Kurtzman and Scanlon 2011 and Baram et al. 2014, respectively). This exhibit very small recharge calculated both through the CMB approximation (Table 1) and by an unsaturated flow model calibrated to water content data (Fig. 1).

Data from the pre-intensive cultivation period (1930s) show chloride concentrations of 200–400 mg/l in the aquifer under these vertisols whereby the unsaturated zone above them shows concentrations of an order of magnitude higher. Kurtzman and Scanlon (2011) show, through a mobile–immobile flow and transport model, how high salinity in the deep unsaturated zone can coexist with rather fresh water in the underlying aquifer. Under natural (non-cultivated) vertisols, the recharge consists of a small component (~5%) of saline matrix recharge and a large component of preferential flow recharge (95%) with low salinity. Using data from Goldenberg et al. (1996), Kurtzman and Scanlon (2011) show that salinity of the aquifer raised after the rainy winter of 1991–92 under loamy sand (Hamra) areas, whereas under vertisols in the southern part of the aquifer salinity decreased. This follows: dominant matrix flow under the Hamra soils (cultivated or not) that pushes down the saline water from the deep unsaturated zone to the aquifer, whereas under the vertisols, mostly under non-cultivated areas, freshwater recharge through preferential flow made the difference.

After some years of observations of transient water content of the deep unsaturated zone with VMS, Baram et al. (2012a) came out with the surprising result that the wetting front propagates faster in clays (vertisols) than under other soils including sand dunes (Fig. 4). The fast rise in water contents at larger depths is associated with non-Darcy flow in cracks. Physical measurements in cracks and crack networks show that they do not clog fully during the rainy winter and the preferential flow recharge can carry on later in winter and should not be considered as a phenomenon related to the first rains only (Baram et al. 2012a).

Deep preferential flow in soil cracks is one of the important characteristics of unsaturated zone under vertisols, which are relatively abundant, soils overlying alluvium valleys that commonly host an aquifer below them. Kurtzman et al. (2016) summarized the soil-aquifer phenomenon under vertisols in a review, based mainly on studies from Australia and the vertisols overlying the Israeli coastal aquifer.

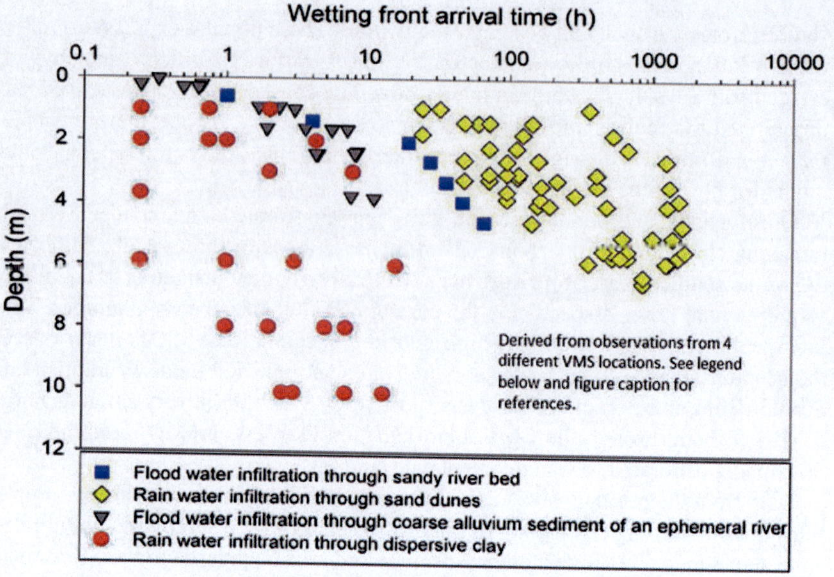


**Fig. 4** Wetting-fronts arrival time at different depths in different soils and infiltration conditions (Rimon et al. 2007; Dahan et al. 2007, 2008; Baram et al. 2012a). Note via the logarithmic horizontal scale, the orders of magnitudes faster travel time of the wetting fronts under vertisols (red points) (modified after Baram et al. 2012a)

## *4.5 Unsaturated Zone Flow in Rocky Vadose Zone*

In this subsection, I will mention three topics all dealing with flow in unsaturated carbonate rock and confront the reality of fast versus slow flows in these complex porous mediums. These include: (1) recharge in karst through cave-drip monitoring; (2) preferential recharge through fractures in the chalk of Ramat Hovav; (3) basin scale and more local estimates of recharge to the western mountain carbonate aquifer (Yarkon-Taninim).

Drips in the famous karstic Soreq Cave were analyzed for major ions and tritium by Even et al. (1986). They concluded that most drips originate in a large reservoir above the cave, that can hold water for decades, and recent rain events are not reflected in the drip dynamics, which is quite constant. Water age and chemistry vary a lot between cave seeps. Sheffer et al. (2011), investigated drips in a shallower cave at Samaria (near Barkan) with cave ceiling 3–10 m from ground level. They concluded that no recharge would occur before 100 mm of rain raised the water content of the overlaying soil. After that, drips discharge, as fraction of rain, will rise through the rainy season up to 70% of precipitation in March–April. Arbel et al. (2010) investigated cave drips in the Orenin cave in Mount. Carmel (Nahal Oren). They show that in this system, which is more seasonal than that of the Sorek cave, the correlation of drip intensity and rain events is clear, whereby most of the water of the drips is "old water" and only about 20% of the dripping water were "young" water that flowed fast along preferential paths through the unsaturated zone above the cave.

Water pollution problems, caused by disposal of liquid industrial wastes in the area of the chemical industry and waste-depository park at Ramat Hovav, resulted in research on flow and transport in the Eocene chalk aquitard of that area. The low permeability of the matrix of this chalk was considered when situating the site there, nevertheless, the upper 50 m of it are noticeably fractured. The fast recharge of this aquitard through the fractures was demonstrated by Nativ et al. (1995). Using tritium, stable water isotope, and bromide data from the unsaturated chalk sequence, they showed that significant flow is recharging the local groundwater body through preferential paths utilizing the fractures. Acknowledging the importance of recharge through the fractures in Ramat Hovav, Dahan et al. (2000) described a controlled field flow experiment on a vertical fracture plain. This experiment enabled to further understand the controls on flow along a single (1 m) scale fracture. The important conclusions were as follows: flow along the fracture plain is uneven and discharge along the fracture plain span a few order of magnitudes, including zero. Filling materials, within the fracture control the paste of the flow. Measurements of fracture aperture do not teach much as to the flow through it. Dissolution features along the fracture plain facilitate most of the flow, and these tend to develop at fracture intersections. The latter feature was found to control flow also in the saturated fracture network scale (10–100 m) in this area as concluded from pumping and tracer tests in the Ramat Hovav chalk by Kurtzman et al. (2005).

Sheffer et al. (2010) developed a daily recharge model of the entire Yarkon–Taninim outcrop area in order to feed a groundwater model. They used a tank with 3-drains model for the soil corresponding to three water content levels: above wilting point—the ET drain starts, above field capacity (FC) the recharge drain joins, and above saturation, the surface runoff will add to the other two drains. The spatial variability was reflected in three soil types (that vary in their aforementioned water contents) and two rock types (chalk/marl vs. limestone/dolomite) which were modeled as resistors to recharge (i.e., chalk will let less water from soil to become recharge than limestone). This type of model results in discrete recharge days (when soil water content is above FC) while in data-based Richards Equation recharge estimates recharge never goes down to 0 (Figs. 1 and 2).

Dvory et al. (2016) analyzed recharge in the upper-Soreq valley (Ein Karem well field) using a dual permeability variably saturated flow model (e.g., two coupled 1D Richards equations) of the ~100 m scale carbonate medium between ground surface and water table. Recharge in wells connected to conduit-dominated unsaturated zone was up to 50% of precipitation and showed correlation in time between recharge rates and precipitation, whereas wells in areas of matrix-dominated unsaturated zone showed steady recharge rates accumulating to ~20% of annual precipitation.

## 5 UZH and the Investigation of Aquifer Salinization

The high storage of chloride and nitrate in the deep unsaturated zone, below the root zone, was termed as a "ticking bomb" by Mercado and his colleagues in Tahal in different reports on the future of the coastal aquifer, written in the 1970s. Regarding salinity, which is measured in Israel's water resources mainly by the concentration of chloride, this is certainly a bomb that for some parts of the coastal aquifer already exploded as described herein.

Plant roots reject most of the chloride that comes with the irrigation water (all of the chloride approximated in many calculations). Water fluxes under the root zone (recharge rates) are significantly smaller than surface inputs (Table 1); hence, the concentration of chloride in the deep unsaturated zone is high and water moves slow. Heavy rains increases the recharge fluxes, which pushes to the water table some of this high salinity bulk and causes salinization of the aquifer. An example of the enhanced salinization under the central coastal aquifer after the extremely rainy winter of 1991/92 is described by Goldenberg et al. (1996).

Another type of salinization due to accumulation of chloride in the unsaturated zone is the salinization of the southeastern coastal aquifer, under wide alluvial valleys of the Judea mountains foothills, where vertisols (cracking clays) dominate the upper unsaturated zone (Fig. 5a, d). The deep unsaturated zone there under non-cultivated vertisols of the Judea-Lowland (southeastern coastal aquifer) is saline, with chloride concentration that exceed 3000 mg/l in pore water (Kurtzman and Scanlon 2011; Baram et al. 2013, 2014). Evapotranspiration through natural vegetation and accompanied evaporation through cracks (Baram et al. 2013), left very little water to percolate through the matrix of the deep unsaturated zone under the vertisols. Rain-water fluxes of 400–500 mm/yr containing an average 15 mg/l chloride on surface were remained with most of the chloride, carried by a water flux of 1–2 mm/yr, under the root and cracking zone (Table 1), causing evapo-concentration of 200 folds and more, with an end result of the aforementioned chloride concentrations of over 3000 mg/l. The small fluxes (and corresponding velocities) caused a buildup of a large storage of chloride in the unsaturated zone under the vertisols for centuries and millennia. The change of land use to intensive irrigated agriculture that started in the 1930s in the Beer Tuvia area and continued increasingly until the 2000s, caused a change in recharge through the matrix (~150 mm/yr under irrigated cropland, Table 1). These new high fluxes pushed the saline unsaturated storages, formed prior to cultivation, down to the aquifer, causing the well-known salinization of the coastal aquifer in these areas (Fig. 5) (Vengosh and Ben-Zvi 1994; Livshitz and Zentner 2009; Kurtzman 2011; Kurtzman et al. 2016).

Based on the above, the salinization of the coastal aquifer under these soils in the decades following the land-use change is understood and obvious (Kurtzman et al. 2016). The more difficult question is how did a relatively freshwater aquifer exist under a very saline unsaturated zone in the 1930s?

The answer to the question above is that recharge to the coastal aquifer under non-cultivated vertisols is mainly through preferential pathways along soil cracks at

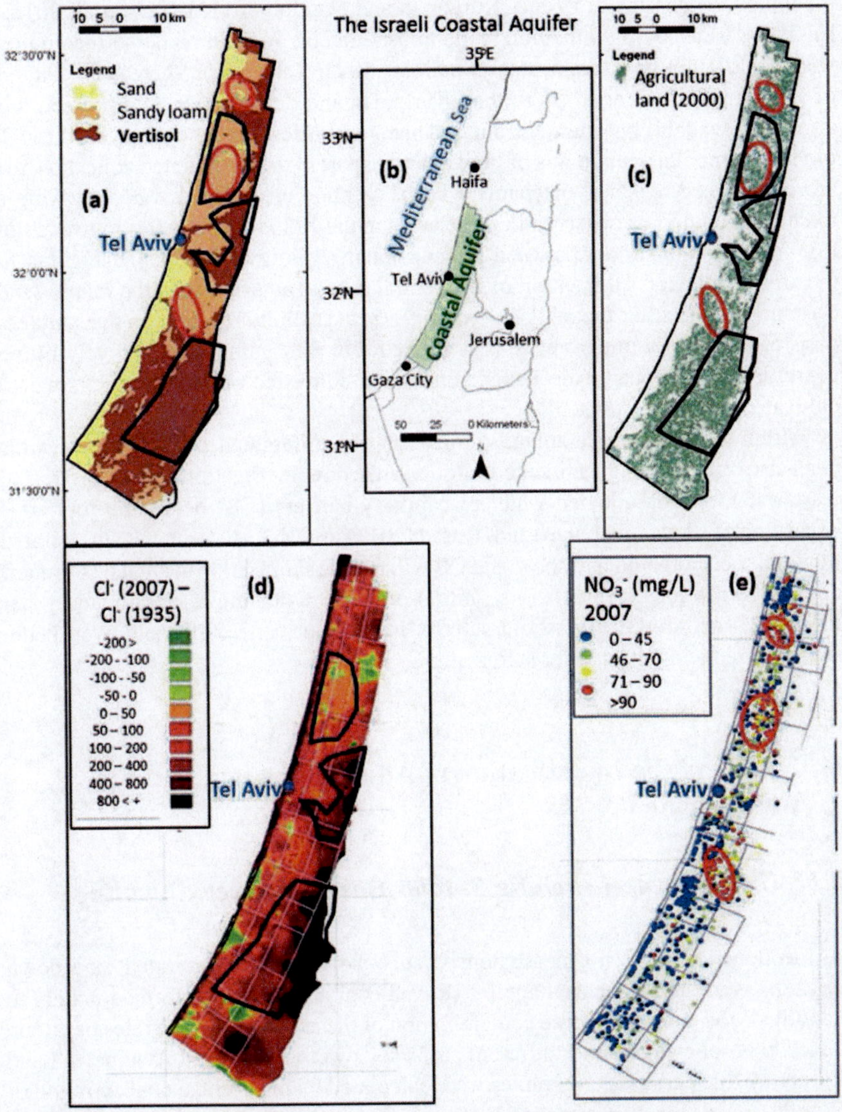


**Fig. 5** Plan views of the Israeli coastal aquifer: **a** soil type (black polygons and red ellipses also in panels d and e, respectively); **b** location map; **c** cultivated land in the year 2000; **d** difference in chloride concentrations between 2007 and 1935 (modified after Livshitz and Zentner 2009). Black polygons are characteristic cultivated areas that were salinized severely (southern polygons) and hardly salinized (northern polygon), compared with soil type in panel a; **e** nitrate concentration in groundwater wells in 2007 (modified from Hydrological Service 2008). Red ellipses—areas with many wells contaminated with nitrate, compared with soil type in panel a (modified from Kurtzman et al. 2016)

the top of the unsaturated zone (Kurtzman and Scanlon 2011; Baram et al. 2012a, b). These water by-pass the bulk saline unsaturated zone and recharge the aquifer with 20–120 mm/yr of freshwater (containing 10 mg/l chloride), whereas 1–3 mm/yr of saline matrix recharge ads to that. Kurtzman and Scanlon (2011) discussed this issue with available unsaturated zone and historic aquifer data. They showed, through mobile–immobile simulations of flow and transport in the unsaturated zone, how the aforementioned recharge mechanism keeps a saline unsaturated zone overlying a freshwater aquifer as observed in these areas in the 1930s whereby the aquifer chloride concentration mostly between 100 and 400 mg/l (Kurtzman 2011). In contrast to the aforementioned salinization of the central coastal aquifer under the loamy sand ("Hamra") soils, after the rainy year of 1991/2, the salinity response in the southern parts of the aquifer under vertisols was neutral or freshening—qualitatively fitting the difference in recharge mechanism under non-cultivated vertisols and Hamra both cultivated and non-cultivated.

Within the soil-science arena, salinization of aquifers under agricultural land is dealt also by examining engineered solutions like double irrigation systems of treated wastewater, and desalinated water or capillary barriers (CB) below the root zone consisting of a layer of course sand. Russo (2017) used a 3D heterogenic unsaturated flow and transport model to show that CB will reduce salinization in annual (summer) crops whereas for perennials (e.g., citrus orchard) switching to desalinated water irrigation, according to threshold salinity in the root zone, will yield even better results in reducing the aquifer salinization.

# 6 UZH and the Investigation of Aquifer Contamination with Nitrate

## *6.1 UZH Nitrate-Leaching Studies Based on Deep Sampling*

As mentioned in Sect. 2 the most extensive survey of the deep unsaturated zone, down to 12 m, overlying the coastal aquifer (Raveh et al. 1972) was performed mainly to investigate the potential nitrate contamination of the aquifer. Average rates of nitrate fluxes to groundwater under different crops/soils were reported in Avnimelech and Raveh (1976). These $NO_3$–N fluxes were calculated by the average concentration of $NO_3$–N from sediment samples of 2–10 m depth and recharge rates based on regional groundwater budget (Table 2). Avnimelech and Raveh (1976) found that these fluxes are considerably smaller than those estimated by N input and N removal by crop and explained the difference by high denitrification rates.

Mercado (1976) used a simplified model of the unsaturated zone (not based on a solution of Richards and ADE equations) where the nitrate flux to the water table is a linear fraction of the total nitrogen surface input, the nitrogen consumed by crops and denitrification. The time lag between the surface input and aquifer contamination, namely the delay in the unsaturated zone, was calculated according to average water

**Table 2** Annual $NO_3$–N fluxes to groundwater based on deep unsaturated-zone $NO_3$–N data

| Soil type texture | Land-use crop | # of boreholes | $NO_3$–N deep concentration (mg N/L) | $NO_3$–N flux to GW kg/(ha * yr) | Citation |
|---|---|---|---|---|---|
| Sandy loam | Rangeland | 4 | 11 ± 5 | 13 ± 6 | Avnimelech and Raveh (1976) |
| Clay loam | Rain-fed crop field | 2 | 23 ± 8 | 50 ± 36 | Avnimelech and Raveh (1976) |
| Clay loam | Irrigated crop field | 2 | 11 ± 4 | 37 ± 12 | Avnimelech and Raveh (1976) |
| Sandy loam | Citrus orchard | 10 | 35 ± 33 | 127 ± 124 | Avnimelech and Raveh (1976) |
| Clay loam | Citrus orchard | 4 | 13 ± 15 | 50 ± 48 | Avnimelech and Raveh (1976) |
| Sandy loam and loam | Field crop irrigated with TWW | 4 | 37 ± 13 | 117 ± 81 | Avnimelech and Raveh (1976) |
| Sandy loam | Irrigated crop field | 4[a] | | 64 ± 21 | Hadas et al. (1999) |
| Sandy loam | Oranges | 3[a] | | 77 ± 33 | Hadas et al. (1999) |
| Sandy loam | Oranges | 4[b] | | 68 ± 39 | Hadas et al. (1999) |
| Hamra sandy loam | Oranges | 3 | | 80–130 | Kurtzman et al. (2013) |
| Hamra loamy sand | Oranges | 3 | | 50–220 | Kurtzman et al. (2013) |
| Hamra sandy loam | Grapefruit | 3 | | 70–140 | Kurtzman et al. (2013) |
| Sandy loam | Tomato greenhouse | 250 simulations[c] | | 179 ± 10 | Turkeltaub et al. (2015) |
| Sandy loam | Lettuce greenhouse | 250 simulations[c] | | 243 ± 25 | Turkeltaub et al. (2015) |
| Hamra loamy sand | Potato | 3 | 73 ± 17 | 233 ± 49 | Levy et al. (2017) |
| Hamra loamy sand | Persimmon | 2 | 32 ± 9 | 125 ± 21 | Levy et al. (2017) |
| Hamra loamy sand | Strawberry | 3 | 59 ± 15 | 223 ± 61 | Levy et al. (2017) |

[a]Data from 1972
[b]Data from 1988
[c]Calibrated to VMS pore-water concentration

content (Raveh et al. 1972), the thickness of the unsaturated zone, the irrigation fluxes, the estimates of the fraction of irrigation that leaches and the recharge flux (estimated externally). These steady-state estimations that lack the effect of precipitation are probably partly balanced by neglecting the high reduction of hydraulic conductivity in dryer periods, gave transit time of 2–24 years for the nitrate contamination in the 20–25 m unsaturated zone in Rehovot–Rishon region in the 1970s.

Drilling to a depth of 12 m under agricultural land in order to estimate the nitrate flux to groundwater was also performed in a research originating in the agricultural science. The effect of fertilization and irrigation, in the scope of leached nitrate, rather than only yield, was taken into account in the 1990s (Hadas et al. 1999). However, as mentioned in the introduction to this chapter, nitrate contamination of groundwater was less an issue in the water resources community in Israel after the dis-alarm study of Ronen et al. (1983), until the beginning of the twenty-first century.

Two studies dealing with groundwater (nitrate) contamination under cultivated land that were lead by the Geological Survey of Israel, brought back the nitrate problem to the front (Oren et al. 2004; Kass et al. 2005). These studies focused on the geochemical and hydrogeological aspect of the problem. Both Oren et al. (2004) who investigated the Arava aquifer under the Ein Yahav intensive agriculture complex and Kass et al. (2005) who studied the coastal aquifer under cultivated land in Nitzanim, realized that high fertilizer and irrigation rates are causing the local groundwater contamination. Both of them used indirect methods such as ion ratios, stable isotopes, etc., whereas Hadas et al. (1999) arrived at similar conclusions using a direct description of water flow and nitrogen transport from land surface through the unsaturated zone. Direct observation of the deep unsaturated zone coupled with mechanistic modeling of flow and the complex nitrogen-species transport became the more common practice in nitrate contamination studies in the last decade.

Kurtzman et al. (2013) used laboratory methods (e.g., nitrification potential) and lysimeter experiments to help parameterizing the transport model of the nitrogen species in the unsaturated zone, which was ultimately calibrated to the deep unsaturated-zone nitrate data from the field. They showed the significance of precipitation in controlling the transient flux of nitrate to groundwater under citrus orchards in the Sharon area. The rainy years of the 1990s contributed a significant mass of nitrate to the aquifer (Fig. 6). This study suggested that a 50% reduction in farmers nitrogen application, with no change in nitrogen in the irrigation water, is required to reduce the nitrate concentration in the deep unsaturated zone below the Israeli drinking water standard (70 mg/l $NO_3$), which will make the practice sustainable (from water resources point of view). This reduction in fertilization was predicted to reduce the crop nitrogen uptake, a proxy for agricultural yield, by about 20%.

Levy et al. (2017) expanded the deep unsaturated-zone surveys in the Sharon area to cover the main three types of crops namely, citrus orchards, vegetables (e.g., potato, strawberry), and deciduous orchards (e.g., persimmon; Fig. 7). Calibrated flow and transport models to the deep unsaturated-zone concentrations of nitrate under the typical crops fed a flow and transport model in the aquifer. This combination of models completed a full mechanistic description of nitrogen transport from application in field to arrival of nitrate to the wells' perforation. This work revealed that

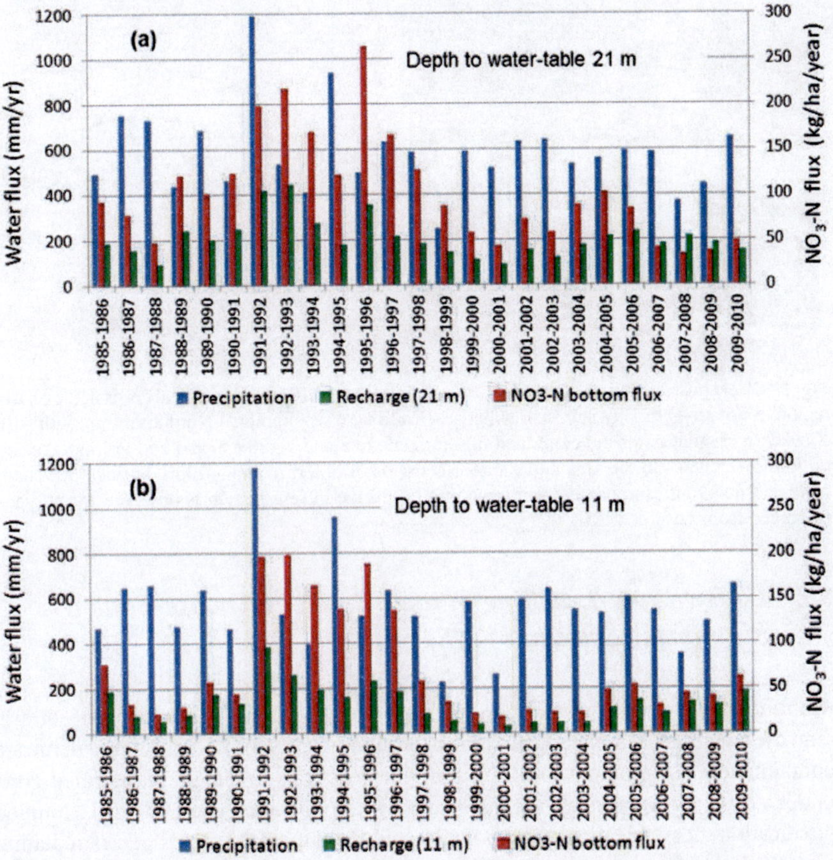


**Fig. 6** Simulated annual water and $NO_3$–N fluxes to groundwater (green and red bars, respectively) compared to precipitation measured in the nearest rain gage (blue bars): **a** orange grove at Bnei Dror, unsaturated-zone thickness of 21 m, **b** grapefruit grove at Tel Yitzhak, unsaturated-zone thickness of 11 m (modified after Kurtzman et al. 2013)

the nitrate concentration in the most contaminated wells ($NO_3$ > 100 mg/l) could not be explained by any rational agricultural practice and "normal" flow and transport. These high concentrations are a result of either extreme heterogeneity in the well area, or more likely, some mall-function(s) in the vicinity of the well (e.g., "short-cut" flow through the well annulus, leakage from fertilizer tank, compost piles left undistributed in the rainy season, etc.). Nevertheless, Levy et al. (2017) calibrated the $NO_3$–N flux to groundwater by multipliers in some areas of the model domain to fit nitrate concentrations in the wells. The calibrated model served for the simulation of 40 years ahead and revealed that a 50% reduction in fertilization will result in the reduction of the contamination in wells within 40 years (Fig. 7).

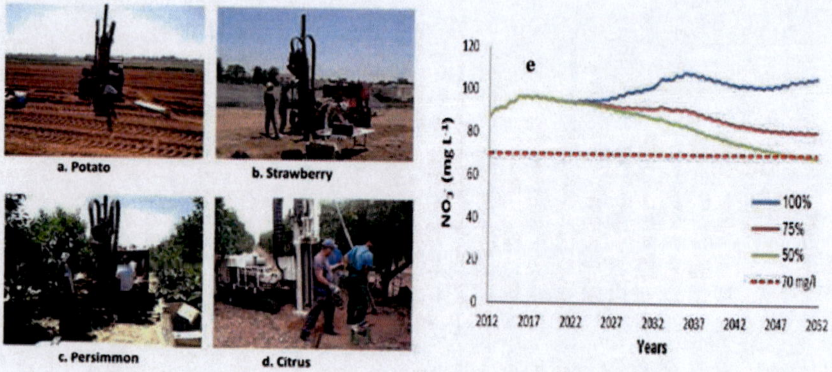


**Fig. 7** Direct push sampling of the unsaturated zone down to 10 m under typical crops in the Sharon region: **a** potato—Nir Eliyahu; **b** strawberry—Kadima; **c** persimmon—Nordia; **d** grapefruit—Tel Yitzhak; **e** simulation of the combined unsaturated zone and aquifer model and average concentrations in 15 wells in the area under three scenarios of N fertilization [100% business as usual]. Note that no significant average well response to change in fertilization is observed for 10 years (modified from Levy et al. 2017)

## 6.2 UZH Nitrate-Leaching Studies Based on Vadose-Zone Monitoring Systems (VMS)

Monitoring water content and continuous pore-water sampling under agricultural land use with the VMS method has advanced significantly the understanding of nitrate contamination of groundwater. The nitrogen reactions in clayey unsaturated zone under a dairy farm waste lagoon were investigated by Baram et al. (2012b). Coupled nitrification–denitrification was shown to reduce the mass of nitrogen percolating directly under the lagoon by more than 90%. This local data corresponds well with the observation of less nitrate contamination under vertisols on the basin scale as was generally discussed by Kurtzman et al. (2016) (Fig. 5a, e). On the other hand, incidental overflows caused deep percolation of nitrate on the margins of the lagoon and delivery channel, along the soil cracks (vertisols), bypassing the reactive matrix clay (Baram et al. 2012b). Baram et al. (2014) studying an area of ~10 $km^2$ in the Beer Tuvia region, in view of the unsaturated-zone observations, shows that the many dairy farms (point sources) in this agricultural area contribute some 25% of the groundwater nitrate.

Dahan et al. (2014), using VMS under conventional and organic greenhouses for vegetables in the Zikim area, showed how intensive organic farming causes groundwater contamination with nitrate, especially in the years of conversion of the farm to organic practice. The high loads of organic nitrogen, mostly composts, followed by establishment irrigations for reducing salinity cause mineralization, nitrification and high leaching of nitrate.

Turkeltaub et al. (2015b) monitored a conventional-practice greenhouse for vegetables with VMS for 5 years. During this time, the owners changed as well as crop from tomato to lettuce. The short season of lettuce (6–8 seasons a year) is accompanied by high annual loads of nitrogen, in comparison to nitrogen uptake by lettuce plants, and high irrigation rates, including many establishment irrigations. A unique observation of the VMS here is the extremely high concentration of nitrate under the greenhouse after the switch to lettuce down to 15 m below surface (Fig. 8).

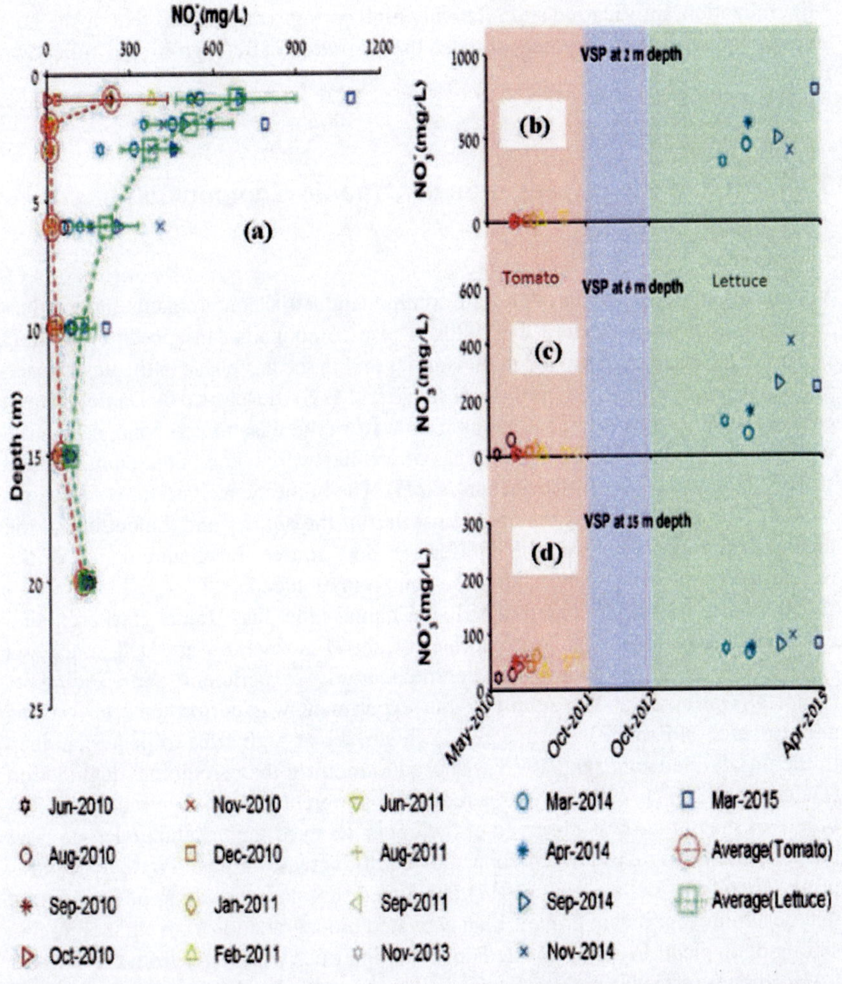


**Fig. 8** Nitrate concentrations in pore water from the unsaturated zone beneath a greenhouse. **a** Nitrate profiles along the unsaturated zone under the greenhouse as obtained by continuous sampling of the pore water with the VMS. The continuous lines are the averaged profiles. The red color represents the tomato cultivation period, and green represents the lettuce cultivation period. **b–d** Change of nitrate concentration with time from three vadose-zone sampling ports (VSP) at different depths (pink—tomato, purple—no crop, green—lettuce) (modified after Kurtzman et al. 2013)

Further model simulations show the effects of different agronomical practices on nitrate leaching.

Turkeltaub et al. (2016) reported another unique observation, facilitated by VMS. Six years of monitoring of an 18.5-m-thick unsaturated-zone sequence under a crop field in Nir Galim (east of Ashdod) revealed the advance of a pulse of high nitrate through the unsaturated zone for 5 years. The pulse was initiated by a single winter legume-cropping season, which left excess nitrogen in the root zone by fixation. Mineralization, nitrification and relatively high spring seasons rains that pushed the nitrogen to underneath the roots initiated the migration of the high-nitrate bulk down towards the water table.

## 7 UZH and the Investigation of Organic-Compounds Contamination

Volatile organic compounds (VOC) are common industrial contaminants under industrial areas that operated until the 1990s. These compounds can appear as separate phases in the aquifer, dissolved in the aquifer and in the gas phase in the unsaturated zone. The passive diffusion multi-level sampler (MLS) (developed by Daniel Ronen) enabled observations of a continuous profile from the unsaturated zone, across the water table into the aquifer. These types of profiles were taken in the contaminated site of Magen in Tel-Aviv (Ronen et al. 2005). The highest concentrations of various VOCs (TCE, PCE, etc.,) at the uppermost part of the aquifer and the bottom of the unsaturated zone reveal that the aquifer can be a source for contamination of the unsaturated zone even decades after the contamination occurs.

In another industrial contaminated site, namely the Taas Ramat Hasharon site, enhanced biodegradation of perchlorate in the 40-m-thick unsaturated zone, was observed possible by adding acetate in laboratory experiments with sediments samples (Gal et al. 2008). A full in situ experiment was performed later on this site. Ethanol and water were injected from surface at high rates to penetrate deep in the unsaturated zone, and VMS was used to monitor the perchlorate degradation (Dahan et al. 2017). Within a year, a reduction of perchlorate concentrations by 2–4 orders of magnitude was observed at the upper 13 m of the unsaturated zone. The deeper parts of the unsaturated zone showed an increase in perchlorate concentrations. Therefore, Dahan et al. (2017) concluded that down leaching of perchlorate maybe a more significant process than activated biodegradation with ethanol, in this setting of an electron donor injected down at high rates with water from the surface.

Hormones (estrogen and testosterone) were found in levels of ng/kg (dry soil) all the way down into the 40-m-thick unsaturated zone, and in ng/l levels in the aquifer underneath a dairy farm in the Kiryat Malachi area (Arnon et al. 2008). Equilibrium flow and transport modeling in the unsaturated zone could not reproduce the distribution of the hormones in the deep unsaturated zone, suggesting that preferential pathways starting in the vertisol may be the explanation for the deep penetration of these sorptive compounds in this low matrix-permeability unsaturated zone (Arnon et al. 2008).

Two other microcontaminants (ng/l concentrations), carbamazepine (pharmaceutical) and caffeine, are typical of non-treated and treated sewage. Dvory et al. (2018) compared the use of these microcontaminants as tracers for sewage following a sewage spill into the upper-Soreq valley. Carbamazepine and caffeine were detected at the depth of 100 m below surface in groundwater samples of the Ein Karem 11 well located downstream of the spill. Using a dual permeability model for the carbonate unsaturated zone (Dvory et al. 2016) for the transport of the contaminants and calibrating of the distribution and first-order-degradation coefficients, Dvory et al. (2018) characterized their tracer performance. According to them, carbamazepine hardly degrades and is of low sorption, hence will be detected for a long time and long distances away and downstream from the spill. Nevertheless, a great deal of carbamazepine will diffuse from the fast-flow fracture domain to the low permeability matrix. Back diffusion to the permeable domain during increased rainy periods will result in a long and fussy tail of the breakthrough curve in groundwater. Caffeine degrades fast (half-life ~7.6 days, Dvory et al. 2018) and leaves no background concentration; hence, it is suitable for tracing the time of contamination but not the possible spatial spread and residence time of some sewage pollutants, which can be indicated, in turn, by the carbamazepine (Dvory et al. 2018).

## 8 Unsaturated-Zone Geochemistry

The exclusion of anions such as chloride and sulfate, from the fast flowing parts of the porous medium of a clayey unsaturated zone, was claimed to double their flow velocity in comparison to water, in the 10-m-scale-transport distance above the coastal aquifer (Gvirtzman et al. 1986).

A 30 m thick, calibrated variably saturated flow and reactive transport model of cation exchange and dissolution of calcite, that investigated infiltration of desalinated seawater into the Menashe MAR system, was presented by Ganot et al. (2018) (Fig. 9). The unsaturated-zone observations showed that the Hedera-plant desalinated water, enriched with $CaCO_3$, gain ~3 mg/l magnesium at the top 3 m of the unsaturated zone in a process which looks like an advancing front. The calibrated model showed that this process is sustainable only for a few decades in a scenario of ~20 m desalinated water infiltration per year (the depth of desalinated water that was applied in 2015 and 2018). Calcite dissolution (or precipitation) of the enriched desalinated water was found to be not a significant process. Nevertheless, the simulations show that if non-enriched desalinated water will be introduced to the infiltration pond the same enrichment level with $CaCO_3$ that is gained today in the desalination plant will be achieved through dissolution of in situ calcite during infiltration. This process is cheaper environmentally and is sustainable for centuries. Moreover, the enrichment with magnesium will probably be similar to the current one since the dissolution of calcite is expected to occur close to the surface, enabling calcium to exchange with

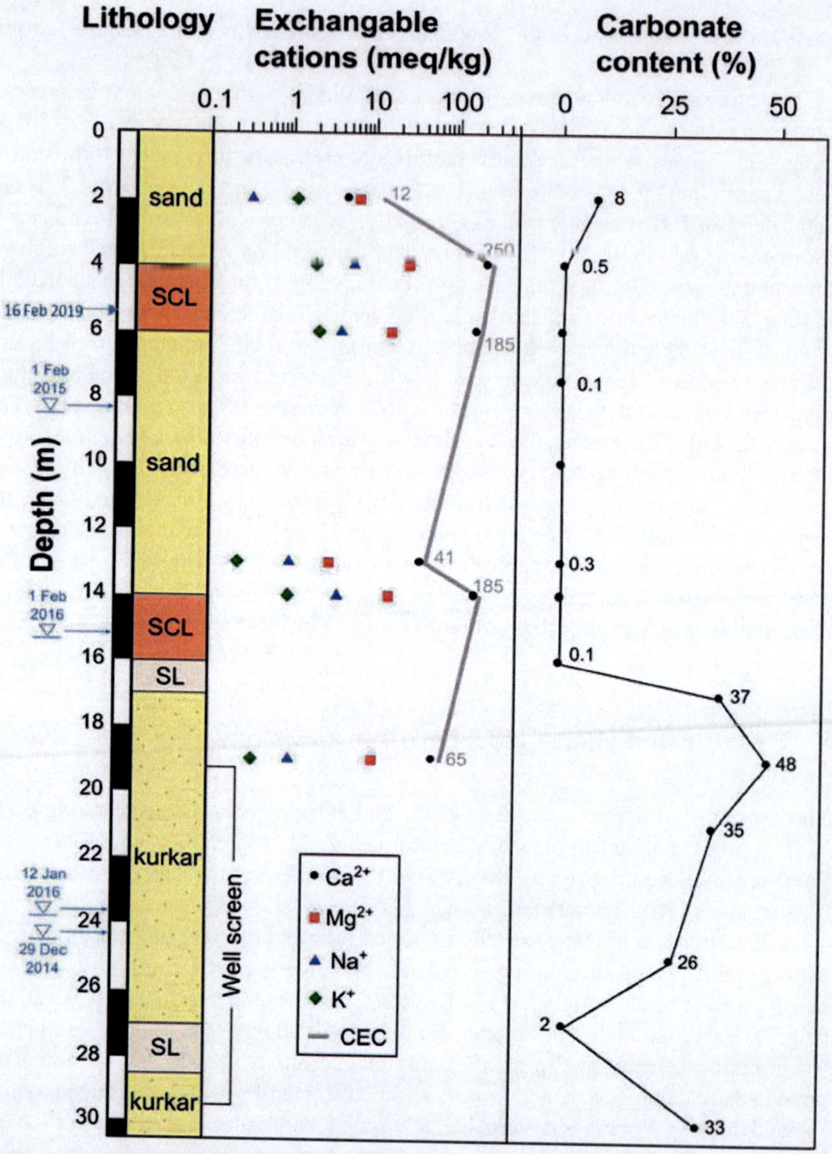


**Fig. 9** Variably saturated zone data used for calibrating a flow and reactive-transport model under the southern infiltration pond of the Menashe MAR system. Profiles of: texture class/sediment (SCL—sandy clay loam, SL—sandy loam, Kurkar—local calcareous sandstone); exchangeable cations ($Ca^{2+}$, $Mg^{2+}$, $Na^{+}$, $K^{+}$) and cation exchange capacity (CEC) in the gray profile; carbonate content. Sediment samples were obtained in 2014 before the studied MAR events. The well screen and water levels before and after MAR events are shown for reference [note 2019 water level at 5.4 m below surface after a large MAR event (modified from Ganot et al. 2018)]

magnesium cations from the exchanging medium deeper in the profile (Ganot et al. 2018).

Zilberbrand et al. (2001) describe the main geochemical processes in the unsaturated zone below the fast growing city of Tel Aviv during 1930–1948 when only infiltration of pit-latrine effluents occurred (before a piped sewage system was installed). Nitrification of ammonium caused: (1) nitrate leaching to groundwater; and (2) calcite dissolution from the unsaturated-zone sediment due to acidification associated with nitrification. The mineralization of organic matter infiltrating from pits caused an increase in carbon dioxide (up to 7.6%) and decrease in oxygen concentrations in the unsaturated-zone gas phase to 4–13%. The impermeable surface of the city constraints diffusion of oxygen from the air and preserved low levels of oxygen in the unsaturated zone.

Kass et al. (2005) described the difference in the geochemical processes that occur in the unsaturated zone under a field irrigated with fresh water and a field irrigated with treated wastewater in the vicinity of Nitzanim. $CaCO_3$ precipitated under the wastewater-irrigated field, whereas under the freshwater irrigation, there was dissolution of $CaCO_3$. Cation exchange also exhibited a mirror image; namely, under the wastewater irrigated, Na is adsorbed and Ca desorbed, whereas under the freshwater irrigation Na desorbed and Ca is adsorbed.

## 9 Concluding Remarks

This chapter tried to give a picture of the effort made to look on Israeli aquifers from the zone from which they are recharged. This torch lights fundamental groundwater resources issues in a bright and straightforward beam, with many examples brought in Sects. 4–8 above. Groundwater resources in Israel today are investigated with a better balance between approaches originating from geological sciences and attitudes coming from the soil science. This off-course lead to the bloom of the unsaturated-zone hydrology in Israel in the last 15 years.

In some aspects, Israeli unsaturated-zone hydrology is in the world front (e.g., monitoring, heterogeneous field-scale modeling), and in others we are well-appreciated contributors. Achieving better association between non-intrusive sensing technologies and unsaturated-zone hydrology important variables (e.g., water content, nitrate concentration) is a challenge that can take unsaturated zone monitoring a great deal forward, worldwide. In our local Israeli interests, continuing understanding impacts of land use, soils, unsaturated zone on our aquifers and transforming these understandings to quantitative tools that can help us decide is the never-ending effort of the Israeli unsaturated-zone hydrology.

## References

Allison GB, Hughes MW (1978) The use of environmental chloride and tritium to estimate total recharge to an unconfined aquifer. Aust J Soil Res 16:181–195

Arbel Y, Greenbaum N, Lange J, Inbar M (2010) Infiltration processes and flow rates in developed karst vadose zone using tracers in cave drips. Earth Surf Process Landf 35:1682–1693

Arnon S, Dahan O, Elhanany S, Cohen K, Pankratov I, Gross A, Ronen Z, Baram S, Shore LS (2008) Transport of testosterone and estrogen from dairy-farm waste lagoons to groundwater. Environ Sci Technol 42:5521–5526

Avnimelech Y, Raveh J (1976) Nitrate leakage from soils differing in texture and nitrogen load. J Environ Qual 5:79–82

Baram S, Kurtzman D, Dahan O (2012a) Water percolation through a clayey vadose zone. J Hydrol 424–425:165–171

Baram S, Arnon S, Ronen Z, Kurtzman D, Dahan O (2012b) Infiltration mechanism controls nitrification and denitrification processes under dairy waste lagoons. J Environ Qual 41:1623–1632

Baram S, Ronen Z, Kurtzman D, Külls C, Dahan O (2013) Desiccation-crack-induced salinization in deep clay sediment. Hydrol Earth Syst Sci 17:1533–1545

Baram S, Kurtzman D, Ronen Z, Peeters A, Dahan O (2014) Assessing the impact of dairy waste lagoons on groundwater quality using a spatial analysis of vadose zone and groundwater information in a coastal phreatic aquifer. J Environ Manage 132:135–144

Dafny E, Simunek J (2016) Infiltration in layered loessial deposits: revised numerical simulations and recharge assessment. J Hydrol 538:339–354

Dahan O, Nativ R, Adar EM, Berkowitz B, Weisbrod N (2000) On fracture structure and preferential flow in unsaturated chalk. Ground Water 38:444–451

Dahan O, McDonald EV, Young MH (2003) Flexible time domain reflectometry probe for deep vadose zone monitoring. Vadose Zone J 2:270–275

Dahan O, Shani Y, Enzel Y, Yechieli Y, Yakirevich A (2007) Direct measurements of floodwater infiltration into shallow alluvial aquifers. J Hydrol 344:157–170

Dahan O, Tatarsky B, Enzel Y, Kulls C, Seely M, Benito G (2008) Dynamics of flood water infiltration and ground water recharge in hyperarid desert. Ground Water 46:450–461

Dahan O, Babad A, Lazarovitch N, Rusak EE, Kurtzman D (2014) Nitrate leaching from intensive organic farms to groundwater. Hydrol Earth Syst Sci 18:333–341

Dahan O, Katz I, Avishai L, Ronen Z (2017) Transport and degradation of perchlorate in deep vadose zone: implications from direct observations during bioremediation treatment. Hydrol Earth Syst Sci 21:4011–4020

Dvory NZ, Livshitz Y, Kuznetsov M, Adar E, Yakirevich A (2016) The effect of hydrogeological conditions on variability and dynamic of groundwater recharge in a carbonate aquifer at local scale. J Hydrol 535:480–494

Dvory NZ, Livshitz Y, Kuznetsov M, Adar E, Gasser G, Pankratov I, Lev O, Yakirevich A (2018) Caffeine vs. carbamazepine as indicators of wastewater pollution in a karst aquifer. Hydrol Earth Syst Sci 22:6371–6381

Eriksson E, Khunakasem V (1969) Chloride concentration in groundwater, recharge rate and rate of deposition of chloride in the Israel coastal plain. J Hydrol 7:178–197

Even H, Carmi I, Magaritz M, Gerson R (1986) Timing the transport of water through the upper vadose zone in a karstic system above a cave in Israel. Earth Surf Proc Land 11:181–191

Gal H, Ronen Z, Weisbrod N, Dahan O, Nativ R (2008) Perchlorate biodegradation in contaminated soils and the deep unsaturated zone. Soil Biol Biochem 40:1751–1757

Ganot Y, Holtzman R, Weisbrod N, Nitzan I, Katz Y, Kurtzman D (2017) Monitoring and modeling infiltration–recharge dynamics of managed aquifer recharge with desalinated seawater. Hydrol Earth Syst Sci 21:4479–4493

Ganot Y, Holtzman R, Weisbrod N, Russak A, Katz Y, Kurtzman D (2018) Geochemical processes during managed aquifer recharge with desalinated seawater. Water Resour Res 54:978–994

Goldenberg LC, Melloul AJ, Zoller U (1996) The "short cut" approach for the reality of enhanced groundwater contamination. J Environ Manage 46:311–326

Gvirtzman H (2002) Israel water resources, Chapters in hydrology and environmental sciences. Yad Ben-Zvi Press, Jerusalem, 287 p (in Hebrew)

Gvirtzman H, Ronen D, Magaritz M (1986) Anion exclusion during transport through the unsaturated zone. J Hydrol 87:267–283

Gvirtzman H, Shalev E, Dahan O, Hatzor YH (2008) Large-scale infiltration experiments into unsaturated stratified loess sediments: monitoring and modeling. J Hydrol 349:214–229

Haaken H, Furman A, Weisbrod N, Kemna A (2016) Time-lapse electrical imaging of water infiltration in the context of soil aquifer treatment. Vadose Zone J 15(11):1–12

Hadas A, Hadas A, Sagiv B, Haruvy N (1999) Agricultural practices, soil fertility management modes and resultant nitrogen leaching rates under semi-arid conditions. Agric Water Manag 42:81–95

Hydrological Service (2008) Current condition and development af water resources in Israel until autumn 2007. Annual report. Israel Water Authority, Jerusalem (in Hebrew)

Kass A, Gavrieli I, Yechieli Y, Vengosh A, Starinsky A (2005) The impact of freshwater and wastewater irrigation on the chemistry of shallow groundwater: a case study from the Israeli coastal aquifer. J Hydrol 300:314–331

Kurtzman D (2011) The soil, the unsaturated zone, and groundwater-salinization in the south-eastern part of the (Israeli) coastal aquifer. Peer reviewed through the reviewing system for publications in Hebrew that operated in ARO, article # 603/11 (in Hebrew)

Kurtzman D, Scanlon BR (2011) Groundwater recharge through vertisols: irrigated cropland vs. natural land, Israel. Vadose Zone J 10:662–674

Kurtzman D, Nativ R, Adar E (2005) The conceptualization of a channel network through macroscopic analysis of pumping and tracer tests in fractured chalk. J Hydrol 309:241–257

Kurtzman D, Shapira RH, Bar-Tal A, Fine P, Russo D (2013) Nitrate fluxes to groundwater under citrus orchards in a Mediterranean climate: observations, calibrated models, simulations and agro-hydrological conclusions. J Contam Hydrol 151:93–104

Kurtzman D, Baram S, Dahan O (2016) Soil-aquifer phenomena affecting groundwater under vertisols: a review. Hydrol Earth Syst Sci 20:1–12

Levy Y, Shapira RH, Chefetz B, Kurtzman D (2017) Modeling nitrate from land surface to wells' perforations under agricultural land: success, failure, and future scenarios in a Mediterranean case study. Hydrol Earth Syst Sci 21:3811–3825

Livshitz Y, Zentner E (2009) Changes in water and salt storage of the coastal aquifer during the years 1933–2007: development and application of a method and comparison to previous methods. A hydrological service report. Israel Water Authority (in Hebrew)

Mercado A (1976) Nitrate and chloride pollution of aquifers: a regional study with the aid of a single-cell model. Water Resour Res 12:731–747

Nativ R, Adar E, Dahan O, Geyh M (1995) Water recharge and solute transport through the vadose zone of fractured chalk under desert conditions. Water Resour Res 31:253–261

Oren O, Yechieli Y, Böhlke JK, Dody A (2004) Contamination of groundwater under cultivated fields in an arid environment, central Arava Valley, Israel. J Hydrol 290:312–328

Raveh J, Saliternik CH, Avnimelech Y (1972) The distribution of nitrogen compounds in the soil column above the groundwater table in the coastal plain HG/70/073 P 149 Tahal, Tel-Aviv (in Hebrew)

Rimon Y, Dahan O, Nativ R, Geyer S (2007) Water percolation through the deep vadose zone and groundwater recharge: preliminary results based on a new vadose zone monitoring system. Water Resour Res 43:1–12

Ronen D, Kanfi Y, Magaritz M (1983) Sources of nitrates in groundwater of the coastal plain of Israel—evolution of ideas. Water Res 17:1499–1503

Ronen D, Graber ER, Laor Y (2005) Volatile organic compounds in the saturated–unsaturated interface region of a contaminated phreatic aquifer. Vadose Zone J 4:337–344

Russo D (1995) Contaminant transport below an orchard irrigated with sewage water. In: Fine P, Haruvi N, Shainberg I (eds) The Ra'anana sewage purification project: recommendations of criteria for the sewage water quality for different purification alternatives. Prelim. Rep. Institute of Soils, Water and Environmental Sciences, The Agricultural Research Organization, Bet Dagan, Israel, pp 35–49 (in Hebrew)

Russo D (2017) On the control of solute mass fluxes and concentrations below fields irrigated with low-quality water: a numerical study. Water Resour Res 53:8925–8940

Russo D, Bouton M (1992) Statistical analysis of spatial variability in unsaturated flow parameters. Water Resour Res 28:1911–1925

Sheffer NA, Dafny E, Gvirtzman H, Navon S, Frumkin A, Morin E (2010) Hydrometeorological daily recharge assessment model (DREAM) for the Western Mountain Aquifer, Israel: model application and effects of temporal patterns. Water Resour Res 46:W05510

Sheffer NA, Cohen M, Morin E, Grodek T, Gimburg A, Magal E, Gvirtzman H, Nied M, Isele D, Frumkin A (2011) Integrated cave drip monitoring for epikarst recharge estimation in a dry Mediterranean area, Sif Cave, Israel. Hydrol Processes 25:2837–2845

Turkeltaub T, Dahan O, Kurtzman D (2014) Investigation of groundwater recharge under agricultural fields using transient deep vadose zone data. Vadose Zone J 13:1–13

Turkeltaub T, Kurtzman D, Bel G, Dahan O (2015a) Examination of groundwater recharge with a calibrated/validated flow model of the deep vadose zone. J Hydrol 522:618–627

Turkeltaub T, Kurtzman D, Russak EE, Dahan O (2015b) Impact of switching crop type on water and solute fluxes in deep vadose zone. Water Resour Res 51. https://doi.org/10.1002/2015WR017612

Turkeltaub T, Kurtzman D, Dahan O (2016) Real-time monitoring of nitrate transport in the deep vadose zone under a crop field-implications for groundwater protection. Hydrol Earth Syst Sci 20:3099–3108

Vengosh A, Ben-Zvi A (1994) Formation of a salt plume in the coastal plain aquifer of Israel: the Be'er Toviyya region. J Hydrol 160:21–52

Yaron B, Dagan G, Goldshmid J (eds) (1984) Pollutants in porous media, the unsaturated zone between soil surface and groundwater. Springer, Berlin. https://doi.org/10.1007/978-3-642-69585-8

Zilberbrand M, Rosenthal E, Shachnai E (2001) Impact of urbanization on hydrochemical evolution of groundwater and on unsaturated-zone gas composition in the coastal city of Tel Aviv, Israel. J Contam Hydrol 50:175–208

# Dating of Groundwater in Israeli Aquifers and Determination of Groundwater Flow Velocities

**Yoseph Yechieli**

## 1 Introduction

The primary aims of groundwater dating are to determine their sources and to estimate their velocity and direction of their flow. Age and flow velocity of groundwater are necessary for estimation of recharge rate and of the risk of pollution. Older ages usually signify slow recharge rate and no risk of pollution, while the opposite is true for young groundwater. Therefore, older ages are significant regarding management whereby pumping should be done with caution since flow rate is small and slow replacement of the pumped water is expected. In such case, it is difficult to maintain sustainable condition of groundwater exploitation where pumped water is equal to recharge.

In general, age is defined as the time from penetration into the subsurface and loss of contact with the atmosphere or other water body until arrival at the discharge zone or sampling wells.

## 2 The Dating Tools

Dating of groundwater is usually done by analysis of radioactive isotopes with known decay rates. Other methods use radiogenic elements which are produced by radioactive decay in the subsurface whereby the rate of accumulation can be estimated (e.g., helium) albeit usually not with a great accuracy. The best known and most widely used radioactive isotopes for groundwater dating are radiocarbon and tritium (half-lives of 5730 and 12.4 yrs, respectively). Radiocarbon dating of groundwater is usually done on its dissolved inorganic carbon (DIC) portion although several studies

Y. Yechieli (✉)
Geological Survey of Israel, Jerusalem, Israel
e-mail: yechieli@gsi.gov.il


U. Kafri and Y. Yechieli (eds.), *The Many Facets of Israel's Hydrogeology*,
Springer Hydrogeology, https://doi.org/10.1007/978-3-030-51148-7_13

were conducted on dissolved organic carbon (DOC) with limited success. Dating by tritium has become ambiguous due to the low values in recent precipitation and its short half-life.

Radiocarbon dating of old groundwater can be problematic due to contamination with modern atmospheric carbon. The addition of radiocarbon from the atmosphere results in incorrect age calculations (Aggarwal et al. 2014). One way to solve this problem is to date the gaseous phase of $CO_2$ which is obtained during the regular sampling for the analyses of noble gases (Yokochi et al. 2018). Moreover, the translation of radiocarbon activities into ages could be complicated and inconclusive due to chemical processes, such as water–rock interaction and isotopic exchange (e.g., Mook 1980; Gonfiantini and Zuppi 2003). Thus, dating by $^{14}C$ is performed with the aid of mass balance equations, using computer codes such as the Netpath and PhreeqC, which take into account the known chemical reactions. An estimated initial value of 60 pmc can be assumed for many cases, due to chemical processes in the unsaturated zone in a freshwater aquifer (e.g., Carmi et al. 2007), and oxidation of old organic matter in a seawater component (Sivan et al. 2005).

Another tool for dating old groundwater is $^{36}Cl$ (half-life of 301 kyr), which is problematic in the case of brackish or saline water. Noble gas radionuclides, with chemical inertness, simple source function, and limited subsurface production, are emerging as ideal tracers of groundwater movement. Indeed, dating of fresh groundwater with noble gases has been performed recently in many parts of the world, using $^{85}Kr$ (half-life of 10.7 yr) and $^{3}He$ (decay product of $^{3}H$, whose half-life is 12.4 yr) for short-term hydrological processes (Althaus et al. 2009; Visser et al. 2013) $^{39}Ar$ (269 yr) for intermediate and $^{81}Kr$ (229 kyr) (Fig. 1) for long-term processes (Gerber et al. 2017; Aggarwal et al. 2015; Purtschert et al. 2013; Suckow et al. 2013). In some cases, especially where groundwater is saline and therefore absorption is limited, radium isotopes can also be used for estimation of ages (e.g., Kiro et al. 2013, and further discussion below). These include four radium isotopes: $^{226}Ra$ (half-life of 1600 yr), $^{228}Ra$ (5.8 yr), $^{224}Ra$ (3.6 days), $^{223}Ra$ (11.4 days). In the

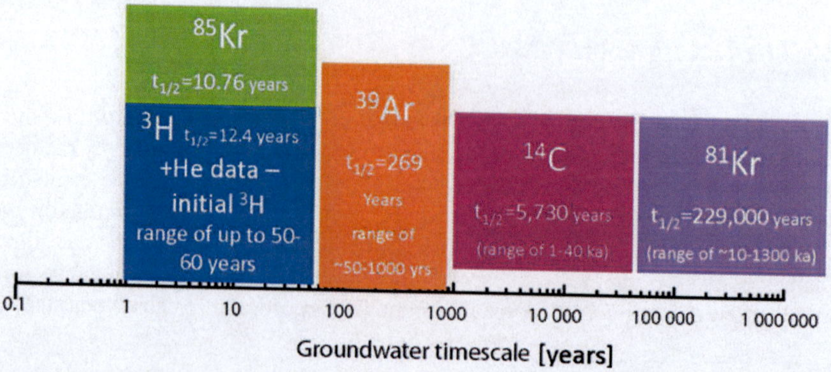


**Fig. 1** Half-lives and timescales of radioactive isotopes that are used for dating of groundwater (modified after Aggarwal 2013)

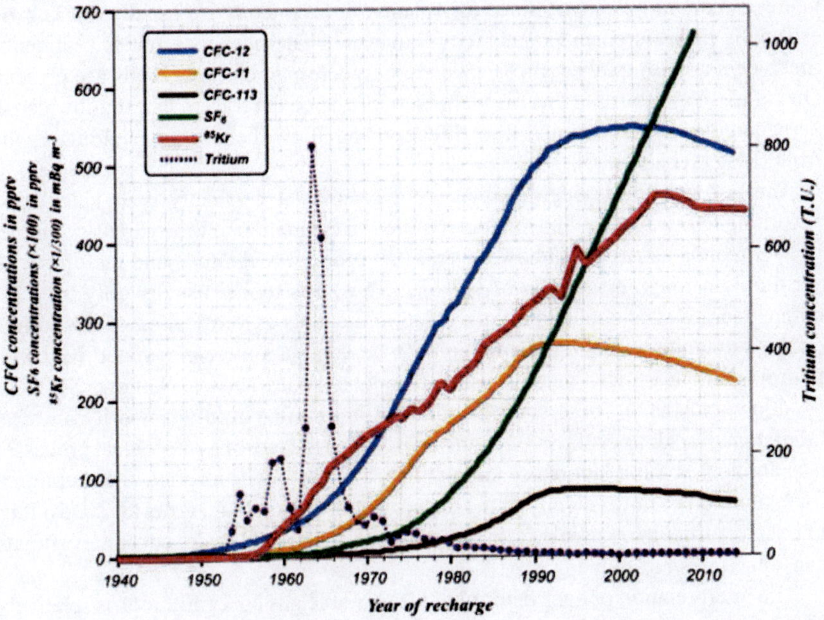


**Fig. 2** Atmospheric values of several tracers that are used for groundwater dating (after Kagabu et al. 2017)

cases of very young water, $^{222}Rn$ can also be used (3.8 days). The accumulation rate of $^{4}He$ can also provide a rough age estimation of old groundwater in cases where the production rate can be estimated.

Other dating options are tracers whose concentrations in their source are known along their relevant history. Some of the best tracers of this kind are CFCs and SF6, since the variations in their concentrations are known (Busenberg and Plummer 1992) (Fig. 2). Moreover, the appearance of any specific chemical parameter can be used for dating provided that the location and the time of its introduction into the hydrological system are known. Therefore, tracking of any pollution at a specific well can provide age of groundwater and therefore also the velocity and direction of flow.

Another tool is the introduction of artificial dye tracer, such as uranine. The tracer is injected at a specific part of the hydrological system and is monitored down-gradient in other parts of the system. Once it is observed in the monitoring locations, an immediate simple calculation of the flow direction and velocity can be performed. The advantage of using dye tracers is relative low cost and simple analytical detection by a spectrophotometer, allowing a very dense sampling and analyzing. The difficulty in dye tracers, however, is that they are not conservative enough in hydrological system and thus can be absorbed on the aquifer's matrix and never reach the monitoring wells. Also, they might be diluted while flowing to a level below detection. Moreover, since the specific flow conduits are usually not known, especially in fractured and karstic rocks but also in heterogeneous deltaic sediments, the tracers are not tracked in cases where only a few monitoring wells are present. One of the ways to solve this problem is to introduce the tracers in a shallow canal, perpendicular to the assumed flow direction and allow them to move along a wide front (Magal et al. 2010).

The last tool for a rough estimation of groundwater velocity is its calculation according to Darcy law. This method requires an estimation of the hydraulic conductivity which is usually not known even as an order of magnitude and therefore can yield only a rough estimation of velocity. Thus, this tool is usually only used for comparison to ages obtained by other dating methods as verification of the correct order of magnitude. Similar estimation can be done in a more elaborated way with hydrological numerical models using computer codes such as Feflow, SUTRA, Modflow, or others. These computer simulations can be accompanied by physical laboratory modeling in 2D or 3D sand box in which the rate of hydrological processes is examined. In these laboratory simulations, a tracer can be inserted and monitored for estimation of flow velocity in the experimental setting (e.g., Oz et al. 2015). It is, of course, difficult to upscale from the lab to the actual field scale, but some estimate can sometimes be attained.

The interpretation of ages determined from noble gas concentrations is relatively simple since noble gases do not chemically interact with the aquifer's rocks and therefore do not require specific corrections for chemical processes (as opposed to the case of $^{14}$C, see below). However, physical processes such as diffusive exchange with stagnant water bodies for all tracers (Sanford 1997) or nucleonic production due to neutron activation of potassium for $^{39}$Ar may affect the tracers' concentrations (Lehmann et al. 1993; Purtschert et al. 2004). Practically, diffusion is more pronounced for tracers with longer half-life (e.g., $^{14}$C) than for fast decaying tracers. $^{39}$Ar, for example, does not "see" the total inactive system volume since the time to diffusively reach the inactive volume exceeds the mean lifetime of the tracer. Therefore, $^{14}$C is more sensitive to the diffusion process than is $^{39}$Ar (half-life of 269 years) resulting in an apparent overestimation of $^{14}$C ages as compared to $^{39}$Ar ages (Mayer et al. 2014). $^{81}$Kr ages are similarly affected by this process despite the slightly reduced diffusion coefficient due to the larger mass of the Kr atom. Uncorrected $^{81}$Kr ages, therefore, represent only the upper age limits if diffusive exchange is ignored.

The effect of diffusive exchange between young water in permeable units and old groundwater in less permeable units can increase the apparent radiocarbon age by ~50% (Sanford 1997). This effect can also be significant when the volume ratio between fractures and rock matrix is relatively low. The high degree of heterogeneity of a karstic system can lead to a highly dispersive flow and thus wide age distributions (Ghasemizadeh et al. 2012).

The dating tools above described can be problematic in cases where a specific water sample is a result of mixing of several water bodies of different ages. In such cases, the codes estimate one age for the mixed sample and do not provide the distinct age of each component in the mixed sample.

The following chapter is divided into two parts: The first deals with dating of fresh groundwater and the second with dating of saline water. Each part is divided according to the different dating methods.

## 3 Dating of Fresh Groundwater

As explained above, the age of groundwater indicates the time it took the water to flow from the recharge point or intake area (as precipitation in the case of fresh groundwater) until reaching the sampling point. Attempts to date fresh groundwater were conducted in Israel using several methods, as described below:

### *3.1 Radiocarbon and Tritium*

The most usable dating methods of groundwater are, by far, those of radiocarbon and tritium.

**Judea Group Aquifer (JGA)**

As explained in Chapter "The Yarkon-Taninim Basin—An Example of a Major Carbonate Aquifers in Israel", the JGA (also named the Mountain aquifer) is the largest aquifer in Israel, consisting of Cretaceous carbonates of the Judea Group. In Israel, as elsewhere in the world, most dating of groundwater in such aquifers was done with radiocarbon and tritium. The first extensive dating was done by Kroiteru et al. (1989) who studied the ages of groundwater along a cross section from the Mediterranean Sea to the Dead Sea using radiocarbon and tritium. The ages were found, as expected, to increase from the recharge zone at the Judea Mountain toward both the eastern and western base levels, reaching values of several hundreds or thousands of years. A later study distinguishes between two sub-aquifers within the JGA and show a difference in radiocarbon ages between them, suggesting younger ages at the lower sub-aquifer (Avisar et al. 2001).

A more recent study (Dafny 2009) did further radiocarbon dating of the western part of the JGA, named the Yartan aquifer. According to this study, groundwater ages generally increase toward the northern part of the aquifer, in agreement with the general flow direction in this aquifer. Relatively old groundwater was found in the Negev desert, in the southern part of the Yartan aquifer (Kronfeld et al. 1993). However, as mentioned above, several samples of younger water were found also in the central and northern part of the aquifer, indicating another source of water from recharge of younger water in the mountain area in the east.

The eastern Mountain aquifer, from the recharge zone in the Jerusalem area to the discharge zone near the Dead Sea, was first studied by Kroiteru et al. (1989). A recent study (Avrahamov et al. 2018) used radiocarbon and tritium combined with newer tools of noble gases ($^{39}Ar$ and $^{85}Kr$), as discussed below.

## The Kurnub Aquifer and Shallower Aquifers in the Negev and the Arava Valley

The detailed description of the Kurnub aquifer in the Negev desert in the southern part of Israel is given by Ram et al. in Chapter "The Nubian Sandstone Aquifer in the Sinai Peninsula and the Negev Desert". The first dating efforts were conducted by Issar et al. (1972) and Gat and Issar (1974), using the radiocarbon method. According to their study, wherein very low values of radiocarbon were found (<5 pmc), the age of this groundwater was estimated as 20,000–30,000 years, namely that they were recharged during the last ice age. This age led to the interpretation that the aquifer was recharged during a wetter period and is currently not recharged. Similar age ranges were also given by Vengosh et al. (2007). A later study argued that radiocarbon dating cannot be trusted for such low values (Burg et al. 2013; Aggarwal et al. 2014) and that this groundwater should be older than 25,000 years, still without attributing a specific age (Burg et al. 2013). A more complex picture is given by Ram et al. (Chapter "The Nubian Sandstone Aquifer in the Sinai Peninsula and the Negev Desert") supported by dating with the tool of $^{81}Kr$.

A more accurate radiocarbon dating was conducted on the carbon in the gaseous phase (Yokochi et al. 2018). This sampling was conducted with a unique equipment for noble gases, which is more scrupulous with regards to contamination of atmospheric gases. Indeed, the results of this sampling yielded much lower radiocarbon values (mostly <1 pmc) indicating ages of >40 kyrs, more in agreement with the $^{81}Kr$ results (Yokochi et al. 2018).

The age of groundwater of shallower aquifers than the Kurnub aquifer in the Arava was found to be younger, as expected. A most pronounced difference was from the groundwater ages of the Miocene-Pliocene aquifers (Hazeva and Arava aquifers), where some of the water samples were found to contain tritium and relatively high radiocarbon values (e.g., ~60 pmc) indicating ages of less than 60 years (Burg et al. 2013). This water was recharged by recent floods in the mainstreams, such the Arava stream, indicating that recharge occurs even in such extremely arid conditions typified by annual precipitation below 50 mm. These shallow aquifers are also laterally fed from the east via the alluvial fans at the foothills of the high and more rainy Edom mountains.

## The JGA in Galilee Near Lake Kinneret

Other dating surveys were conducted in smaller areas including the JGA in the northeastern part of Israel, adjacent to Lake Kinneret. The main dating study was done using radiocarbon and tritium dating. As expected, the fresh groundwater at the recharge area in Central Galilee was found to be young (mostly 50–70 pmc and containing some tritium), while the saline water near the Kinneret was found to be older with values of <10 pmc (Bergelson et al. 1999).

## Dating of Fresh Groundwater in the Mediterranean Coastal Aquifer

Fresh groundwater was dated with radiocarbon and tritium in many parts of the coastal aquifer, showing in most places relatively young ages of less than 50 years (Bruce et al. 2001, 2007). The specific variabilities in radiocarbon values in this aquifer were partly explained by the difference in flow paths through the unsaturated zone which could be related to open versus closed systems (Bruce et al. 2001). According to the peaks in radiocarbon and tritium concentrations, due to the effect of the nuclear bomb testing, the age interpretation of relatively young water could be problematic since similar concentration can yield different ages, due to the similar initial values prior to and after the peak. One way to overcome this problem was suggested to be the method of repetitive analyses, in which samples from the same wells were analyzed with a time gap of ~20 years or more (Bruce et al. 2007). Such an exercise indeed provided a better and more specific solution for age determination.

The area of the coastal aquifer adjacent to the shoreline is more unique since it is divided to several sub-aquifers, separated by clay layers. The groundwater ages in this part of the aquifer, mostly in the lower sub-aquifers, were found to be much older than in the entire aquifer (Nativ and Weisbrod 1994; Yechieli et al. 2001, 2009). For example, very low radiocarbon values of ~2 pmc were found in the deeper sub-aquifer (Fig. 3) indicating an age of more than 10,000 years. This old age implies that there is no simple flow from this aquifer to the sea, and therefore, the connection is probably blocked by units of lower permeability, as was suggested by Kolton (1988).

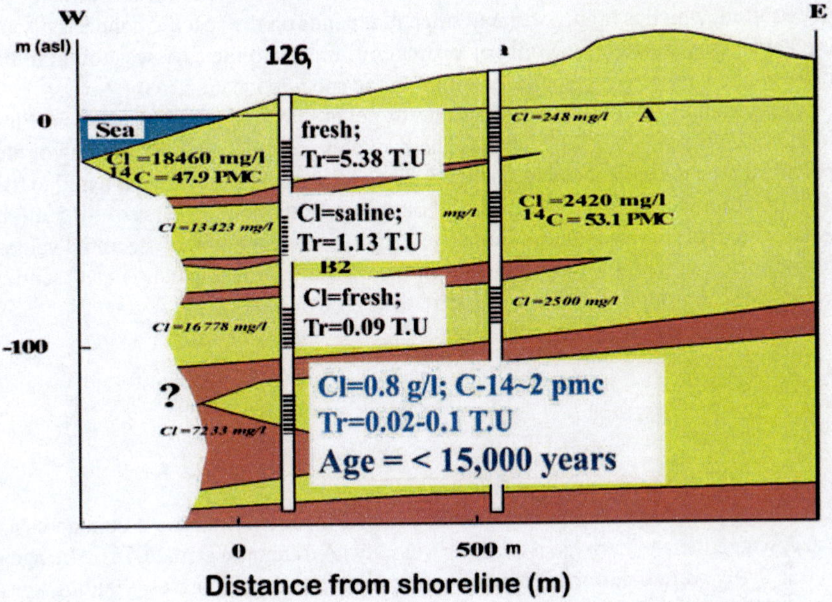


**Fig. 3** Dating of fresh groundwater in the vicinity of the shoreline, 10 km south of Tel Aviv

Another study of dating with radiocarbon was conducted in the Mount Carmel coastal aquifer, located north of the main coastal aquifer, indicating that a significant source of water to the coastal aquifer is originating from the Mountain aquifer of Mount Carmel (Guttman et al. 2011). The apparent difference in ages between the two aquifers was explained by a difference in initial values, and both aquifers have young water (containing tritium, Guttman et al. 2011).

**Dating of Groundwater in the Unsaturated Zone**

Several dating studies were conducted in Israel in order to estimate the rate of water infiltration through the unsaturated zone (USZ) above the coastal and Mountain aquifers, using the tritium and radiocarbon methods. Tritium dating is more efficient for the USZ since the water usually reaches the groundwater table within several tens of years. Tritium profiles were conducted in different parts of the coastal area of Israel showing that the downward flow velocity is about 0.7–2.3 m/yr. Thus, the water can reach the groundwater table, usually several tens of meters deep, within less than 50 years (Gvirtzman et al. 1986; see also Chapter "Unsaturated-Zone Hydrology in Israel: A Subjective Review"). A better time resolution was obtained when taking into account the specific values of the different sources of infiltrating water, namely natural rain or irrigation water (Gvirtzman and Magaritz 1986).

The rate of water penetration into the JGA was also examined using tritium in the mountain area, from the surface toward Soreq cave, located some 40 m below surface (Kaufman et al. 2003). Although the cave is considerably (a few hundred m) above the regional water table, this study yields an order of magnitude for the velocity of penetration, which is highly variable since it depends on the specific connectivity of the karstic flow system. The time of vertical infiltration to the cave was found to be usually a few to tens of years, depending on the specific location in the cave.

Other studies tried to estimate the rate of penetration using radiocarbon profiles in the USZ (Carmi et al. 2007, 2004). These studies required an understanding of the chemical processes that occur in the USZ. The radiocarbon data was compared to the data of tritium at the same location (in Nitzanim), both yielding rates of ~0.5 m/yr. Another part of this research was aimed toward the determination of the initial values for radiocarbon dating of groundwater in the coastal aquifer, which is an essential parameter for groundwater dating (Carmi et al. 2004, 2007).

## 3.2 *Noble Gases*

**Eastern (JGA) Mountain Aquifer**

The dating revealed two separate groundwater flow routes from the mountain eastward, where the southern, is of a slower velocity (Avrahamov et al. 2018). The ages in this study were determined mainly by the analysis of noble gases, yielding ages below 1000 years, since $^{39}Ar$ (half-life of 269 years) was found in all samples. Moreover, the $^{39}Ar$ data was used for correction of the radiocarbon ages for the effect of diffusion and water–rock interaction with the aquifer's matrix (Avrahamov et al.

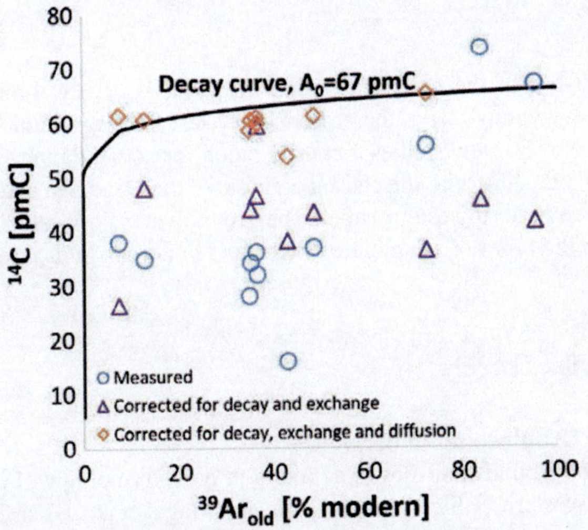


**Fig. 4** Decay curve for the system $^{14}C$ versus $^{39}Ar$ assuming initial 14C value of A0 = 67 pmc (used by permission of Elsevier from Avrahamov et al. 2018, Fig. 11). Blue circles-measured, uncorrected values; purple triangles—$^{14}C$ for A0 = 67 corrected on the assumption of isotope exchange; orange diamonds—corrected $^{14}C$ values in which radioactive decay, isotope exchange and matrix diffusion, were taken into account

2018). Indeed, the radiocarbon ages were found to be considerably older according to the Ar data (Fig. 4). Together with the Ar analysis, the noble gases of $^{3}He$ and $^{85}Kr$ were analyzed to determine whether there is a young component in this groundwater. The $^{3}He$ together with tritium data yielded a better resolution than just the tritium data, especially since in the last 20 years the tritium values are quite low.

**The Kurnub Aquifer in Southern Israel**

Noble gases dating ($^{81}Kr$) showed that the ages in this aquifer range from several tens of thousands to ~600,000 years, extremely different from the quite uniform radiocarbon ages of 20–30 kyr, obtained in previous studies (Ram et al., see also Chapter "The Nubian Sandstone Aquifer in the Sinai Peninsula and the Negev Desert"). The high variability of the $^{81}Kr$ ages, however, indicates that there is a mixing of several water bodies of different ages within this aquifer. The ages were compared to ages obtained by other less reliable methods (e.g., $^{36}Cl$ and $^{4}He$) in order to understand the limitations of the other methods. Indeed, the $^{36}Cl$ is problematic in this case due to the abundance of Cl from other sources, such as halite and other, post halite deposition residues of the Dead Sea brines. Provided that the $^{81}Kr$ data is indicating the true ages, only then the $^{36}Cl$ data could be calibrated and employed for dating of groundwater (Ram et al. 2018). The process of the $^{4}He$ dating is still in progress with no conclusion as to the efficiency of this dating method in the Kurnub aquifer.

## 3.3 Radon

The short half-life of the noble gas radon of 3.8 days usually limits its potential to dating of very young water, therefore to very fast flowing groundwater systems. Thus, there are only a few studies that used radon for actual dating, one of which is by Kafri (2001) dealing with the coastal aquifers of the Dead Sea and the JGA in the Western Galilee in northwestern Israel. The groundwater flow velocities there were found to be in the order of magnitude of tens and hundreds m/day, respectively.

## 3.4 Artificial Tracers

**Dye Tracers (Uranine and Others)**

The velocity of groundwater flow was found to be extremely fast in the JGA karst system of the Western Galilee, using dye tracers (Magal et al. 2013). The dye tracers traveled from a sinkhole, where they were inserted, to discharging springs down-gradient at a distance of several kilometers within a few days.

Dye tracers (uranine) were also used in the JGA in Mount Carmel, indicating a very fast recharge rate in which the dominant flow was of the same order of magnitude for all drip types (22–41 cm/h; Arbel et al. 2010). Another study with dye was conducted in a cave in the JGA in the Judea mountains, showing also fast flow from the surface down to the karstic system (Sheffer et al. 2011).

A more comprehensive tracer study, using several dye tracers simultaneously, was conducted in the coastal aquifer of the Dead Sea (Magal et al. 2010). Due to several unsuccessful attempts of inserting tracer into a specific borehole and tracking it in other boreholes, a different approach was taken, in which the tracers were inserted into a shallow channel. This study allowed the tracking of the different tracers in several boreholes (Fig. 5), indicating an extremely fast flow of tens of meters per day and managed to determine the specific direction of flow in this preferential flow system (Magal et al. 2010). The above described velocities are indeed in agreement with those obtained for the same region, using radon dating by Kafri (2001).

Tracers were also used in order to study recharge processes in the shallow aquifer in the Arava valley, indicating the potential for fast recharge in the alluvial aquifer (Dahan et al. 2009).

**Pollutants**

As indicated above, pollutants in groundwater can also be used as tracers for groundwater flow direction and velocity. One example for such research was the indication of the fast flow exhibited by the arrival of several pollutants, traveling a long distance within a short time, as was also indicated by the arrival of dye tracers (Magal et al. 2013). Another example was described related to the recharge zone, near Jerusalem, where the velocities of arrival of several organic compounds such as caffeine and carbamazepine were studied, one of which is less conservative that the other (Dvory et al. 2018). The arrival of these organic tracers within less than one year (~120 days),

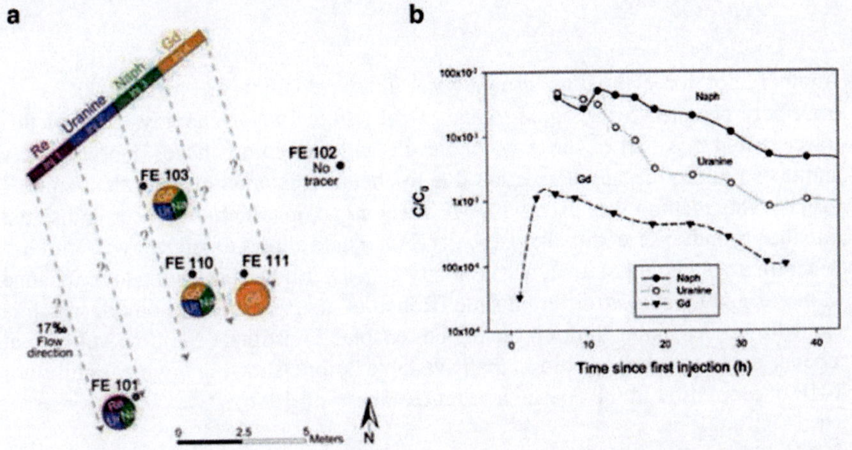


**Fig. 5** **a** Location map of the four segment (Inj 1 to Inj 4) line-source injection system and the observation wells (FE101, FE102, FE103, FE110, and FE111). The dashed lines represent the assumed direction of the groundwater flow lines derived from the water head in the observation wells. Marked in colors are the tracers injected through the segments and their arrival at the observation wells. **b** C/C0 breakthrough curves in observation well FE110 (4.5 m depth): uranine injected at a distance of 6.8 m, naphthionate injected at a distance of 7.3 m, and Gd injected at a distance of ~6.5 m (modified after Magal et al. 2010)

as indicated both in the field evidence and in computer simulations, indicate that the recharge through the unsaturated zone all the way to groundwater table is fast as could be predicted in such karstic system with large conduits.

## 3.5 *CFCs and* $SF_6$

As stated above, the knowledge of the concentrations of CFCs and $SF_6$ in the atmosphere (Fig. 2) enables their use as a natural tracer for estimation of groundwater ages. However, this method was only used in a preliminary manner in a few samples in the JGA. The results indeed indicate that these tools are reasonable, yielding a young age (<50 years) for groundwater in the recharge zone in a Jerusalem well, where anthropogenic pollution also indicates its relatively young age, and an older age (>50 years) for groundwater down-gradient which was found to be older also by other tracers (e.g., tritium) (Avrahamov et al. 2018).

## 3.6 Depleted $^{18}O$

Another tool for estimating groundwater flow velocities was using a tracer of extremely depleted $\delta^{18}O$ signal (Ronen et al. 1986, 1991). The advantage of this tracer is that it is part of the water molecule and therefore behaves conservatively and does not require any correction due to chemical interaction with the aquifer's matrix. This method was used with MLS technique, in which the tracer is inserted into the groundwater within dialysis cells (Fig. 6) and allows to interact with groundwater for a set amount of time, with velocity then determined according to the change in the original signal at different time (Ronen et al. 1991). This method, which is named modified point dilution technique, enables to estimate the flow velocity at a better depth interval resolution than available before, showing large variability of 1–10 or more m/yr in the coastal aquifer (Ronen et al. 1986).

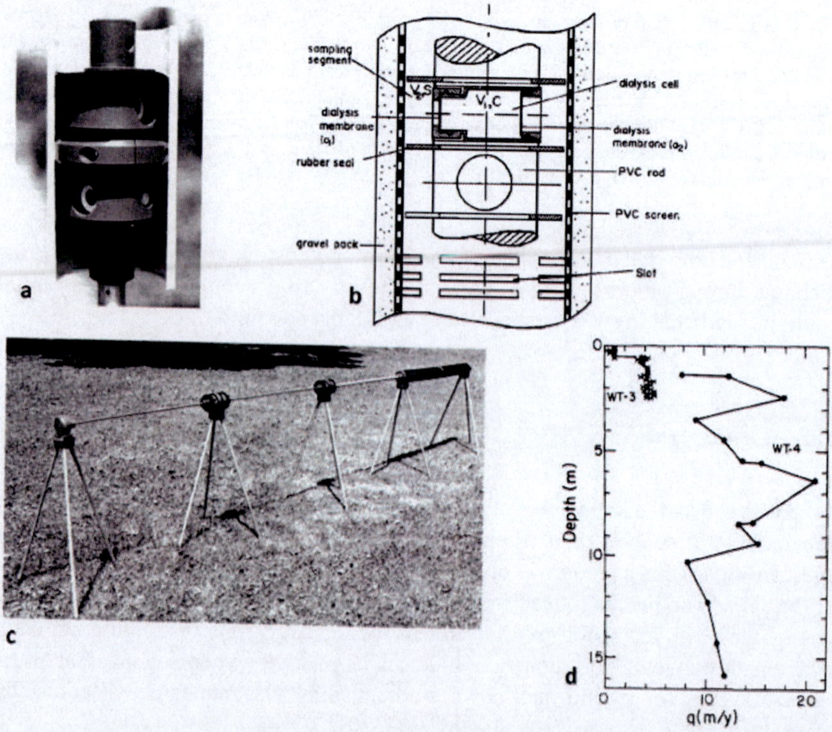


**Fig. 6** Use of the MLS technique (modified after Ronen et al. 1986, 1991). **a** Picture of a segment of the MLS; **b** sketch of MLS; **c** the full length of MLS; **d** profile of specific discharge obtained by MLS

### 3.7 *Uranium Isotopes*

The effectiveness of the value of uranium isotopes as a tool for dating has been debated in the literature for some time with no specific evidence for it. Several studies were conducted in Israel, using the uranium isotopes for understanding the chemical processes in the aquifer (e.g., Kronfeld and Rosenthal 1981; Rosenthal and Kronfeld 1982). A later study (Rogojin et al. 1998) exhibited a decrease in $^{14}C$ activity that is concomitant to the growth in $^{234}U$-excess, suggesting that under specific conditions, $^{234}U$-excess dating of groundwater can be used to corroborate $^{14}C$ dates, possibly exceeding the range of dating of groundwater beyond that of $^{14}C$.

## 4 Dating of Saline Groundwater and Brines—Determination of Rate of Seawater Intrusion

### 4.1 *General*

Coastal aquifers are potentially prone to salinization via seawater intrusion or via brines from saline lakes. The issue of coastal aquifer degradation has become increasingly acute in the past few decades due to over exploitation that induces saline water into adjacent freshwater aquifers. In most cases, seawater intrusion rates are inferred from increase in salinity (Acworth and Dasey 2003; Linderfelt and Turner 2001), change of chemical composition (Russak et al. 2016; Mercado 1985), or movement of the fresh–saline water interface (Melloul and Zeitoun 1999). This chapter deals with the determination of the rate of seawater intrusion by a direct method of groundwater dating.

Saline water usually mixes with freshwater in wells, thereby complicating the chronological interpretation. In coastal aquifers, samples contain both fresh and saline water components, each with its distinct flow path and residence time. Hence, it is required to define separately the age for each component as the time since penetration or recharge of the water into the aquifer. Concerning the saline water, its age, or residence time, is considered to be from the time of penetration of the seawater into the bottom sediment (Yechieli et al. 2009). In such a case, there are three potential relevant processes: (1) long-term seawater intrusion due to natural density-driven circulation near the sea shore; (2) the effect of sea level rise; (3) seawater encroachment landwards due to the short-term effect of pumping. The rate of seawater intrusion, and therefore the age of the saline water, would be the result of the superposition of these three possible processes. The initial values of the penetrating seawater can be assumed to be similar to the current ones since the age of the bottom Mediterranean Sea water probably does not exceed ~100 years according to their radiocarbon values (Broecker and Gerard 1969).

## 4.2 Radiocarbon and Tritium

**Mediterranean Coastal Aquifer**

Direct determination of the seawater intrusion rate in the coastal aquifer has been seldom done and was based only on radiocarbon and tritium tracers (e.g., Sivan et al. 2005; Yechieli et al. 2001, 2009). A detailed study was conducted on the upper phreatic sub-aquifer, indicating a rate of seawater intrusion of several meters per year, even though the radiocarbon value in groundwater was found to be significantly lower than that of the seawater (60 and 100 pmc, respectively) due to the effect of oxidation of organic matter during the seawater penetration inland (Sivan et al. 2005). The presence of tritium in this seawater is a clear evidence for young seawater and therefore for a relatively fast penetration (Sivan et al. 2005).

Saline groundwater was also sampled in the lower sub-aquifers in a few locations where seawater intrusion was observed, indicating ages of thousands of years according to the very low values of tritium and radiocarbon (Yechieli et al. 2001, 2009). In the Nitzanim area, the age of the saline water was found to become younger upward (Fig. 7), implying that it represents the penetration of younger seawater parallel to the sea level rise after the last glacial period (Fig. 8). (Yechieli et al. 2009).

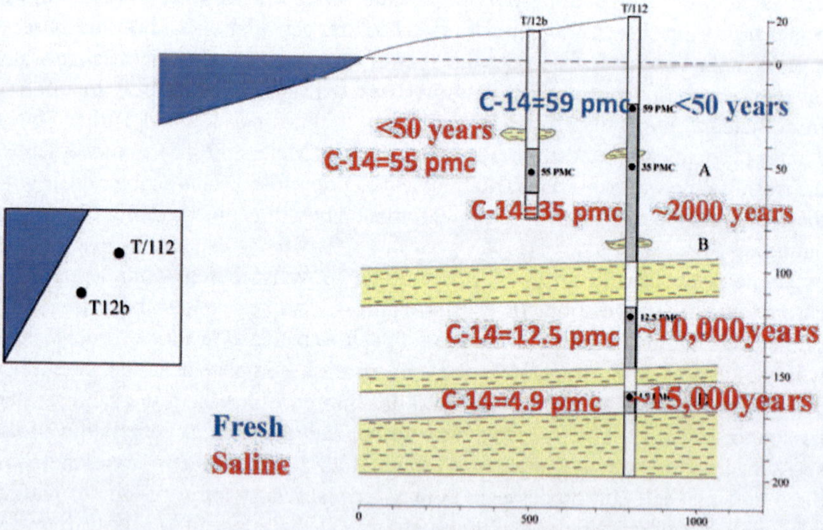


**Fig. 7** Schematic hydrogeological cross section of the coastal aquifer in the Nitzanim area, where the different sub-aquifers are denoted A, B, C, and D (after Yechieli et al. 2009). Borehole 112 is located in the inserted map. The $^{14}C$ of the dissolved inorganic carbon (DIC) data and ages are shown. Ages are corrected due to reaction processes, as explained in the text. Clay layers are shown in yellow. Elevation is in meters relative to sea level. The $^{14}C$ value of borehole T/12b was taken from Yechieli et al. (2001)

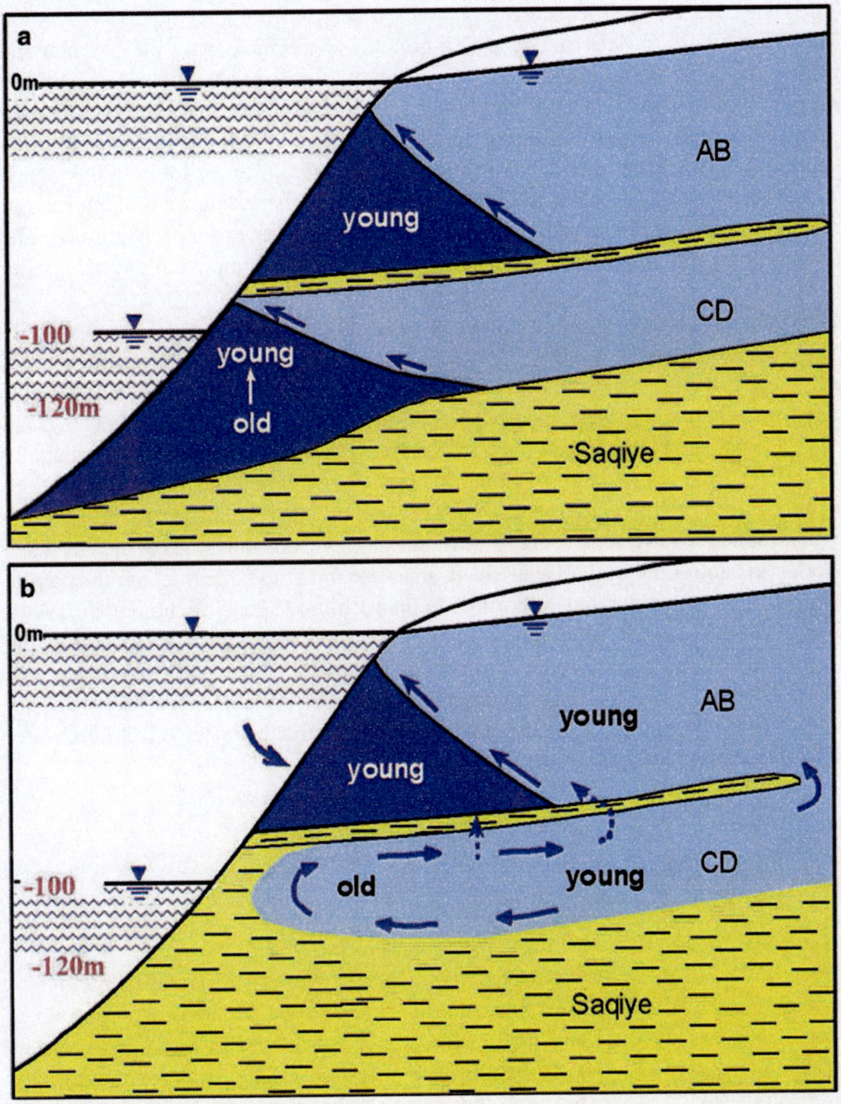


**Fig. 8** Conceptual models of groundwater flow and ages and their connection with the sea (modified after Yechieli et al. 2009). **a** The upper sub-aquifers are connected to the sea, while the lower ones are not. **b** Both the upper and the lower sub-aquifers are connected to the sea. The sea level of 120 below sea level (bsl) is shown because it is the lowest level in recent sea level history (in the last glacial period)

**The Dead Sea Coastal Aquifer**

The first attempt to date the Dead Sea groundwater brines using radiocarbon and tritium was done by Yechieli et al. (1996) showing that the brines do not constitute a simple mixture of current Dead Sea water and fresh groundwater, even though their salinity is usually about half that of the Dead Sea. According to their radiocarbon values (<10 pmc), the age of the brine was estimated to exceed ~10,000 years. It should be noted that due the complexity of their chemistry and the problematics of radiocarbon dating, these values should be taken only as an order of magnitude. The low tritium values also indicate that there is no significant component of young water in these brines.

The drilling of new boreholes in the Arugot stream allowed a more detailed study of saline groundwater in several boreholes near the Dead Sea (Avrahamov et al. 2009). The dating of the saline water in these boreholes, which are located at short distances from the shoreline, mostly up to 100 m from the shore, implies that the Dead Sea brine penetrated in recent time into the alluvial fan of Arugot stream (Avrahamov et al. 2009). The presence of tritium in these waters, although in low concentration, also indicates a relatively fast rate of penetration of Dead Sea brine.

It is interesting to note that also the brines of the evaporation ponds of the Dead Sea Works (located south of the Dead Sea itself) were found to penetrate into the coastal aquifer adjacent to the ponds according to their tritium values (Yechieli et al. 1994). This penetration was further supported by the unique low Na/Cl ratio in the saline groundwater whose only source could be the brine of the ponds. A similar rationale was also used for tracking the intrusion of the Dead Sea brine into its adjacent coastal aquifer (Kiro et al. 2013; see also Chapter "Geochemical Aspects of Groundwater in the Dead Sea Coastal Aquifer").

## *4.3 Noble Gases*

**Saline Groundwater in the JGA**

The existence of saline groundwater in the JGA in Western Israel has been known for many years. The source of this saline water has been debated for a long time, suggested to be from the Neogene time (e.g., Bein and Burg 2001) or through current intrusion (Kafri 1967; Kafri and Arad 1979; Paster et al. 2006). Some preliminary radiocarbon analyses yielded very low value implying relatively old ages, but no specific ages could be determined. The unique combination of analyses of two noble gases ($^{81}$Kr and $^{39}$Ar) allowed a better age resolution than available before and even a separate estimation for the fresh and saline water components (Fig. 9) (Yechieli et al. 2019). This dating indicated that the age of the saline water in this very deep aquifer (~1000 m deep) is younger than 40 kyr, implying that the seawater intrusion started following the sea level rise after the last glacial period.

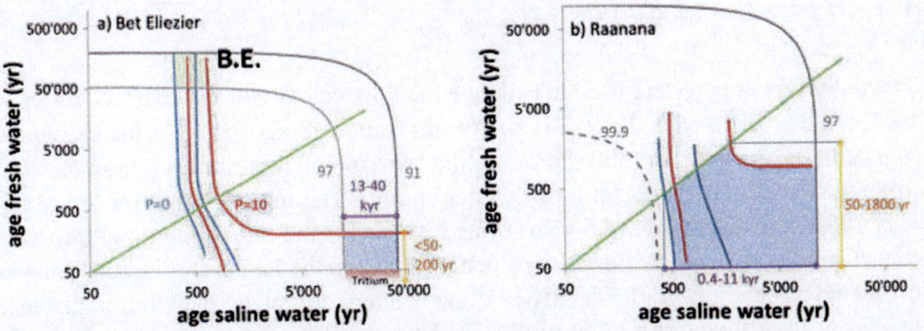


**Fig. 9** Age ranges of saline and fresh groundwater determined by $^{39}Ar$ and $^{81}Kr$ data (modified after Yechieli et al. 2019). The gray areas indicate possible ages based on the $^{81}Kr$ data and the mixing ratios according to Cl concentration. The bandwidth is defined by the analytical uncertainty. The red areas define constraints from the $^{39}Ar$ data including the analytical uncertainties and a range for underground production between 0 and 10% modern. Age ranges that fulfill the constraints for both tracers correspond to the shaded areas. The criterion tf < ts is therefore included for each plot by the green line. The final concluded age bands for the saline (violet numbers) and freshwater (orange numbers) are also given

## Mediterranean Coastal Aquifer

The dating of seawater intrusion into the Mediterranean coastal aquifer was preliminarily conducted with noble gases of $^{39}Ar$, $^{85}Kr$, and $^{3}He$. The value of ~50% 39Ar (half-life of 269 years) in the saline water at a borehole at a distance of 700 m from the sea yields an age of ~300 years, implying an average rate of seawater intrusion of ~2 m/yr (Reznik et al. 2018). The relatively high values of tritium and He in the boreholes closer to the sea also indicate a velocity of intrusion of >2 m/yr.

## Dead Sea Coastal Aquifer

As stated above, the age of the DS groundwater brines has been an issue for research for a long time with no specific resolution. The chemical similarity between the Qedem spring brine and pore-water from a Dead Sea deep drilling record, dated as ~30–50 kyr, hinted that the Qedem brine could have penetrated from the sea during that time. The preliminary $^{81}Kr$ dating of one sample yielded an age of ~30 kyr, supporting the above result. It should be noted that penetration occurred at the time of the high-stand of Lake Lisan (the precursor of the DS) allowing the brine of the lake to penetrate, as expected, due to its higher pressure compared to that of the on-land groundwater (Weber et al. 2018). The reverse process, of drainage of the residual brines toward the Dead sea, is occurring at present due to the fast drop in the DS level.

## 4.4 Radium and Radon

The radium isotopes were used to estimate the flow velocity in the coastal aquifer of the Dead Sea (Kiro et al. 2012, 2013). The advantage of this aquifer is the extremely high salinity in its water, and therefore, the radium behaves almost conservatively, different from its behavior in a freshwater aquifer. The main isotope for this study was $^{228}Ra$ whose half-life (5.5 years) was suitable for the expected rate of intrusion in this aquifer. Indeed, the velocity of penetration and the lake water circulation was estimated to be ~3 m/yr, the Dead Sea brine reaching an inland penetration distance of ~70 m in ~20 years (Kiro et al. 2013). This circulation rate was later used also for mass balance calculation of the contribution of water and salts by groundwater toward the Dead Sea proper (Kiro et al. 2015).

The radium and radon isotopes in the Dead Sea coastal aquifer were also used to test another concept, regarding the age of a groundwater system rather than the usual age of a specific water molecule (Kiro et al. 2015). This system age means the time at which a groundwater system is existing without a change in the flow regime. As expected, the system age at the Arugot stream is young due to the drop in the DS level, while the system at the Qedem spring is older (Kiro et al. 2015).

An attempt to use the same method of radium isotopes was conducted in the Mediterranean coastal aquifer, assuming that the salinity of seawater is enough to cause the radium to behave conservatively (somewhat similar to its behavior near the Dead Sea). However, this preliminary study showed that the picture is more complicated here (Weinstein, personal communication) and requires further studies.

## 4.5 Artificial Tracers

Several studies were conducted near the Dead Sea to estimate flow velocity with artificial tracers. The first study was done in the salt layer next to the Dead Sea, using the tracer of extremely depleted $\delta^{18}O$ values, showing that the salt is not necessarily an impermeable layer (Yechieli et al. 1995) as was suggested before. The flow velocity was measured with the MLS technique as explained above. It should be noted that the calculation of the flow velocity in this system took into account a required lower diffusion rate of the Dead Sea brine (Yechieli and Ronen 1997) and a higher viscosity and density (Weisbrod et al. 2016), which affect significantly the value of hydraulic conductivity, and therefore the flow velocity.

Another attempt was done adjacent to the Dead Sea shoreline with dye tracers to estimate the inflow and outflow of the Dead Sea. While the inflow of the Dead Sea brine could not be estimated, the flow toward the Dead Sea was found to be very fast, of several meters per day (Yechieli et al. 2012). Prior to the implementation of the dye tracer, a research was conducted on several tracers, examining which one is relatively conservative and thus more suitable for the Dead Sea environment (Magal et al. 2008). A somewhat similar study was conducted also in the Mediterranean coastal aquifer, showing also a fast flow (2 m/day) of the saline water toward the sea (Yechieli et al. 2012).

**Acknowledgements** The studies in this chapter could not be done without the help of the late Haim Hemo from the Geological Survey of Israel, who did most of the field work with great care and knowledge. I also want to thank all the colleagues that cooperate with me in this research during the past 30 years, from Israel and abroad.

## References

Acworth RI, Dasey GR (2003) Mapping of the hyporheic zone around a tidal creek using a combination of borehole logging, borehole electrical tomography and cross-creek electrical imaging, New South Wales, Australia. Hydrogeol J 11(3):368–377

Aggarwal PK (2013) Introduction, Chapter 1. In: Suckow, A., Aggarwal, P.K., Araguas-Araguas LJ (eds) Isotope methods for dating old groundwater. International Atomic Energy Agency (IAEA), Vienna, pp 16–19

Aggarwal PK, Araguas-Araguas L, Choudhry M, van Duren M, Froehlich K (2014) Lower groundwater $^{14}$C age by atmospheric $CO_2$ uptake during sampling and analysis. Groundwater 52(1):20–24

Aggarwal PK, Matsumoto T, Sturchio NC, Chang HK, Gastmans D, Araguas-Araguas LJ, Jiang W, Lu Z-T, Mueller P, Yokochi R, Purtschert R, Torgersen T (2015) Continental degassing of $^{4}$He by surficial discharge of deep groundwater. Nat Geosci 8:35–39

Althaus R, Klump S, Omnis A, Purtschert R, Kipfer R, Stauffer F, Kinzelbach W (2009) Noble gas tracers for characterisation of flow dynamics and origin of groundwater: a case study in Switzerland. J Hydrol 370:64–72

Arbel Y, Greenbaum N, Lange J, Inbar M (2010) Infiltration processes and flow rates in developed karst vadose zone using tracers in cave drips. Earth Surf Process Landforms 35:1682–1693

Avisar D, Kronfeld J, Kolton J, Rosenthal E, Weinberger G (2001) The sources of the Yarkon Springs. Radiocarbon 43:793–799

Avrahamov N, Sivan O, Yechieli Y, Lazar B, Levenberg O (2009) Characterization and dating of saline groundwater in the Dead Sea area. Radiocarbon 52(3):1123–1140

Avrahamov A, Yechieli Y, Purtschert R, Levy Y, Sültenfuß J, Vergnaud V, Burg A (2018) Characterization of a carbonate karstic aquifer flow system using multiple radioactive noble gases ($^{3}$H-$^{3}$He, $^{85}$Kr, $^{39}$Ar) and $^{14}$C as environmental tracers. Geochim Cosmochim Acta 242:213–232

Bein A, Burg A (2001) Quantitative, 3-D hydrogeological model for the Yarkon-Tanninim aquifer as a tool for assessment of operational limitation ("red lines") and planning of optimal exploitation, opening report; Rep GSI/28/2001. Geological Survey of Israel, Jerusalem, Israel (in Hebrew)

Bergelson G, Nativ R, Bein A (1999) Salinization and dilution history of ground water discharging into the Sea of Galilee, the Dead Sea transform, Israel. Appl Geochem 14:91–118

Broecker WS, Gerard R (1969) Natural radiocarbon in the Mediterranean Sea. Limnol Oceanogr 14(6):883–888

Bruce DL, Friedman GM, Kaufman A, Yechieli Y (2001) Spatial variations of radiocarbon in the coastal aquifer of Israel—indicators of open and closed systems. Radiocarbon 43(2B):735–743

Bruce DL, Yechieli Y, Zilberbrand M, Kaufman A, Friedman GM (2007) Delineation of coastal aquifer of Israel based on repetitive analysis of $^{14}$C and tritium. J Hydrol 343:56–70

Burg A, Zilberbrand M, Yechieli Y (2013) Radiocarbon variability in groundwater in an extremely arid zone—the Arava Valley, Israel. Radiocarbon 55(2–3):963–978

Busenberg E, Plummer LN (1992) Use of chlorofluorocarbons ($CCl_3F$ and $CCl_2F_2$) as hydrologic tracers and age-dating tools: the alluvium and terrace system of central Oklahoma. Water Resour Res 28(9):2257–2283

Carmi I, Kronfeld J, Yechieli Y, Boaretto E, Bar-Matthews M, Ayalon A (2004) A direct estimate of the initial concentration of 14C in the Mountain Aquifer of Israel. Radiocarbon 46(2):497–500

Carmi I, Kronfeld J, Yechieli Y, Yakir D, Stiller M, Boaretto E (2007) Quantitative extraction of dissolved inorganic carbon (as $CO_2$) and water by vacuum distillation from sediments of the unsaturated zone for carbon isotope analysis ($^{13}C$ and $^{14}C$). Radiocarbon 49(1):83–94

Dafny E (2009) Groundwater flow and solute transport within the Yarqon-Tanninim Aquifer, Israel. Ph.D. thesis, Hebrew University, Jerusalem (in Hebrew), English Abstract

Dahan O, Talbi R, Yechieli Y, Adar E, Enzel Y (2009) Insitu monitoring of water percolation and solute transport using a vadose zone monitoring system. J Vadose Zone 8:1–10

Dvory NZ, Livshitz Y, Kuznetsov M, Adar E, Gasser G, Pankratov I, Lev O, Yakirevich A (2018) Caffeine vs. carbamazepine as indicators of wastewater pollution in a karst aquifer. Hydrol Earth Syst Sci 22:6371–6381

Gat JR, Issar A (1974) Desert isotope hydrology: water sources of the Sinai Desert. Geochim Cosmochim Acta 38(7):1117–1131

Gerber C, Vaikmäe R, Aeschbach W, Babre A, Jiang W, Leuenberger M, Lu Z-T, Mokrik R, Müller P, Raidla V, Saks T, Waber HN, Weissbach T, Zappala JC, Purtschert R (2017) Using $^{81}Kr$ and noble gases to characterize and date groundwater and brines in the Baltic Artesian Basin on the one-million-year timescale. Geochim Cosmochim Acta 205:187–210

Ghasemizadeh R, Hellwege F, Butscher C, Padilla I, Vesper D, Field M, Alshawabkeh A (2012) Review: groundwater flow and transport modeling of karst aquifers, with particular reference to the North Coast Limestone aquifer system of Puerto Rico. Hydrogeol J 20(8):1441–1461

Gonfiantini R, Zuppi GM (2003) Carbon isotope exchange rate of DIC in karst groundwater. Chem Geol 197:319–336

Guttman J, Kronfeld J, Carmi I (2011) Dating of groundwater recharge in two small adjacent aquifers in Israel and their initial $^{14}C$ activities. Radiocarbon 53(1):137–149

Gvirtzman H, Magaritz M (1986) Investigation of water movement in the unsaturated zone under an irrigated area using environmental tritium. Water Resour Res 22(5):635–642

Gvirtzman H, Ronen D, Magaritz M (1986) Anion exclusion during transport through the unsaturated zone. J Hydrol 87:267–283

Issar A, Bein A, Michaeli A (1972) On the ancient water of the Upper Nubian sandstone aquifer in central Sinai and southern Israel. J Hydrol 17:353–374

Kafri U (1967) Facies changes in the southwestern Carmel (Israel) and their influence on groundwater regime. Isr J Earth Sci 16:206–214

Kafri U (2001) Radon in groundwater as a tracer to assess flow velocities: two test cases from Israel. Environ Geol 40:392–398

Kafri U, Arad A (1979) Current subsurface intrusion of Mediterranean seawater—a possible source of groundwater salinity in the Rift Valley system, Israel. J Hydrol 44:267–287

Kagabu M, Matsunaga M, Ide K, Momoshima N, Shimada J (2017) Groundwater age determination using $^{85}Kr$ and multiple age tracers ($SF_6$, CFCs, and $^3H$) to elucidate regional groundwater flow systems. J Hydrol Reg Stud 12:165–180. https://doi.org/10.1016/j.ejrh.2017.05.003

Kaufman A, Bar-Matthews M, Ayalon A, Carmi I (2003) The vadose flow above Soreq Cave, Israel: a tritium study of the cave waters. J Hydrol 273:155–163

Kiro Y, Yechieli Y, Voss C, Starinsky A, Weinstein Y (2012) Modeling radium distribution in coastal aquifers during sea level changes: the Dead Sea case. Geochim Cosmochim Acta 88:237–254

Kiro Y, Weinstein Y, Starinsky A, Yechieli Y (2013) Groundwater ages and reaction rates during seawater circulation in the Dead Sea aquifer. Geochim Cosmichim Acta 122:17–35

Kiro Y, Weinstein Y, Starinsky A, Yechieli Y (2015) Application of radon and radium isotopes to groundwater flow dynamics: an example from the Dead Sea. Chem Geol 411:155–171

Kolton Y (1988) Examination of the connection between groundwater and seawater in the Pleistocene aquifer in the continental shelf of the Mediterranean sea coast in Central Israel. Tahal Rept 01/88/31 (in Hebrew)

Kroiteru L, Carmi I, Mazor E (1989) Groundwater as affected by initial water-rock interactions in the Judea Mountains, Israel. Chem Geol (isotope geoscience section) 79:259–274

Kronfeld J, Rosenthal E (1981) Uranium isotopes as a natural tracer in the waters of the Bet-Shean Harod Valleys, Israel. J Hydrol 50:179–190

Kronfeld J, Rosenthal E, Weinberger G, Flexer A, Berkowitz B (1993) The interaction of two major old water bodies and its implication for the exploitation of groundwater in the multiple aquifer system of the central and northern Negev, Israel. J Hydrol 143(3–4):169–190

Lehmann B, Davis S, Fabryka MJ (1993) Atmospheric and subsurface sources of stable and radioactive nuclides used for groundwater dating. Water Resour Res 29:2027–2040

Linderfelt WR, Turner JV (2001) Interaction between shallow groundwater, saline surface water and nutrient discharge in a seasonal estuary: the swan-canning system. Special issue: integrating research and management for an urban estuarine system. The Swan-Canning Estuary, Western Australia. Hydrol Proc 15(13):2631–2653

Magal E, Weisbrod N, Yakirevitz A, Yechieli Y (2008) Artificial tracer in saline groundwater. J Hydrol 358:124–133

Magal E, Weisbrod N, Yakirevich A, Kurtzman D, Yechieli Y (2010) Line-source multi-tracer test for assessing high groundwater velocity. Ground Water 48(6):892–897

Magal E, Arbel Y, Caspi S, Glazman H, Greenbaum N, Yechieli Y (2013) Determination of pollution and recovery time of karst springs, an example from a carbonate aquifer in Israel. J Contam Hydrol 145:26–36

Mayer A, Sültenfuß J, Travi Y, Rebeix R, Purtschert R, Claude C, Le Gal La Salle C, Miche H, Conchetto E (2014) A multi-tracer study of groundwater origin and transit-time in the aquifers of the Venice region (Italy). Appl Geochem. https://doi.org/10.1016/j.apgeochem.2013.10.009

Melloul AJ, Zeitoun DG (1999) A semi-empirical approach to intrusion monitoring in Israeli coastal aquifer. In: Bear J et al (eds) Seawater intrusion in coastal aquifers—concepts, methods and practices. Kluwer Academic Publishers, The Netherlands, pp 543–558

Mercado A (1985) The use of hydrogeochemical patterns in carbonate sand and sandstone aquifers to identify intrusion and flushing of saline water. Ground Water 23:635–645

Mook WG (1980) Carbon-14 in hydrogeological studies. In: Fritz P, Fontes C (eds) Handbook of environmental isotope geochemistry, vol 1. The terrestrial environment. Elsevier Scientific Publishers, New York, pp 49–74

Nativ R, Weisbrod N (1994) Hydraulic connections among sub-aquifers of the coastal plain aquifer, Israel. Ground Water 32:997–1007

Oz I, Shalev E, Yechieli Y, Gvirtzman H (2015) Saltwater circulation patterns within the freshwater–saltwater interface in coastal aquifers: laboratory experiments and numerical modeling. J Hydrol 530:734–741

Paster A, Dagan G, Guttman J (2006) The salt-water body in the Northern part of Yarkon-Tanninim aquifer: field data analysis, conceptual model and prediction. J Hydrol 323:154–167

Purtschert R, Loosli HH, Corcho JA (2004) How reliable are 39Ar ages? In: International workshop on the application of isotope techniques in hydrological and environmental studies. UNESCO, Paris

Purtschert R, Sturchio NC, Yokochi R (2013) Krypton-81 dating of old groundwater. In: Suckow A, Aggarwal P, Araguas-Araguas L (eds) Isotope methods for dating old groundwater. IAEA, Vienna, pp 91–124

Ram R, Purtschert R, Vockenhuber C, Yechieli Y, Burg A, Adar EM, Yokochi R, Zappala JC, Lu Z-T, Mueller P, Jiang W, Bernier R (2018) Calibration of $^{36}$Cl dating with $^{81}$Kr ages in the brackish Nubian Aquifer of the Negev Desert (Israel). Abstract in 45th IAH, Daejeon, Korea

Reznik IJ, Yechieli Y, Purtschert R, Sültenfuß J, Vergnaud V, Avrahamov A, Wollman S, Weinstein W (2018) Direct determination of the rate of seawater intrusion with noble gases. Abstract, SWIM, Gdansk, Poland, p 250

Rogojin V, Carmi I, Kronfeld K (1998) $^{14}C$ and $^{234}U$ excess dating of groundwater in the Haifa Bay region, Israel. Radiocarbon 40(2):945–951

Ronen D, Magaritz M, Paldor N, Bachmat Y (1986) The behavior of groundwater in the vicinity of the water table evidenced by specific discharge profiles. Water Resour Res 22:1217–1224

Ronen D, Magaritz M, Molz FJ (1991) Comparison between natural and forced gradient tests to determine the vertical distribution of horizontal transport properties of aquifers. Water Resour Res 27:1309–1314

Rosenthal A, Kronfeld J (1982) $^{234}U/^{238}U$ disequilibria as an aid to the hydrological study of the Judea Group aquifer in eastern Judea and Samaria, Israel. J Hydrol 58:149–158

Russak A, Sivan O, Yechieli Y (2016) Trace elements (Li, B, Mn and Ba) as sensitive indicators for salinization and freshening events in coastal aquifers. Chem Geol. https://doi.org/10.1016/j.chemgeo.2016.08.003

Sanford WE (1997) Correcting for diffusion in carbon-14 dating of groundwater. Groundwater 35:357–361

Sheffer NA, Cohen M, Morin E, Grodek T, Gimburg A, Magal E, Gvirtzman H, Nied M, Isele D, Frumkin A (2011) Integrated cave drip monitoring for Epikarst recharge estimation in a dry Mediterranean Area, Sif Cave—Israel. Hydrol Process 25(18):2837–2845

Sivan O, Yechieli Y, Herut B, Lazar B (2005) Geochemical evolution and timescale of seawater intrusion into the coastal aquifer of Israel. Geochim Cosmochim Acta 69:579–592

Suckow A, Aggarwal PK, Araguas-Araguas L (2013) Isotope methods for dating old groundwater. IAEA, Vienna

Vengosh A, Hening S, Ganor J, Mayer B, Weyhenmeyer CE, Bullen TD, Paytan A (2007) New isotopic evidence for the origin of groundwater from the Nubian Sandstone Aquifer in the Negev, Israel. Appl Geochem 22:1052–1073

Visser A, Broers HP, Purtschert R, Sültenfuß J, de Jonge M (2013) Groundwater age distributions at a public drinking water supply well field derived from multiple age tracers ($^{85}Kr$, $^{3}H/^{3}He$, and $^{39}Ar$). Water Resour Res 49:7778–7796

Weber N, Yechieli Y, Stein M, Yokochi R, Gavrieli I, Zappala J, Mueller P, Lazar B (2018) The circulation of the Dead Sea brine in the regional aquifer. Earth Planet Sci Lett 493:242–261

Weisbrod N, Yechieli Y, Shandalov S, Lensky N (2016) On the viscosity of natural hyper-saline solutions and its importance: The Dead Sea brines. J Hydrol 532:46–51

Yechieli Y, Ronen D (1997) Early diagenesis of highly saline sediments after exposure. Chem Geol 138:93–106

Yechieli Y, Ronen D, Carmi I, Kaufman A (1994) Tritium concentrations in fresh and saline water in the Dead Sea area. Isr J Earth Sci 43(3–4):203–212

Yechieli Y, Ronen D, Berkovitz B (1995) Are sedimentary salt layers always impermeable? Geophys Res Lett 22:2761–2764

Yechieli Y, Ronen D, Kaufman A (1996) The source and age of groundwater brines in the Dead Sea area, as deduced from $^{36}Cl$ and $^{14}C$. Geochim Cosmochim Acta 60:1909–1916

Yechieli Y, Sivan O, Lazar B, Vengosh A, Ronen D, Herut B (2001) Radiocarbon in seawater intruding into the Israeli Mediterranean coastal aquifer. Radiocarbon 43:773–781

Yechieli Y, Kafri U, Sivan O (2009) The inter-relationship between coastal sub-aquifers and the Mediterranean Sea, deduced from radioactive isotopes analysis. Hydrogeol J 17:265–274

Yechieli Y, Lutzky H, Magal E, Weinstein Y (2012) Direct measurement of groundwater flow velocity near the shoreline using chemical tracers. In: SWIM conference, Buzios, Brazil

Yechieli Y, Yokochi R, Zilberbrand M, Lu Z-T, Purtschert R, Sueltenfuss J, Jiang W, Zappala J, Mueller P, Bernier R, Avrahamov N, Adar E, Talhami F, Lifshitz Y, Burg A (2019) Recent seawater intrusion into deep aquifer determined by the radioactive noble-gas isotopes $^{81}Kr$ and $^{39}Ar$. Earth Planet Sci Lett 507:21–29

Yokochi R, Bernier R, Purtschert R, Zappala JC, Yechieli Y, Adar E, Jiang W, Lu ZT, Mueller P, Olack G, Ram R (2018) Field degassing as a new method for $^{14}C$ analyses in old groundwater. Radiocarbon 60:349–366

# Engineering and Infrastructural Aspects, Related to Groundwater Regime and Processes

# Soil Aquifer Treatment System Performance: Israel's Shafdan Reclamation System as an Ultimate Case Study

**Roy Elkayam, Ovadia Lev, Ido Negev, Oded Sued, Lilach Shtrasler, Dalit Vaizel-Ohayon, and Yoram Katz**

## 1 Introduction

Soil Aquifer Treatment (SAT) is a well-known method for treating water and polishing effluent for water reuse (Wilson et al. 1995; Peter 1998; Bouwer 2002; Drewes et al. 2003; Amy and Drewes 2006; Hernández et al. 2012; Lian et al. 2013).

Generally, the method is based on artificial recharge of effluent via infiltration ponds into the vadose (unsaturated) zone of the aquifer, where substantial contaminant removal takes place. The treated water is further stored in the saturated zone of the aquifer and later reclaimed for various reuse purposes.

Despite the simplicity of the SAT system, organizations such as the US Environmental Protection Agency (US Environmental Protection Agency 2012) and

R. Elkayam · I. Negev · O. Sued · L. Shtrasler · D. Vaizel-Ohayon · Y. Katz (✉)
Mekorot—National Water Company of Israel, 9 Lincoln St., 6713402 Tel Aviv, Jaffa, Israel
e-mail: Ykatz@MEKOROT.CO.IL

R. Elkayam
e-mail: relkayam@MEKOROT.CO.IL

I. Negev
e-mail: Nido@MEKOROT.CO.IL

O. Sued
e-mail: osved@MEKOROT.CO.IL

L. Shtrasler
e-mail: lshtrasler@MEKOROT.CO.IL

D. Vaizel-Ohayon
e-mail: Dvaizel@MEKOROT.CO.IL

O. Lev
Casali Center of Applied Chemistry, The Institute of Chemistry, Edmond J. Safra Campus, The Hebrew University, 9190401 Jerusalem, Israel
e-mail: ovadia@mail.huji.ac.il


U. Kafri and Y. Yechieli (eds.), *The Many Facets of Israel's Hydrogeology*,
Springer Hydrogeology, https://doi.org/10.1007/978-3-030-51148-7_14

the World Health Organization (World Health Organisation 2006) recommend this method, along with other advanced technologies such as reverse osmosis, advance filtration and oxidation processes, as a proper solution for water reuse.

Beginning in 1970 and developing ever since, the Dan Region Wastewater Treatment system (Shafdan) collects, treats and reclaims municipal wastewater of the Dan metropolitan of Israel, now numbering about 2½ million residents. The system satisfies two main goals: 1. It minimizes environmental hazards and health risks using advanced sewage collection and disposal facilities and prevents discharge of raw sewage into the Mediterranean Sea. 2. It preserves Israel's water resources, enabling the use of recycled water for agricultural irrigation of all types of crops without limitation.

This present chapter describes the Shafdan SAT system structure, the actions and principles governing its management and the means and methods developed to ensure the chemical and biological quality of the reclaimed water.

## 2 The Shafdan Plant Main Components

The overall recycling process involves water purification by a central mechanical–biological wastewater treatment facility (Shafdan Waste Water Treatment Plant—WWTP) and a following SAT in the aquifer. The WWTP is a conventional activated sludge process of nutrient removal (general layout in Figs. 1 and 2) which produces secondary effluent of 10:10, i.e., 10 mg/L of total suspended solids and 10 mg/L of biochemical oxygen demand (Icekson-Tal 2014).

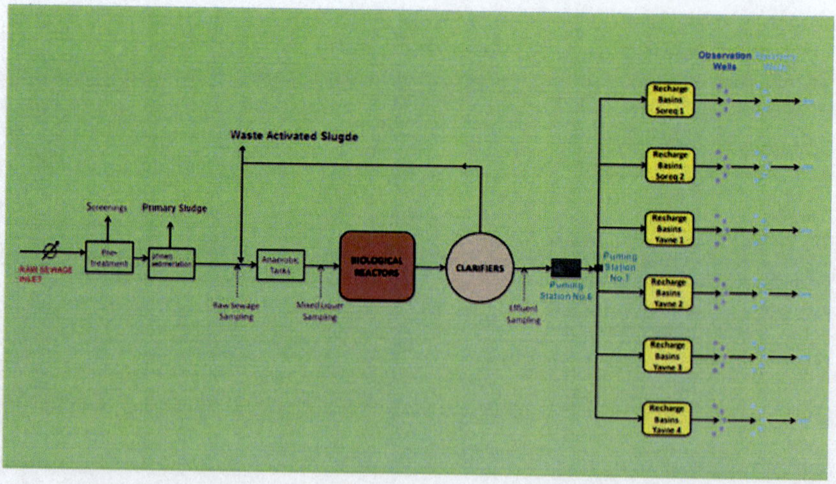


**Fig. 1** Process flow diagram of the Shafdan WWTP and reclamation system. Modified after Icekson-Tal (2014)

**Fig. 2** Shafdan WWTP (partial view). The raw sewage inlet pipe, the biological reactors and the clarifiers can be seen from back to front

Next, the secondary effluents are recharged via some 50 infiltration ponds into a hydrologically separated portion of the Israeli coastal aquifer (termed the Shafdan basin). During this activity, the effluents undergo a complementary, tertiary SAT treatment and enrich the underground reservoir for seasonal and annual storage. The water with complementing 20–30 $Mm^3$ of freshwater from the Coastal Plain aquifer is later recovered by a gallery of ~150 recovery wells, which provide annually 130–160 million cubic meters (MCM) of recycled water. The excellent quality water is delivered to the Negev region for unrestricted agriculture use, irrigating over 60% of its total crops. The SAT system layout is presented in Fig. 3.

The SAT of such a large scale demands comprehensive hydrological management and quality monitoring, in order to secure the aquifer storage, prevent effluent leakage to the fresh aquifer and maintain the Shafdan as a reliable, stable, sustainable and high-quality water provider.

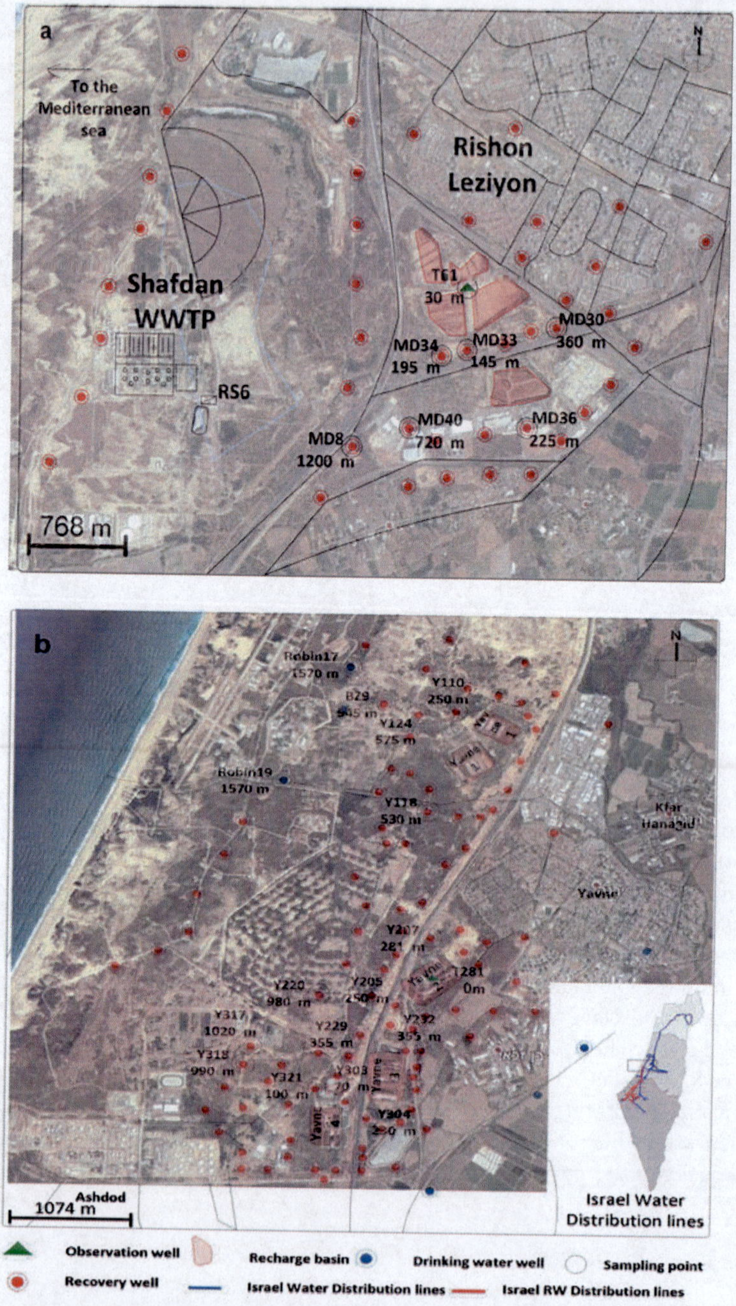


**Fig. 3** **a** The northern part (Soreq) of the Shafdan SAT system and the location of Shafdan WWTP. **b** The southern part (Yavne) of the Shafdan SAT system. A location map including Israel's major water distribution system is shown at the bottom right corner. Modified after Elkayam et al. (2018)

## 3 Hydrogeological Setting and the Related SAT System Structure

The Israeli coastal aquifer, extending along the Mediterranean Sea on the Western part of the country, consists of Pleistocene-age coastal environment rocks dominated by calcareous sandstone (The Kurkar group). The aquifer was formed during a series of depositional cycles and is characterized by interbedding of sandstone layers with marine and continental clay, silt and shale lenses. These represent the varying sea level stages during the glacial and interglacial periods of the Pleistocene (Issar 1968; Bruce et al. 2001). The sandstone suite of the Kurkar Group is still accumulating today on the shallow continental shelf (Gvirtzman and Buchbinder 1978).

The coastal aquifer overlays the late Eocene to early Pleistocene Saqiye group, a thick, impermeable clayish unit which slopes westward. Consequently, the overlying aquifer reaches its maximum thickness (180–200 m) along the coastline and gradually wedges out eastward, until it vanishes some 10–15 km inland. The vadose zone below the infiltration ponds is approximately 30–40 m thick.

Three north–south trending, sub-parallel calcareous sandstone (kurkar) ridges, situated roughly 0.5–0.7, 1.3–2 and 3–3.5 km east of the coastline and separated by clayey elongated basins, dominate the aquifer morphology. The recharge ponds are constructed mostly on the Eastern Ridge. In some places, thicker marine clay layers divide the aquifer into sub-horizontal sections. This division is more dominant toward the coast and affects the western parts of the aquifer. Although the division is irregular, four main sub-aquifers situated one on top of the other can be identified and are named from top to bottom by the letters A–D (Fig. 4).

The Shafdan SAT system produces most of the water from sub-aquifer B—it is usually 60–80 m thick and is characterized mostly by sandstone and conglomerate of fairly high hydraulic conductivities. Typically, the conductivity is on the order of 4–10 m/s (Kloppmann et al. 2009). The RWs are arranged roughly in a double-ring structure around the percolation ponds; the inner ring pumps 100% treated effluent, whereas the outer ring pumps 70–85% of treated effluent and a complementary amount of natural, regional aquifer water. The number, locations, capacities, screen positions and other features of the wells were designed to create a closed SAT system, which is isolated from the surrounding freshwater aquifer and does not affect its chemical quality.

The first infiltration basins (Soreq) were set up west of Rishon Lezion in 1977 (Icekson-Tal 2014) and were designed to treat 50 $Mm^3$/year of effluent. The SAT system has expanded ever since to meet demand, and more facilities were constructed west of Yavne and north of Ashdod (Fig. 3). Presently, overall recharge of secondary effluent is approximately 130 $Mm^3$/year; whereas, reclaimed water totals approximately 145 $m^3$/year and is pumped by some 150 recovery wells.

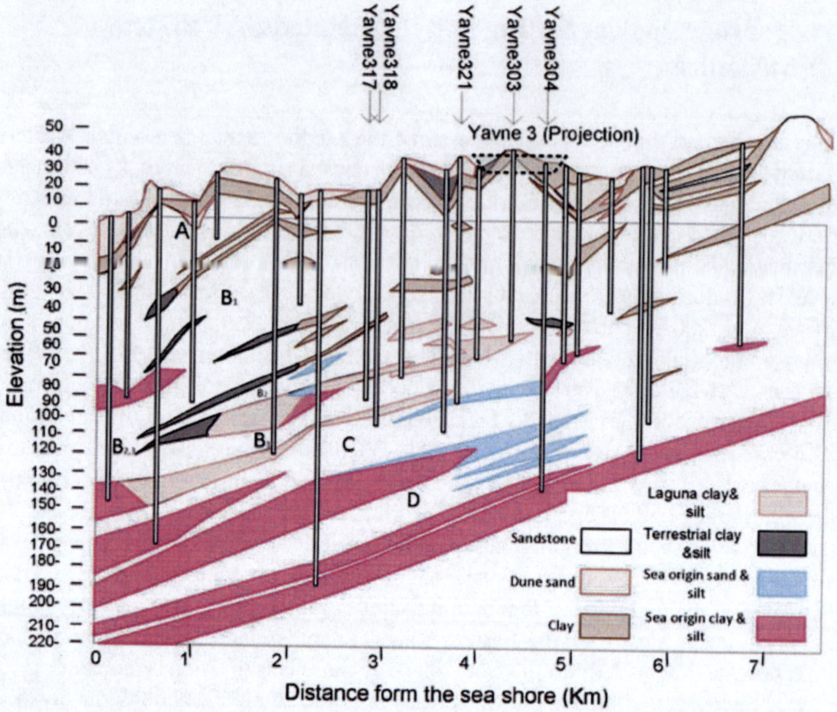


**Fig. 4** Representative geological cross section of the Israeli coastal aquifer; recharge basin Yavne 3 and some of the surrounding wells are projected over the geological cross section; capital letters indicate sub-aquifers; the base of sub-aquifer D is the Saquie formation (Elkayam et al. 2018, based on TAHAL Atlas XX)

# 4 Hydrological Management of the SAT System

## 4.1 Goals and Constrains

The primary goal of the SAT management is to supervise the system's operation according to a balanced, sustainable plan, while securing the aquifer storage, preventing effluent leakage into the regional aquifer and avoiding salinization and any deterioration of the Shafdan water quality. Among further objectives are the improvement and adjustment of the recharge—recovery arrays to the current and predicted hydrological status, tracking and prediction of water quality shifts (especially spreading of the effluent and Mn dissolution trends in the aquifer).

Several factors, sometimes contradictory, affect the operational plan and production regime of the Shafdan basin. Three of the main factors may be defined as constrains:

1. The amount of effluent that is available and can be recharged: In the absence of alternative use for the secondary level effluents flowing from the biomechanical facility, the main challenge is to recharge the entire effluent amounts into the aquifer for complementary SAT . Higher effluents amounts, or alternatively poor utilization of the available recharge capacity, result in the dismiss of significant amounts of effluent to the sea.
2. Minimal recovery amounts: Recovery amounts can vary according to hydrological considerations and different operational constrains. However, a minimal recovery estimated to be ~120% of the recharge amount is required in order to prevent leakage of effluent to the surrounding, fresh aquifer. The 20% over-pumping is assumed to claim also the natural rain recharge over the pond areas and to stabilize the hydrological barrier between the Shafdan and the surrounding regional aquifer.
3. Water consumption demand: The amounts of recovered water which is supplied for irrigation are predetermined by the Water Authority, according to water quotas given to the farmers. Moreover, the agricultural water demand regime may vary significantly in time due to shifts in agronomical parameters such as climate, seeding seasons and plant types. For example, continuous and high precipitation during a rainy year can significantly limit the irrigation season to six months only (from April to October). As a result, the recovery amounts may dramatically drop due to the limitation of production capacity of the recovery wells and due to hydraulic restrictions of the third line.

In addition to these constraints, other factors and operational considerations may affect the production regime: the availability of recovery wells, hydraulic considerations, energy-saving issues and water quality regulations (especially relating high manganese concentrations and sand amounts).

Above all, there are the long-term hydrological considerations, intending to preserve the aquifer and its water quality over time. Therefore, a skillful hydrological management is a central tool for routine operation, necessary for settling various constraints and factors and for the future planning of the plant.

### *4.2 Means and Principles for the Shafdan Plant Hydrological Management*

Hydrological management of the Shafdan plant is based on a detailed operational plan. The plan is set at the beginning of each (calendar) year and is based on the current hydrological conditions.

The hydrological condition analysis is based on up-to-date infiltration, production and rainfall data, as well as on groundwater levels recorded during the former fall season (usually during September–November). These data are used both for water balance calculations and for reliable, up-to-date evaluation of groundwater levels in the area.

Considering the abovementioned analysis, the annual infiltration prediction and the water budgets allocated by the Israel Water Authority, a detailed production plan for the coming year is prepared. The hydrological management principles are based on number of prime hydrological parameters and unique theoretical tools developed over the years.

Implementation of the production plan is complemented by close, monthly monitoring of the water balance major components, namely the infiltration in the ponds and the water amounts reclaimed by the production wells. Arrays of ~150 Shafdan recovery wells and 75 Shafdan inspection wells, as well as some 200 additional private and research, production and inspection wells, all provide further data on a monthly/yearly basis, regarding the aquifer water quality and water levels in the Shafdan vicinity. Occasional additional specific water level measurements are conducted at sensitive regions (especially at shoreline vicinity). All the gathered data are used to update the production plan and to prepare ahead for the coming year.

### 4.3 Water Balance and Groundwater Levels

The Shafdan water balance is based on the following principles:

1. A clear, although not sealing, boundary separates the Shafdan treatment volume from the fresh, regional aquifer, allowing to relate to the Shafdan aquifer volume separately.
2. The division of the Shafdan aquifer is into three sub-basins. Each basin contains a set of infiltration ponds, series of surrounding recovery wells and a series of coastal drain wells located to the west. The sub-basins are characterized by high groundwater level zones under the infiltration ponds and surrounding low-level areas created by pumping. Each sub-basin is hydrologically constrained by the sea toward west and by low head (water level) zones created by the outer recovery well rings, toward all other directions (Fig. 5).
3. A unique water balance is calculated for each sub-basin, based on its specific effluent infiltration, reclaim and estimated natural recharge.
4. The ponds evaporation component is small (<0.5%) and therefore is neglected.
5. Pumping rates by private wells adjacent to the Shafdan basins are small and are neglected as well.

The water balance components are evaluated as follows:

**Effluent infiltration** is based on monthly and annual sums of effluent volumes introduced into the infiltration ponds.

**Natural recharge** is evaluated from rainfall measurements (at the Shafdan meteorological station), assessment of the areas where effective infiltration takes place and estimation of the infiltration coefficient. The total recharge area for each sub-basin includes the ponds, the recovery wells area and some further 500–1000 m strip

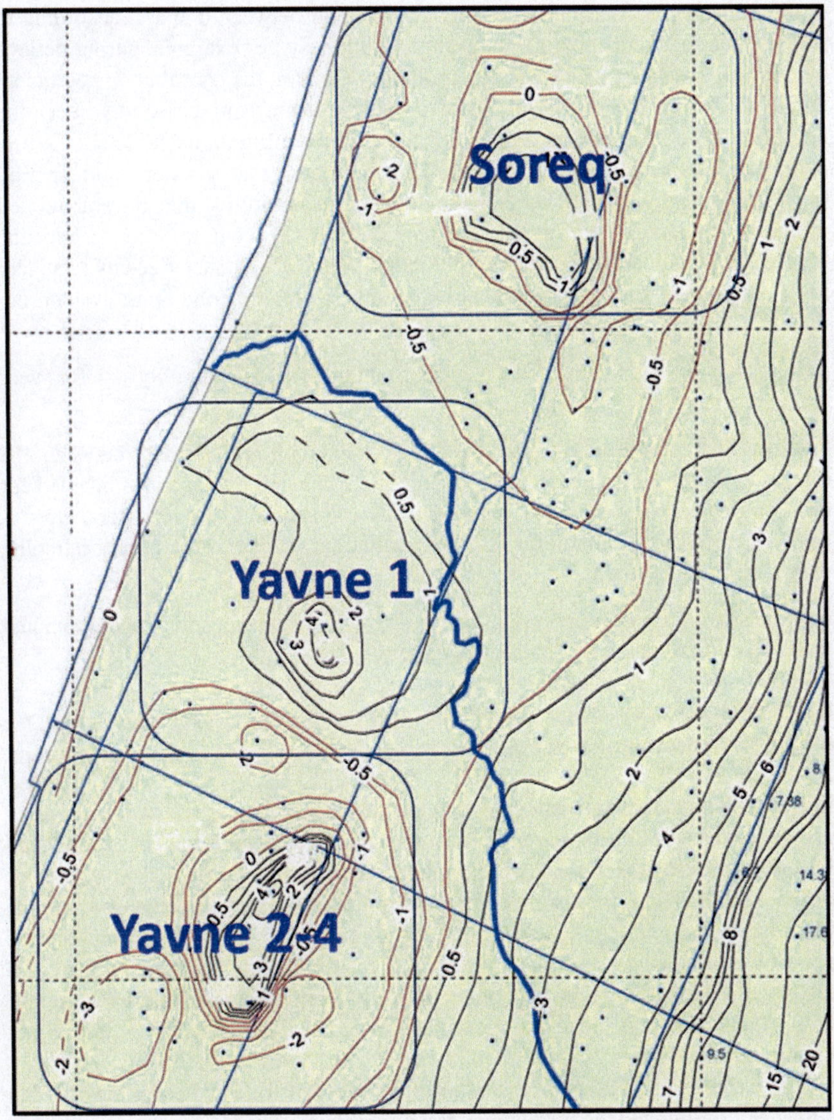


**Fig. 5** Typical groundwater level map of the coastal aquifer—the Shafdan area

beyond the outer well ring. Toward west, the area is bounded some 750 m inland, roughly estimated as the fresh–salt water interface location.

The infiltration coefficient evaluation is based on the hydrological survey data, on elaborated analysis of soil characteristics and uses and on the spatial distribution of clay layers underlying a local perched aquifer. Furthermore, assumptions regarding the recharge contribution of single rain events were made: The infiltration coefficient is multiplied by a contribution coefficient, ranging between 0 and 1, according to the amount and frequency of rain events. In this way, for example, during periods of frequent events (once a week at minimum) in winter, the contribution coefficient equals 1. During spring and fall seasons (or after a long rain cease in winter), the contribution coefficient may drop to 0.2 and alter the infiltration coefficient. During summer, the contribution coefficient is given 0, assuming that occasional summer rain events have negligible influence. It should be emphasized that the abovementioned attitude is conceptual and was not calibrated. However, it seems to better describe the local Shafdan climate and recharge conditions, which are characterized by relatively low precipitation and long periods of high temperature, dry conditions, even during winter.

**Production amounts** are derived from constant measure and monthly sum for every single recovery well of the Shafdan.

**Lateral flow components**, namely the exchange of groundwater between the Shafdan and the surrounding fresh aquifer, are evaluated by equations adapted for each sub-basin, as well as by numerical flow modeling. The net lateral flow is directed from the fresh aquifer into the Shafdan area, due to the excess pumping over infiltration policy.

**Annual storage change** in the Shafdan aquifer is evaluated via comparison of annual water level maps and water balance equations.

### 4.3.1 Groundwater Level

The hydrological management of the Shafdan aquifer includes preparation and analysis of water level maps, based on updated measurements during the former fall season. In addition, multi-year hydrographs are used for monitoring sensitive portions of the aquifer. These means are especially important as they directly associate the water balance calculations and the present aquifer state, thus enabling simultaneous update of the management plan. Water level measurements are done within ~400 recovery, production and inspection wells at the Shafdan aquifer and its surroundings. Much effort is dedicated to the collection and analysis of valuable data during a short period of time.

The water level maps are prepared via the Surfer software and cover the Shafdan influenced aquifer and the surrounding fresh aquifer, bounded from shoreline to 9 km inland. As a first step, the maps are used to locate sensitive areas where the target water levels were not reached, to track storage changes and flow directions. The maps are further compared to maps of previous years using grid volume calculations, allowing to assess the operational plan and to update it as necessary.

### 4.3.2 Groundwater Balance Equation and Its Implementation for the Sub-basins

The general groundwater balance equation is given by

$$R_C + R + \Delta f - P - \Delta V = 0 \tag{1}$$

where $R_c$ is the effluent infiltration, $R$ is the rain recharge, $\Delta f$ is the net flow from the fresh to the Shafdan aquifer, $P$ is the production of the recovery wells, and $\Delta V$ is the Shafdan aquifer storage change. All components represent water volumes.

Let B be the sum of the measurable components:

$$B = R_C + R - P \tag{2}$$

Next, the Shafdan aquifer storage delta may be evaluated using the specific yield ($S_y$), the aquifer area ($A$) and the change in hydraulic head ($H$):

$$\Delta V = A \cdot S_y \cdot \Delta H \tag{3}$$

After rearranging:

$$B = A \cdot S_y \cdot \Delta H - \Delta f \tag{4}$$

Assuming that the lateral flow is not dependent on infiltration, production or water levels in the aquifer, Eq. 4 suggests a linear relationship between $B$ and $\Delta H$. This allows to evaluate the effective porosity of the aquifer (the ratio between the curve inclination and the aquifer area) and to assess the flow of water into the Shafdan aquifer when no water level change occurs between one year to another (interception value when $\Delta H = 0$ in Fig. 6).

Figure 6 demonstrates the measured water balance $B$ versus the average annual water level change $\Delta H$, at the three sub-basins Sorek, Yavne 1 and Yavne 2–4. All component data were collected and analyzed during 2008–2017 according to the scheme described above. $\Delta H$ was calculated by water level map comparison of sequential years.

The summary of evaluated values of specific yields and net flow into each sub-basin is given in Table 1. Figure 6 suggests a good linear relationship between the measurable annual balance and $\Delta H$, demonstrating a correlation coefficient of 0.7–0.8 for all sub-basins. This correlation further supports the assumption that given the average water levels variety measured during the recent years (−1.7 to +1.1 m MSL), the net flows are not dependent on water level state nor on the infiltration and production amounts. The effective porosity values evaluated for all sub-basins are 24–27% and are typical for the main rock types composing the coastal aquifer, namely calcareous sandstone, coarse- and medium-grained sand. In general, the relatively

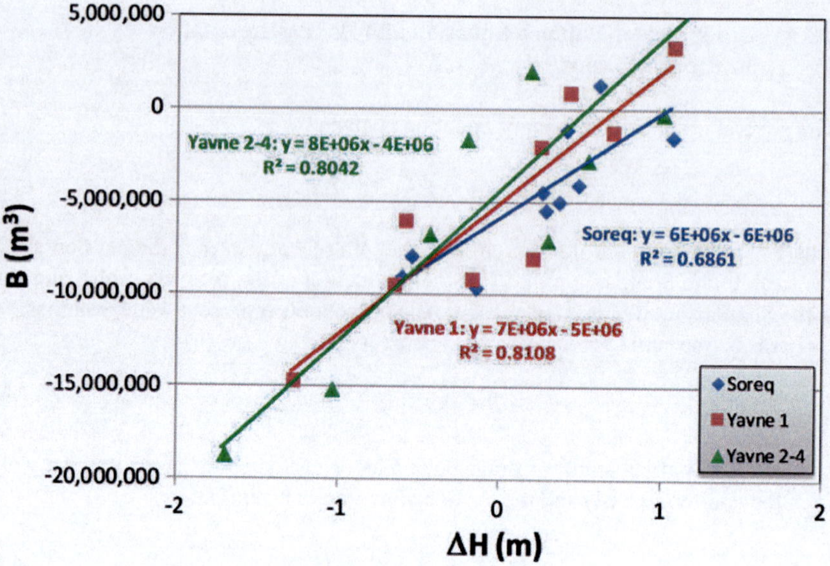


**Fig. 6** Measured water balance ($B$ = Recharge + Rain − Recovery) versus calculated water level differences (DH) for the Shafdan sub-basins

**Table 1** Calculated specific yield and net fluxes to the Shafdan sub-basins

| Sub-basin | Specific yield (%) | Yearly net flux to the basin (MCM) |
|---|---|---|
| Soreq | 27 | 5.9 |
| Yavne 1 | 24 | 5.1 |
| Yavne 2–4 | 25 | 4.4 |

good correlation between the measured balance components and $\Delta H$ enables a short-term prediction (1–3 years) of water level states expected as a result of infiltration and production policies.

### *4.4 Regions of High Hydrological Sensitivity and Relating Production Priorities*

The recovery wells are spatially dispersed and located in variable distances from the infiltration ponds, the watersheds and the shoreline. Therefore, the key for an appropriate hydrological management is a skillfull adjustment of the water production to the aquifers spatial conditions. In general, four indicative areas of different hydrological sensitivity are distinguished and are the basis for operation priorities:

1. The shoreline strip—includes the coastal drainage wells (Rishon and Rubin series) which produce water west of the infiltration ponds, some 1.5 km inland. Water production is taking place close to the fresh–saline interface, hence classifies this area as highly sensitive. Accordingly, the production in this area is limited and aims at balancing the local natural recharge, while maintaining water levels around (+)1 MSL.
2. The low water head areas—bounding the Shafdan aquifer toward north, east and south and separating it from the surrounding fresh aquifer. The production from this area is of the highest priority and aims at the delineation of the tertiary treatment activity while maintaining water levels of 0 to (−)1 MSL.
3. The infiltration ponds vicinities—characterized by the water mounding caused by the infiltration activity. Production in these areas is generally high, maintaining water levels at (+)2 to (+)7 MLS. Due to some limitations (as temporal lows in agriculture water demand) and pumping adjustment requirements, these areas are further subdivided into subsections (east, west and "manganese") for fine-tuning production.
4. High density wells areas—various reasons have led to an uneven well distribution in the Shafdan aquifer and to consequent low hydraulic head areas. A typical example is the "Yavne low," situated SW of Yavne 2–4 ponds and exhibits a relatively wide, low water level aquifer section.

The ongoing operation of the wells follows an annual operating plan and is coordinated with the Shafdan control room. The total group of 150 recovery wells is subdivided into 11 subgroups of variable operation priority, varying from high (group 1) to low (group 10). This allows to match the operation measures consumption needs to the spatial hydrological sensitivities.

Figure 7 shows the Sorek region water level map during fall 2011. For various reasons, the operational plan was not properly implemented during that year: Excess of production was only 4% ($B = +1.3$ MCM), over-pumping of 0.5 MCM took place within the coastal drainage wells, and priority-wise production was very limited. As a result, the low head hydrological boundaries had weakened and unwanted low head areas evolved near the shoreline.

On the contrary, Fig. 8 shows the water level map of Sorek region at the end of the year 2017, during which an operational plan was properly implemented, and resulting water levels and hydrological boundaries were reached and stabilized. A proper operational plan not only enables to reach the hydrological goals, but also positively affects the maintenance and rehabilitation of wells and the company's long-term planning in general.

### *4.5 Recharged Effluent Migration in the Aquifer*

Reliable monitoring of the recharged effluent is essential for efficient management of the infiltration and recovery systems. Data regarding the location and expansion

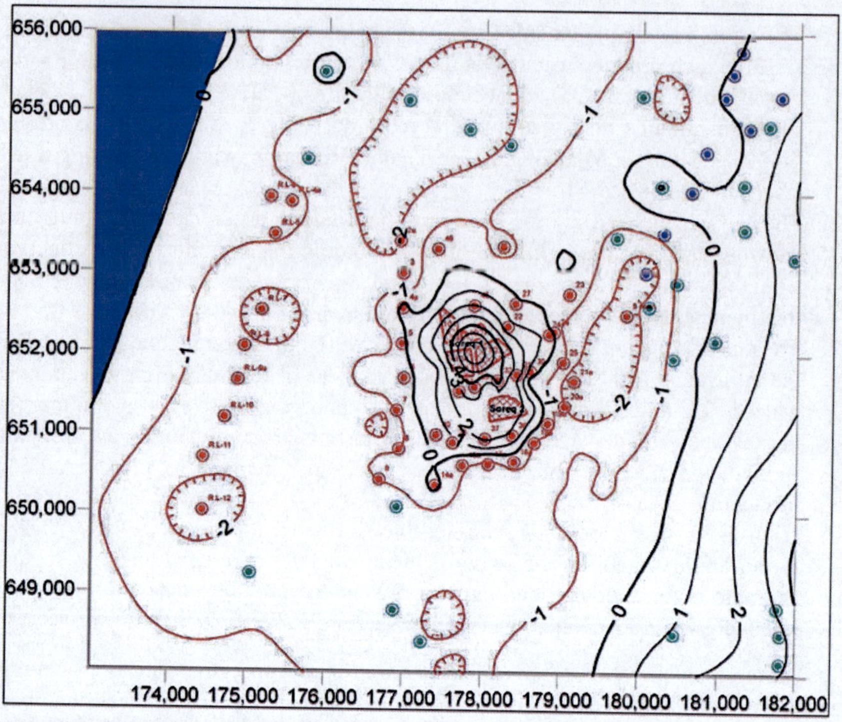


**Fig. 7** Groundwater level map of the Soreq sub-basin at the end of the year 2011. The operation regime during 2011 did not match hydrological principals of water balances and production priorities according to hydrological sensitive areas

trends of the effluent frontiers are required to keep the Shafdan aquifer well defined and prevent effluent leakage to the surrounding fresh aquifer.

The effluent expansion rate is assessed by calculating the mixing ratios of effluent and freshwater in wells (inspection, recovery and production) of the Shafdan area and its vicinity. In general, calculations are based on the measured concentrations of a representative compound (marker) dissolved in the water, as follows:

$$MR = \frac{[X]_i - [X]_b}{[X]_{ef} - [X]_b} \tag{5}$$

where MR is the calculated effluent mixing ration at a specific location (well), and $[X]_i$, $[X]_b$ and $[X]_{ef}$ are the marker concentrations in the well, in the fresh regional aquifer and in the effluent, respectively. For years, the chloride ion ($Cl^-$) was used as a primary marker due to its inert nature, its measuring simplicity and the large difference between its concentrations in the effluent (200–300 mg/l) and in the regional aquifer (20–80 mg/l). However, considerable spatial diversity of background concentrations in the regional aquifer, additional local Cl ion sources and temporal changes in

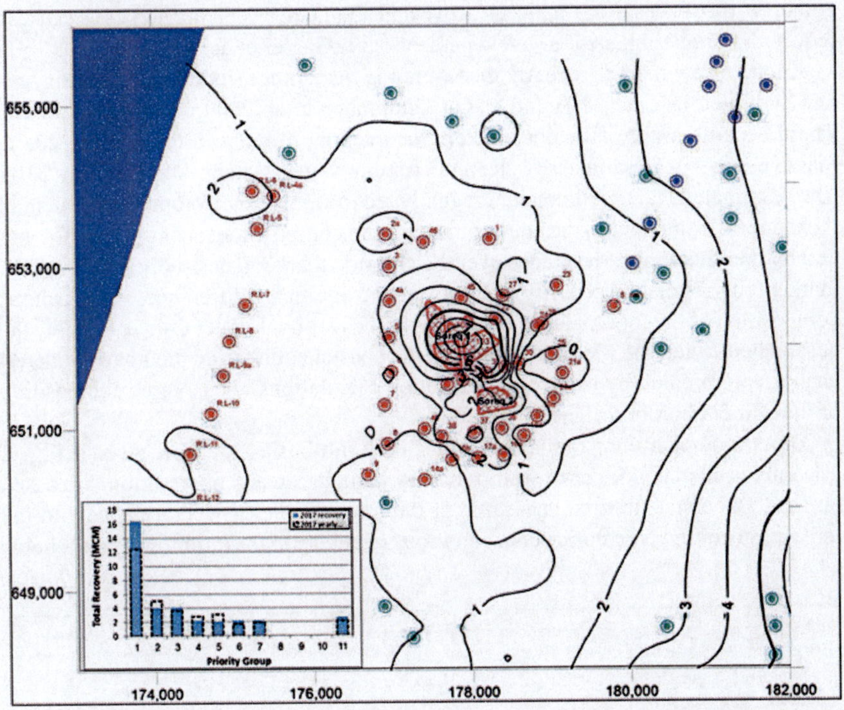


**Fig. 8** Groundwater level map of the Soreq sub-basin at the end of the year 2017. The operation regime during 2017 followed the hydrological principals of water balances and production priorities according to hydrological sensitive areas (inset)

the effluent salinity (due to massive desalinated water consumption starting ~2010) led to uncertainties and statistical errors up to 40% in mixing ratio calculations (Gasser et al. 2011).

In recent years, two new markers identified in the Shafdan SAT system enable a quantum leap of the mixing equation sensitivity, and hence, gradually substitute the chloride as a tracking marker of the effluent distribution in the aquifer.

The first marker is carbamazepine (CBZ), an organic residue of an antiepileptic drug. CBZ is dissolved in the effluents; it does not decompose or undergo absorption or deposition during the various water treatments and is being influenced only by dilution. Unlike chloride, the spatial CBZ background concentration variance in the aquifer is small, and there is a considerable three order of magnitude difference between the CBZ background concentration and its concentration in the effluent. Technological advances in analytical analysis over the last decade allow measurement of concentrations of CBZ in the range of down to 0.1 ng/l, thus enabling the calculation of mixing ratios to the resolution of a few percent. However, the precision of the analysis of organic compounds at trace levels is low, and therefore, multiple measurements can reduce the intrinsic error considerably. In addition, the use of a combination of acesulfam, a biorefractive and abundant sweetener and CBZ can also reduce the prediction error to a few percent only (Gasser et al. 2014).

Another new marker recently discovered is the isotope compositions of oxygen and hydrogen in water, $\delta^2H$ and $\delta^{18}O$ (Kloppmann et al. 2009; Negev et al. 2017). These became highly efficient markers for tracking effluent in the aquifer, due to massive entry of desalinated water into Israel's water system, beginning in 2010. The desalinated water component gradually led to significant isotopic enrichment of the effluents compared to the aquifer background composition. Using water isotopes as a marker has some advantages over the chloride marker (sensitivity) and the CBZ marker (environmental stability). Moreover, the significant differences in the isotope composition of the four main water sources of the National Water Carrier (namely the desalinated water, the Mountain Aquifer, the coastal aquifer and the Lake Kinneret) enable us to predict, by using simple dilution calculations, the isotopic composition of the effluent before recharge.

The computed mixing ratios within the Sorek sub-basin (2012) are given in Fig. 9. The ratios calculated for each well (recovery, production and observation) are based on the CBZ concentrations measured at the most distanced wells relatively to the infiltration ponds and chloride concentrations relatively to the infiltration ponds fields

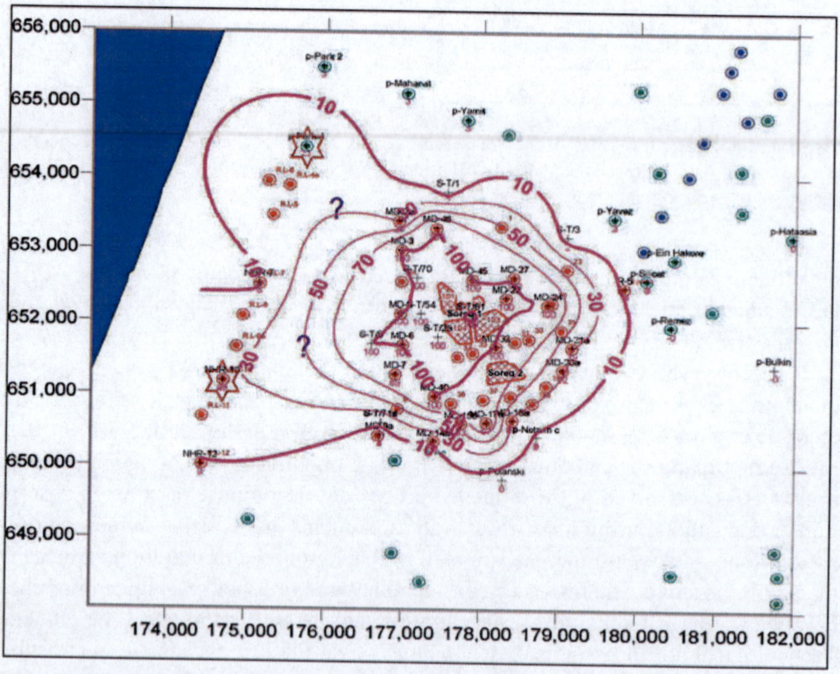


**Fig. 9** Mixing ratio (MR) between the recharge effluents and the native water at the Soreq sub-basin, 2012. The MR map is based on CBZ measured concentrations at the distant wells (MR < 70%) and on Cl concentration at the wells near the recharge fields (MR > 70%)

and on the basis of chloride concentrations at the closest wells, in which calculated mixing ratio is greater than 70%. Analysis of the spatial trends of calculated mixing ratios was performed using Surfer 8.0 software and the Kriging method.

## 4.6 *Numerical Modeling as a Management Tool*

In order to test operational scenarios, to evaluate flow directions and velocities and to estimate retention times, a numerical flow and transport model was constructed for the saturated zone.

The model base was defined by the impermeable Saquie formation underlying the aquifer. Representing a phreatic aquifer, the model top was defined by the water level. Boundary conditions along the northern, eastern and southern model edges were set to be of transient head type, based on periodic water level measurements. The western edge was set to be a constant head type dictated by the sea. Initial conditions were based on static heads measured at dozens of wells included in the model (Fig. 10).

The geological data processed from well logs, geological and structural maps served as the basis for the conceptual model, constructed via the GMS software package. The variety of rock types was represented by four material categories, each characterized by a set of hydrological properties. Hundreds of well logs were analyzed using the T-PROGS (stochastic subsurface modeling system developed by

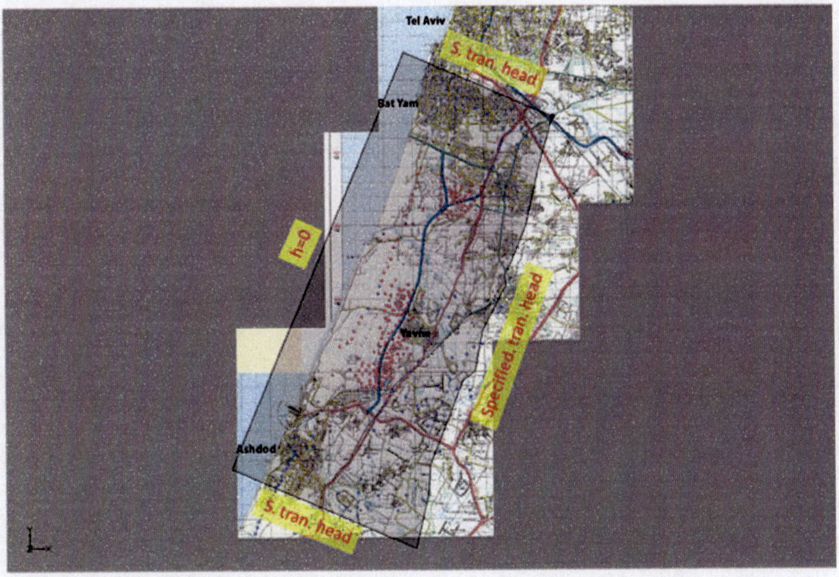


**Fig. 10** Numerical model extent and boundary settings

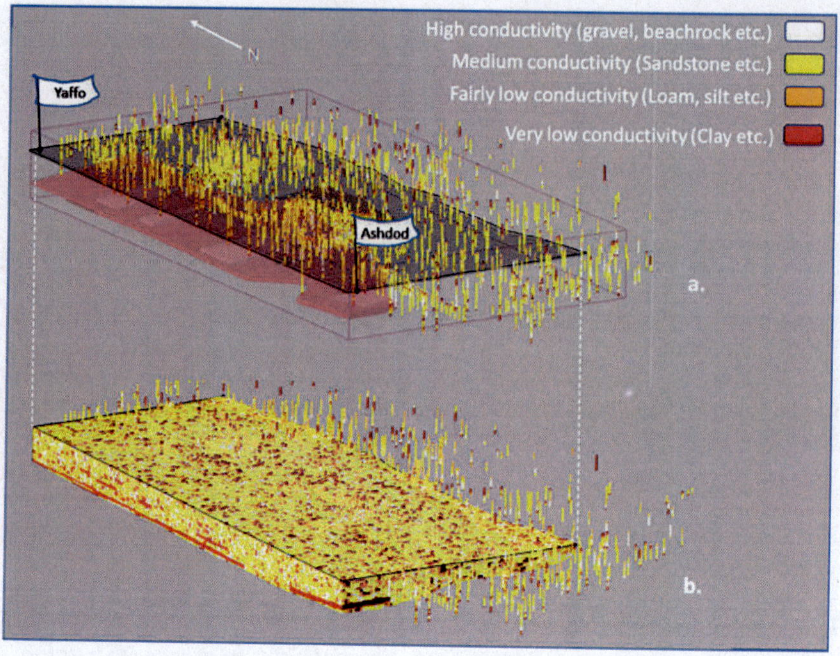


**Fig. 11** Numerical model structure: Borehole logs and marine clay structural data (**a**) and the integrated geological configuration (**b**)

S. Carle, University of California, Davis) and provided the material spatial occurrence. This stochastically generated material array, conditioned to the borehole logs, was combined with structural map data of the major marine clay lenses present in the aquifer. The resulting model hence reflects the material proportions and transition trends, as well as the division into sub-aquifers by marine clay within the western part of the aquifer (Fig. 11).

Calibration of the model was done by matching the calculated heads and breakthrough curves to field data measured at various inspection wells. Using the model abilities, various factors as diverse geology and complicated flow regime conditions, which vastly affect the retention time and cannot be accounted for by simpler methods, were considered.

# 5 Water Quality Performance of the Shafdan SAT System

## 5.1 Overview

Since 1977, annual reports consistently prove that the Shafdan reclaimed water complies with all chemical drinking water quality parameters. Elkayam et al. (2015a, b) showed that most of the DOC and all of the ammonia are removed during filtration through the unsaturated zone. Recent research (Elkayam et al. 2015a, b; Sopliniak et al. 2017a, b) showed a large removal of DOC and a complete removal of ammonia, both taking place at the uppermost aquifer layers within a few hours. Moreover, Sopliniak et al. (2015) showed that appropriate operating conditions allow oxygen to penetrate a few meters into the unsaturated zone and enhance intermittent aerobic–anaerobic conditions, thus creating a unique effluent polishing reactor.

Routine water sampling conducted over decades has proven that the reclaimed water is clean from bacteriological pollution. A detailed examination and data summary done in 2015 (Fig. 12) showed that no fecal coliforms have reached the recovery wells in the Soreq area (Fig. 13) for over three decades. Moreover, the data imply that the entire bacteriologic contamination reduction occurs within the unsaturated zone.

The role of retention times of the water quality is well known, both, the vadose zone and the saturated aquifer influence the SAT efficiency and consequent water quality. Elkayam (2012) found that the retention time in the unsaturated zone is $15 \pm 2$ days for a 3 d cycle in the Yavne 2 basin. The retention times for various wells were evaluated by the numerical model and are given along with further data in Table 2.

Elkayam et al. (2018) analyzed the Shafdan water quality as a function of the distance from the basin and of retention time proved that most of the chemical and microbial removal takes place after the percolation through the unsaturated zone.

Figure 14 presents several measured water quality indicators as a function of retention time and flow distance. The water quality improvement with time and distance is noticeable but it is still much less significant than pollutant removal in the unsaturated zone, e.g., DOC removal from about 11 mg/L to about 1.5 mg/L, complete ammonia oxidation, 17% total nitrogen removal by denitrification and complete elimination of most Emerging Organic Contaminants, EOC (Elkayam et al. 2018).

Although conventional bacteriological tests have being done since the origin of the Shafdan SAT, a comprehensive survey to assess diverse pathogen removal during the process was conducted for the first time in 2018 (Elkayam et al. 2018). The survey aimed to provide a better insight regarding SAT pathogen removal efficiency and further unfold the effect of retention time and the distance from the recharge basin on reclaimed water quality, with a focus on microbial survival. Therefore, Elkayam et al. (2018) chose a combination of identification and enumeration methods of indicator and pathogenic bacteria and viruses, using growth techniques and qPCR techniques with state-of-the-art DNA probes. Elkayam et al. (2018) performed a

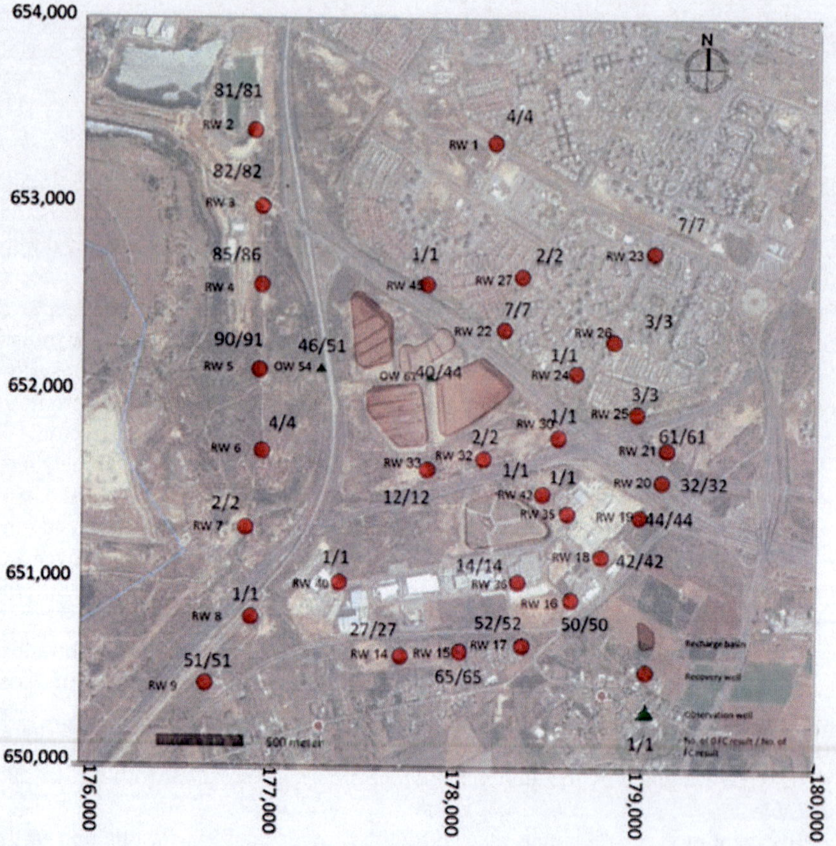


**Fig. 12** Northern part (Soreq) of the Shafdan SAT system: location of the recharge basins and the surrounding water reclamation wells (RW) is shown. The number of fecal coliform analyses during 1980–2012 and the number of negative findings are denoted near each RW (Modified after Elkayam et al. 2015)

comprehensive survey in order to obtain a representative view of the SAT controlled closed system (CCS), by examining 25 sampling sites: 1. RS6 site represented the secondary effluent feed to the recharge basins; 2. Two observation wells (OWs) that sample the water beneath the recharge basins, just below the water table. The sampled water in the OWs represented the recharged water after percolation through the vadose zone, with little retention time in the saturated zone; 3. Twenty recovery wells (RWs) located at various distances from the recharge basins and 4. Two RW sites that represented reclaimed water from the regional aquifer, reclaiming less than 3% effluent origin water (EOW). We examined the water from these 25 sites, searching for four types of waterborne pathogenic viruses (adenovirus, enterovirus, norovirus, parechovirus), together with coliphage. The viruses are known to be of human origin and to have high severity of impact, long persistence in water and high

**Fig. 13** Aerial view of several infiltration ponds of the Soreq 1 basin

relative infectivity (World Health Organisation 2006). In addition, Elkayam et al. (2018) performed microbial source tracking (MST) analysis, a molecular method that can determine the source of bacterial contamination (human, cattle, pig, etc.) (Ohad et al. 2015). The results are presented as the number of PCR cycles required to obtain identification (Cq), the fewer cycles it requires to identify the DNA, the higher is the level of the contamination. All results were transformed into ΔCq, which is the difference between the number of cycles required to obtain identification of DNA and the total cycles used, and thus, the higher the number, the higher the level of the contamination is. Since theoretically, in each PCR cycle, the amount of DNA is multiplied by 2, when comparing the results of different tests, and every additional PCR cycle indicates that the initial amount of DNA was higher by a factor of two. The difference in "X" units of ΔCq indicates exponential difference from the initial amount of the DNA sample ($2^X$). As mentioned above, the samples were tested in triplicate, and their average is presented even in cases where one or two of the triplicates tested negative ($\Delta Cq < 5$).

Elkayam et al. (2018) also incorporated into this study monitoring by means of quantitative PCR (qPCR) of two antibiotic resistance genes (16S rRNA was used as a proxy for overall bacterial abundance and blaTEM and *qnrS* which encode resistance to β-lactam and quinolone antibiotics, respectively). Antibiotic resistance genes are often targeted to assess levels of anthropogenic impact in the environment.

**Table 2** Hydrological data regarding selected wells

| Well tag | Near recharge basin | Minimal distance from basin* [m] | Calculated minimal retention time in the saturated zone [month] | Dilution ratio effluent/background [%] | Sub-aquifer | Top screen and water level height difference [m] | Bottom screen and water level height difference [m] |
|---|---|---|---|---|---|---|---|
| T 281 | Yavne 2 | 0 | 0 | 100 | AB | (+) 6 | (−) 3 |
| T 61 | Soreq 1 | 30 | 0.3 | 100 | AB | (−) 12 | (−) 36 |
| Y 303 | Yavne 4 | 70 | 0.5 | 100 | B | (−) 16 | (−) 53 |
| Y 321 | Yavne 4 | 100 | 1.5 | 100 | B | (−) 31 | (−) 57 |
| MD 33 | Soreq 1 | 145 | 2 | 100 | B (C) | (−) 45 | (−) 72 |
| MD 34 | Soreq 1 | 195 | 3 | 100 | B (C) | (−) 53 | (−) 75 |
| MD 36 | Soreq 2 | 225 | 2 | 100 | B (C) | (−) 33 | (−) 55 |
| Y 304 | Yavne 4 | 230 | 1 | 100 | B | (−) 24 | (−) 47 |
| Y 110 | Yavne 1 | 250 | 2 | 100 | B | (−) 49 | (−) 67 |
| Y 205 | Yavne 2 | 250 | 2 | 100 | B | (−) 26 | (−) 45 |
| Y 207 | Yavne 2 | 281 | 2 | 100 | B | (−) 28 | (−) 60 |
| Y 229 | Yavne 3 | 335 | 4 | 90 | B | (−) 27 | (−) 59 |
| Y 232 | Yavne 3 | 355 | 3 | 100 | B | (−) 23 | (−) 54 |
| MD 30 | Soreq 1 | 360 | 4 | 100 | B | (−) 35 | (−) 61 |
| Y 118 | Yavne 1 | 530 | 6 | 100 | B | (−) 31 | (−) 65 |
| Y 124 | Yavne 1 | 575 | 5 | 100 | B | (−) 37 | (−) 77 |
| MD 40 | Soreq 2 | 720 | 14 | 100 | C | (−) 59 | (−) 80 |
| BZ 9 | Yavne 1 | 945 | 48 | 85 | B | (−) 25 | (−) 91 |
| Y 220 | Yavne 3 | 980 | 30 | 100 | B | (−) 38 | (−) 77 |
| Y 318 | Yavne 4 | 990 | 60 | 85 | B | (−) 37 | (−) 67 |
| Y 317 | Yavne 4 | 1020 | 60 | 100 | B | (−) 33 | (−) 77 |
| MD 8 | Soreq 1 | 1200 | 60 | 100 | B | (−) 31 | (−) 59 |
| Robin 17 | Yavne 1 | 1570 | Out of range | 0 | AB | (−) 34 | (−) 65 |
| Robin 19 | Yavne 1 | 1930 | Out of range | 3 | AB | (−) 31 | (−) 60 |

The table indicates the distance from the nearest recharge basin and the computed minimal retention time in the saturated aquifer (Elkayam et al. 2018). The water retention time in the vadose zone is estimated to be about two weeks for the different percolation basins (Elkayam et al. 2012)
*Distances are measured from the edge of the nearest basin

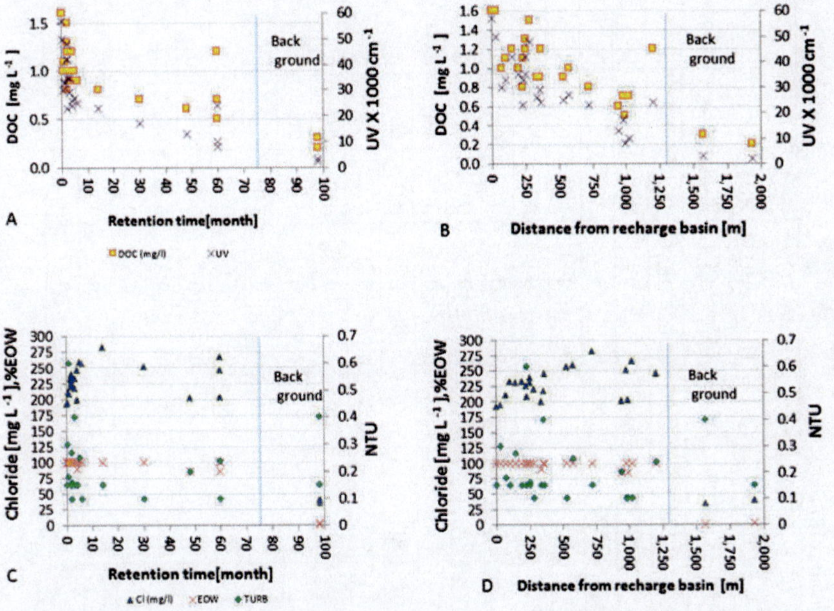


**Fig. 14** Chemical measurements of samples from various sites. **a**, **b** DOC and UV in the sampled wells as a function of the distance from the basin and of retention time. **c**, **d** Chloride, turbidity and the effluent origin water (EOW) in the sampled wells as a function of the distance from the basin and of retention time (Modified after Elkayam et al. 2018)

Finally, Elkayam et al. (2018) included in the survey examination of classical indicator bacteria [total coliforms (Coli), fecal coliforms (fecal Coli), fecal streptococcus (STREP) and total bacteria counts (TOTB)]. Table 3 shows the bacteriological and MST results. The bacterial levels in the influent to the recharge basin were similar to the annual average (Icekson-Tal 2014); Coli, fecal Coli and STREP levels were 240,000 MPN/100 mL, 160,000 MPN/100 mL and 9400 MPN/100 mL, respectively. No fecal Coli or STREP was found in any other sites; however, a single Coli (1 CFU/100 mL) was found in one sample (Y110). Site Y110 is located 250 m from the edge of the nearest spreading basin and the minimal water retention time to reach this point is 2 months (Table 2). Large counts of total heterotrophic bacteria were found in RS6 and the two OWs (T61 and T281) with approximately 71,000, 5800 and 5700 CFU/mL, respectively. Sampling site MD8, approximately 1200 m from the nearest recharge basin, had approximately 220 CFU/mL. Other sampling sites had low levels of total heterotrophic bacteria plate counts.

Elkayam et al. (2018) showed the quantification of microbial gene indicators in the sampled wells. qnrS was detected in the effluent water before infiltration, but not in any of the infiltrated groundwater samples. Additionally, it was not detected in the background water adjacent to the sampled infiltrated groundwater wells. In addition, the antibiotic resistance gene blaTEM was above the detection limit (>2 log

**Table 3** Bacteriological and MST results in all sampling sites (modified after Elkayam et al. 2018)

| Location | Minimal distance form basin [m] | Minimal retention time in the saturated zone [month] | Classic bacteriological test | | | | MST | | | | | |
|---|---|---|---|---|---|---|---|---|---|---|---|---|
| | | | Coli | Fecal coli | STREP | TOTB | GB1 | HU4 | HU5 | HU6 | EC10 | CNT8 |
| | | | CFU/100 ml | CFU/100 ml | CFU/100 ml | CFU/1 ml | ΔCq | ΔCq | ΔCq | ΔCq | ΔCq | ΔCq |
| RS6* | – | – | 240,000 | 160,000 | 9400 | 71,000 | 23.80 | 12.60 | 18.70 | 16.30 | 15.80 | 0 |
| T 281 | 0 | 0 | <1 | <1 | <1 | 5700 | 6.00 | 0 | 0 | 0 | 0 | 0 |
| T 61 | 30 | 0.3 | <1 | <1 | <1 | 5800 | 8.20 | 0 | 0 | 0 | 5 | 0 |
| Y 303 | 70 | 0.5 | <1 | <1 | <1 | 1 | 6.60 | 0 | 0 | 1.67 | 0 | 0 |
| Y 321 | 100 | 1.5 | <1 | <1 | <1 | 3 | 8.60 | 0 | 0 | 0 | 0 | 0 |
| MD 33 | 145 | 2 | <1 | <1 | <1 | 1 | 6.70 | 0 | 0 | 0 | 0 | 0 |
| MD 34 | 195 | 3 | <1 | <1 | <1 | 4 | 3.80 | 0 | 0 | 0 | 0 | 0 |
| MD 36 | 225 | 2 | <1 | <1 | <1 | 13 | 7.00 | 0 | 0 | 0 | 1.67 | 0 |
| Y 304 | 230 | 1 | <1 | <1 | <1 | 30 | 5.60 | 0 | 0 | 0 | 0 | 0 |
| Y 110 | 250 | 2 | 1 | <1 | <1 | 4 | 4.23 | 0 | 0 | 0 | 0 | 0 |
| Y 205 | 250 | 2 | <1 | <1 | <1 | 30 | 5.70 | 0 | 0 | 0 | 0 | 0 |
| Y 207 | 281 | 2 | <1 | <1 | <1 | 9 | 5.50 | 0 | 0 | 0 | 0 | 0 |
| Y 229 | 335 | 4 | <1 | <1 | <1 | 0 | 6.60 | 0 | 0 | 1.67 | 0 | 0 |
| Y 232 | 355 | 3 | <1 | <1 | <1 | 20 | 8.80 | 0 | 1.67 | 0 | 0 | 0 |
| MD 30 | 360 | 4 | <1 | <1 | <1 | 8 | 3.33 | 0 | 0 | 0 | 0 | 0 |
| Y 118 | 530 | 6 | <1 | <1 | <1 | 0 | 3.93 | 0 | 0 | 0 | 0 | 0 |
| Y 124 | 575 | 5 | <1 | <1 | <1 | 0 | 5.50 | 0 | 1.67 | 0 | 0 | 0 |
| MD 40 | 720 | 14 | <1 | <1 | <1 | 18 | 1.67 | 0 | 0 | 0 | 1.67 | 0 |
| BZ 9 | 945 | 48 | <1 | <1 | <1 | 14 | 3.33 | 0 | 0 | 0 | 0 | 0 |

(continued)

**Table 3** (continued)

| Location | Minimal distance form basin [m] | Minimal retention time in the saturated zone [month] | Classic bacteriological test | | | | MST | | | | | |
|---|---|---|---|---|---|---|---|---|---|---|---|---|
| | | | Coli | Fecal coli | STREP | TOTB | GB1 | HU4 | HU5 | HU6 | EC10 | CNT8 |
| | | | CFU/100 ml | CFU/100 ml | CFU/100 ml | CFU/1 ml | ΔCq | ΔCq | ΔCq | ΔCq | ΔCq | ΔCq |
| Y 220 | 980 | 30 | <1 | <1 | <1 | 0 | 1.67 | 0 | 0 | 0 | 0 | 0 |
| Y 318 | 990 | 60 | <1 | <1 | <1 | 1 | 3.97 | 0 | 0 | 0 | 0 | 0 |
| Y 317 | 1020 | 60 | <1 | <1 | <1 | 1 | 4.07 | 0 | 0 | 0 | 1.67 | 0 |
| MD 8 | 1200 | 60 | <1 | <1 | <1 | 220 | 5.20 | 0 | 0 | 0 | 0 | 0 |
| Robin 17 | 1570 | – | <1 | <1 | <1 | 0 | 5.50 | 0 | 0 | 0 | 0 | 0 |
| Robin 19 | 1930 | – | <1 | <1 | <1 | 2 | 6.20 | 0 | 0 | 0 | 0 | 0 |

*Results of classic bacteriological test in MPN/100 ml units copies/mL) in effluent water and was classified as DBNQ in infiltrated water (Y110 and Y118) and in the control groundwater (Robin 17 and Robin 19).

Elkayam et al. (2018) also showed the results of the tests for adenovirus, enterovirus, norovirus and parechovirus in the Shafdan SAT. These viruses were found in RS6 site at approximately 115,000 ± 7000, 65,000 ± 78,000, 13,000 ± 18,000 and 55,000 ± 7000 copies/1000 L, respectively.

Surprisingly, no pathogenic viruses were found in all sampling sites throughout the SAT system, and even more unexpected, no pathogenic viruses were found after the percolation through the unsaturated zone.

## 5.2 Retention Time, Regulation and Water Quality

For the Shafdan SAT, the regulator sets a guideline of 60 days retention time in the aquifer before supplying the reclaimed water for irrigation of crops eaten raw, and this regulation was set in order to assure all pathogenic hazards before unlimited irrigation. For decades, the notion was that the retention time in the aquifer at the Shafdan SAT varied from 6 months to more than a year, but a recent hydraulic model concluded that the retention time in the nearest recovery wells can be even shorter than one month and the longest retention time extends well beyond 3 years. Thus, some of the Shafdan SAT recovery wells may not fully comply with the guideline concerning minimal retention time. Nevertheless, as the recent study shows, the unsaturated zone plays a major role in the removal mechanism, and as shown the unsaturated zone alone is a barrier between the infiltration basins and the aquifer. It is important to note that the unsaturated thickness in the Shafdan area is between 30 and 40 m while according to Water Reuse EPA, 2012 for Indirect Potable Reuse, tertiary effluent (i.e., secondary after filtration and chlorination) should be infiltrated through a minimum of 2 m of the unsaturated zone only (and then must be "stored" 60 days in the aquifer before use). It might be more reasonable to grant a separate disinfection credit to deep unsaturated zone depending on its thickness, retention time, structure and proven microbial (and chemical) removal efficiency.

## 5.3 Suitability of the Shafdan Reclaimed Effluent for Unlimited Irrigation

The current use of the Shafdan water is unrestricted irrigation of crops eaten raw. Nowadays, more than 150 MCM treated effluents are supplied to the Northern Negev by the Shafdan system. Although some of the recovery wells retention time may be shorter than 60 days, the water quality has been high and stable for decades. The demand for 60 days retention time before the supply for unrestricted irrigation should, therefore, be carefully re-examined because it is pertinent to the use of the Shafdan effluent for unrestricted irrigation.

The U.S. Food and Drug Administration's Food Safety Modernization Act (FMSA), which was finalized in 2016 (US Food and Drug Administration 2016), specifies microbial quality criteria for crop irrigation. It does not specify the type of treatment required (to be applied to the water) but just the water quality requirements.

The American requirements (along with Israeli and other leading regulatory bodies) for unrestricted irrigation of crops eaten raw include pre-chlorination and post-chlorination (Table 4). However, as Elkayam (2019) concluded, the microbial "risk" of the use of the Shafdan effluent for irrigation, even without pre-disinfection is very low, and the Israeli demand for 60 days retention time in the aquifer before being supplied can be reduced—as the accumulated microbial observation in Elkayam et al. (2015a, b), Elkayam et al. (2018) prove that 30 days of retention in the aquifer (in addition to approximately 14 days in the unsaturated zone) are enough to achieve 0 *E. coli* consistently. In addition, pre-chlorination will generate (perhaps unnecessarily) disinfection by products and is likely to interfere with the microbial diversity in the unsaturated zone, where most of the removal of pollutants takes place.

Table 4 summarizes the leading regulatory bodies and their guidelines for bacteriological quality for unrestricted irrigation of crops eaten raw.

**Table 4** List of regulatory bodies and water reuse guidelines for unrestricted irrigation of crops eaten raw (Elkayam 2019)

| Country/regulatory body | Regulation/guidelines | Treatment processes | List of bacterial regulatory demands and disinfection | Source |
|---|---|---|---|---|
| USA | EPA water reuse | Secondary treatment<br>Filtration<br>Disinfection | No detectable fecal coliform/10 ml<br>1 mg/l $Cl_2$ residual (min.), contact time of at least 90 min | US Environmental Protection Agency (2012) |
| USA | California | Tertiary<br>Filtration (media of membrane):<br>**Media**, including coagulation—Turbidity ≤ 2 NTU<br>**Membrane:** UF, MF, NF, RO—Turbidity ≤ 0.5 NTU<br>Disinfection | Total coliform < most probable number (MPN) of 2.2 per 100 mL<br>Disinfection:<br>chlorine disinfection: CT < 450 milligram-minutes per liter, modal contact time of at least 90 min<br>Disinfection process that, when combined with the filtration process, has been demonstrated to inactivate and/or remove 99.999 percent of the plaque forming units of F-specific bacteriophage MS2, or poliovirus in the wastewater | California Department of Public Health (2014) |
| UN Food and Agriculture Organization (FAO) | | Secondary treatment<br>Filtration<br>Disinfection | <1000 fecal coliform/100 mL<br><1 Intestinal nematodes eggs/l | Pescod, Engineering, Newcastle-upon-tyne, Nations and Division (1992) |

(continued)

**Table 4** (continued)

| Country/regulatory body | Regulation/guidelines | Treatment processes | List of bacterial regulatory demands and disinfection | Source |
|---|---|---|---|---|
| JRC SCIENC | EU | Secondary treatment<br>Filtration<br>Disinfection<br>(advanced water treatments) | <10 fecal coliform/100 ml<br>Legionella spp.: ≤ 1000 cfu/l<br>helminth eggs: ≤ 1 egg/l<br>Log removal in disinfection:<br>*E. coli* ≥ 5<br>Total coliphages/F-specific coliphages/somatic coliphages ≥ 6<br>Clostridium perfringens spores/spore-forming sulfite-reducing bacteria ≥ 5<br>Alternatively, Campylobacter ≥ 5<br>Rotavirus: ≥ 6 Cryptosporidium: ≥ 5 | Laura and Bernd (2017) |
| Australia | Victoria | Tertiary<br>Pathogen reduction with sufficient log reductions to achieve:<br><10 *E. coli* org/100 mL<br><1 helminthes per liter<br><1 protozoa per 50 L<br><1 virus per 50 L | <10 *E. coli* org/100 mL<br>Turbidity < 2 NTU4<br>1 mg/L $Cl_2$ residual after contact time of 30 min (or equivalent disinfection) | VEPA (2003) |
| Israel | Inbar committee | Secondary treatment<br>Filtration<br>Disinfection | 10 fecal coliform/100 mL<br>1 mg/l $Cl_2$ residual (min) | Inbar (2007) |

(continued)

**Table 4** (continued)

| Country/regulatory body | Regulation/guidelines | Treatment processes | List of bacterial regulatory demands and disinfection | Source |
|---|---|---|---|---|
| Israel | Halperin committee | Secondary treatment<br>Filtration/high retention time in the reservoir<br>Disinfection | 10 fecal coliform/100 mL<br>1 mg/l $Cl_2$ residual (min)<br>Contact time of at least 30 min | Halperin (2002) |

## References


Amy G, Drewes J (2006) Soil aquifer treatment (SAT) as a natural and sustainable wastewater reclamation/reuse technology: fate of wastewater effluent organic matter (EfOM) and trace organic compounds. Environ Monit Assess 129:19–26. https://doi.org/10.1007/s10661-006-9421-4

Bouwer H (2002) Artificial recharge of groundwater: hydrogeology and engineering. Hydrogeol J 10:121–142. https://doi.org/10.1007/s10040-001-0182-4

Bruce D, Friedman GM, Kaufman A, Yechieli Y (2001) Spatial variations of radiocarbon in the coastal aquifer of Israel—indicators of open and closed systems. Radiocarbon 43:783–791. https://doi.org/10.1017/s003382220004145x

California Department of Public Health (2014) Regulations related to recycled water. http://www.waterboards.ca.gov/drinking_water/certlic/drinkingwater/documents/lawbook/RWregulations_20140618.pdf

Drewes JE, Reinhard M, Fox P (2003) Comparing microfiltration-reverse osmosis and soil-aquifer treatment for indirect potable reuse of water. Water Res 37:3612–3621. https://doi.org/10.1016/s0043-1354(03)00230-6

Elkayam R (2012) Evaluation and identification of subsurface processes at the Shafdan SAT. M.Sc. thesis, The Hebrew University of Jerusalem (in Hebrew)

Elkayam R (2019) Shafdan soil aquifer treatment system; Process assessment & improvement. Ph.D. thesis, The Hebrew University of Jerusalem, Israel

Elkayam R, Michail M, Mienis O, Kraitzer T, Tal N, Lev O (2015a) Soil aquifer treatment as disinfection unit. J Environ Eng 141:05015001. https://doi.org/10.1061/(asce)ee.1943-7870.0000992

Elkayam R, Sopliniak A, Gasser G, Pankratov I, Lev O (2015b) Oxidizer demand in the unsaturated zone of a surface-spreading soil aquifer treatment system. Vadose Zo J 14. https://doi.org/10.2136/vzj2015.03.0047

Elkayam R, Aharoni A, Vaizel-Ohayon D, Sued O, Katz Y, Negev I, Marano RBM, Cytryn E, Shtrasler L, Lev O (2018) Viral and microbial pathogens, indicator microorganisms, microbial source tracking indicators, and antibiotic resistance genes in a confined managed effluent recharge system. J Environ Eng 144:05017011. https://doi.org/10.1061/(asce)ee.1943-7870.0001334

Gasser G, Rona M, Voloshenko A, Shelkov R, Lev O, Elhanany S, Lange F, Scheurer M, Pankratov I (2011) Evaluation of micropollutant tracers. II. Carbamazepine tracer for wastewater contamination from a nearby water recharge system and from non-specific sources. Desalination 273:398–404. https://doi.org/10.1016/j.desal.2011.01.058

Gasser G, Pankratov I, Elhanany S, Glazman H, Lev O (2014) Calculation of wastewater effluent leakage to pristine water sources by the weighted average of multiple tracer approach. Water Resour Res 50:4269–4282. https://doi.org/10.1002/2013wr014377

Gvirtzman G, Buchbinder B (1978) The late Tertiary of the coastal plain and continental shelf of Israel and its bearing on the history of the Eastern Mediterranean. In: Initial reports of the Deep Sea drilling project, vol 42. U.S. Government Printing Office, Washington, D.C., pp 1195–1222

Halperin R (2002) Halperin Committee Report, Principles for granting permits to irrigate effluents [WWW Document]. Minist. Heal. Isr. https://www.health.gov.il/hozer/bsv_Halperin.doc

Hernández M, Magarzo C, Lemaire B (2012) Degradation of emerging contaminants in reclaimed water through soil aquifer treatment (SAT). J Water Reuse Desalin 2:157–164. https://doi.org/10.2166/wrd.2012.016

Icekson-Tal N (2014) groundwater recharge with municipal effluent: recharge basins Soreq, Yavne 1, Yavne 2 and Yavne 3 : 2014. Mekorot Water Company Limited, Central District, Dan Region Unit

Inbar Y (2007) Inbar [WWW Document]. Minist. Environ. Prot. Isr. http://www.sviva.gov.il/InfoServices/ReservoirInfo/DocLib2/Publications/P0301-P0400/P0321.pdf

Issar A (1968) Geology of the central coastal plain of Israel. Isr J Earth Sci 17:16–29

Kloppmann W, Chikurel H, Picot G, Guttman J, Pettenati M, Aharoni A, Guerrot C, Millot R, Gaus I, Wintgens T (2009) B and Li isotopes as intrinsic tracers for injection tests in aquifer storage

and recovery systems. Appl Geochem 24:1214–1223. https://doi.org/10.1016/j.apgeochem.2009.03.006
Laura AS, Bernd G (2017) Minimum quality requirements for water reuse in agricultural irrigation and aquifer recharge. Publications Office of the European Union. https://doi.org/10.2760/887727
Lian J, Luo Z, Jin M (2013) Transport and fate of bacteria in SAT system recharged with recycling water. Int Biodeterior Biodegrad 76:98–101. https://doi.org/10.1016/j.ibiod.2012.06.012
Negev I, Guttman J, Kloppmann W (2017) The use of stable water isotopes as tracers in soil aquifer treatment (SAT) and in regional water systems. Water 9:73. https://doi.org/10.3390/w9020073
Ohad S, Vaizel-Ohayon D, Rom M, Guttman J, Berger D, Kravitz V, Pilo S, Huberman Z, Kashi Y, Rorman E (2015) Microbial source tracking in adjacent karst springs. Appl Environ Microbiol 81:5037–5047. https://doi.org/10.1128/aem.00855-15
Peter H (1998) Wastewater treatment and groundwater recharge for reuse in agriculture: Dan Region Reclamation Project, Shafdan. Artificial recharge of groundwater. A.A. Balkema, Rotterdam
Pescod T, Engineering EC, Newcastle-upon-tyne T, Nations U, Division P (1992) Wastewater treatment and use in agriculture - FAO irrigation and drainage paper 47. Food and agriculture organization of the united. Nat. Resour. Manag. Environ. Dep. vol 4.https://www.fao.org/docrep/T0551E/t0551e06.htm
Sopilniak A, Elkayam R, Lev O, Elad T (2015) Oxygen profiling of the unsaturated zone using direct push drilling. Environ Sci Process Impacts 17:1680–1688. https://doi.org/10.1039/c5em00270b
Sopliniak A, Elkayam R, Lev O (2017a) Nitrification in a soil-aquifer treatment system: comparison of potential nitrification and concentration profiles in the vadose zone. Environ Sci Process Impacts 19:1571–1582. https://doi.org/10.1039/c7em00402h
Sopliniak A, Elkayam R, Lev O (2017b) Quantification of dissolved organic matter in pore water of the vadose zone using a new ex-situ positive displacement extraction. Chem Geol 466:263–273. https://doi.org/10.1016/j.chemgeo.2017.06.017
U S Environmental Protection Agency (2012) 2012 guidelines for water reuse, development. EPA Environmental Protection Agency, Cincinnati, OH
US Food and Drug Administration (2016) Food safety modernization act (FSMA). Public Law
VEPA (2003) Guidelines for environmental management: use of reclaimed water. EPA Victoria, Southbank, Victoria, pp 1–101
Wilson LG, Amy GL, Gerba CP, Gordon H, Johnson B, Miller J (1995) Water quality changes during soil aquifer treatment of tertiary effluent. Water Environ Res 67:371–376. https://doi.org/10.2175/106143095x131600
World Health Organisation (2006) Guidelines for the safe use of seawater, excreta and graywater. Wastewater use in agriculture, vol 2

# Hydrological and Geological Controls on the Evolution of the Dead Sea Sinkholes

**Meir Abelson**

## 1 Introduction

### *1.1 Background*

The Dead Sea (DS) is a terminal lake that is situated in the deepest basin along the Dead Sea Transform (Fig. 1), the deepest place on Earth continents. The desiccation of the DS in the last 40 years has led to a rapid decline of its level, more than a 1 m/year during the last 13 years (Fig. 2). This desiccation caused rapid decline of the lake level and retreat of the DS shoreline toward the lake center, and exposed broad new coastal areas at regions of gentle slope topography. This shrinkage of the DS caused the entire surrounding areas to become unstable in two ways, (1) by incision of water channels due to increase of the gradient of stream flow, and (2) by formation of thousands of collapse sinkholes within the last 25 years (Fig. 1). These instabilities have caused severe damage to human activity along the DS coast in both countries, Israel and Jordan (e.g., Abou-Karaki et al. 2019).

This chapter presents some results from an interdisciplinary research on the development of sinkholes along the western coast of the DS, conducted by the Geological Survey of Israel in the last 20 years. In this research, a broad variety of exploration methods were used (e.g., Abelson et al. 2006, 2017a; Baer et al. 2018; Yechieli et al. 2006). For instance, in order to understand the mechanism of sinkhole formation, boreholes were used to identify the lithology around sinkhole sites as well as groundwater chemistry, conductivity/salinity, and level variations (Yechieli et al. 2006, 2010). The lithological identification was assisted by seismic refraction in order to achieve a broader coverage (Yechieli et al. 2006; Ezersky 2010; Frumkin et al. 2011). Seismic reflection was used to identify concealed faults, buried within the sediments with no surface expression (Abelson et al. 2003, 2006). In the efforts to identify

M. Abelson (✉)
Geological Survey of Israel, Jerusalem, Israel
e-mail: meira@gsi.gov.il



U. Kafri and Y. Yechieli (eds.), *The Many Facets of Israel's Hydrogeology*,
Springer Hydrogeology, https://doi.org/10.1007/978-3-030-51148-7_15

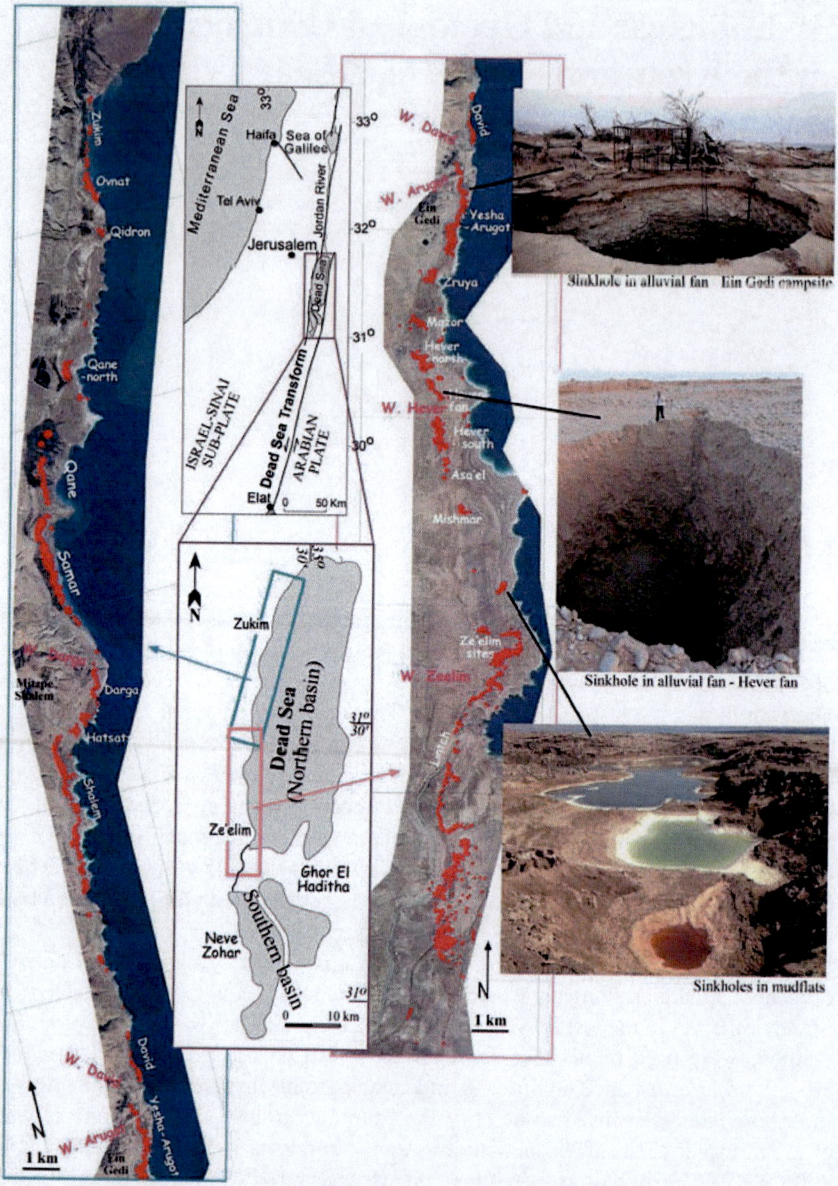


**Fig. 1** Distribution of sinkholes (red dots) along the west coast of the Dead Sea (DS) north basin, as observed airborne LiDAR from April 2017 and presented on rectified collage of aerial photographs (modified from Abelson et al. 2017a). Note the skewed map of the northernmost part of the DS coast. Photographs on left are examples of sinkholes in alluvial fans (upper and intermediate) and mudflats (lower). Names of sinkhole clusters/sites are in white

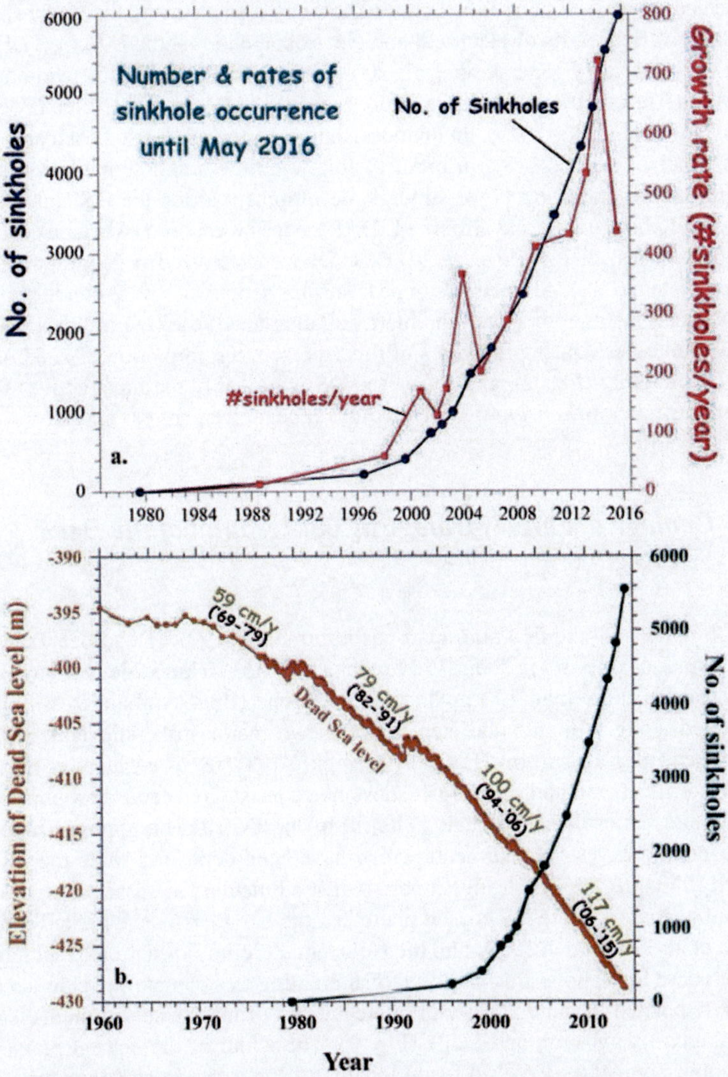


**Fig. 2** Evolution of the DS sinkholes since their formation and temporal variations of the DS level. **a** Temporal variation of sinkhole number (black line) and sinkhole occurrence rate (red line) in number of sinkholes per year. **b** Number of sinkholes (dark green line) versus decline of the DS level (brown line and dots) (after Abelson et al. 2017b)

concealed subsurface cavities prior to sinkhole collapse, many methods were applied, including various sorts of electrical and electromagnetic methods (e.g., TDEM and GPR), microgravity (Rybakov et al. 2001), seismic diffraction (Shtivelman et al. 2005), interferometric synthetic aperture radar (InSAR) from satellites (Baer et al. 2018; Nof et al. 2013, 2019), and nanoseismic monitoring (Wust-Bloch and Joswig 2006; Abelson et al. 2018). In order to follow-up the sinkhole evolution and the mapping of potential levels for sinkhole development along the DS sinkholes, we used aerial photographs and airborne LiDAR images were used (Abelson et al. 2004, 2017b, Avni et al. 2016; Filin et al. 2011), which were analyzed by geographical information system (GIS). All methods of near-surface geophysics were conducted by the Geophysical Institute of Israel, the main collaborator of the GSI in the research on the DS sinkholes. During the years of this research, the formation of sinkholes has been accelerating (Fig. 2), and today, the DS is probably the most active sinkhole site on Earth in terms of number of sinkhole appearance per year and area.

## 1.2 *Geological and Hydrological Background of the Dead Sea Region*

The DS is located within a pull-apart basin formed along the Dead Sea Transform. The DS pull apart basin is about 150 km long, and 15–20 km wide, with a maximal subsidence of more than 10 km since the Neogene (Ben-Avraham et al. 2008). It developed along a dilatational step between two major strike-slip fault segments of the Dead Sea Transform (Fig. 1) (Garfunkel 1997; Ben-Avraham et al. 2008). The western rift wall and the pre-rift substance consist Cretaceous carbonates, such as dolomite, limestone, and chalk. The fill of the DS rift is composed mainly by lacustrine and terrestrial sediments which have been deposited since the Miocene (Picard 1943; Zak 1967). The upper part of this sedimentary sequence is made of two formations that form the DS coastal plain, i.e., the late Pleistocene Lisan Formation (Begin et al. 1974; Sneh 1979) and the Holocene Ze'elim Formation (Yechieli et al. 1993). These formations are composed of alternating clastic material, clay, sand, and gravel, deposited in fan deltas, with intercalations of lacustrine sediments, such as clay, aragonite, gypsum, and halite (Fig. 3). The width of the coastal plain varies from about 5 km at the Ze'elim fan to less than 1 km in the Enot Zuqim area, at the northern basin of the Dead Sea (Fig. 1). The sinkholes develop within the sediments of the Ze'elim Formation.

The DS is a hyper-saline terminal lake, which gets most of its water from the north through the Jordan River system. The DS brine, with a salinity of 340 g/L and an overall composition and ion ratio, which differs markedly from that of normal seawater, is saturated with respect to halite. The lake acts as the regional base level and as such any change in its water level is reflected in the surrounding surface drainage and groundwater systems (Kafri and Yechieli 2010) (Figs. 4 and 5). The Lisan Lake, which its climatic-induced shrinkage formed the present DS, started

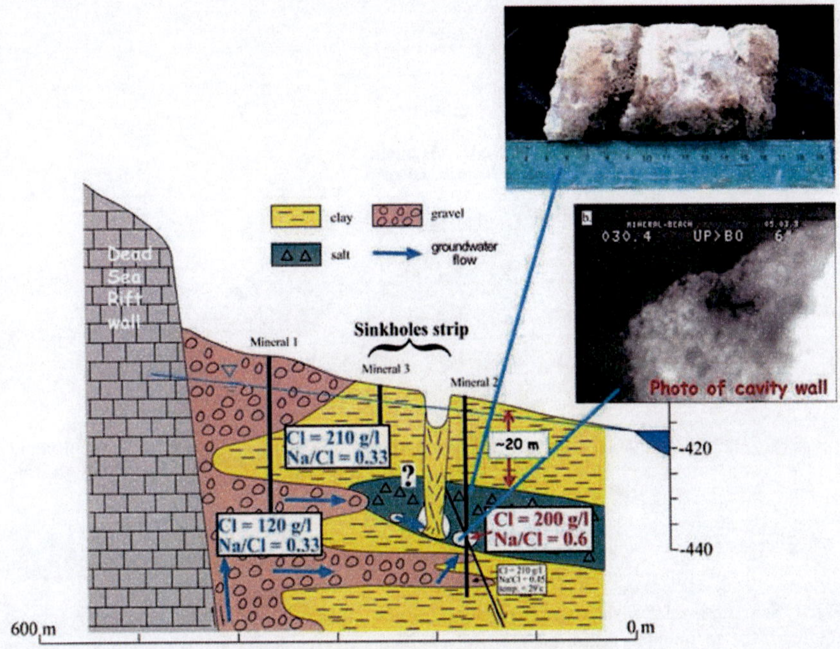


**Fig. 3** Representative observations indicating sinkhole formation by dissolution of a buried salt layer within sediments of mudflats. The hydrogeological cross section across the sinkhole cluster near Mineral Beach is based on findings from boreholes (modified from Yechieli et al. 2006). The pertinent observations are (1) presence of a salt layer beneath the sinkholes (upper photo), (2) borehole encountered cavity in the salt layer, as displayed in the lower photograph showing the salt crystals of the cavity wall (Abelson et al. 2006), (3) high Na/Cl ratio indicating groundwater the dissolved salt, and (4) incursion of groundwater sub-saturated with respect to halite (Yechieli et al. 2006). Note depth of salt layer top is ~20 m in mudflats, while in the alluvial fans is ~50 m (see Fig. 5)

some 70,000 years BP and attained a maximal level of 160 m bsl around 25,000 years BP (Bartov et al. 2002). Since the rapid shrinkage of the Lisan Lake, the lake have reached the present DS dimensions during the early Holocene (Neev and Emery 1967; Sneh 1979; Yechieli et al. 1993) and even smaller (Stein et al. 2010). Bookman (Ken-Tor) et al. (2004) show that the DS water level over the last 4000 years was intermittently below 400 m bsl.

The drop in the DS levels from 392 to 430 m bsl over the last few decades is the prime factor affecting the infrastructure and the groundwater systems around the lake. This drop results from the intensive exploitation of the freshwater flow in the DS drainage basin by surrounding countries, Israel, Jordan, Syria, and Lebanon. The level drop has been accelerating, and since 2006, it has reached nearly 1.2 m/year (Fig. 2). The rate of the DS level drop is expected to continue at similar rates for the next tens of years. However, this level drop is expected to decrease when approaching

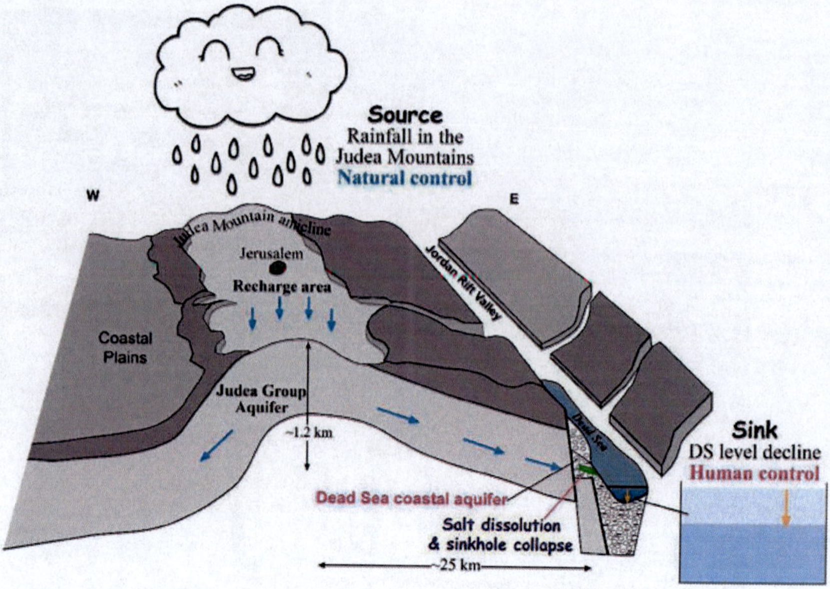


**Fig. 4** Schematic relationship between the eastern Judea Group aquifer and the Quaternary coastal aquifer of the sedimentary fill of the DS rift (modified after Yechieli et al. 2006). The salt layer is embedded at the base of the Holocene sediments. The two major controls on the supply of fresh groundwater into the DS coastal aquifer are the decline of the lake level, a sink drawing fresh groundwater due to human activity, and precipitation in the recharge area near the ridgeline of the Judea Mountains (JM), the natural source of fresh groundwater. The vertical 1.2 km interval is stretched between the salt layer in the Dead Sea rift and top of the Judea Mountains

a new equilibrium in the water balance at a level of 100–150 below the present level (Yechieli et al. 1998; Krumgalz et al. 2000).

The subsurface hydrology of the DS rift is dominated by two main aquifers (Fig. 4): the Judea Group aquifer and the DS coastal aquifer (Arad and Michaeli 1967; Yechieli et al. 1995). The former is composed mainly of the Cretaceous carbonates that constitute the Judea Mountains (JM) and the DS rift margin (Fig. 4). The latter includes the fluvial and lacustrine sediments of the Ze'elim formation. Alternating clays and gravels within the alluvial sequence create several sub-aquifers with different heads and water composition (Yechieli et al. 2006) (Fig. 3). The Judea Group aquifer is recharged around the ridgeline of the JM, some 20–30 km to the west (Fig. 4). Most of the water supply to the Quaternary coastal aquifer is from the Judea Group aquifer, by lateral flow across the rift faults. In addition, the DS coastal aquifer is recharged occasionally by flash floods in major stream channels (e.g., Avni et al. 2016). Direct rainfall recharge to the aquifer over the alluvial sediments is negligible due to the extreme aridity of the region, with rainfall less than 70 mm annually, as compared to 500–700 mm/year over the JM intake area to the west. The groundwater discharge from this aquifer converges toward three springs in the northern part of the northern basin of the DS, namely the Zukim, Qane, and Samar springs (Laronne Ben-Itzhak

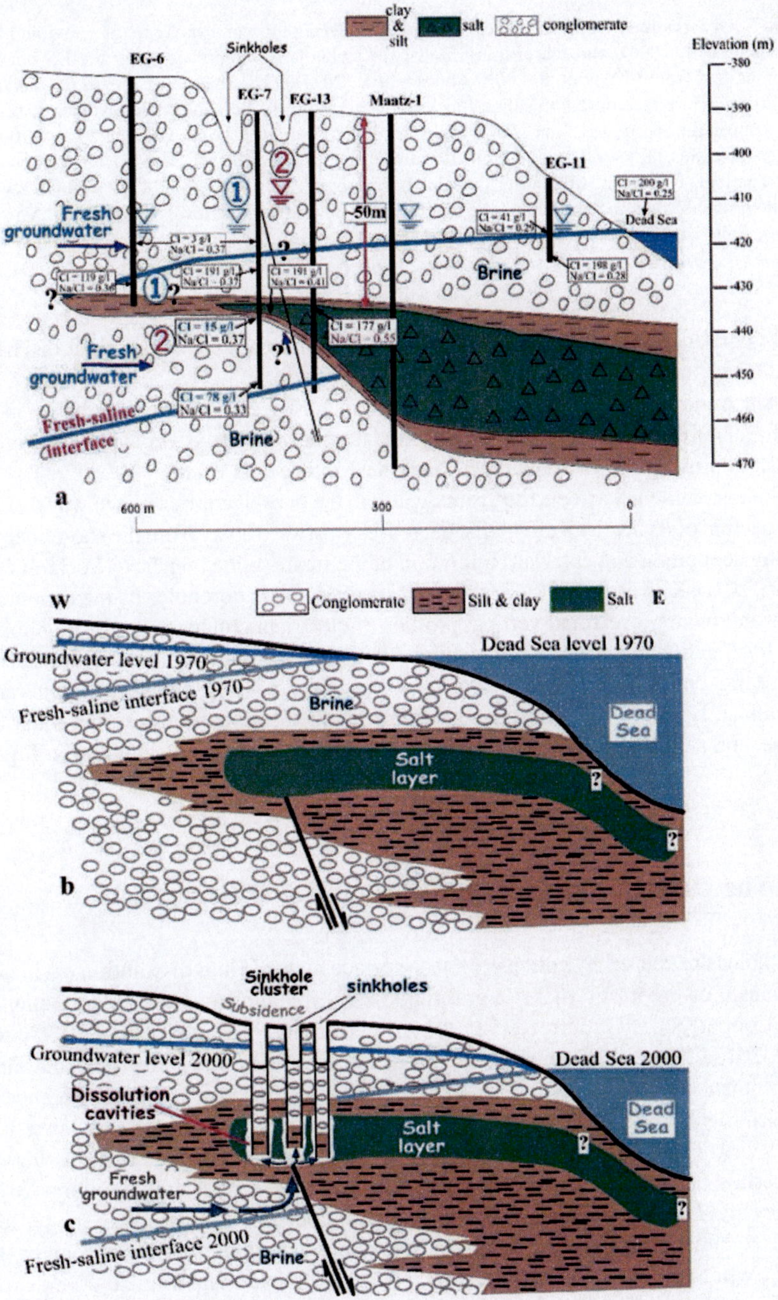
clay & silt
salt
conglomerate
Elevation (m)
Sinkholes
EG-6
EG-7
EG-13
Maatz-1
EG-11
Fresh groundwater
Fresh-saline interface
Brine
Dead Sea
~50m
a
600 m
300
0
W
E
Conglomerate
Silt & clay
Salt
Groundwater level 1970
Dead Sea level 1970
Fresh-saline interface 1970
Salt layer
b
Sinkhole cluster
sinkholes
Subsidence
Groundwater level 2000
Dead Sea 2000
Dissolution cavities
Fresh groundwater
c
Fresh-saline interface 2000

◀**Fig. 5** Observations from boreholes in Arugot alluvial fan near Ein-Gedi (**a**) (modified from Yechieli et al. 2006), and the ensuing model for sinkhole development along the DS coast (**b, c**) (modified from Shalev et al. 2006; Abelson et al. 2017a). The thick blue line in (**a**) marks the fresh-saline interface based on salinity measurements from the boreholes. Note values of salinity (Cl concentrations in g/L) and Na/Cl ratios are marked on each borehole, where $C_{Cl}$ ~200 g/L is for Dead Sea brine. **a** The fine-grained sedimentary unit and the salt layer divide between two sub-aquifers. The lower sub-aquifer contains fresh groundwater ($C_{Cl}$ ~15, 78 g/L), which is under higher pressure than the upper phreatic aquifer (note in sketch *a* that level "2" is higher than "1"). The salinity of the groundwater of the upper phreatic sub-aquifer is similar to that of the DS water (**a**)

and Gvirtzman 2005; Yechieli et al. 2006) (Fig. 1), accompanied by minor discharge that takes place along other parts of the coast.

Since the density of the Dead Sea brine is 1.24 g/cc, the interface between the saline and the fresh groundwater along the lake coastal plain is very shallow (Fig. 5), about 10 times shallower than that found near the ocean (Yechieli 1993, 2000). The shrinkage of the DS affects the groundwater in the coastal aquifer in two ways, (1) by the decline of its level, an effect that decreases westward away from the shoreline, and (2) by deepening and eastward migration of the fresh-saline interface (Yechieli et al. 1995, 2010; Kiro et al. 2008). Both effects are detected in boreholes by measurements of groundwater levels and vertical profiles of electrical conductivity for monitoring the fresh-saline interface (Yechieli et al. 2010). The spatial configuration of salinity in the DS coastal groundwater system is extremely heterogeneous, where it varies significantly among the different hosted end members, namely fresh spring water, DS brine, and saline springs issuing close to the shoreline (Gavrieli et al. 2001) (Figs. 3 and 5).

## 2 The Dead Sea Sinkholes—Mechanism of Formation

Sinkholes are caused by collapse of near-surface material into dissolution cavities in carbonate or evaporitic rocks (e.g., Frumkin 2013; Galloway et al. 1999; Gutierrez and Cooper 2002; Gutiérrez et al. 2014; Johnson 1997; Martinez et al. 1998; Nof et al. 2013). The rate of sinkhole occurrence is controlled by (1) the rock solubility, i.e., control on growth rate of the subsurface cavity, and by (2) the competence of the sediments overlying the dissolution cavity, which controls their collapse into the cavity. The sediments competence is related to their thickness and rheological properties (e.g., Baer et al. 2018; Maimon et al. 2009; Parise and Lollino 2011; Shalev and Lyakhovsky 2012). The high solubility of evaporitic rocks, e.g., gypsum and salt, enhances sinkhole formation when exposed to undersaturated groundwater (e.g., Avni et al. 2016; Frumkin 2013; Galve et al. 2009; Yechieli et al. 2006).

The evolution of the DS sinkholes results from dissolution of a salt layer embedded within the Holocene Ze'elim formation along the DS shores (Abelson et al. 2003, 2006; Ezersky and Frumkin 2013; Frumkin et al. 2011; Shalev et al. 2006; Yechieli et al. 2006). This inference is based on observations from boreholes (Figs. 3 and 5) showing (1) the presence of salt layer in all boreholes adjacent to sinkholes as well as from seismic refraction, (2) findings of cavities in the salt layer (three drilling sites), (3) high Na/Cl ratios, an indication of salt dissolution in the DS groundwater, and (4) presence of sub-saturated groundwater with respect to halite below the salt layer (see summary in Figs. 3 and 5) (Yechieli et al. 2006). As seen worldwide, that sinkholes are usually karst feature, the DS sinkholes are also of karstic origin, but are formed at very high rates due to dissolution of the highly dissolvable salt (halite) layer. The salt layer is dated to ~10k years (Yechieli et al. 1993; Stein et al. 2010) and therefore defines the base of the Holocene Ze'elim formation in the DS basin. The elevation of the top of the salt layer varies between 415 m and 443 m bsl.

The sinkholes are concentrated in a narrow strip along the DS shoreline, a strip ~65 km long with width varying from several meters to 1200 m (Fig. 1). The rate of sinkhole occurrence along the Israeli (western) side of the DS has been increasing dramatically since 2000 (Fig. 2), with more than 700 sinkholes appearing between April 2014 and April 2015. The sinkholes appear in the two main sedimentary environments of the DS coast (Arkin and Gilat 2000), the mudflats, consisting of fine-grained lacustrine sediments, and the alluvial fans, composed of coarse-grained fluvial sediments (Figs. 3 and 5). The depth of the salt layer top in the alluvial fans is in between 35 and 65 m, whereas in the mudflats is usually ~20 m deep and in cases shallower (Avni et al. 2016; Yechieli et al. 2006). The sediments of these two environments differ also in their mechanical properties: the alluvial fan sediments deform mostly as brittle material, whereas the fine-grained sediments in the mudflats deform in a more viscous manner (Baer et al. 2018; Shalev and Lyakhovsky 2012). The differences in salt depth and sediment rheology affect sinkhole collapses in two ways. First, sinkholes in the alluvial fans tend to be deeper and narrower (the deepest is ~27 m deep with an average diameter of ~4 m) than those in the mudflats (Fig. 1). Secondly, field observations (Abelson et al. 2006) and mechanical models (Shalev and Lyakhovsky 2012; Baer et al. 2018) indicate that the thick and stiff overburden in the alluvial fans impedes the final collapse into the awaiting cavity in the salt layer (further discussed below).

The formation of the DS sinkholes is related to the DS shrinkage, i.e., the lake level decline and the eastward migration of its shoreline. This shrinkage drives the deepening and eastward migration of the interface between the fresh and saline groundwater (Fig. 5), which in turn brings fresh groundwater from the Judea Group aquifer to approach the salt layer, hence causing its dissolution (Figs. 4 and 5) (Yechieli et al. 2006, 2010). The final path of the fresh groundwater to the salt layer occurs through fault lines, as also attested by the linear form (lineaments) of the sinkhole clusters (Abelson et al. 2003, 2006; Closson and Abou Karaki 2009) (Fig. 5).

According to the results from the Ein-Gedi area, it appears that groundwater in the lower sub-aquifer, below the salt layer, is over-pressurized, as attested by its higher groundwater level as detected in boreholes (Fig. 5). Therefore, this fresh groundwater can rise via fault plain toward the salt layer causing its dissolution. This may also explain the difference between the lower and the upper phreatic aquifers in the location of the fresh-saline interface as inferred from the boreholes (Fig. 5), where the higher pressure within the lower aquifer pushes the interface further to the east.

In the Jordanian side of the DS, sinkholes are much scarcer, and mainly concentrated near the site of Ghor Al-Hadita (Taqieddin et al. 2000; Closson et al. 2005; Al-Halbouni et al. 2018; Polom et al. 2018; Watson et al. 2018). In this site, ~1150 sinkholes have formed between 1985 and 2017 (Watson et al. 2018). Polom et al. (2018) conducted a survey of shear-wave seismic reflection and did not identify a massive salt layer in the Ghor Al-Hadita area. This led Polom et al. (2018) and Al-Halbouni et al. (2018) to suggest that the mechanism of sinkhole formation in this area is dominated by subrosion and some dissolution of salt pockets dwelling in the silty sediments, rather than a dissolution of a massive salt layer as found in the Israeli site. However, as found in the Israeli side, many of the sinkhole clusters are located above the edge of the salt layer (Abelson et al. 2004; Ezersky and Frumkin 2013), where the sinkhole cluster area display a thin salt layer, perhaps due to the salt dissolution. Therefore, we estimate that a massive salt layer indeed exists also in this area and could be identified by boreholes and shear-wave reflection profile parallel to, and west of the sinkhole line of Ghor Al-Hadita, similarly found in the Israeli side.

## 3 Controls on Rates of Sinkholes Propagation Along the Dead Sea Coast

Whilst the trigger for the formation of the DS sinkholes is the shrinkage of the DS and the consequent replacement of brine by fresh groundwater, it has been shown that variation in the rate of sinkholes formation could have been controlled by other factors rather than the rate of level drop of the DS (Abelson et al. 2017a). Three prominent factors controlling the sinkholes propagation rate are discussed herein: (1) the rheological properties of the sedimentary cap above the dissolution cavities in the salt layer (Abelson et al. 2006, 2018; Baer et al. 2018; Shalev and Lyakhovsky 2012), (2) variations in rate of fresh groundwater supply from rainfall in the recharge area of the JM that control salt dissolution rates (Abelson et al. 2017a), and (3) episodes of floods in some of the stream tributaries, where the floodwater penetrates through sinkholes, initially formed by salt dissolution by fresh groundwater (Avni et al. 2016; Shviro et al. 2017).

### 3.1 *Control of Sediments Properties and Thickness on Final Sinkhole Occurrence and Technics of Precursors for Sinkhole Collapses*

As mentioned above, the sediments overlying the cavities in the salt layer have a crucial role in the time lag between formation of sufficiently large subsurface cavity and final collapse of sinkholes (Abelson et al. 2006, 2017a, 2018; Atzori et al. 2015; Baer et al. 2018). It was first realized when during the follow-up to document, the development of the DS sinkholes that sinkholes appeared in the mudflats years before those in the adjacent alluvial fan. For instance, at the alluvial fan of wadi Hever, the mudflats sinkholes were noticed first at 1987 (Site Asa'el in Fig. 6). Ten years later, the more northerly Hever-south sinkhole site, appeared in the fringe of the alluvial fan, and in 2000, the first sinkhole appeared in the core of the Hever alluvial fan (Site Hever-central, Fig. 6), suggesting a time lag of about 13 years. This propagation of sinkhole appearance toward north-west occurred despite the fact that the fresh groundwater flows from the west, and presumably salt dissolution started first at the more westerly sites (Fig. 6). Such a propagation of sinkholes appearance from the mudflats into the adjacent alluvial fan is also seen in the sinkhole site of Mineral Beach (Abelson et al. 2017a, 2018) in a similar time-scale in the lag of occurrence of sinkholes between these two environments (Fig. 7). The salt layer is deeper in the alluvial fans due to more rapid accumulation rate of sediments (Abelson et al. 2006), and alluvial fan sediments are much more competent to stresses than the mudflat sediments (Abelson et al. 2006; Shalev and Lyakhovsky 2012; Baer et al. 2018). Therefore, it was suggested then that the thick and competent sediments of the alluvial fan impeded sinkhole collapses.

One of the goals of the research on the DS sinkholes was to develop a geophysical tool to predict sinkhole formation before their final collapse. The requirement was that this tool should actually find the presence of subsurface cavities and concealed activity related to the subsurface cavities. A variety of geophysical tools was tested for this purpose since the early 2000s but only two were found applicable for this purpose, the interferometry synthetic aperture radar (InSAR) from satellites (Nof et al. 2013, 2019) and nanoseismic monitoring (Abelson et al. 2018). The InSAR was found as very efficient for this purpose due to its ability to cover broad areas by frequent measurements (every two weeks), and was proven as a useful means to predict sinkhole collapses in vulnerable sites, e.g., under the regional main road (Fig. 8) (Nof et al. 2013, 2019). Additional advantage was the ability to document the time lag between the onset of subsidence, as recorded by the InSAR, and the related sinkhole final collapse. Therefore, this enabled a quantitative detection of this time lag with dependency on sediments properties, i.e., rheology and thickness (Baer et al. 2018). Accordingly, Baer et al. (2018) showed (Fig. 9) that the time lag increases with the increase in the gravel content relative to fine-grained material, i.e., more alluvial deposits; when the sediments are purely lacustrine, i.e., mudflats, the time lag is 0–50 days. On the other hand, for alluvial sediments of >80% gravel, the time lag can be more than 300 days (Fig. 9). In one case, in the Hever alluvial fan, the

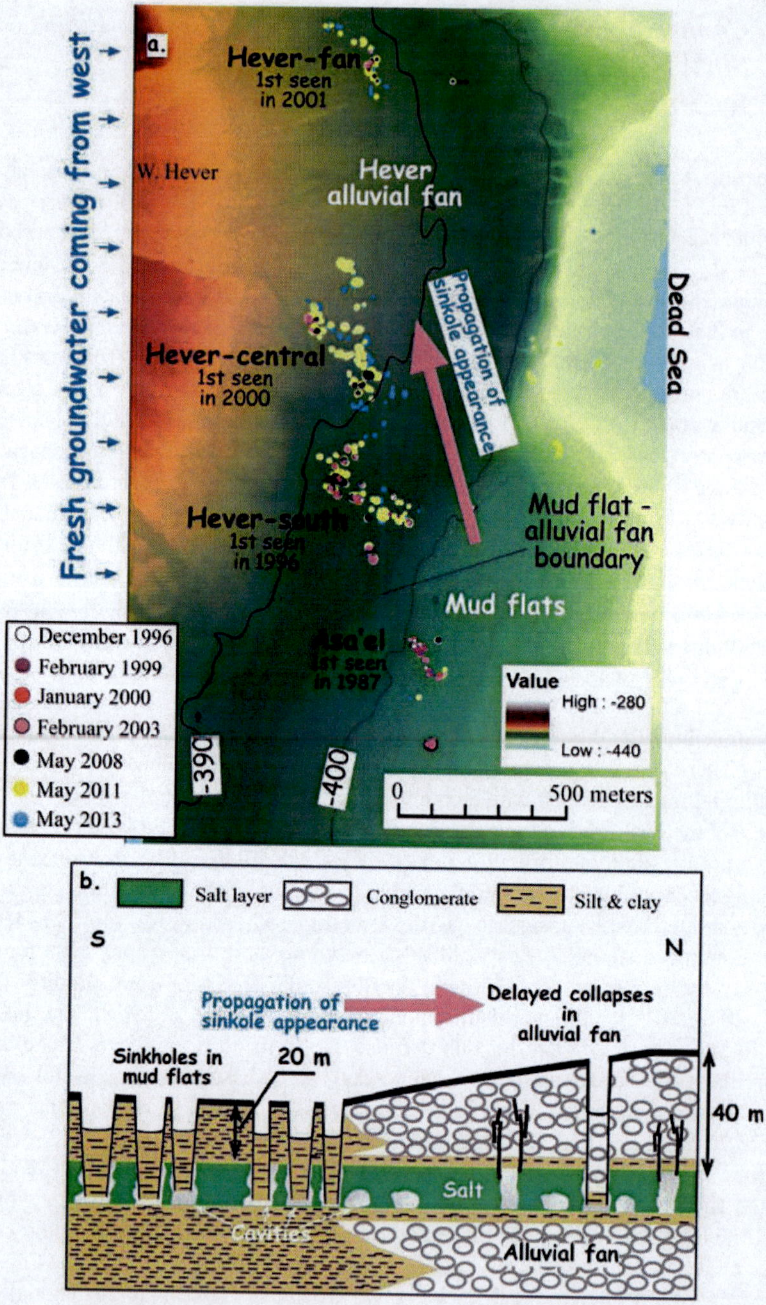
a.
Fresh groundwater coming from west
Hever-fan
1st seen in 2001
W. Hever
Hever alluvial fan
Propagation of sinkole appearance
Dead Sea
Hever-central
1st seen in 2000
Hever-south
Mud flat - alluvial fan boundary
Mud flats
Asa'el
1st seen in 1987
December 1996
February 1999
January 2000
February 2003
May 2008
May 2011
May 2013
Value
High : -280
Low : -440
-390
-400
0 500 meters
b.
Salt layer
Conglomerate
Silt & clay
S
N
Propagation of sinkole appearance
Delayed collapses in alluvial fan
Sinkholes in mud flats
20 m
40 m
Salt
Cavities
Alluvial fan

◀**Fig. 6** Propagation of sinkhole appearance from mudflats into adjacent alluvial fan, **a** in map view from airborne LiDAR image around the Hever alluvial fan, and **b** a schematic N-S cross section based on results from boreholes (the latter is modified after Abelson et al. 2018). The boundary between mudflats and the alluvial fan is marked. Direction of entrance of fresh groundwater is marked by blue arrows in (**a**). First sinkhole cluster (sinkholes sites Asa'el) noticed by aerial photographs from 1987 is located in mudflats. The second, Hever-south site, the fringe of the alluvial fan, occurred in 1997. First sinkhole in the Site Hever-central, which is located in the fan core, occurred in 2000 (Abelson et al. 2006)

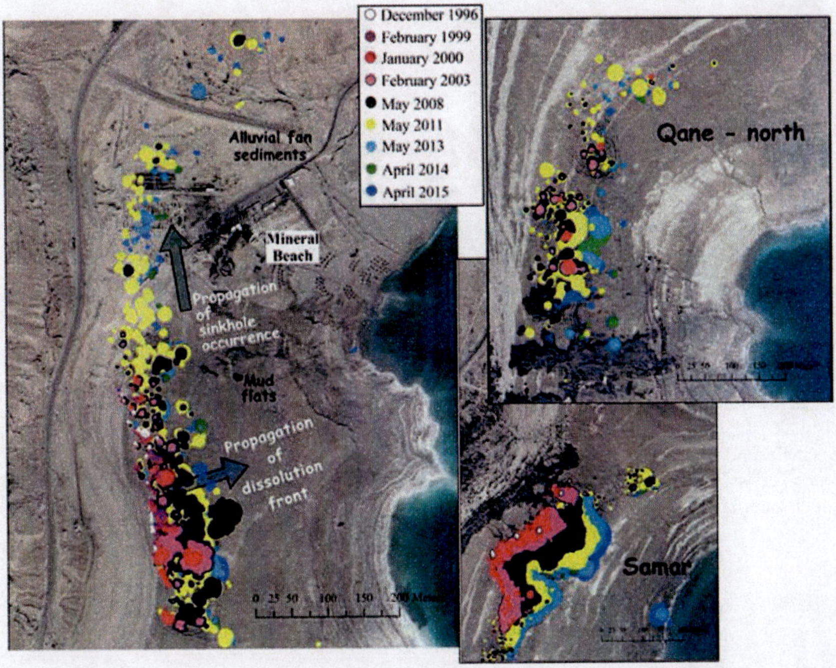


**Fig. 7** Examples of patterns of sinkhole clusters evolution (Sites Mineral Beach, Qane, and Samar). These sinkhole clusters show relative long history and located in mudflats. Therefore, the growth rates of sinkholes area multiplied by salt layer thickness could serve as a proxy for salt dissolution rates in these sites (Abelson et al. 2017a)

time lag even exceeded 5 years. They also showed that the time lag increases with depth of the salt layer where the cavities are actually formed (Fig. 9). The damage viscoelastic model, developed by Baer et al. (2018), infers from these observations the viscosity difference of nearly two-order of magnitude between the mudflats and alluvial fan with high content of gravel.

Still, the time lag between subsidence onset recorded by InSAR and sinkhole collapse is short relative to the difference in occurrences of sinkholes in the neighboring alluvial fans and mudflats, a year or even five years (Baer et al. 2018) versus

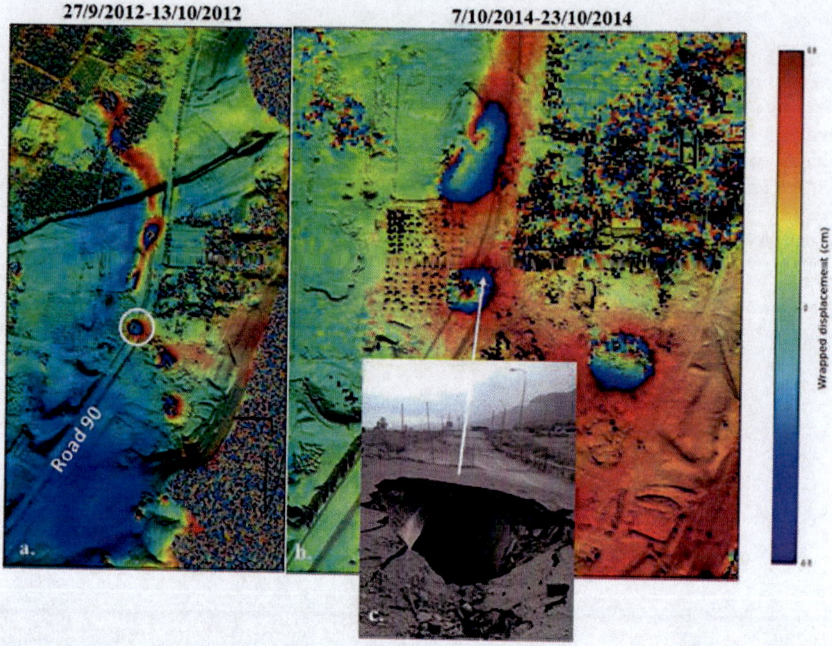


**Fig. 8** Precursory subsidence recorded by InSAR predicted the presence of a concealed sinkhole just below the regional main road (Road 90) near Ein-Gedi date plantation (modified from Nof et al. 2013, 2019). **a** Subsidence recorded by the InSAR in October 2012 with no surface indications such as cracks and prominent subsidence. **b** Two years later, a broader InSAR-recorded subsidence is seen at the same location. Signs of concealed sinkhole were seen on the surface in February 2015, which later developed to a sinkhole (**c**)

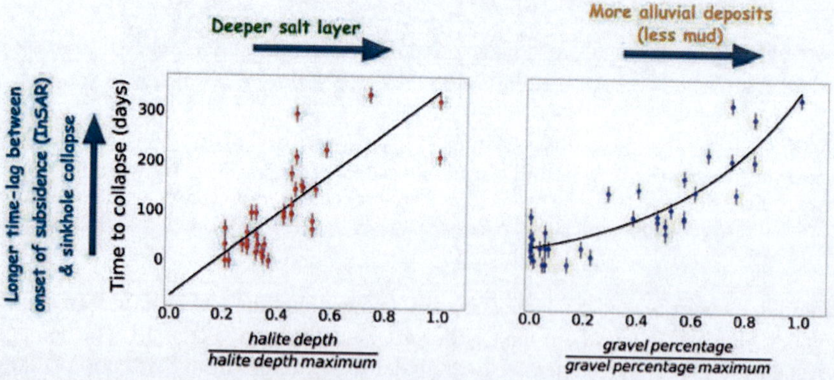


**Fig. 9** Time lag between onset of subsidence seen by the InSAR and the subsequent sinkhole collapse as dependence on depth of salt layer (left panel) and content of alluvial fan deposits (gravel percentage in right panel) (modified from Baer et al. (2018))

13 years accordingly (Fig. 6). Such a gap in the estimated time lag raises the possibility of presence of subsurface cavities prone to collapse long before any expression, even subtle as InSAR-recorded subsidence, is seen on the surface of the alluvial fans. This inference was tested by nanoseismic monitoring in the sinkhole site near the resort site of Mineral Beach in a purpose to identify brittle deformation related to subsurface cavities with no surface expression of subsidence (Fig. 10) (Abelson et al. 2018). A two-month long nanoseismic monitoring identified 75 events with magnitude ranging between (−3) and (−2) located between the sinkhole line and eastward (Fig. 10), as well as at a depth range between surface and bottom of the salt layer (Abelson et al. 2018). Both locations, in plan-view and depth, suggest that these nanoseismic events occur due to activity on short faults and joints above concealed cavities. The surface above these events, east of the sinkhole line, indicates no signs of subsidence as seen by the InSAR images from the time of the seismic monitoring

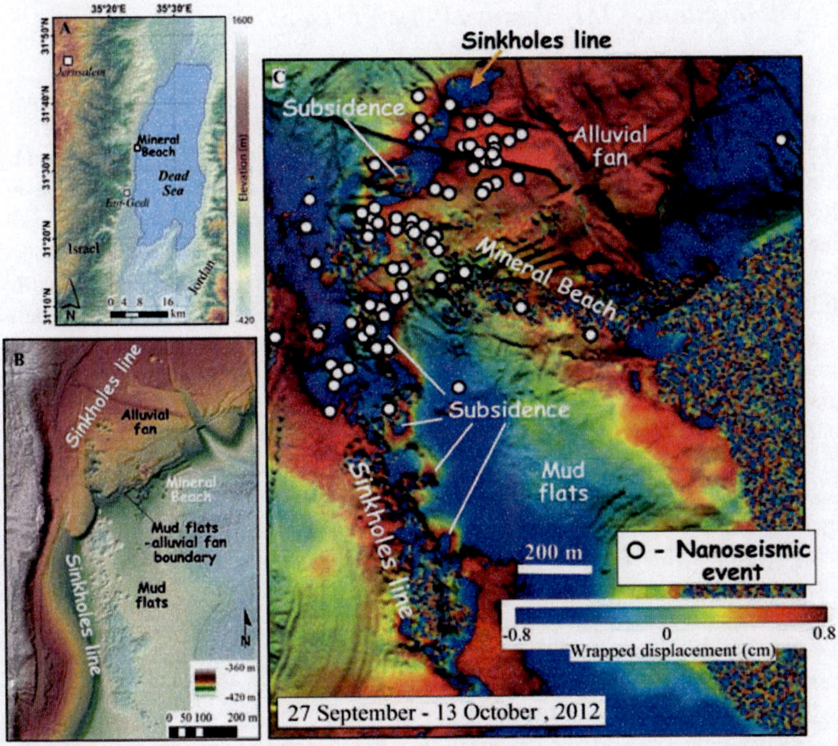


**Fig. 10** Nanoseismic events indicating brittle deformation above concealed subsurface cavities versus subsidence recorded by InSAR near the sinkhole site of Mineral Beach (modified after Abelson et al. 2018). **a** Location of Mineral Beach site. **b** The sinkhole cluster of Mineral Beach on a LiDAR image that strands on mudflats and Wadi Hatsatson alluvial fan. **c** Epicenters of nanoseismic events are projected upon InSAR-detected subsidence. Interferogram shows land subsidence ~1 month after conduction of the seismic monitoring. Note the absence of subsidence in a nanoseismic active zone east of the sinkhole-line

(Fig. 10c). Therefore, it is possible that the alluvial fan sediments can impede sinkhole collapse and occurrence of subtle subsidence by more than 10 years, despite the presence of subsurface cavities that later can develop to sinkholes. On the other hand, collapses in the lacustrine sediments in the mudflats occur once a dissolution cavity has reached a sufficient extent (several meters). Several years after the nanoseismic survey, several sinkholes appeared in the parking lot of the Mineral Beach resort site, within the "cloud" of the nanoseismic events east of the sinkhole line. These collapses led to final closure of the Mineral Beach as a touristic resort sites, and this "cloud" of nanoseismic events is defined as an area of highly prone to sinkhole development (Abelson et al. 2018; Nof et al. 2019).

## 3.2 Controls on Dissolution Rates of the Salt Layer Along the Dead Sea Coast—Precipitation in the Judea Mountains (JM) Versus DS Level Drop

The most pertinent factor controlling propagation rate of the DS sinkholes is the dissolution rate of the salt layer. According to the inferences in the previous sub-section, the final occurrence of collapse sinkhole on the alluvial fan surface is controlled by the dissolution rate of the salt layer impeded by the stiffness and thickness of the overlying sediments. On the other hand, at the mudflats, where the weak overlying sediments do not impede the sinkhole collapse, sinkhole formation is mainly controlled by dissolution rate of the salt layer. In addition, the subsidence area recorded by InSAR in the mudflats which reflects the width of the subsurface cavity is usually similar to subsequent sinkhole diameter. Therefore, Abelson et al. (2017a) demonstrated that for a given sinkhole cluster in the mudflats (Fig. 7), the growth rate of the area of the sinkholes multiplied by the thickness of the salt layer as observed from borehole can be a proxy for dissolution rates of the salt layer (Fig. 11c). However, the subsequent challenge was to find out what was the main factor controlling these dissolution rates. As mentioned above, the two principle factors that control the hydrological regime of groundwater flow in the aquifer of the eastern Judea Mountains (hereafter JM) are (1) the rainfall and recharge quantities on the intake area of the Judea Group aquifer in the JM, and (2) rate of the DS level drop (Fig. 4). The former is a natural factor, whereas the latter is controlled by human activity upstream the drainage basin of the DS. The influence of these two factors on salt dissolution rates could have been detected by the history of groundwater level and its response to these two factors (Abelson et al. 2017a).

Abelson et al. (2017a) showed that the variations of the flux of incursion of undersaturated groundwater with respect to halite are controlled by long-term variations in rainfall quantities in the recharge area in the JM rather than variations in DS level drop. The long-term variations are cycles of ~10 years changes in precipitation ranging between 200 and 400 mm/year (Fig. 11a). Changes in the flux of undersaturated groundwater were determined by variations of groundwater level in boreholes

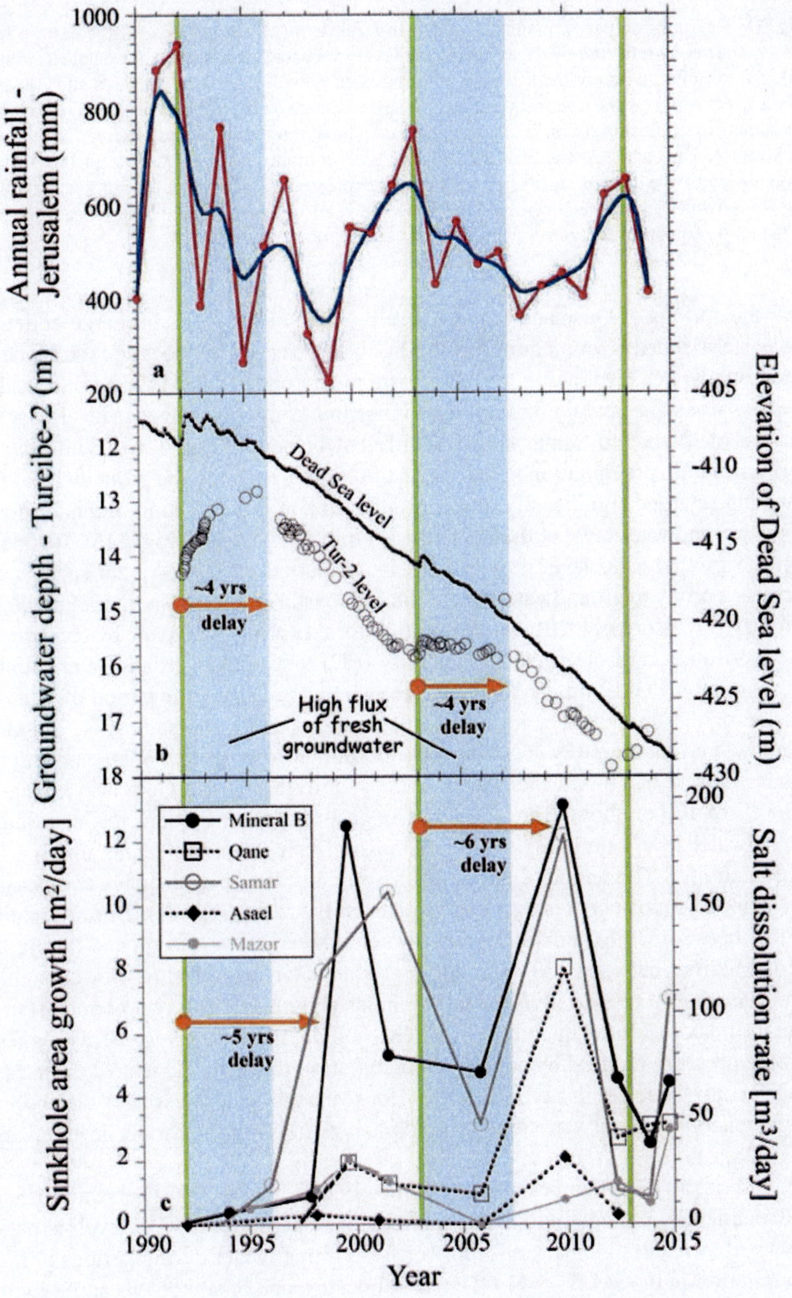

Annual rainfall - Jerusalem (mm)
1000
800
600
400
200
a
Groundwater depth - Tureibe-2 (m)
12
13
14
15
16
17
18
Dead Sea level
Tur-2 level
~4 yrs delay
~4 yrs delay
High flux of fresh groundwater
b
Elevation of Dead Sea level (m)
-405
-410
-415
-420
-425
-430
Sinkhole area growth [m²/day]
Mineral B
Qane
Samar
Asael
Mazor
~6 yrs delay
~5 yrs delay
c
Salt dissolution rate [m³/day]
200
150
100
50
0
1990
1995
2000
2005
2010
2015
Year

◀**Fig. 11** Relationship between annual rainfall in Jerusalem (**a**), the response of the eastern Judea Group aquifer as reflected by the groundwater level in Qane-Samar springs (borehole Tureibe-2) (**b**), and dissolution rate of the salt layer inferred from growth rate of the sinkhole sites (**c**) (after Abelson et al. 2017a; see examples in Fig. 9). The coordinate on the right-hand side of (**c**) is the left coordinate multiplied by 15 m, the approximate thickness of the salt layer as observed in boreholes at Mazor (15 m) and Mineral Beach (14.5 m). Blue columns mark years of high flux of fresh groundwater in the eastern Judea Group aquifer, as expressed by increase or lowering retardation of the groundwater level. Green vertical lines mark the years of peak rainfall in the Judea Mountains in the long-term episodicity

near the springs of Qane and Samar (Tureibe), one of the main discharge outlets of the eastern Judea Group aquifer in the DS valley (Fig. 1 for location). In general, the groundwater level along the DS coastline responds to the DS level drop, thus displays constant decline through time. Dissimilar, the groundwater level near the springs of Qane and Samar shows a different behavior (Fig. 11b). While during periods of high precipitation in the JM, in the 10-years-cycles, the groundwater level lowering retarded significantly or even started to rise. The duration of this response of the groundwater level at the DS rift was typically ~4 years (Fig. 11b). It appears that the groundwater level near this discharge outlets of Qane-Samar springs can serve as a proxy to groundwater flux to the DS coastal aquifer from rainfall in the JM (Fig. 10) (Abelson et al. 2017a). Verification for this proxy was made by comparison to measurements of electrical conductivity (EC) that reflects groundwater salinity, in a borehole (EG-19) in the Ein-Gedi area (Fig. 12). During the period of retarded decline of the groundwater level in Qane-Samar springs, the salinity in Ein-Gedi was low, i.e., measured conductivity was 60 mmho/m (Fig. 12). Whereas with the increasing rate of groundwater decline, the salinity/EC in Ein-Gedi also increased to levels of 100 mmho/m (Fig. 12). This suggests that the variations in Qane-Samar groundwater level mimic regional variations in fresh groundwater flux into the DS coastal aquifer. The temporal profile of the groundwater level displays two periods of enhanced supply of fresh groundwater from the west (Fig. 11). During the first period, between 1992 and 1996 groundwater level rose by ~1.5 m, and during the second period, between 2003 and 2007, groundwater level decline slowed significantly, from ~40 to ~7.5 cm/year in boreholes (Fig. 10b). The two periods started with the peak in rainfall quantities in the JM, and both lasted ~4 years. This suggests that the pressure induced by enhanced precipitation in the recharge area diffuses to the DS coastal aquifer in a typical time of four years. This also explains the absence of influence of annual variations of precipitation in the groundwater level change (Fig. 11a, b).

Consequently, the two-peak pattern of the salt dissolution (Fig. 11c) which cannot be explained by the continuous and monotonous decline of the DS level, is better explained as result of enhanced supply of fresh groundwater during periods of high precipitation in the JM (Fig. 11). The actual occurrence of sinkholes caused by the higher rates of salt dissolution is in a lag of 6–7 years after the long-term peak in the JM precipitation (Fig. 11). This conclusion may also explain the marked difference in rates of salt dissolution between the northern sinkhole sites (Qane, Samar, Mineral)

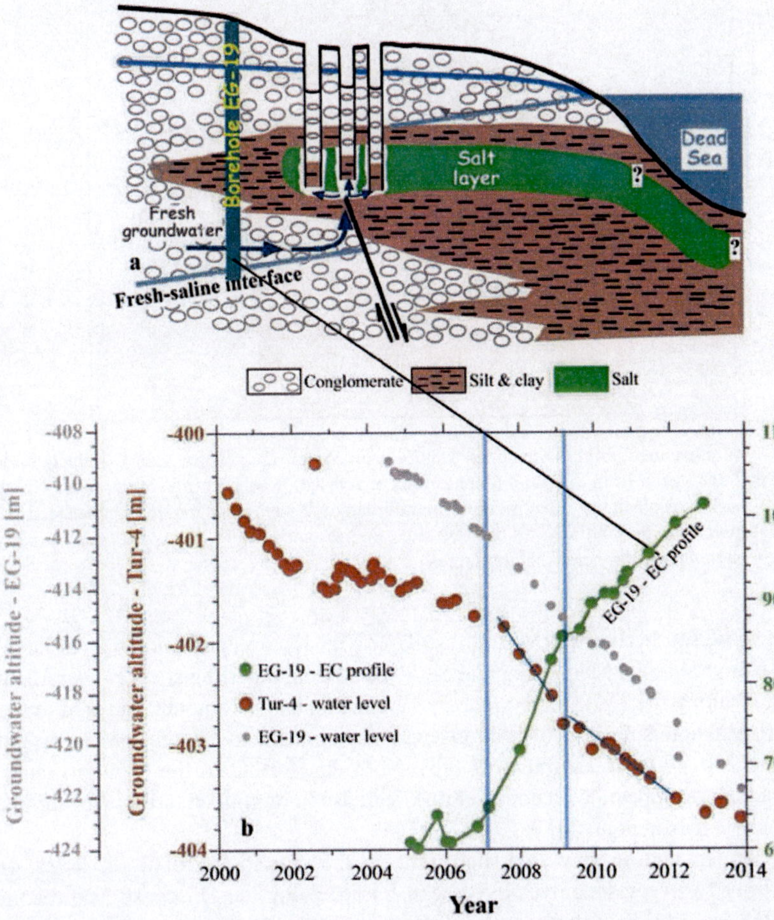


**Fig. 12** Variations in Tureibe groundwater level versus changes in salinity of fresh groundwater component from Ein-Gedi borehole EG-19 (see Fig. 1 for location) (Abelson et al. 2017a). **a** Schematic hydrogeological section of the area of borehole EG-19 that demonstrates the location of borehole measurements of electrical conductivity. **b** Record of electrical conductivity (EC) measurements in EG-19 made between February 2005 and December 2012 record a step increase in EC values. The level of the profile of EC temporal variations in the borehole is marked schematically in (**a**) and located above the fresh-saline interface and within the range of the pipe perforation. Variations in groundwater level in the Tureibe-4 and EG-19 boreholes are also presented in (**b**). Note the correlation between the salinity/EC variations and groundwater depth in Tureibe, corroborating the Tureibe groundwater level as proxy for fluxes of fresh groundwater from the Judea Mountains

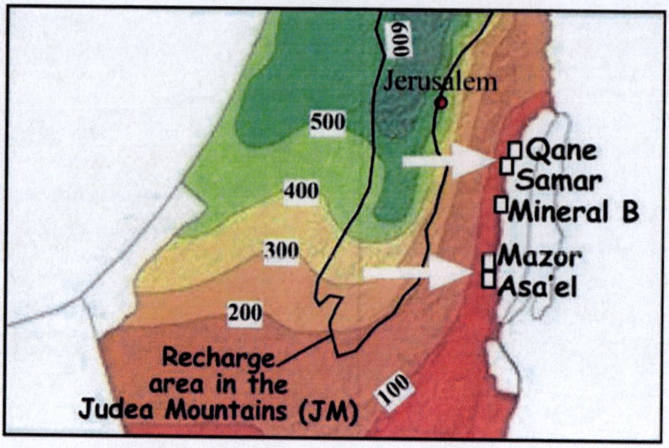


**Fig. 13** Average annual rainfall quantities (numbers on contours) in central Israel, in the latitude of the Dead Sea (taken from the Israel Meteorological Service). Recharge area of the Judea Group aquifer is marked by black line. The five representative sinkhole sites are marked along the Dead Sea coast. Estimated salt dissolution rates of these sites are presented in Fig. 11c, where the northerly sites show significantly higher dissolution rates

and the more southerly sites (Mazor, Asaél). In Fig. 13, a precipitation map of Israel (taken from the Israel Meteorological Service) demonstrates the steep N-S gradient in rainfall along the JM, i.e., the recharge area. The annual amount opposite to the northern sinkhole sites is 500–600 mm, whereas opposite to the sites of Mazor and Asael, is 200–400 mm rainfall (Fig. 13). Note that further to the east, this sharp N-S gradient disappears with considerable decrease in rainfall quantities (for further details see Abelson et al. 2017a).

This understanding may also shed light on the scarcity of fossil sinkholes, of which only two representatives have been found, along the DS coast. The record of the DS level prior to the human intervention in the 1960s served as a regional rain gauge (Enzel et al. 2003, 2008; Bookman (Ken-Tor) et al. 2004, 2014) and reflected climatic conditions of the Levant. Thus, there was a climatic coherency between fresh groundwater supply from the JM and the lake level; when the DS level was low, the supply of fresh groundwater was low, and vice versa. Since the human-induced decline of the DS level, the fresh groundwater supply has remained high, while the rapid decline of the DS level simulates dry climatic conditions. These unique conditions generated such an intensive salt dissolution and the formation of the present thousands of sinkholes along the DS coast (Abelson et al. 2017a).

### 3.3 *Acceleration of Salt Layer Dissolution and Sinkhole Appearance in Stream Tributaries by Penetration of Floodwater Through Existing Sinkholes*

Another mode of sinkhole formation has recently been identified in some of the active tributaries of alluvial fans (Avni et al. 2016; Shviro et al. 2017). There, sinkhole formation was accelerated by flood waters draining into the subsurface through existing sinkholes which were formed by the groundwater that dissolved the salt layer in advance. Penetration of flood waters enhanced significantly the salt dissolution and the formation of new sinkholes (Fig. 14). As a result, the sinkhole sites at the outlet of wadi Zeelim (Fig. 1) became the most active site along the DS coast, where flood water was found to penetrate sinkhole in western clusters and discharging at eastern sinkholes after dissolving marked amount of the salt layer (Avni et al. 2016). This acceleration of sinkhole formation in Zeelim fan began in 2010. Prior to 2010, two clusters of sinkholes, eastern and western (Fig. 14) developed across the active stream channels by groundwater salt dissolution. Since 2010, floodwater penetrated some of these already existing sinkholes as also documented by Avni et al. (2016) by time-lapse camera. The floodwater penetration dissolved the salt layer very rapidly and formed sinkholes along its subsurface flow route, where sinkholes occurred after several hours to several days after the flood event (Avni et al. 2016). Some of the water that dissolved the subsurface salt layer discharged in the more eastern sinkholes, identified by time-lapse camera. The salt dissolution by the floodwater was attested by measurements of $\delta^{18}O$ (typical for the Jerusalem rainwater) and by Na/Cl ratio (corroborating the salt dissolution) (Avni et al. 2016).

## 4 Summary

Since their early appearance in the mid-1990s, the DS sinkholes accelerated their rate of appearance dramatically (Figs. 1 and 2) with increasing damage to the regional infrastructure (Figs. 1, 8 and 10). The primary driver for the sinkhole formation is the DS shrinkage leading to incursion of fresh groundwater from the west that dissolves a subsurface salt layer. After understanding the mechanism of the formation of the DS sinkholes in early stages of our research (e.g., Abelson et al. 2006; Shalev et al. 2006; Yechieli et al. 2006), there was not enough data and relative limited technology to understand the major controls on the rates of development of the DS sinkholes. Accumulation of observations and application of new technologies (e.g., LiDAR and InSAR, Figs. 6, 7 and 8) during eighteen years of research and monitoring enabled a further understanding of the main factors controlling the rates of appearance of sinkholes along the DS coast (Abelson et al. 2017a; Avni et al. 2016; Baer et al. 2018). These factors can be summarized as part of chain of events leading to the sinkhole formation. First, the dissolution rate of the salt layer and the resultant formation of subsurface cavities were controlled by long-term variations in precipitation in the

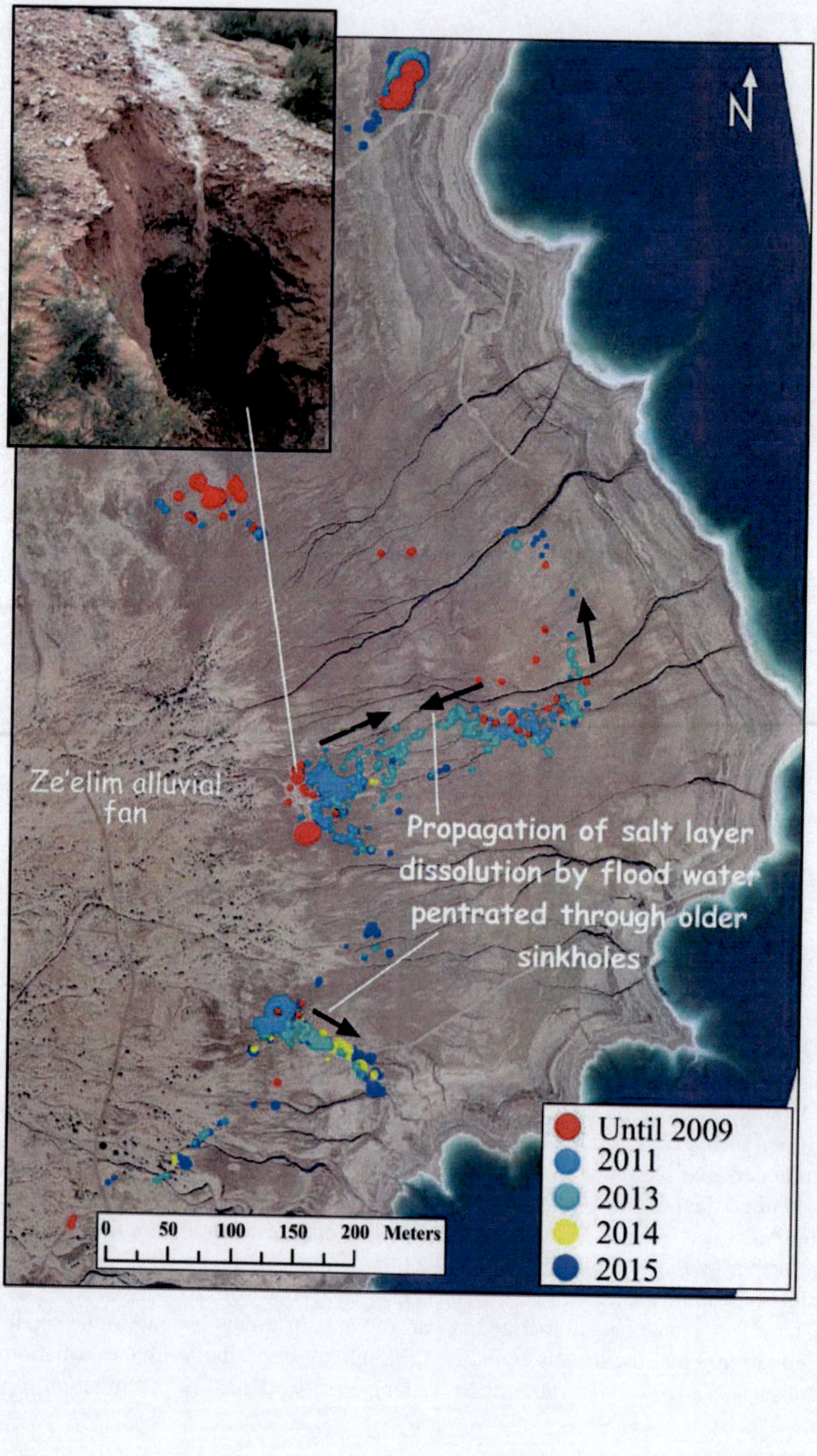
N
Ze'elim alluvial fan
Propagation of salt layer dissolution by flood water pentrated through older sinkholes
Until 2009
2011
2013
2014
2015
0 50 100 150 200 Meters

◀**Fig. 14** Evolution of sinkholes in the outlet of Zeelim stream (see location in Fig. 1) that indicates its significant acceleration since 2011 due to penetration of flood waters through older sinkholes (modified from Avni et al. 2016). The photograph (upper left; Avni et al. 2016) shows stream flowing into an existing sinkhole. This water dissolves the salt layer through the pre-existing cavities and accelerates sinkholes formation along a dissolution line expressed on the surface by propagation of sinkhole occurrence

Judea Mountains (Figs. 11, 12 and 13), rather than by the rates of the DS level drop and shoreline retreat (Abelson et al. 2017a). The influence of precipitation variations is seen on groundwater level in the DS coast aquifer on a typical time lag of 4–5 years, and actual sinkhole occurrence after 6–7 years (Fig. 11). Following the formation of subsurface cavities, the timing of sinkhole collapses was controlled by the properties of the overlying sediments (Abelson et al. 2006, 2018; Baer et al. 2018; Shalev and Lyakhovsky 2012). At the alluvial fans, due to thick and stiff sediments, large cavities can exist several years before their final collapse (Figs. 9 and 10). On the other hand, at the mudflats, once a cavity exceeds a width of several meters, the delay until final collapse can be several days or weeks (Fig. 9). Consequent to the formation of sinkholes by salt dissolution by groundwater at active stream channels, sinkhole formation is accelerated markedly by penetration of floodwater through the open sinkholes. This water flow through the salt layer enhances the rate of salt dissolution, as found in the Ze'elim delta, Hever and Rahaf alluvial fans (Shviro et al. 2017), and leading to abrupt increase in sinkhole formation in these areas (Fig. 14).

## References

Abelson M, Baer G, Shtivelman V, Wachs D, Raz E, Crouvi O, Kurzon I, Yechieli Y (2003) Collapse-sinkholes and radar interferometry reveal neotectonics concealed within the Dead Sea basin. Geophys Res Lett 30:1545. https://doi.org/10.1029/2003GL017103

Abelson M, Yechieli Y, Bein A, Shtivelman V, Crouvi O, Calvo R, Steinberg J, Baer G, Wachs D (2004) Potential levels for sinkholes collapse along the western coast of the Dead Sea. Report GSI/11/2013. Geological Survey of Israel, Jerusalem (in Hebrew, English abstract)

Abelson M, Yechieli Y, Crouvi O, Baer G, Wachs D, Bein A, Shtivelman V (2006) Evolution of the Dead Sea sinkholes. In: Enzel Y, Agnon A, Stein M (eds) New frontiers in Dead Sea paleoenvironmental research, vol 401. Geological Society of America Special Papers, pp 241–253

Abelson M, Boroshevsky A, Baruch O, Mushkin A, Yechieli Y (2013) Potential levels for sinkholes collapse along the western coast of the Dead Sea—2nd update. Report GSI/11/2013. Geological Survey of Israel, Jerusalem (in Hebrew, English abstract)

Abelson M, Yechieli Y, Baer G, Lapid G, Behar N, Calvo R, Rosensaft M (2017a) Natural versus human control on subsurface salt dissolution and development of thousands of sinkholes along the Dead Sea coast. J Geophys Res Earth Surf 122. https://doi.org/10.1002/2017JF004219

Abelson M, Yechieli Y, Baer G (2017b) Advancement in research and susceptibility assessment for development of sinkholes along the Dead Sea west coast—2017 update. Report GSI/24/2017. Geological Survey of Israel (in Hebrew, English abstract)

Abelson M, Aksinenko T, Kurzon I, Pinsky V, Baer G, Nof R, Yechieli Y (2018) Nanoseismicity forecasts sinkhole collapse in the Dead Sea coast years in advance. Geology 46:83–86

Abou-Karaki N, Fiaschi S, Paenen K, Al-Awabdeh M, Closson D (2019) Exposure of tourism development to salt karst hazards along the Jordanian Dead Sea shore. Hydrol Earth Syst Sci 2323:2111–2127

Al-Halbouni D, Holoan E, Taheri A, Schopfer MPJ, Emam S, Dahm T (2018) Geomechanical modelling of sinkhole development using distinct elements: model verification for a single void space and application to the Dead Sea area. Solid Earth 9:1341–1373

Arad A, Michaeli A (1967) Hydrogeological investigations in the western catchments of the Dead Sea. Isr J Earth Sci 16:181–196

Arkin Y, Gilat A (2000) Dead Sea sinkholes: an ever-developing hazard. Environ Geol 39:711–722

Atzori S, Antonioli A, Salvi S, Baer G (2015) InSAR-based modeling and analysis of sinkholes along the Dead Sea coastline. Geophys Res Lett 42

Avni Y, Lensky N, Dente E, Shviro M, Arav R, Gavrieli I, Yechieli Y, Abelson M, Lutsky H, Filin S, Haviv I, Baer G (2016) Self-accelerated development of salt karst during flash floods along the Dead Sea Coast, Israel. J Geophys Res Earth Surf 121. https://doi.org/10.1002/2015JF003738

Baer G, Magen Y, Nof R, Raz, E, Lyakhovsky V, Shalev E (2018) InSAR measurements and viscoelastic modeling of sinkhole precursory subsidence: implications for sinkhole formation, early warning and sediment properties. J Geophys Res Earth Surf 123. https://doi.org/10.1002/2017JF004594

Bartov Y, Stein M, Enzel Y, Agnon A, Reches Z (2002) Lake levels and sequence stratigraphy of Lake Lisan, the late Pleistocene precursor of the Dead Sea. Quat Res 57:9–21

Begin B, Ehrlich A, Nathan Y (1974) Lake Lisan, the Pleistocene precursor of the Dead Sea. Geol Surv Isr Bull 63:30

Ben-Avraham Z, Garfunkel Z, Lazar M (2008) Geology and evolution of the southern Dead Sea fault with emphasis on subsurface structure. Annu Rev Earth Planet Sci 36:357–387

Bookman (Ken-Tor) R, Enzel Y, Agnon A, Stein M (2004) Late Holocene lake levels of the Dead Sea. Geol Soc Am Bull 116:555–571

Bookman (Ken-Tor) R, Filin S, Avni Y, Rosenfeld D, Marco S (2014) Possible connections between large volcanic eruptions and level rise episodes in the Dead Sea Basin. Quat Sci Rev 89:123–128

Closson D, Abou Karaki N (2009) Salt karst and tectonics: sinkholes development along tension cracks between parallel strike-slip faults, Dead Sea, Jordan. Earth Surf Proc Land 34:1408–1421

Closson D, Abou Karaki N, Klinger Y, Jad Hussein M (2005) Subsidence and sinkhole hazard assessment in the southern Dead Sea area, Jordan. Pure Appl Geophys 162:221–248

Enzel Y, Bookman (Ken-Tor) R, Sharon D, Gvirtzman H, Dayan U, Ziv B, Stein M (2003) Late Holocene climates of the Near East deduced from Dead Sea level variations and modern regional winter rainfall. Quat Res 60:263–273

Enzel Y, Amit R, Dayan U, Crouvi O, Kahana R, Ziv B, Sharon D (2008) The climatic and physiographic controls of the eastern Mediterranean over the late Pleistocene climates in the southern Levant and its neighboring deserts. Global Planet Change 60:165–192. https://doi.org/10.1016/j.gloplacha.2007.02.003

Ezersky M (2010) Improvement of the seismic refraction method for mapping of the buried salt layers along the Dead Sea shoreline. Report GII 211/538/10. Geophysical Institute of Israel

Ezersky M, Frumkin A (2013) Fault—dissolution front relations and the Dead Sea sinkhole problem. Geomorphology 201:35–44

Filin A, Baruch A, Avni Y, Marco S (2011) Sinkhole characterization in the Dead Sea area using airborne laser scanning. Nat Hazards 58:1135–1154. https://doi.org/10.1007/s11069-011-9718-7

Frumkin A (2013) Salt karst. In: Shroder J, Frumkin A (eds) Treatise on geomorphology. Karst geomorphology, vol 6. Academic Press, San Diego, CA, pp 407–424

Frumkin A, Ezersky M, Al-Zoubi A, Akkawi E, Abueladas A (2011) The Dead Sea sinkhole hazard: geophysical assessment of salt dissolution and collapse. Geomorphology 134:102–117

Galloway D, Jones DR, Ingberitsen SE (1999) Land subsidence in the United States. United States Geological Survey Circular 1182, 177 pp

Galve JP, Gutiérrez F, Remondo J, Bonachea J, Lucha P, Cendrero A (2009) Evaluating and comparing methods of sinkhole susceptibility mapping in the Ebro Valley evaporite karst (NE Spain). Geomorphology 111:160–172

Garfunkel Z (1997) The history and formation of the Dead Sea basin. In: Niemi TM, Ben-Avraham Z, Gat JR (eds) The Dead Sea, the lake and its setting. Oxford University Press, New York, pp 36–56

Gavrieli I, Yechieli Y, Halicz L, Spiro B, Bein A, Efron D (2001) The sulfur system in anoxic subsurface brines and its implication in brine evolutionary pathways: the Ca-chloride brines in the Dead Sea area. Earth Planet Sci Lett 186:199–213

Gutierrez F, Cooper AH (2002) Evaporite dissolution subsidence in the historic city of Calatayud, Spain: damage appraisal and prevention. Nat Hazards 25:259–288

Gutiérrez F, Parise M, De Waele J, Jourde H (2014) A review on natural and human-induced geohazards and impacts in karst. Earth Sci Rev 138:61–88

Johnson KS (1997) Evaporite karst in the United States. Carbon Evapor 12:2–14

Kafri U, Yechieli Y (2010) Groundwater base level changes and adjoining hydrogeological systems. Springer, Heidelberg, Germany, 171 pp

Kiro Y, Yechieli Y, Shalev E, Lyakhovsky V, Starinsky A (2008) Time response of the water table and saltwater transition zone to a base level drop. Water Resour Res 44:1–15

Krumgalz BS, Hecht A, Starinsky A, Katz A (2000) Thermodynamic constraints on Dead Sea evaporation: can the Dead Sea dry up? Chem Geol 165:1–11

Laronne Ben-Itzhak L, Gvirtzman H (2005) Groundwater flow along and across structural folding: an example from the Judean Desert, Israel. J Hydrol 312:51–69. https://doi.org/10.1016/j.jhydrol.2005.02.009

Maimon O, Abelson M, Rybakov M, Lyakhovsky V (2009) An examination whether microgravity reflects underground cavities along the Dead Sea shoreline: a 3-D numerical model of rock mass stability. Report GSI/09/2009. Geological Survey of Israel, 22 pp

Martinez JD, Johnson KS, Neal JT (1998) Sinkholes in evaporite rocks. Am Sci 86:38–51

Neev D, Emery KO (1967) The Dead Sea. Geol Surv Isr Bull 41:1–147

Nof RN, Baer G, Ziv A, Raz E, Atzori S, Salvi S (2013) Sinkhole precursors along the Dead Sea Israel, revealed by SAR interferometry. Geology. https://doi.org/10.1130/G34505.1

Nof R, Abelson M, Raz E, Magen Y, Atzori S, Salvi S, Baer G (2019) SAR interferometry for sinkhole early warning and susceptibility assessment along the Dead Sea, Israel. Remote Sens 11:89. https://doi.org/10.3390/rs11010089

Parise M, Lollino P (2011) A preliminary analysis of failure mechanisms in karst and man-made underground caves in Southern Italy. Geomorphology 134:132–143. https://doi.org/10.1016/j.geomorph.2011.06.008

Picard L (1943) Structure and evolution of Palestine with comparative notes on neighboring countries. Hebr Univ Geol Depart Bull 4(2–4):1–187

Polom U, Alrashdan H, Al-Halbouni D, Holoan E, Dahm T, Sawarieh A, Atallah MY, Krawczyk CM (2018) Shear wave reflection seismic yields subsurface dissolution and subrosion patterns: application to the Ghor Al-Haditha sinkhole site, Dead Sea, Jordan. Solid Earth 9:1079–1098

Rybakov M, Goldschmidt V, Fleischer L, Rotstein Y (2001) Cave detection and 4-D monitoring: a microgravity case history near the Dead Sea. Lead Edge 20:896–900

Shalev E, Lyakhovsky V (2012) Viscoelastic damage modeling of sinkhole formation. J Struct Geol 42:163–170

Shalev E, Lyakhovsky V, Yechieli Y (2006) Salt dissolution and sinkhole formation along the Dead Sea shore. J Geophys Res 111(B3). https://doi.org/10.1029/2005jb004038

Shtivelman V, Keydar S, Pelman D (2005) Detecting subsurface cavities at sinkhole sites using 3D diffraction summation (stage 2). Report GII 845/118/05. GeophysIsrael

Shviro M, Haviv I, Baer G (2017) High-resolution InSAR constraints on flood-related subsidence and evaporate dissolution along the Dead Sea shores: interplay between hydrology and rheology. Geomorphology 293:53–68

Sneh A (1979) Late Pleistocene fan-deltas along the Dead Sea Rift. J Sed Petrol 49:541–552

Stein M, Torfstein A, Gavrieli I, Yechieli Y (2010) Abrupt aridities and salt deposition in the post-glacial Dead Sea and their North Atlantic connection. Quat Sci Rev 29:567–575

Taqieddin SA, Abderahman NS, Atallah M (2000) Sinkhole hazards along the eastern Dead Sea shoreline area, Jordan: a geological and geotechnical consideration. Environ Geol 39:1237–1253

Watson RA, Holoan E, Al-Halbouni D, Saberi L, Sawarieh A, Closson D, Alrashdan H, Abou-Karaki N, Siebert C, Walter TR, Dahm T (2018) Sinkholes, stream channels and base-level fall: a 50-year record of spatio-temporal development on the eastern shore of the Dead Sea. Solid Earth Discuss. https://doi.org/10.5194/se-2018-10

Wust-Bloch GH, Joswig M (2006) Pre-collapse identification of sinkholes in unconsolidated media at Dead Sea area by 'nanoseismic monitoring' (graphical jackknife location of weak sources by few, low-SNR records). Geophys J Int 167:1220–1232

Yechieli Y (1993) The effects of water level changes in closed lakes (Dead Sea) on the surrounding groundwater and country rocks. PhD thesis, Weizmann Institute of Science, Rehovot, Israel, 197 pp

Yechieli Y (2000) Fresh-saline groundwater interface in the Western Dead Sea area. GroundWater 38:615–623

Yechieli Y, Magaritz M, Levy Y, Weber U, Kafri U, Woelfli W, Bonani G (1993) Late Quaternary geological history of the Dead Sea area, Israel. Quat Res 39:59–67

Yechieli Y, Ronen D, Berkovitz B, Dershovitz WS, Hadad A (1995) Aquifer characteristics derived from the interaction between water levels of a terminal lake (Dead Sea) and an adjacent aquifer. Water Resour Res 31(4):893–902

Yechieli Y, Gavrieli I, Ronen D, Berkovitz B (1998) Will the Dead Sea die? Geology 26:755–758

Yechieli Y, Abelson M, Bein A, Crouvi O, Shtivelman V (2006) Sinkholes "swarms" along the Dead Sea coast: reflection of disturbance of lake and adjacent groundwater systems. Geol Soc Am Bull 118:1075–1087

Yechieli Y, Shalev E, Wollman S, Kiro Y, Kafri U (2010) Response of the Mediterranean and Dead Sea coastal aquifers to sea level variations. Water Resour Res 46:W12550. https://doi.org/10.1029/2009WR008708

Zak I (1967) The geology of Mount Sedom. PhD thesis, The Hebrew University, Jerusalem, 208 pp (in Hebrew, English summary)

# Geoelectric, Geoelectromagnetic and Combined Geophysical Methods in Groundwater Exploration in Israel

Mark Goldman and Uri Kafri

## 1 Introduction

Among all geophysical methods, applied in groundwater exploration, geoelectric (GE) and geoelectromagnetic (GEM) ones are most successful for characterization of groundwater salinity (e.g., Fitterman and Stewart 1986; Goldman et al. 1988) and for delineating geometrical features of different hydrogeological targets, including the configuration of the fresh–saline groundwater interface, the thickness of alluvial fill and gravel lenses overlying the bedrock, the thickness and depth of sand and gravel lenses in till, etc. (e.g., Albouy et al. 2001; Fitterman and Labson 2005). The methods are less efficient in quantitative evaluation of hydraulic/hydrological parameters by using geoelectric ones, either measured (e.g., apparent resistivities) or interpreted in terms of true physical parameters (e.g., resistivities). In this respect, the most extensive efforts were devoted to studying the empirical relationship between the porosity and/or the pore water salinity/conductivity on one hand and the bulk electrical resistivity recovered by GE/GEM methods on the other hand (e.g., Archie 1942; Kelly and Mareš 1993; Yang et al. 1999). However, due to uncertainties in determining all the parameters used in the above relationship, such as the pore water and bulk electrical resistivities, clay content, or the type of a rock sequence, these attempts proved to be less successful than those devoted to the delineation of geometrical parameters. Somewhat better estimates can be obtained in both clastic and carbonate aquifers saturated with saline water, where the use of the time domain electromagnetic (TDEM) method allows to more accurately determine the pore water and bulk

M. Goldman (✉)
Leon H. Charney School of Marine Sciences, University of Haifa, Haifa 31905, Israel
e-mail: mgol1302@gmail.com

U. Kafri
Geological Survey of Israel, 32 Leibovitz Str., Jerusalem 9692100, Israel
e-mail: uri.kafri@gsi.gov.il



U. Kafri and Y. Yechieli (eds.), *The Many Facets of Israel's Hydrogeology*,
Springer Hydrogeology, https://doi.org/10.1007/978-3-030-51148-7_16

electrical resistivities and thus to reduce uncertainties in estimating the porosity or storativity of the aquifers (Kafri and Goldman 2005).

In the last few decades, the efforts have been also made to develop geophysical methods, which allow a more detailed characterization of aquifers by direct quantitative determination of their hydraulic properties, such as porosity/storativity and/or conductivity/transmissivity. These methods include cross-hole seismic and radar tomography, surface nuclear magnetic resonance (SNMR), seismoelectric and electroseismic methods and spectral induced polarization (SIP) method (e.g., Garambois and Dietrich 2001; Linde et al. 2008; Yaramanci 2003; Thompson and Gist 1993; Kemna et al. 2012).

In Israel, suffering not only from a shortage of freshwater, but also from severe salinization problems of both surface and groundwater resources, the main objective of geophysical surveys was aimed at detecting and accurately delineating saline ground water bodies of different origins. Taking into account the close relationship which exists between water salinity and electrical conductivity (resistivity), it is natural that the principal geophysical methods applied in Israel are the GE/GEM methods. Seismic methods, which are better suited for delineation of different rock units, are applied to discriminate aquifers and aquicludes in the subsurface (e.g., Shtivelman and Goldman 2000). The combined application of the GEM and SNMR methods in the Mediterranean and Dead Sea coastal aquifers allowed a quantitative estimation of fresh groundwater volume within the aquifers as well as the discrimination of the completely saturated brine with regard to halite from the slightly diluted Dead Sea brine (Legchenko et al. 2009). The objective of this chapter is mainly to focus on the application of GE/GEM methods for characterization of groundwater salinity in different aquifers throughout Israel.

## 2 Methods

The following three main GE/GEM methods, which were widely applied in groundwater exploration in Israel, are described below: DC resistivity soundings, electrical resistance tomography (ERT) and the time domain electromagnetic (TDEM) method. In addition, only one non-GE/GEM method, namely the surface nuclear magnetic resonance, SNMR (also known as proton magnetic resonance, PMR, or magnetic resonance soundings, MRS) method, is shortly described below. The description of all other geophysical methods is beyond the scope of this chapter.

### 2.1 *The DC Resistivity Soundings*

In this method, the earth is energized using a direct current. Thus, the only way to inject the current into the earth is by using a pair of current electrodes connected to a DC transmitter. Another pair of electrodes, named potential or measurement electrodes, is used to measure the established potential difference throughout the earth surface. The surficial distribution of the potential depends on spatial resistivity distribution in the subsurface and thus the latter can be, in principle, recovered by proper measurements of the surface potential differences. This is done by moving either the current or potential or all electrodes along straight lines above a subsurface exploration target. Depending on the inter-electrode configurations, there are several methods of DC resistivity soundings, each having different advantages and drawbacks. The most popular configurations are shown in Fig. 1.

In the case of the Wenner method (Fig. 1a), all four electrodes are relocated keeping the distance (separation) between them constant. This method provides the largest signal-to-noise ratio, but is mostly time consuming and requires long cables. In the case of the Schlumberger method (Fig. 1b), only the outer (current) electrodes are relocated, keeping the potential electrodes in the center of the array. The latter are moved only occasionally, when the signal-to-noise ratio drops below some specified

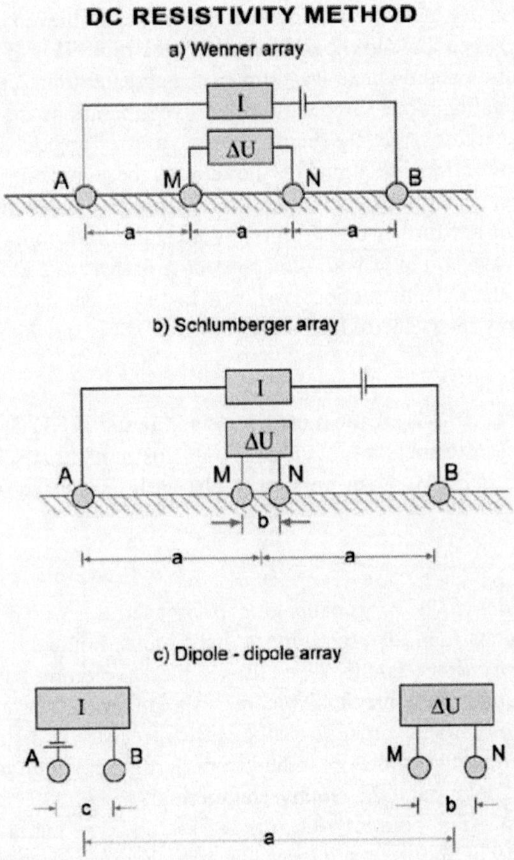


**Fig. 1** DC resistivity arrays

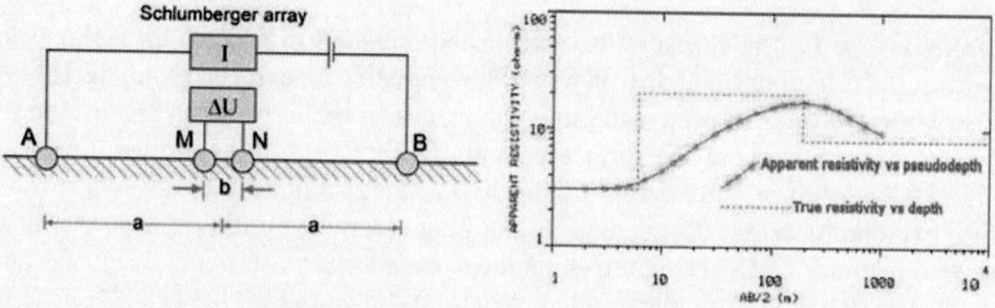


**Fig. 2** Example of the apparent resistivity versus pseudo-depth curve and the true resistivity versus depth section for the Schlumberger array

level. The method provides a larger exploration depth for a given cable length and is less time consuming than the Wenner method. The dipole–dipole method (Fig. 1c) requires minimum cables lengths for a given depth, but suffers from the smallest signal-to-noise ratio.

In all DC resistivity methods, the exploration depth is achieved by increasing the separation (***a***) between the current and potential electrodes (Fig. 1). In general, the maximum separation should be at least three times greater than the required exploration depth. The voltages measured with successive separations are first transformed into apparent resistivities, namely the true resistivities of an equivalent half-space, which would produce the same signals as the actually measured ones. These apparent resistivities versus separations curves are used to build semi-qualitative resistivity pseudo-sections or are further transformed by inversion into true resistivity versus depth sections. An example of both the apparent resistivity pseudo-section and the true resistivity versus depth section is shown in Fig. 2 on the right side for the Schlumberger array inter-electrode configuration, which is shown on the left side.

**Advantages**

The DC resistivity methods are the most simple and intuitive GE/GM methods from both hardware and software points of view. The DC instruments are cheap and widely available. In fact, their most basic version can be easily produced in any electronic lab.

**Drawbacks**

The use of electrodes can be very challenging in some terrains with difficult electrical grounding conditions (e.g., dry sand dunes, hard rocks, boulders). The maximum inter-electrode separations should exceed the required exploration depth by roughly an order of magnitude. This inevitably decreases a lateral resolution of the method and distorts a quantitative estimation of the vertical resistivity distribution by 1-D inversion. In addition, the inversion suffers from particularly high non-uniqueness, as common for all potential field geophysical methods.

From the hydrological point of view, the DC methods are mainly applicable for the delineation the of the rock sequence, whether aquifers or aquicludes, in the vertical dimension. The application is restricted to a rather shallow depth, whereby the sequence includes only a few units. The methods are not suitable in a tectonically disturbed area, where the length of a required measuring array is such that it crosses faults. In such a case, the results often represent a mixture of different tectonic blocks and 1-D inversion results might lead to erroneous hydrogeological interpretation.

## 2.2 *The Electrical Resistance Tomography (ERT)*

In order to significantly improve a resolution of the DC resistivity soundings, a multi-dimensional modification of the method was developed a few decades ago (e.g., Loke 2004). It realizes modern digital technologies, which provide automatic scanning and selection of any two pairs of electrodes among a large number of preinstalled electrodes along a profile (2D imaging) or within a selected area (3D imaging). The selected pairs are used as current and potential electrodes either in conventional arrays, as shown in Fig. 1, or in any other geometrical combination of equally spaced electrodes, required by the survey design.

An example of a field setup and the measurement procedure using the most popular Wenner inter-electrode configuration is exhibited herein (Fig. 3). In this example, 20 electrodes are preinstalled along a profile with the spacing between adjacent electrodes "a". First, all the possible measurements, with an electrode spacing of "1a", are carried out (Fig. 1). Electrode numbers 1, 2, 3 and 4 are used for the first measurement. Electrode 1 is used as the first current electrode C1, electrode 2 as the first potential electrode P1, electrode 3 as the second potential electrode P2 and electrode 4 as the second current electrode C2. Then, electrodes 2, 3, 4 and 5 are used for C1, P1, P2 and C2, respectively, for the second measurement and the measurements are repeated along the entire line of electrodes until electrodes 17, 18,

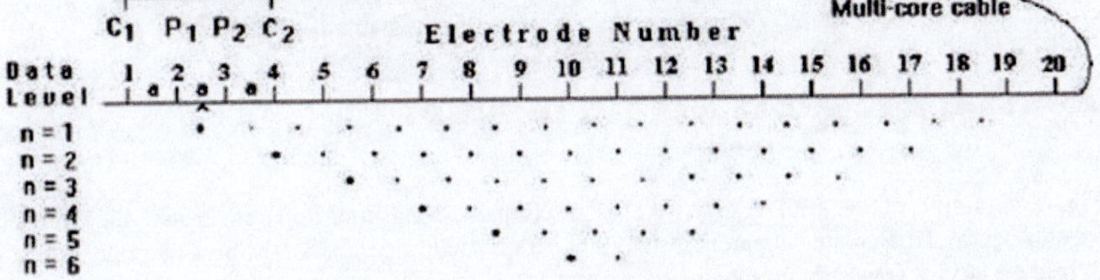


**Fig. 3** Arrangement of electrodes for a 2D electrical survey and the sequence of measurements used to build up a pseudo-section (modified after Loke 2004)

19 and 20 are used for the last measurement with "1a" spacing. For a system with 20 electrodes, there are 17 (20–3) possible measurements with "1a" spacing using the Wenner array.

Upon completing the measurements with "1a" spacing, the next sequence of measurements with "2a" electrode spacing is carried out. Electrodes 1, 3, 5 and 7 are used for the first measurement. The electrodes are selected with "2a" spacing between adjacent electrodes. For the second measurement, electrodes 2, 4, 6 and 8 are used. This process is repeated along the entire line until electrodes 14, 16, 18 and 20 are used for the last measurement with "2a" spacing. For a system with 20 electrodes, there are 14 ($20 - 2 \times 3$) possible measurements with "2a" spacing.

The same procedure is repeated for measurements with "3a", "4a", "5a" and "6a" spacings between electrodes. Each set of measurements "n" corresponds to a certain pseudo-depth, whereby the greater "n" yields the greater pseudo-depth (Fig. 3). Since each measurement is related to the middle of the array, the deeper parts of the pseudo-section are shorter and the resulting pseudo-section is represented by a trapeze (Fig. 3). Depending on the array, there are different methods to calculate the pseudo-depth, but in the case of the conventional Wenner array, the pseudo-depth is normally equal to the spacing (Loke 2004).

The measured data are customarily presented in the form of the apparent resistivity pseudo-section (Fig. 4a), which provides a semi-quantitative picture of the

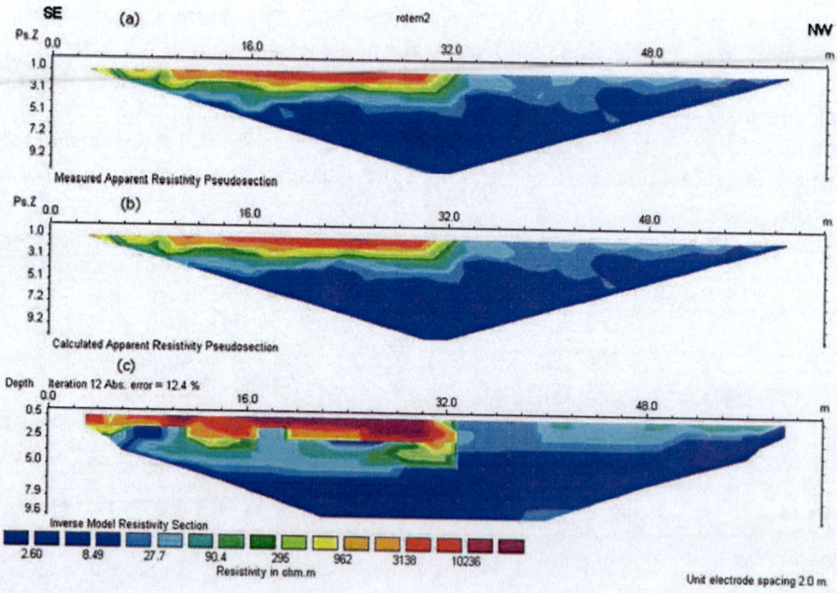


**Fig. 4** Example of the ERT resistivity cross-sections. **a** Measured apparent resistivities versus pseudo-depths, **b** calculated apparent resistivities versus pseudo-depths for the best-fit model, **c** best-fit true resistivity versus depth model

true subsurface resistivity distribution. Such a presentation is feasible for the visualization of the measured apparent resistivities and as an initial model for further quantitative interpretation. In order to obtain quantitative information regarding the true resistivities and depths by using the above apparent resistivity pseudo-section, a further 2D inversion of the measured data is required (Loke 2004). An example of the inverted true resistivity versus depth model and the calculated synthetic response for this model is shown in Fig. 4b, c, respectively. A visual comparison of Fig. 4a, b shows how well the data are fitted to the measurements at different parts of the section. An overall quality of the inversion is controlled by the total RMS (root mean square) error (Fig. 4).

**Advantages**

In principle, the ERT method allows a quantitative determination of the resistivity distribution in all three dimensions, although at present, full 3D surveys are still rare. By increasing the number of electrodes, one can provide a high lateral resolution. The multi-dimensional forward and inverse modeling is well developed.

**Drawbacks**

Similarly to 1-D resistivity soundings, the use of electrodes can be most challenging in some terrains with difficult electrical grounding conditions (sand dunes, hard rocks, boulders). The inversion suffers from non-uniqueness, which is not adequately treated in the available software.

## 2.3 *The Time Domain Electromagnetic (TDEM) Method*

The TDEM sounding system consists of a transmitter (Tx) and receiver (Rx) antennae. Contrary to the DC resistivity methods, the current waveform driven through Tx is not a steady state, but consists of equal periods of time-ons and time-offs (alternative current). This current creates a primary magnetic field, which varies with time. In accordance with Faraday's law, rapid changes of the current and thus of the primary magnetic field during switch-ons and switch-offs induce an electromotive force in the subsurface. As a result, eddy currents are created everywhere in the conductive ground, the intensity of which at each point depends on the ground resistivity or conductivity. These eddy currents generate a secondary magnetic field, the time rate-of-change of which is measured on the earth surface. Thus, contrary to the DC resistivity methods, the use of electrodes as transmitters and/or receivers is only optional and not necessary in the TDEM method. Instead, an ungrounded loop of wire can be used as a transmitter and as an example a small air coil as a receiver. If the signal is only measured during time-offs, i.e., in the absence of the primary field, the receiver can be placed everywhere, both outside and inside the transmitter loop (Fig. 5), or can even be combined with the latter. Unlike the DC resistivity methods, the exploration depth in the TDEM method is not determined by the Tx-Rx separation, but rather by the time after the Tx current is switched off.

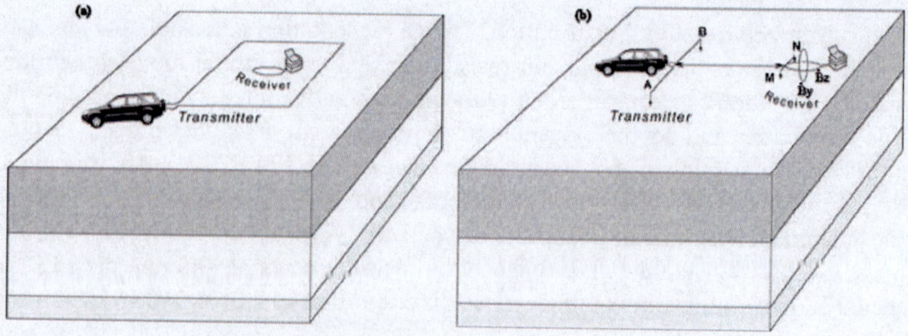


**Fig. 5** Typical TDEM arrays. **a** Central loop, **b** grounded line—Ex, Bz, By

Data processing and interpretation in the TDEM method are fairly similar to those in conventional DC resistivity one. The measured signals are normally first converted to apparent resistivities and then inverted in terms of 1-D layered models. The main difference stems from a more complicated definition of the apparent resistivity, which is equivalent to a true resistivity of the half-space, not in the entire measurement time range, but only at sufficiently late times (Kaufman and Keller 1983). A typical example of the collected data in form of the late-stage apparent resistivities (left side) and the appropriate inverted model (right side) is shown in Fig. 6.

In contrast to all other GEM techniques, the time domain/transient electromagnetic (TDEM/TEM) method was developed in two different modes depending on the

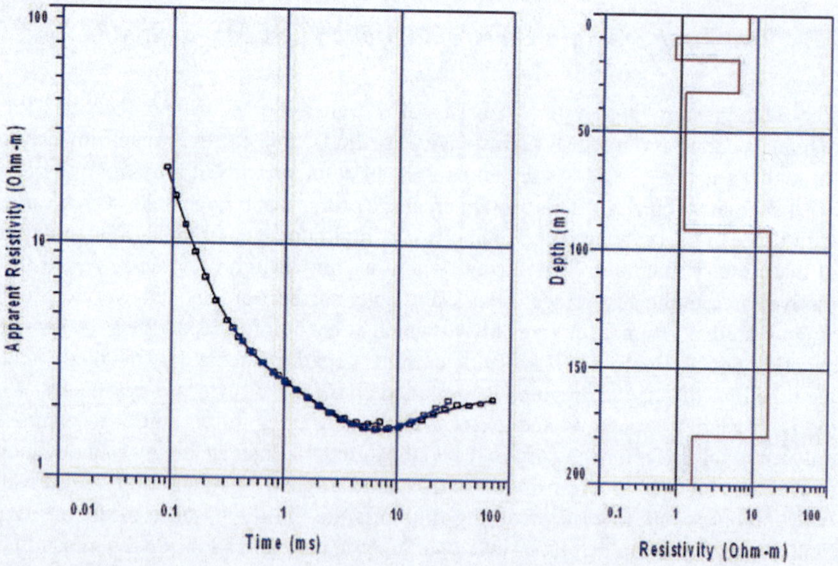


**Fig. 6** An example of the late-stage apparent resistivity TDEM data (open squares) and the best-fit curve (left side) for the best-fit model (right side)

transmitter–receiver separations used. This duality was caused by the specific current waveform, including both time-on and time-off periods employed in the method. If the measurements are only carried out during the time-off periods, the response from the ground is entirely decoupled from the primary field. As a result, the receiver can be located at any distance from the transmitter, and the transmitter and receiver antennae can even be combined in a single antenna. This means that, theoretically, the method provides an exploration depth, which is independent on the Tx-Rx separation (offset). However, in reality, signal-to-noise considerations might force to place the receiver antenna far away from the transmitter (far-field) due to a very fast decay of the signal as a function of time in the close proximity to Tx (near-field). On the other hand, increasing offset causes the deterioration of the resolution and increases undesirable multi-dimensional effects caused by geologic noise. These contradictory considerations actually led to the development of two independent TDEM (TEM) methods. The first one, which uses short offset arrays, significantly shorter than the depth to the target, is named SHOTEM or near-zone TEM. The second one is named LOTEM or far-zone TEM.

**SHOTEM**

The array commonly used in the SHOTEM method is named central loop (CL) array (Fig. 5a). It consists of a transmitter (Tx) antenna represented by a square, sometimes rectangular, or a circular loop of insulated wire laid on the earth surface. The receiver (Rx) antenna can be either another smaller loop of wire or a multi-turn air coil placed in the center of the Tx loop or at any other convenient place inside the Tx loop ("in-loop array"). By using this array, one can detect a target at depths two to three times exceeding the Tx size. Sometimes, mainly for mineral exploration, the Rx coil is placed outside the Tx loop (slingram array) or the signal is picked-up by the same Tx loop (coincident loop or one-loop array).

Less commonly used SHOTEM arrays are those comprising electrical dipoles (Fig. 5b). This is mainly caused by current injection problems, if the dipoles are only used as transmitters. However, if the electrodes are also used to pick up the signal (electrical components), the problem is more complicated. Due to a very fast decay of the transient signal with time, most of the measured responses turn out to be severely distorted by induced polarization (IP) and/or spontaneous or self-potential (SP) effects.

The situation has been changed dramatically, when the SHOTEM method was applied in the marine environment. Both the current injection and measurement problems have been successfully overcome due to extremely low contact resistances, very homogeneous grounding conditions and, particularly, due to the slow decay of the signal in an electrically conductive environment (Goldman et al. 2011).

**LOTEM**

The traditional LOTEM array, which was mainly developed to explore presumably 1-D earth, consisted of a long grounded transmitter line and broadside horizontal electric dipole (Ex) and/or vertical magnetic dipole (dBz/dt) receivers (Vanyan 1967). The modern LOTEM array is also mainly broadside, but generally measures five components of the EM field (Strack 1992), three of which are shown in Fig. 5b. The principal difference between the SHOTEM and LOTEM methods lies in the Tx-Rx separation used. Whereby the SHOTEM offset could be 2–3 times smaller than the depth to the target, the offsets used in LOTEM are normally 3–5 times larger than the depth to the target. Thus, offsets difference between the two methods for the same target depth might be in an order of magnitude. Such a huge difference leads to important geophysical consequences in lateral resolution and multi-dimensional effects.

In contrast to SHOTEM, the LOTEM signals can be measured during either turn-off or turn-on periods. In both cases, the vertical resolution of the electric field is similar to the one exhibited in the DC limit, to which the signal approaches at early times or at late times correspondingly. The lateral resolution is inherently poor due to the large offsets. The LOTEM signals decay with a time much slower than in the case of SHOTEM, thus increasing the signal-to-noise ratio in the LOTEM method as compared to the SHOTEM method for the same exploration depth. The ability to measure the electrical fields makes the LOTEM method superior to the traditional (on-land) SHOTEM one, if the target is resistive and relatively thin. However, the situation has become different in the marine environment in the presence of the discovered coastal effect, which significantly enhances the relative target response of resistive, including thin, structures in the case of the marine SHOTEM method (Goldman et al. 2011).

**Advantages**

The SHOTEM method provides the best lateral resolution of all 1-D GE/GEM sounding methods. With regard to conductive targets, it also demonstrates an excellent vertical resolution. By using ungrounded transmitter loops, it can be applied practically in any terrain.

The LOTEM method is very efficient in deep soundings and, despite the use of current and measurement electrodes, it can be applied practically in any terrain as it requires the grounding of just one pair of current electrodes per sounding.

From the hydrogeological point of view, the SHOTEM method has an advantage as compared to the DC and LOTEM methods. Since the dimensions of the transmitter loop are relatively small with regard to the required exploration depth, the measurements are considerably less affected by neighboring tectonic blocks and other lateral inhomogeneities.

**Drawbacks**

Due to extremely fast signal decay, the SHOTEM method exhibits rather poor signal-to-noise characteristics. The LOTEM method suffers from a fairly low lateral resolution.

The equipment used in both SHOTEM and LOTEM is expensive, and the interpretation software is rather complicated and is still mainly one-dimensional.

## 2.4 The Surface Nuclear Magnetic Resonance (SNMR) Method

In order to avoid undesirable connotations with the term "nuclear," the method is more widely named "proton magnetic resonance" (PMR) and "magnetic resonance soundings" (MRS). Although the method is based on different physical principles than GE/GEM methods, due to a great similarity in almost all aspects of applications (primary field excitation, measurements setup, data processing and interpretation), it is commonly considered among the GEM methods.

The greatest difference between the SNMR method and all GE/GEM, and in fact between all other geophysical methods, is that its inversion does not yield physical parameters, such as electrical resistivity, dielectric permittivity, acoustic velocity, but rather hydrogeological ones, such as water content and transmissivity of the aquifer. The former is directly proportional to the amplitude of the measured signal, whereby the latter is closely related to the speed of its decay with time. This unique feature of the SNMR method greatly reduces an ambiguity of the final hydrogeological interpretation, which in the case of SNMR mainly includes non-uniqueness of the geophysical inversion by itself.

The main principles of the SNMR method are shortly described below (e.g., Goldman et al. 1994a).

1. Hydrogen nuclei (protons) in water possess nonzero magnetic moments.
2. In the presence of geomagnetic field, $\mathbf{B_0}$, the protons in groundwater molecules are aligned in parallel to the field (Fig. 7a). In this equilibrium state, the protons are able to absorb electromagnetic energy that has been emitted at the Larmor frequency, $f_0$, where $f_0 = gB_0$ and $g$ is the gyromagnetic ratio ($g = 4.254597 \times 10^{-2}$ radians per second per nano-Tesla).
3. If a primary electromagnetic (EM) field oscillating at the Larmor frequency is applied, the magnetic moments of the water molecules begin to precess from their initial position (Fig. 7b).

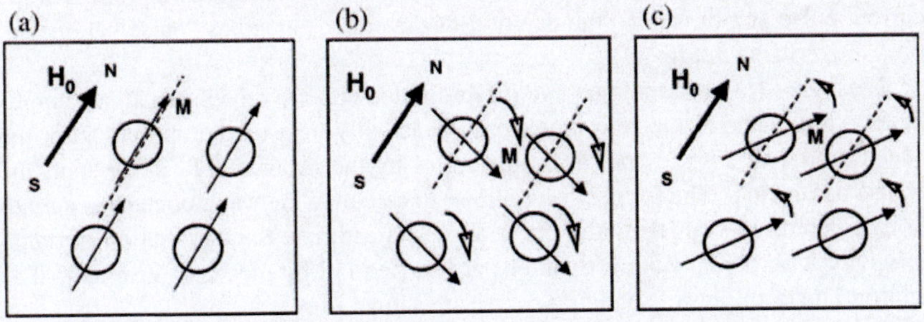


**Fig. 7** General principles of the SNMR method. **a** Water protons in presence of geomagnetic field H0, **b** water protons having magnetic moments M in presence of both geomagnetic and external magnetic field emitted at the Larmor frequency, **c** water protons after the external magnetic field is switched off (modified after Legchenko et al. 2002)

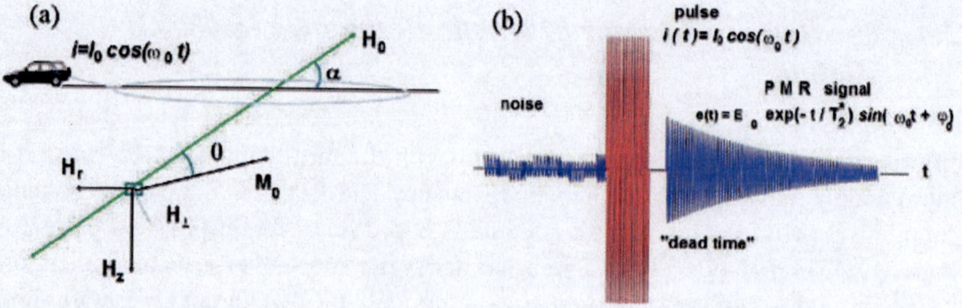


**Fig. 8** **a** Typical SNMR field setup, **b** SNMR signal measured during all three stages shown in Fig. 7a–c, respectively (after Legchenko and Valla 2002)

4. When the applied EM field is stopped, the magnetic moments of the water molecules return to their original positions, generating a secondary electromagnetic field (Fig. 7c). This field also oscillates at the Larmor frequency and exponentially decays with time due to heat losses. The secondary field consists of the important information regarding groundwater and is measured at the earth surface by the geophysical NMR instrument.

A typical SNMR field set up is presented in Fig. 8a. Figure 8b shows EM signals measured by the SNMR instrument during the above-mentioned three stages (Fig. 7a–c).

The initial amplitude of the secondary field is directly proportional to the amount of groundwater, irrespective the hydraulic conductivity/transmissivity of the water-bearing rocks. Fortunately, the measured signal decays differently, depending on the transmissivity. The greater is the time constant $T_2^*$, the slower is the decay and the greater is the transmissivity of the aquifer. Experience shows that $T_2^*$ for most aquicludes is less than 50 ms. The real measurements normally start after this time, named "dead time" whereby the water is detected in aquiferous rocks only.

In order to perform soundings, namely to discriminate signals from different depths, one must to use an appropriate sounding parameter. Unlike the previously described methods, the sounding parameter in the SNMR method is a product of the current pulse amplitude and the duration of the pulse, named as "pulse parameter" (i.e., the red area in Fig. 8b).

Thus, the SNMR sounding curve is represented either by the initial signal amplitude (extrapolated to the zero time from the actually measured amplitude after the "dead time") versus the "pulse parameter" or by the measured $T_2^*$ also versus the "pulse parameter." The former curve allows to calculate the water content as a function of depth, whereas the latter curve allows to estimate the distribution of transmissivity with depth. Both estimations are carried out by means of inversion. The appropriate examples of sounding curves are shown in Fig. 9a, b.

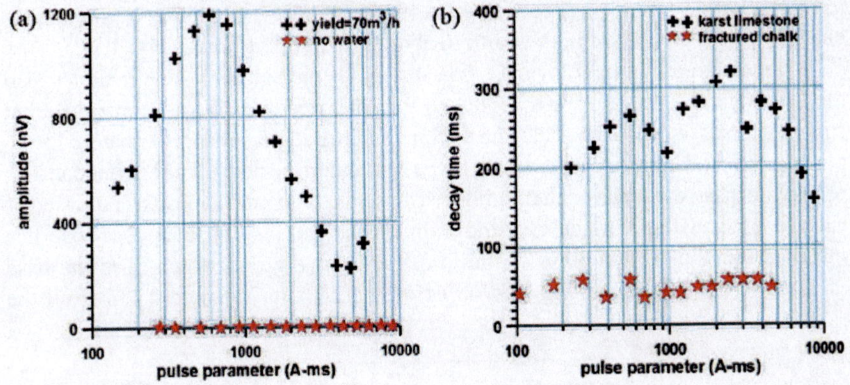


**Fig. 9** Examples of the SNMR data. **a** Initial amplitude versus pulse parameter, **b** time constant (decay time) $T_2^*$ versus pulse parameter (after Legchenko and Valla 2002)

**Advantages**
The SNMR method is the only geophysical tool, which allows a direct detection of groundwater and quantitative estimation of the water content within aquifers and, under favorable conditions, the transmissivity of the aquifer.

**Drawbacks**
Due to the necessity to use a specific and relatively high frequency (roughly 2 kHz), the method has principally a limited exploration depth (roughly 100 m). The instrument is expensive, and measurements are very slow.

## 3 Geoelectric Characterization of Groundwater Salinity in Clastic Coastal Aquifers

### *3.1 The Mediterranean Coastal Aquifer*

The regional coastal aquifer is one of the major water resources of the country. As such, due to the heavy water demand, the aquifer is heavily over-exploited. This in turn leads to water table decline below sea level, resulting in the encroachment of seawater into the aquifer up to approximately 2 km inland. Due to its importance, the hydrogeology of the aquifer was dealt with numerous studies. The following is a concise summary of the configuration of the aquifer, based on Kafri and Goldman (2006) (see also Chapter "General Information and Hydrogeology of the Mediterranean and Dead Sea Coastal Aquifers and Their Relation with Their Base Level").

The aquifer, approximately 110 km long, extends from the Gaza Strip in the south to the Mount Carmel in the north. It is divided into strips, 2 km wide, perpendicular to the shoreline (Fig. 10). The west–east (hereafter W-E) extension of the aquifer changes from some 5 km in the north to about 30 km in the south (Fig. 10).

A generalized hydrogeological cross-section of the aquifer which includes also the lithological units, the water table and the seawater intrusions into the different sub-aquifers is shown in Fig. 11. The thickness of the aquifer is approximately 150 m close to the shoreline, and it decreases to the east and to the north. The aquifer consists of a Quaternary sequence of alternating marine and continental deposits, composed of calcareous sandstones, sands, clay and loam interlayers. The entire sequence overlies impermeable shales and clays of the Pliocene Saqiye group which form the base of the aquifer. In the western portion of the aquifer, up to some 4 km from the

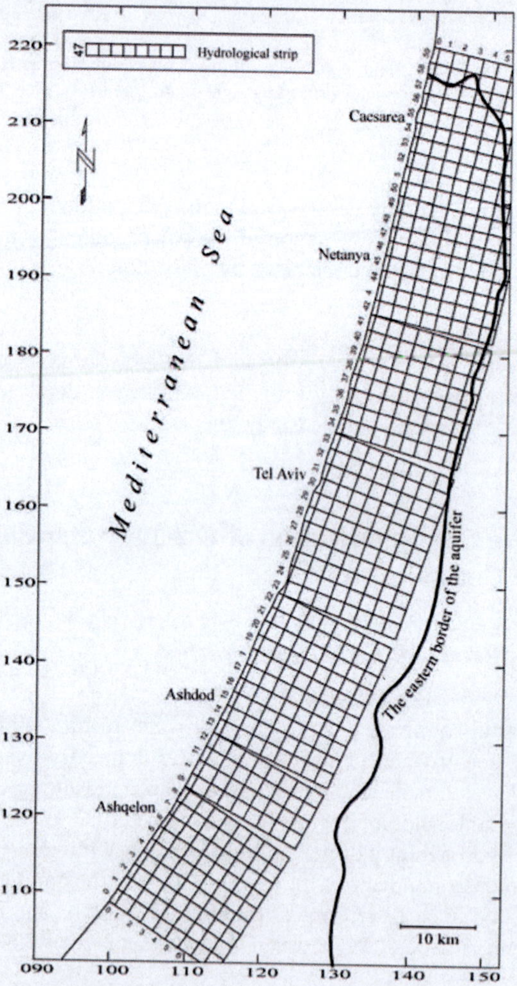


**Fig. 10** Schematic map of the Mediterranean coastal aquifer of Israel

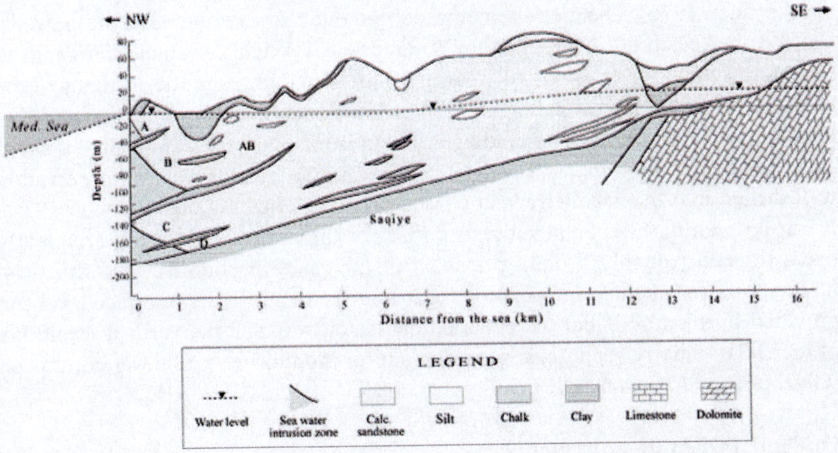


**Fig. 11** Typical hydrogeological cross-section of the Mediterranean coastal aquifer

shoreline, the above-mentioned impermeable, mostly clay, horizons subdivide the aquifer into generally four sub-aquifers from top to bottom (Fig. 11), denoted herein as A (phreatic), B (phreatic or confined), C and D (confined) sub-aquifers.

As of today, the exploitation of the aquifer is being carried out mostly from the upper (A and B) sub-aquifers. However, fresh or brackish water was encountered in the lower C and D, sub-aquifers (hereafter CD), below a seawater intrusion in only a few existing deep observation wells. These sub-aquifers were seldom exploited despite the growing water demand, and the fact that these sub-aquifers are protected from surficial pollution. As a result, very little was known as to this rather deep freshwater body and whether it is subjected to seawater encroachment.

Both problems are crucial for rational exploitation of the lower sub-aquifers, namely the potential of freshwater as well as to its protection from salinization in case that the CD are blocked to the sea, as suggested by Kolton (1988).

**The DC resistivity method**

Since the 50s of the last century, an intensive drilling program for groundwater exploitation was carried out in Israel. The drilling was carried out in both the shallow coastal clastic aquifers and the rather deep mountain carbonate aquifers. In many cases, in order to obtain a drilling forecast of the lithological sequence, DC resistivity (Schlumberger) soundings were carried out at the chosen site prior to the drilling. This was done even in the case of a multi-layered subsurface deep sequence. In general, the success in accurately defining and delineating the different lithological units was found to be poor due to the limitations of the DC method as described above.

The first geoelectric measurements targeted to detect and monitor the fresh–seawater interface in the Mediterranean coastal aquifer were carried out in the 60s and 70s of the last century using the DC resistivity (Schlumberger) soundings (Ginzburg and Levanon 1976). The measurements, carried out near existing observation wells, showed the feasibility of the method to detect and roughly evaluate the depth to the interface and the resistivity of seawater intrusion. However, the accuracy of the measurements and interpretation was rather low due to the above-mentioned drawbacks of the DC resistivity soundings. For example, Fig. 12 shows the apparent resistivities measured in 1960 and 1975 at the same location near an observation well drilled into the Mediterranean coastal aquifer (Ginzburg and Levanon 1976). It can be seen that the obtained resistivities for the seawater intrusion are greatly overestimated (roughly 5 ohm-m instead of the real 1.5 ohm-m). Unfortunately, no quantitative borehole information regarding the depth to the interface has been provided in the article, but the general conclusion was that the vertical resolution of the DC resistivity soundings is insufficient for monitoring the seawater intrusion (Ginzburg and Levanon 1976).

## On-land TDEM measurements

In order to increase the accuracy and both lateral and vertical resolution of the geophysical measurements, the TDEM method was applied for the first time in Israel in the mid-80s of the last century (Goldman et al. 1988). The first feasibility study in the Mediterranean coastal aquifer included just four central loop TDEM stations located in the Nitzanim area, roughly 1 km to the south of observation wells 12-0, 12-B and 12-A (Fig. 13). One of the stations (N4) was located directly on observation well $T_1$. The interpretation of all collected data revealed a very low-resistivity body, gradually dipping from west to east, which was undoubtedly identified with a seawater intrusion. The resistivity of that unit was extremely consistent varying within a narrow range between 1.4 and 1.8 ohm-m. The most exciting result was obtained at station N4. The interpretation of the data collected at this station detected

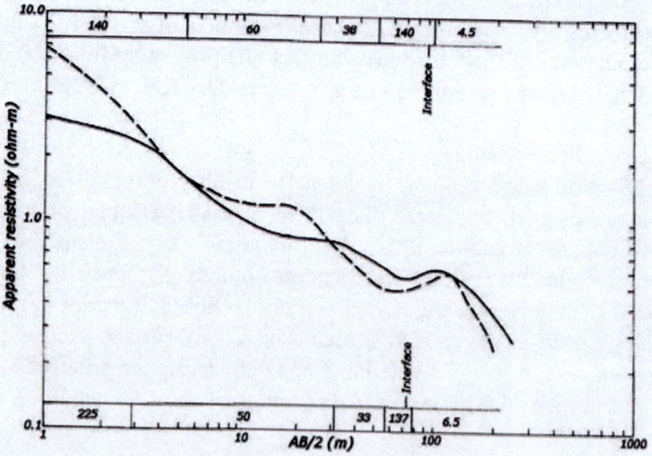


**Fig. 12** DC apparent resistivities measured by the Schlumberger array in the Mediterranean coast of Israel. Solid line, in 1960; dashed line, in 1975 (modified after Ginzburg and Levanon 1976)

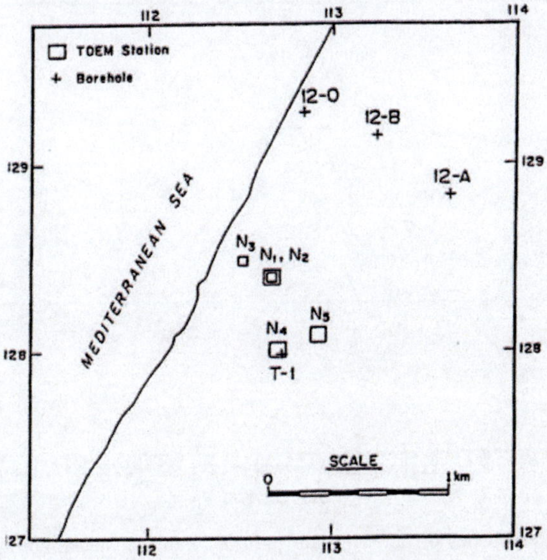


**Fig. 13** Location map of the TDEM feasibility study survey at the Nitzanim area

the top of the very low-resistivity body at a depth of 70 m. According to the measured borehole salinity (conductivity), the fresh–seawater interface on top of a half sea water salinity (17,500 mg/l TDS) was encountered at the depth of 67 m, whereas a seawater salinity of 35,000 mg/l TDS was encountered at a depth of 70 m (Goldman et al. 1988).

Such an exceptionally high accuracy, however, needed further statistically significant verification, especially taking into account that the use of TDEM in ground water exploration was pioneering at that time, not only in Israel but all around the world. Therefore, the next feasibility study, carried out shortly after the first survey, included 100 TDEM stations distributed throughout the entire coastal plain, predominantly in proximity of observation wells (Goldman et al. 1991). The main results of the survey are shown in Fig. 14. It is evident that the resistivities of the seawater-intruded coastal aquifer vary in a narrow range between 1.1 and 2.9 ohm-m, with the mean value, of 1.7 ohm-m. Out of 31 measurements, which detected the resistivities in the above range, 21 showed the discrepancy between the geophysical and the borehole measurements of less than ±5 m, while greater errors were observed in case of a wide transition zone or due to remoteness from an observation well (Goldman et al. 1991). It should be noted that such differences can also stem from the fact that due to temporal groundwater-level fluctuations, the actual depth to the interface at the time of the TDEM measurements has already changed since the borehole salinity was measured.

Based on the above results, it was decided to perform regular TDEM measurements at locations of abandoned observation wells to monitor the sea water intrusion and to map the interface in areas of scarce hydrogeological information.

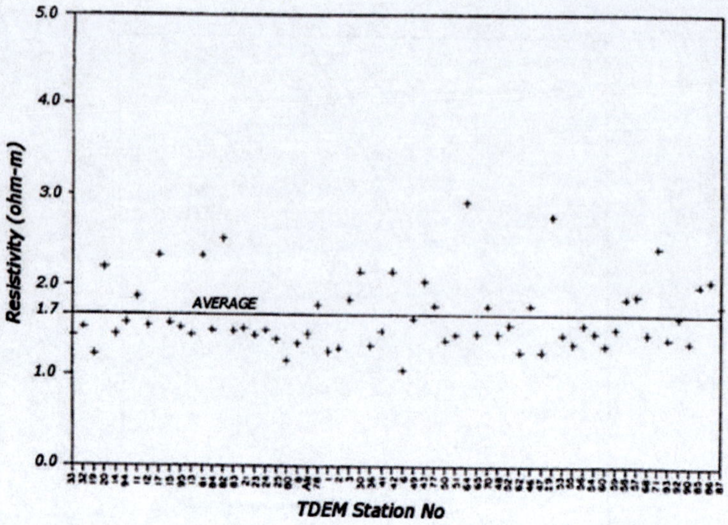


**Fig. 14** TDEM resistivities of the sea water intrusion into the Mediterranean coastal aquifer (modified after Goldman et al. 1991)

During the last decade of the past century, roughly 700 TDEM measurements were carried out throughout the entire coastal aquifer. In the course of these surveys, resistivity "reversals" (higher resistivities below lower ones) were detected in the CD of the coastal aquifer, indicating that they might be saturated with fresher water underneath the seawater-intruded upper sub-aquifers. This assumption has been successfully verified near one of the very few observation wells, which at that time managed to measure the water salinity within the CD. Figure 15 shows the comparison of borehole water salinities (conductivities) measured in Ashdod 112 borehole with the resistivities obtained by the nearby surface TDEM measurements.

One can see that higher TDEM resistivities opposite to CD correspond to lower than seawater salinities measured in the well at appropriate depths. Similar resistivity behavior has been found in many TDEM measurements both isolated and along traverses (Fig. 16).

Based on the above, a TDEM survey aimed to reveal the resistivities of the entire sequence of the aquifer, including those of the underlying Saqiye Group aquiclude, was executed along the entire coastal plain (Kafri and Goldman 2006). A hundred TDEM measurements were done, followed by drilling of ca 200-m-deep observation wells, in which water conductivity logging was performed in all four sub-aquifers. To enable correlation, 14 TDEM measurements have been carried out adjacent to these observation wells. In all cases, the interface that was detected in the wells within the upper sub-aquifers was also accurately detected by the TDEM measurements (Fig. 17). The bulk resistivity of the seawater-intruded portion of the aquifer varied

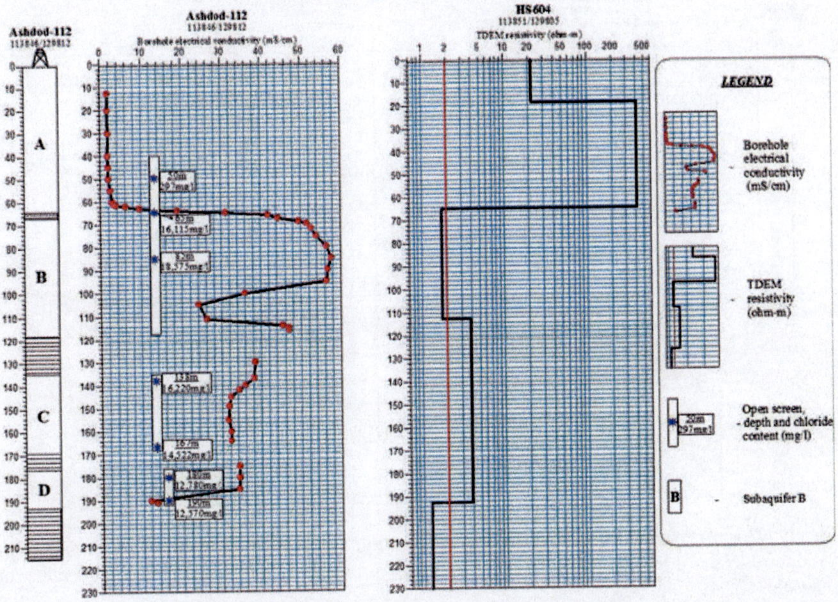


**Fig. 15** Comparison between the EC profile measured in the observation well Ashdod-112 and TDEM resistivity cross-section obtained in the nearby TDEM station HS 604

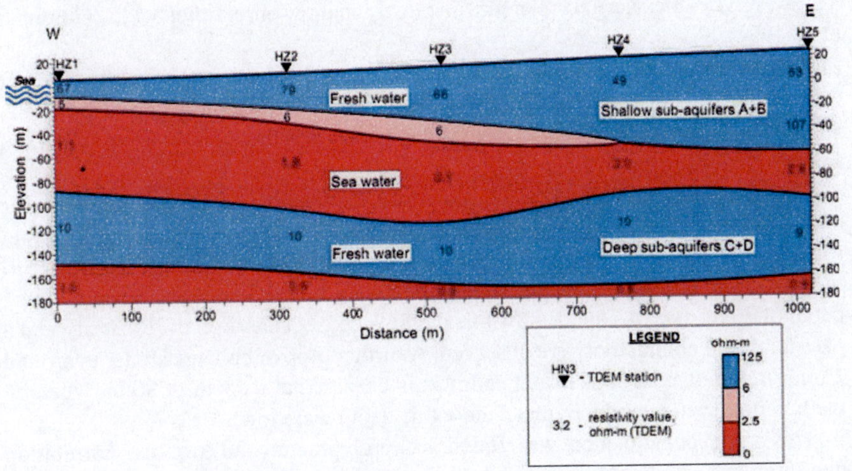


**Fig. 16** Quasi 2D resistivity cross-sections along the TDEM traverse perpendicular to shoreline in the Ziqim area

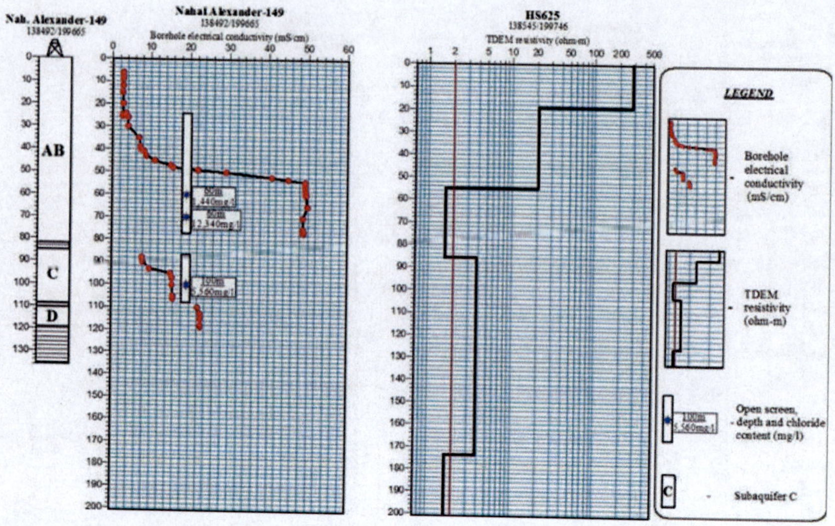


**Fig. 17** Correlation of TDEM resistivities and borehole conductivity measurements in the observation well Nahal Alexander-149. Dashed line roughly represents the threshold resistivity value for seawater

between 1.5 and 1.9 ohm-m, averaging a value of 1.7 ohm-m as found in the previous TDEM surveys that were correlated with mostly shallow observation wells (Goldman et al. 1991).

An increase of the resistivity was observed below the depth of 86 m within the CD sub-aquifer as shown in Fig. 17. This is in accordance with the adjacent borehole, significantly lower water conductivity, than that of the overlying portion identified as seawater intrusion.

In contrast to the above, Fig. 18 shows an example whereby the borehole water conductivity profile does not fit the relevant lateral TDEM resistivities profile within the deep sub-aquifers, despite the good fit in the upper sub-aquifers. High TDEM resistivities and low borehole conductivities were found down to the approximate depth of 70 m, as expected. Below this depth, a consequent drop of resistivity and a parallel rise of conductivity are observed. A further drop of bulk resistivity to around 2 ohm-m (equivalent to seawater salinity) is observed at a depth of 90 m, whereby the borehole water conductivities below this depth were low.

This same phenomenon was found in approximately 20% of the correlation measurements. It should be noticed that in all these cases, the TDEM resistivities within the CD consistently reflect higher salinities than those actually analyzed in the observation wells water samples. It implies that the disagreement is caused by a real hydrogeological setup rather than by measurement errors or an ambiguous geophysical interpretation (Goldman et al. 2003). The most reasonable proposed explanation to the above disagreement is as follows;

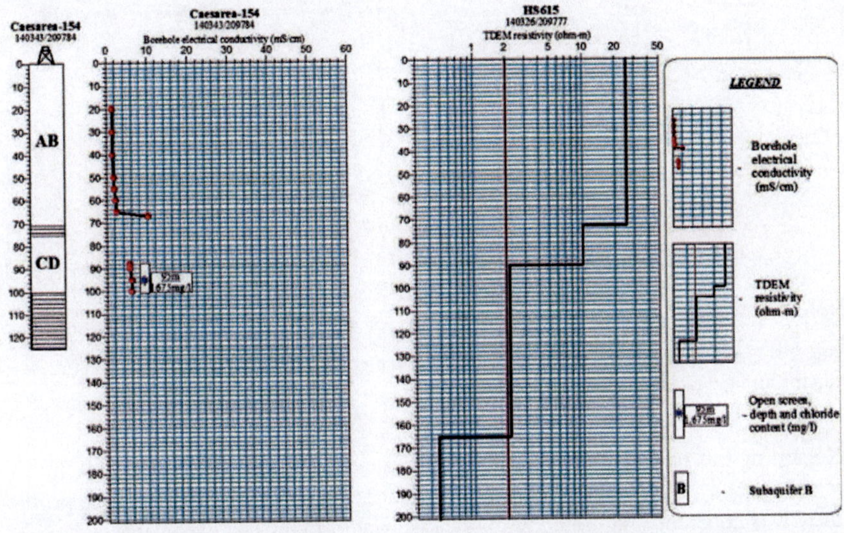


**Fig. 18** Correlation of TDEM resistivities and borehole conductivity measurements in the observation well Caesarea-154. Dashed line roughly represents the threshold resistivity value for seawater

Water salinities that were obtained through analysis of borehole samples reflect only the aquiferous portion of the sequence which contains freshwater. At the same time, the TDEM bulk resistivity measurements of a certain sequence include and integrate also the clay interlayers which seldom contain trapped or interstitial saline water of low resistivity. The latter are more abundant, proportion wise, in the lower sub-aquifers as compared to the upper ones. The end result is that the upper sub-aquifers exhibit higher integrated bulk resistivities, whereas the lower ones exhibit lower bulk resistivities due the low resistivity contribution of the impermeable saline interlayers.

Therefore, the ability of TDEM to accurately characterize the water salinity within the lower sub-aquifers is limited since it strongly depends on the sparse available information regarding aquifer/aquiclude thickness ratio, the depth to the lower sub-aquifers, as well as the salinity distribution within the upper sub-aquifers. However, taking into account the small probability of misinterpretation (three out of fourteen calibration measurements) and the fact that TDEM consistently overestimates the water salinity in the CD, one might expect that the method can be successfully applied to the regional mapping of the entire coastal plain or to its significant parts. This is particularly true regarding those parts, where the resistivities detected in the CD are fairly high and continuous in space.

The distribution of the TDEM resistivities of the CD and those of the underlying Saqiye beds (designated S) is exhibited on a longitudinal profile parallel and close to the seashore (Fig. 19). One can see continuous high resistivities within the CD along the segment between 27 and 47 km of the profile (North Ashdod-Bat Yam).

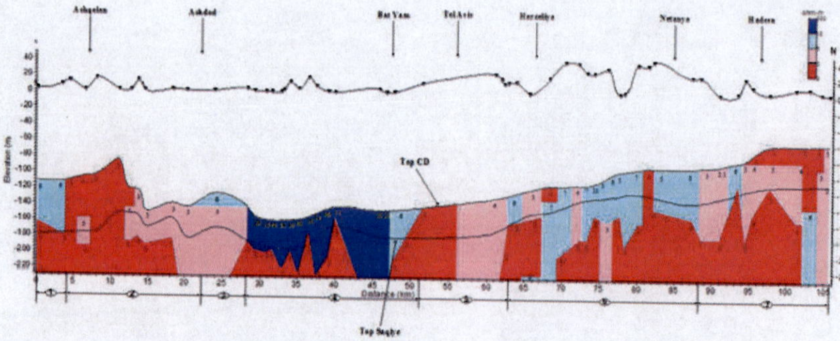

**Fig. 19** Quasi-2D resistivity cross-section along the entire Mediterranean coastal line of Israel (after Kafri and Goldman 2006)

Regarding this part of the coastal aquifer, it is evident, with confidence, that the CD is occupied by freshwater. Along the segment between 70 and 87 km of the profile, the CD is apparently saturated with brackish water. In the rest of the coastal aquifer, the water in the CD is either saline or, if fresh, the thickness of the CD is most likely relatively small.

Since only one measurement is available in the entire metropolitan Tel Aviv area (roughly between kilometers 50 and 60 of the profile), the detected low resistivity of CD could be caused by the above-mentioned artifact and the real extent of the freshwater body within the CD might be more than 30 km. In order to estimate the real south–north and east–west dimensions of the discovered freshwater body within the CD, an offshore continuation of the on-land TDEM measurements was required. These measurements were necessary to estimate the amount of freshwater within the CD, and also to decide whether the CD is laterally blocked to the sea by the Saqiye clays, or alternatively whether it is hydraulically connected to the sea.

**Offshore TDEM measurements**

In the case of the offshore TDEM measurements, the required target is considerably more resistive than the background medium and thus the LOTEM method provides a better detectability than the SHOTEM method, as noted above. On the other hand, accurate detection of the target boundaries requires a better lateral resolution of the SHOTEM method. Therefore, the offshore survey included both methods.

**The marine SHOTEM method**

The use of the above-described conventional SHOTEM central loop array is not feasible in a wavy sea, where laying down a large transmitter loop is a real challenge. The use of the theoretically equivalent so-called AB-loop array (also known as Ex-Bz) consisted of a grounded transmitter line and an equatorial small receiver coil is technically simpler in the sea. Contrary to on-land applications, grounding current electrodes in water is very simple. Calculations show that in a shallow sea (up to 50–100 m deep) the sensitivity of the marine SHOTEM method with regards to a sub-seafloor resistive target (e.g., freshwater) practically does not depend on the location

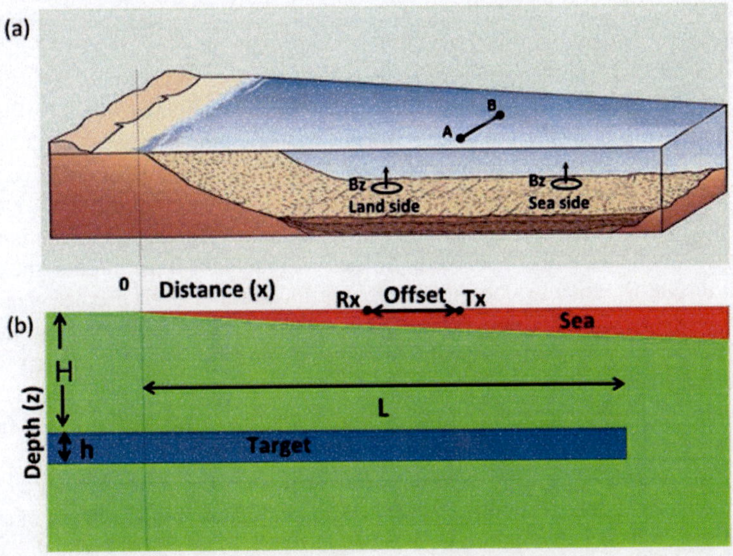


**Fig. 20** **a** Marine TDEM system; **b** conceptual geoelectric model (modified after Levi and Goldman 2017)

of either the transmitter or the receiver within the sea. Therefore, due to technical considerations, the selected array consisted of a floating surface transmitter line (AB) and a bottom located receiver coil (Bz) (Fig. 20a).

In a 1-D Earth, the measured signal is invariant with respect to which side of the transmitter line the receiver coil is located. However, this is not the case in the presence of a strong 2D resistivity contrast, which exists between the very conductive sea water and the relatively resistive coast. Both feasibility study measurements and the follow-up 2D/3D calculations showed that the land-side AB-Bz array (Fig. 20a) is considerably more sensitive to the presence of the sub-seafloor resistive target than either the sea-side AB-Bz (Fig. 20a) or the signal in the 1-D case, where no sea–land resistivity contrast exists. This coastal effect was investigated and discovered during the marine TDEM survey offshore the Palmahim beach in Israel. It provided a new and highly efficient tool for exploration of sub-seafloor resistive targets (Goldman et al. 2011). Indeed, it exhibits a high lateral resolution, as all conventional SHOTEM methods do, but contrary to them, it was found to be also very sensitive to the presence of high-resistivity sub-seafloor targets. The use of the marine LOTEM method remains however important, since the discovered coastal effect only takes place at relatively short distances from the shoreline, up to 3 km under specific bathymetric and the sub-bottom geoelectric conditions of the study area.

The offshore SHOTEM measurements included four marine campaigns, during which a total of 35 measurements, using 20 different transmitter locations, were carried out. The first two campaigns were used as a feasibility study that tested different arrays and acquisition systems. The results of the feasibility study led to the development of a novel highly efficient land-side marine TDEM system, described in details by Goldman et al. (2011). The next two campaigns (Levi et al. 2018) included a 15-km south–north profile (P-1) between Bat Yam and Tel Aviv parallel to the Mediterranean shoreline at an approximate distance of 600 m offshore (Fig. 21). The profile consisted of eight measurements, including those carried out during the previous feasibility study campaigns. In addition, two east–west profiles were run perpendicular to the shoreline up to approximately 3000 m offshore (Fig. 21), up to a water depth of 33 m. The southern profile (P-2) aligned south of Bat Yam included four measurements, and the northern profile (P-3) included three measurements.

All the measurements along the south–north (hereafter S-N) profile were interpreted using a standard 1-D inversion software (TEMIX-XL and IX1D by Interpex Ltd.) and were verified afterward by the appropriate 2D forward modeling, which included the sea–land resistivity contrast. Typical examples of the acquired data and the interpreted 1-D models are shown in Fig. 22 in case of an existing target (a) and in the absence of a target (b).

The interpretation of the E-W measurements was complicated. In order to determine the western boundary of the target as accurate as possible, the initial quasi-2D resistivity cross-section obtained by 1-D inversions was not simply verified, but was rather used as an initial model for a further full 2D trial-and-error interpretation (Levi et al. 2018). The resulting real 2D model, obtained from P-2, is shown in Fig. 23.

Based on all 35 offshore measurements, the interpreted high-resistivity (freshwater) body is roughly 30 km long extending continuously from North Ashdod to North Tel Aviv at the depth range between 100 and 250 meters below sea level. Its

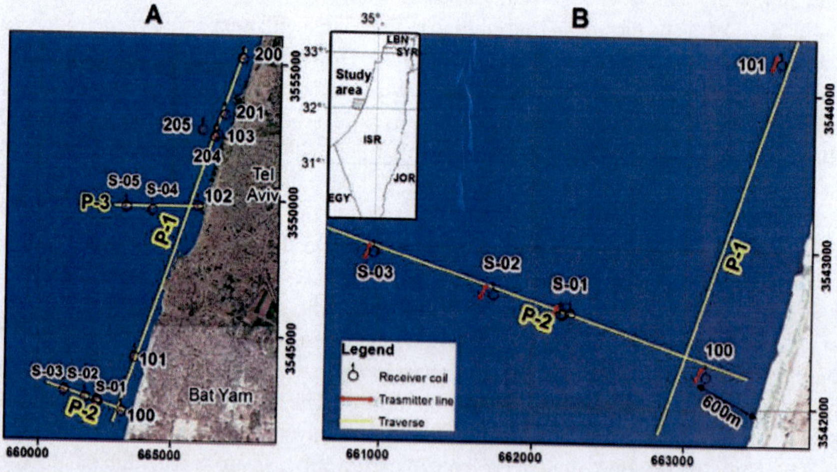


Fig. 21 Location map of the marine TDEM surveys (modified after Levi et al. 2018)

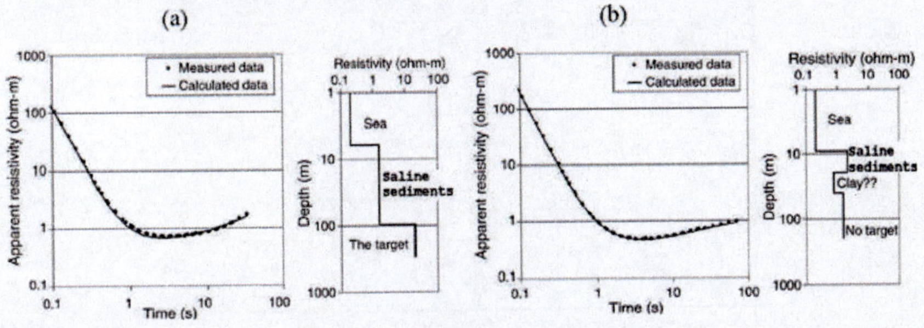


**Fig. 22** Examples of measured and interpreted offshore TDEM data (apparent resistivities), **a** with a target; **b** without a target (modified after Goldman et al. 2011)

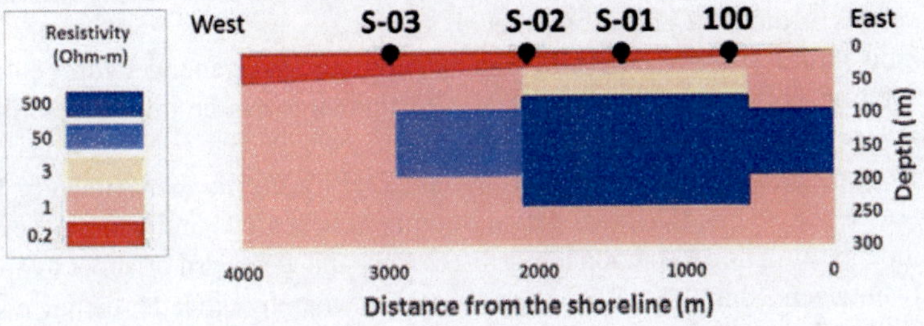


**Fig. 23** The interpreted E-W 2D resistivity cross-section. Freshwater in CD extends to 3 km west of the shoreline (after Levi et al. 2018)

lateral western boundary is determined less accurately and estimated as no less than 3 km west of the sea shoreline.

**The marine LOTEM method**

Due to the more complicated acquisition procedure, the marine LOTEM survey was carried out along only one E-W profile in the vicinity of the city of Bat Yam. It included eight transmitter and fourteen receiver locations using both in-line and mainly broadside transmitter-receiver (T/R) configurations (Fig. 24). The T/R offsets varied between 400 and 1200 m with the first receiver located on-land near the shoreline and the last receiver, located approximately 5 km west of it (Lippert 2015).

Out of 14 measurements, nine were of good quality, which allowed a reliable interpretation, the results of which are shown as a columnar resistivity cross-section in Fig. 25 (Lippert 2015). Both Figs. 23 and 25 show a general similarity between the SHOTEM and LOTEM results with regard to the resistivity, the depth and the thickness of the target. The main difference manifests itself in the location of the western boundary of the target. According to the SHOTEM model, the boundary is located at a distance of approximately 3 km offshore, whereas according to the LOTEM model, it is located some 3700 m offshore. The LOTEM results seem to be more reliable, especially taking into account the recent follow up measurements,

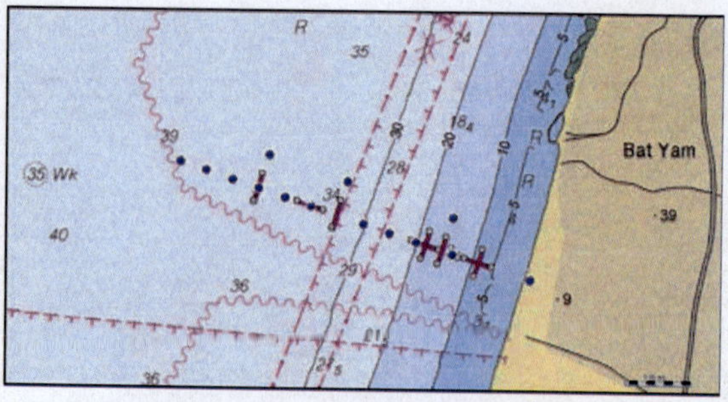


**Fig. 24** Location map of the marine LOTEM survey (modified after Lippert 2015)

using a novel marine differential electric dipole (DED), that led to approximately the same results as did LOTEM (Haroon et al. 2018).

Taking into account all the above considerations, the general hydrogeological interpretation of the marine geophysical measurements can be briefly described as follows:

The top of the freshwater body within the CD is located at approximately 100 m below sea level, and its thickness varies roughly between 100 and 150 m. It extends continuously from north Ashdod to north Tel Aviv and westward offshore to approximately between 3 and 3.7 km from the shoreline, whereby the latter distance seems to be more reliable.

## Combined ERT-TDEM measurements aimed to delineate diluted seawater intrusions from the Alexander stream estuary to the coastal aquifer

The water salinity at the lower reaches of the Alexander stream, as well as at some other coastal streams along the Mediterranean coast of Israel (e.g., Yarkon, Qishon), may attain seawater salinity. Since the riverbeds are below sea level, they are subjected to seawater intrusion along them up to about several kilometers upstream. As a result, a stratified water column is developed with a saline water mass at the bottom. Similar to seawater intrusion into coastal aquifers, the saline water within the streams may

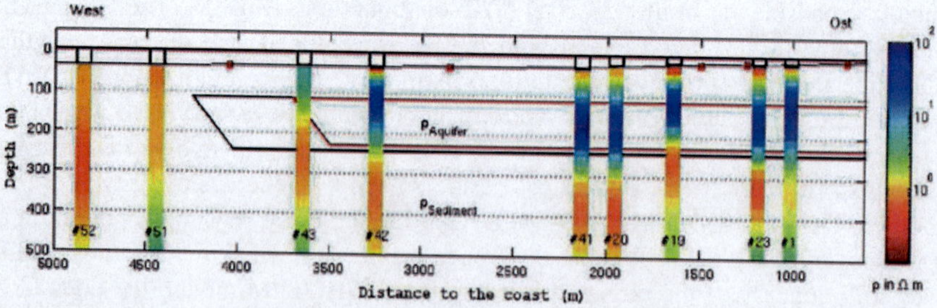


**Fig. 25** Interpreted columnar resistivity cross-section along the profile shown in Fig. 24. Resistivity of the aquifer is 50 ohm-m; resistivity of the sediment is 1 ohm-m (modified after Lippert 2015)

contribute to salinization of the groundwater. Although the process of seawater intrusion into coastal aquifers is well known and properly treated, the local-scale potential salinization from the stream estuaries has never been studied in Israel until recently. A study carried out by Shalem et al. (2014) investigated the salinization mechanism of groundwater along seawater-intruded streams and estuaries. The study, among other methods, included also a combined high-resolution electrical resistance tomography (ERT) and time domain (transient) electromagnetic (TDEM/TEM) methods, which were applied at several measurement sites along the lower Alexander stream within an approximately 1.5 km distance from the shoreline (Fig. 26).

Contrary to the expectations, the geophysical measurements clearly showed a saline water intrusion (resistivity of 1–2 ohm-m) from the river into the aquifer only at the most eastern, and remote from the sea sites located approximately 1.5 km east of the shoreline (lines 03 and 04). The resistivities measured at the western sites, close to the coast, were significantly higher, besides just a few meters from the stream. Thus, contrary to traditional seawater intrusion into coastal aquifers, the saline intrusion from coastal estuaries is site dependent due to hydraulic heterogeneity of the shallow river bed along the stream. Indeed, the eastern sites, that are more remote from the sea, consist of a sandy sequence, whereas the western ones are silty. Therefore, salinization in the east extends laterally to more than 80 m off the stream and only up to 20 m in the west as shown in Fig. 27.

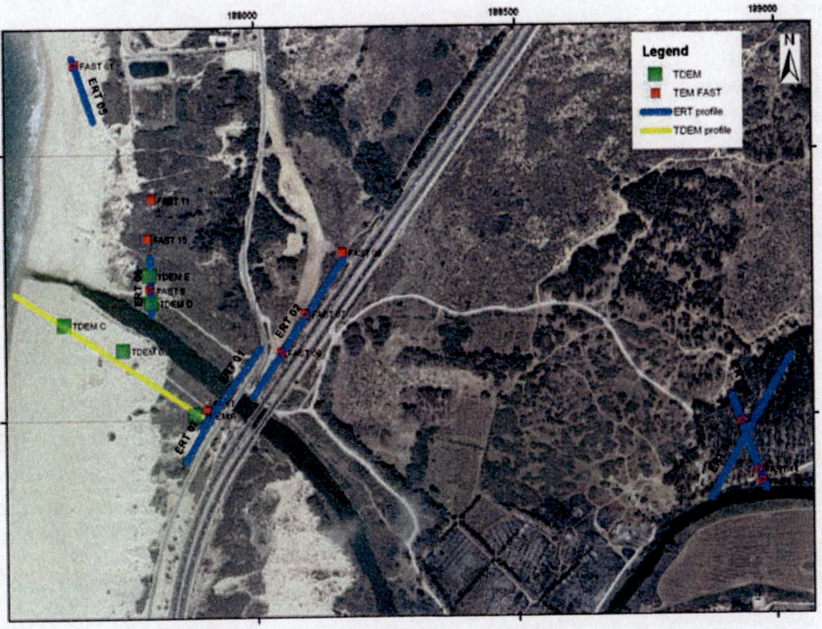


**Fig. 26** Location map of the ERT and TDEM (TEM) measurements at the Alexander stream

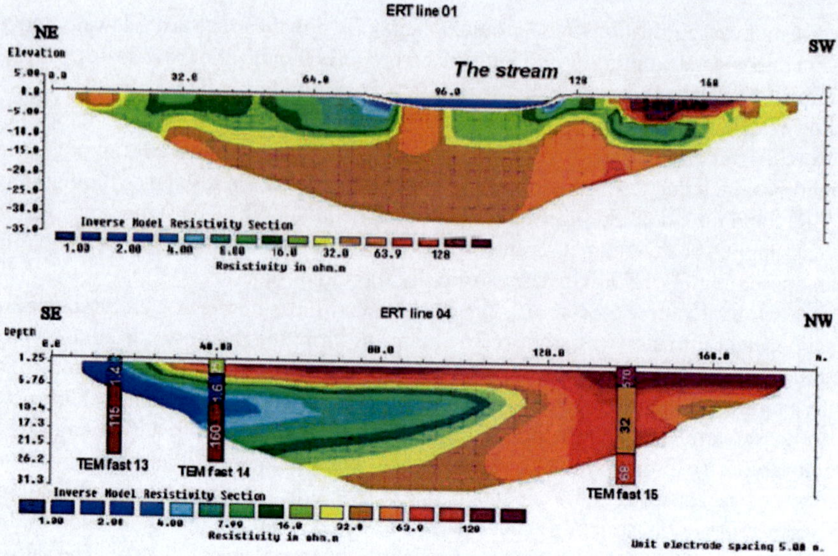


**Fig. 27** ERT resistivity cross-sections along the western line 01 and along the eastern line 04 (see location map in Fig. 26). Line 01 crosses the Alexander stream in the N-S direction. Line 04 runs perpendicular to the stream on its northern bank and is accompanied by TEM soundings (modified after Shalem et al. 2014)

## Geoelectric (ERT) and TDEM measurements aimed to delineate submarine groundwater discharge (SGD)

The SGD phenomenon is known, from all around the world (see also Chapter "Submarine Groundwater Discharge Along the Israeli Eastern Mediterranean Coast and in Inland Basins"). It was studied by geochemical, hydrological as well as of geophysical methods, as were employed in Israel, in the Dor Beach site, which became a major test site for studying SGD since 2006. The first geophysical survey included high-resolution time-series ERT measurements, using a 112-m-long cable (Fig. 28). It provided information regarding the coastal hydrogeology, up to the maximum exploration depth of approximately 20 m (Swarzenski et al. 2006).

In summer 2008, the deeper parts of the site were explored, using roughly two times longer shore diagonal and perpendicular ERT lines combined with onshore TDEM measurements (Fig. 29). These yielded information regarding the distribution of the electrical resistivity up to a maximum depth of roughly 100 m whereby a high-resistivity body was detected at a depth of nearly 63 m, which most likely represents the underlying Judea Group aquifer, apparently saturated with groundwater, with salinity below that of intruding seawater. Two saline/freshwater interfaces were detected at depths of approximately 17 and 49 m (Fig. 30). The upper interface between fresh and brackish water was detected by both the ERT and TDEM methods, whereas the lower fresh and seawater interface was detected only by the TDEM method.

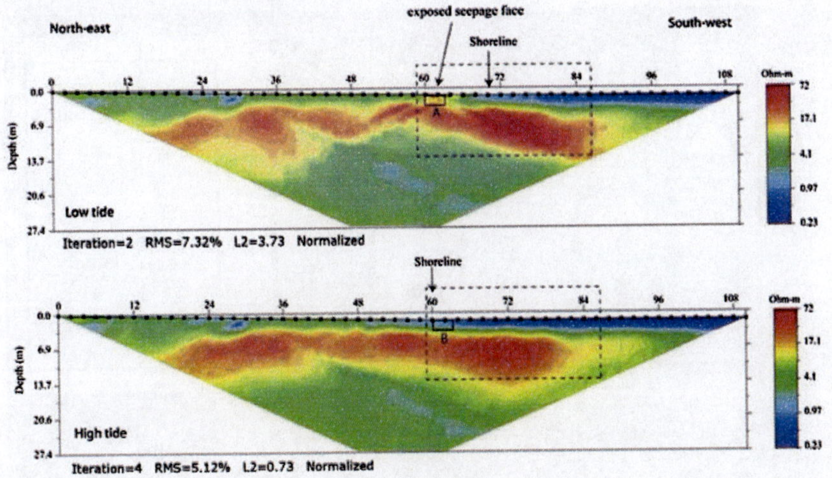


**Fig. 28** ERT resistivity cross-sections at the Dor Beach site during low tide (upper) and high tide (lower) periods (modified after Swarzenski et al. 2006)

## 3.2 *The Gulf of Eilat (Aqaba) Coastal Aquifer*

The Gulf of Eilat (Aqaba) is a base level to the convergent groundwater flow from western and eastern mountain aquifers into the young Rift Valley fill aquifer. The latter consists of intercalations of aquiferous detrital coarse sediments and thick impermeable marls and clays horizons that divide the aquifer into a number of sub-aquifers.

Several TDEM soundings as well as traverses have been carried out in the western and northern coastal strips of the Gulf of Eilat (Aqaba). The results of the northern ones are discussed below.

**TDEM measurements in the northern coastal strip**

The setup of the northern coastal strip is characterized by the following elements that influence groundwater salinity: (1) A gulf consisting of two artificial "fjords" that penetrate several hundred meters inland to the north of the shoreline (Fig. 31).

(2) A deep saline end member, (3) The surficial artificial evaporation ponds of the salt industry, which host concentrated brines and halite deposits (Fig. 31).

The TDEM measurements have been carried out along two lines, perpendicular to each other, namely the F-F′ and G-G′ lines, The F-F′ line runs close to the "fjords" and the evaporation ponds. An additional shorter E-E′ line runs from the seashore inland to the north, adjacent to the state border and crossing the evaporation ponds (Fig. 31).

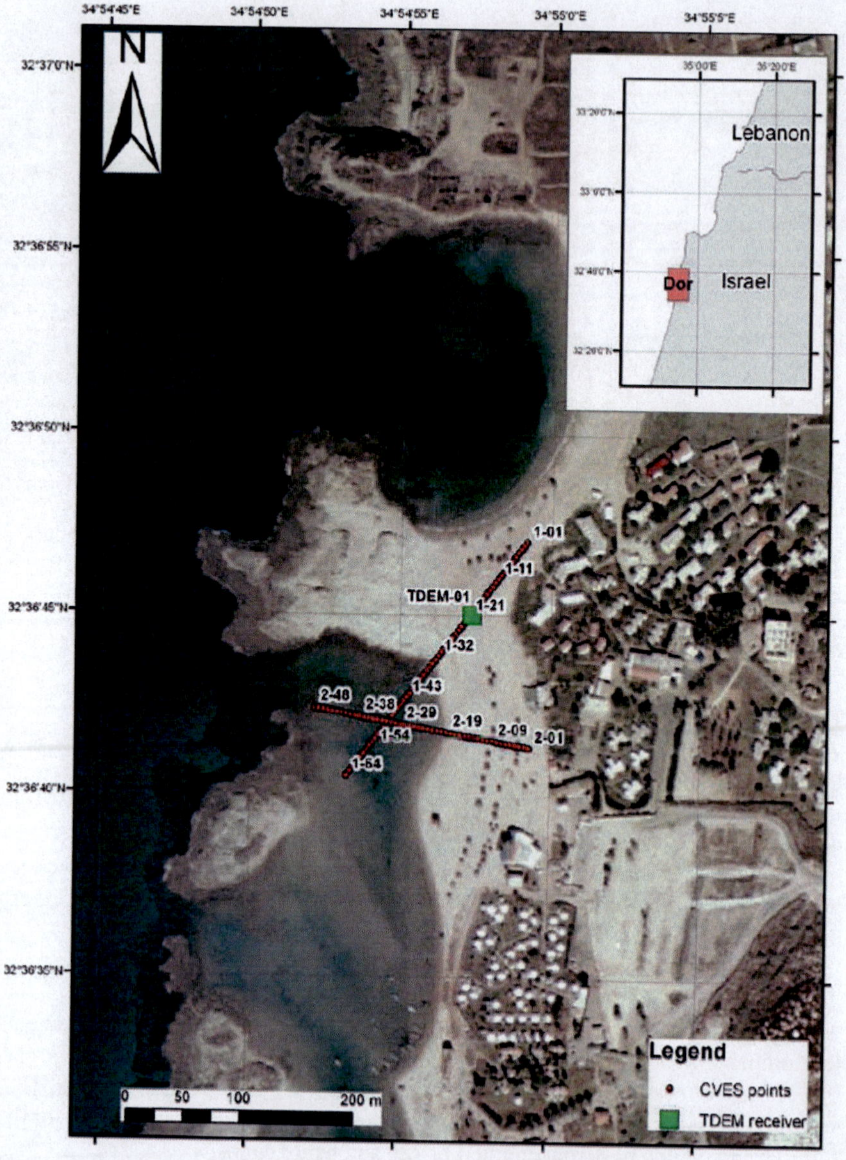


Fig. 29 Location map of the combined ERT-TDEM measurements at the Dor Beach

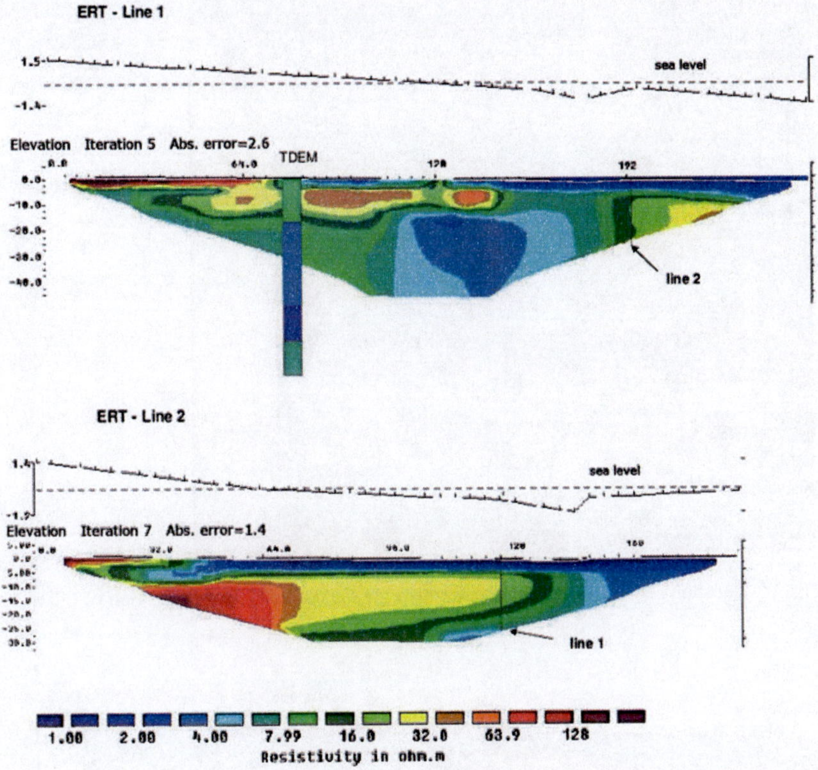


**Fig. 30** 2-D ERT resistivity cross-sections along lines 1 and 2 showing SGD (high resistivities) at shallow depths below the sea. 1-D TDEM columnar resistivity cross-section shows, in addition, freshwater (high resistivity) below the seawater intrusion at greater depth

Figure 32 shows an example of a quasi-2D resistivity cross-section, based on the interpreted 1-D TDEM measurements along line F-F′. To enable the above interpretation, the following resistivity and appropriate hydrogeological units have been taken into account and introduced:

A discrepancy was found between some of the TDEM resistivities and resultant water salinities and water salinities of adjacent boreholes. Lower bulk resistivity values were obtained than the expected for the adjacent appropriate water salinities. For example, groundwater chlorinities between 3100 and 3400 mg/l, from the depth interval of 100–200 m in the Sabkha 19 and 20 wells correspond to low resistivities below 0.5 ohm-m, typical to concentrated brines rather than to the above described brackish waters (Fig. 32). This discrepancy is explained herein by the existence of fairly thick clay intercalations, which assumingly contain unflushed brines which contribute to the integrated lower bulk resistivity, whereas the borehole water salinities represent solely the aquiferous fresher portions of the aquifer.

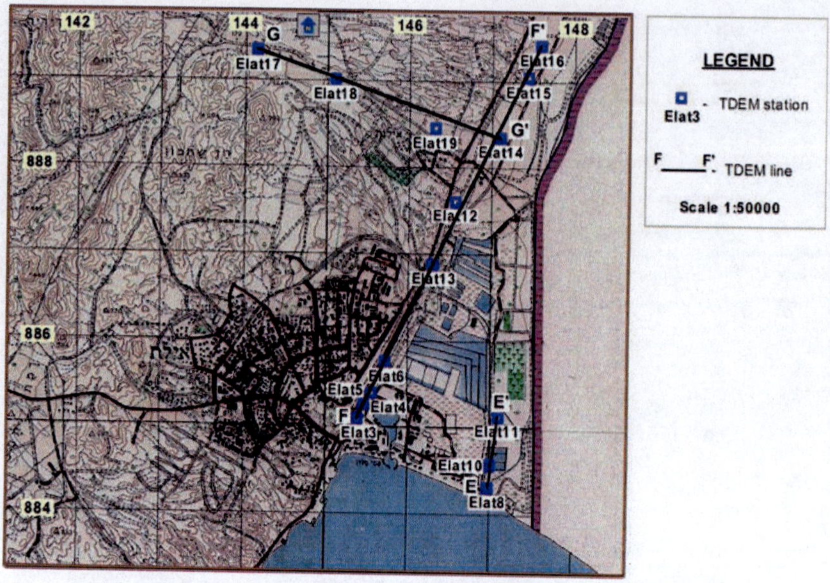


**Fig. 31** Location map of the TDEM measurements at the northern coast of the Gulf of Eilat

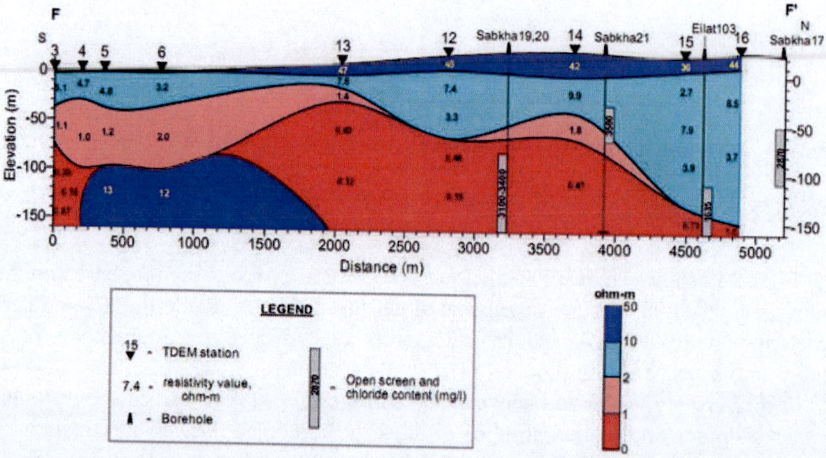


**Fig. 32** Quasi-2D resistivity cross-section along line F-F′ accompanied by the chlorinity data from nearby observation wells

The resistivity results of the above-mentioned lines combined with individual TDEM soundings from the western coastal strip enabled to draw a schematic 3D hydrogeological model of the Gulf of Eilat coastal aquifer (Fig. 33). The above findings can be summarized as follows:

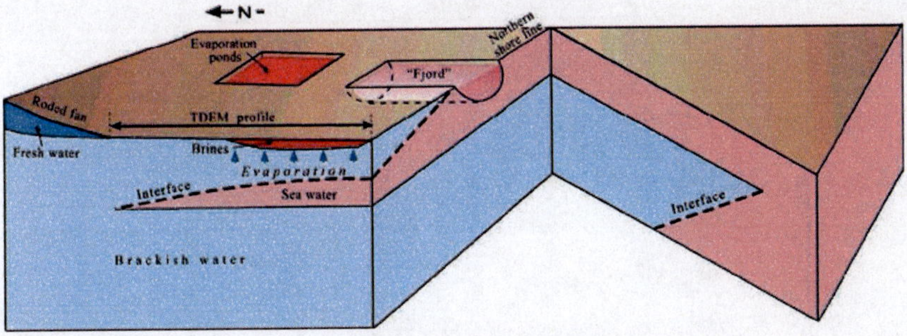


**Fig. 33** Schematic hydrogeological model of the Gulf of Eilat coastal aquifer

- TDEM resistivities below 2 ohm-m uniquely characterize the salinity of seawater intrusion. The depth to the brine–freshwater interface, within the upper sub-aquifers is accurately detected by TDEM measurements and fits the one obtained from borehole measurements.
- In most cases, the TDEM method manages to characterize groundwater salinity within the lower sub-aquifers, except for those sub-aquifers which include saline clay horizons where the resolution of TDEM method is sometimes insufficient to detect the fresh/brackish water bodies.

# 4 Geoelectric Characterization of Groundwater Salinity Along and Close to the Dead Sea Rift (DSR)

## *4.1 The Sea of Galilee*

The Sea of Galilee is located within the DSR and occupies approximately 170 km$^2$ with a maximum water depth of 46 m (Fig. 34). In fact, it is a freshwater lake, which serves as one of the most important water reservoirs of Israel. The lake is subjected to salinization from different sources that results in the average chlorinity of approximately 220 mg/l.

In order to study possible salinization sources, attempts have been made in the past to quantitatively evaluate the amount of salts entering the lake from the bottom. Since drilling within the lake is prohibited by law, the above study has been limited by extracting shallow cores from the lake bottom. Consequently, nineteen short (50 cm) cores and three long (500 cm) ones were taken at various locations on the lake floor (Fig. 34) in order to examine their chlorine content (Stiller 1994). The water in all cores exhibited a roughly linear increase in chlorine concentration from about 220 mg/l at the top of the core (water-sediment interface) to 350–600 mg/l at the depth of 50 cm, and 2000–3500 mg/l at the depth of 500 cm. A rough estimate based on the linear extrapolation of the observed chlorinity gradients in the cores has revealed a diffusion and advection rate of 5.3 mg Cl cm$^{-2}$ yr$^{-1}$ from the lake's

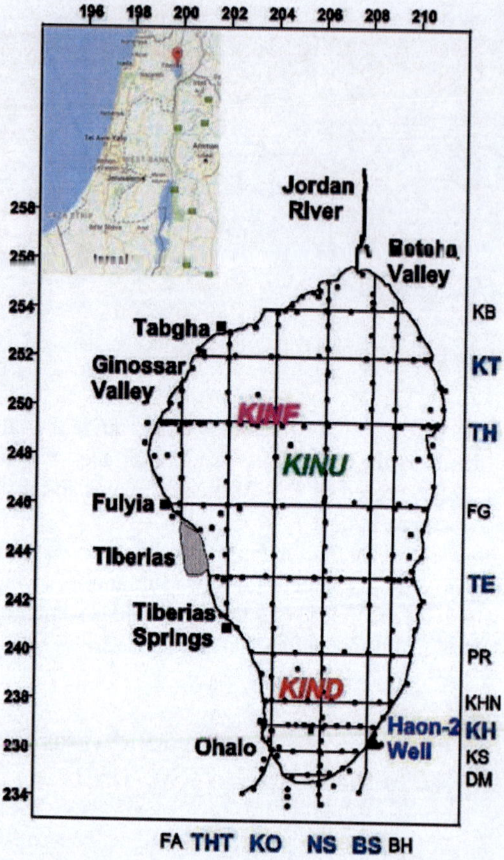


**Fig. 34** TDEM location map of the Sea of Galilee and its vicinity. KINF, KINU and KIND designate the locations of the core samples. KB, KT, TH, etc., are the names of the traverses used in the text (modified after Hurwitz et al. 1999)

bottom sediments, resulting in an annual contribution of 9000 tons of Cl to the lake. (Stiller 1994). These results clearly indicate that saline sources are located at shallow sub-bottom depths over large parts of the lake. However, the spatial distribution of the saline groundwater and actual sub-bottom depth was unknown.

In light of the legal forbiddance to drill boreholes within the lake, the only reasonable way to tackle the above problem was by applying geophysical, more precisely GE/GEM, methods. As a result a novel marine TDEM system has been developed and widely applied in the Sea of Galilee (Hurwitz et al 1999). The system consisted of a 25 m diameter circular transmitter loop, which floated on the water surface and was towed by a motorboat. A 1-m-diameter receiver coil was rigidly connected to the center of the transmitter loop and to another rubber non-motorized boat, which carried an operator and a receiver console (Fig. 35).

**Fig. 35** Central loop marine TDEM system applied at the Sea of Galilee (modified after Hurwitz et al. 1999)

Since the measurements with the described system did not require assembling and disassembling of the measurement array at each point, the productivity of the survey was considerably greater than that of the conventional onshore TDEM (30–40 offshore soundings per day vs. 6–8 onshore soundings). This, in turn, allowed to obtain a dense grid of soundings covering the entire lake and its close vicinity (Fig. 34).

A total of 269 offshore and 33 supplementary onshore TDEM soundings have been carried out along six N-S and thirteen W-E profiles as well as at selected points both offshore and onshore, along the entire coastline (Fig. 34). The interpreted resistivities were calibrated with the salinity measurements in nearby boreholes available in the vicinity of the lake and from deep cores taken from the lake bottom (Hurwitz et al. 1999). The calibration revealed that resistivities below 1 ohm-m are solely indicative of groundwater chlorinity exceeding 10,000 mg/l. Such low resistivities (high salinities) were detected practically everywhere at depths greater than 15 m below the bottom of the lake. At some parts of the lake, particularly in the south, the saline water was detected at shallower depths, sometimes at only a few meters below the bottom (Fig. 36).

A relatively high resistivity identified with fresh groundwater was detected along the margins of the lake to the depths of at least 100 m, which represents the maximum exploration depth of the marine TDEM system. The detected sharp lateral contrasts between high and low resistivities apparently represent a fault controlled lateral separation between carbonate and clastic units (Figs. 36 and 37), where the uplifted block, consisting of the carbonate aquifer is saturated with fresher groundwater. The geometry of the fresh/saline groundwater interface below the central part of the lake approximately follows the shape of the lake's bathymetry.

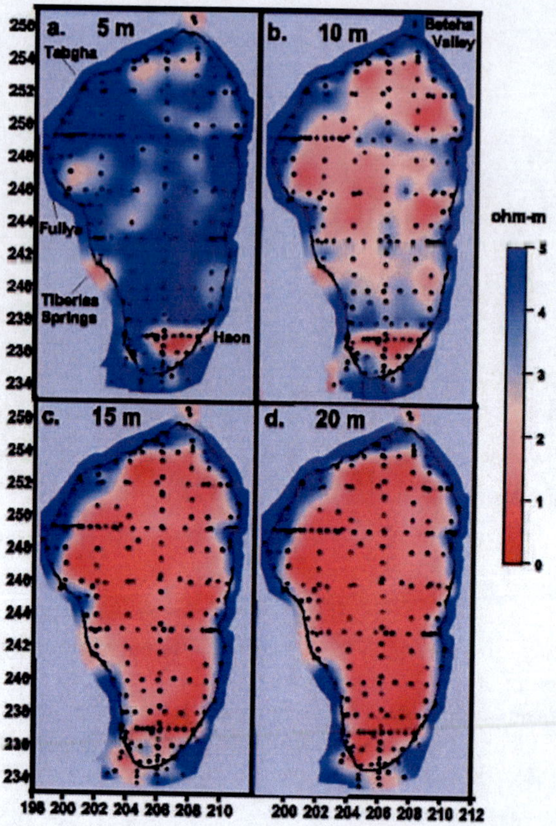


**Fig. 36** Distribution of resistivities at different depths below the lake's bottom (modified after Hurwitz et al. 1999)

The described TDEM results indicate that the distribution of the saline groundwater entrapped within the lake's sediments is highly heterogeneous. The greatest variability, both in concentration and depth, is limited to a narrow strip of 1–2 km wide near the shores. A significant change is also observed between the southern and northern parts of the lake. The interpreted TDEM resistivity maps and cross-sections define two zones with different transport mechanisms. These are the diffusion zone under most parts of the lake and the advective flux zone in the margins, where diffusion is negligible.

The similar shape of the lake's bathymetry and the top of the submarine saline groundwater body in the central part of the lake suggest that in the past, a saline lake covered the area and its water was trapped in the sediments. Consequently, when the Sea of Galilee became a freshwater lake, a diffusion of salinity from the sediments into the lake began to take place.

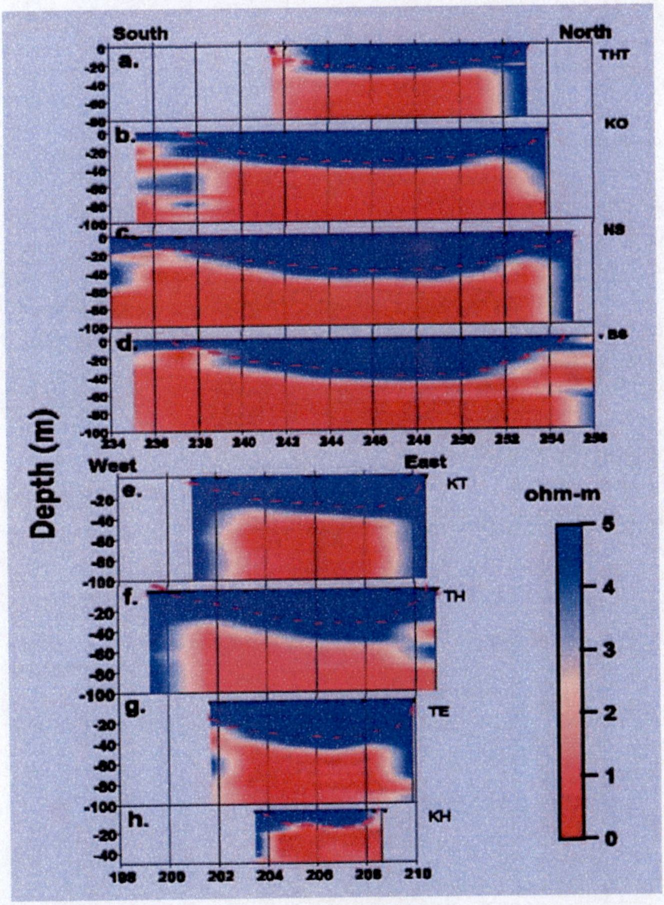


**Fig. 37** TDEM quasi-2D resistivity cross-sections along the SN and WE traverses (modified after Hurwitz et al. 1999)

In the southeastern part of the lake, the depth to the top of the saline groundwater body below the bottom is small and sometimes negligible. This implies that this part of the Sea of Galilee turned to be a freshwater lake later than in the northern part.

The Beteha Valley, in the northeast looks as an isolated pocket (Fig. 36). Since it is practically a delta of freshwater rivers, entering the lake, this observation seems surprising. Delta sediments are known there to be coarse grained with a high permeability. Therefore, either the Beteha Valley was also covered by the same lake as the one below the Sea of Galilee or alternatively that it was an isolated separated lake.

## 4.2 The Dead Sea Coast

The Dead Sea (hereafter DS) is the terminal base level of the DSR system (Fig. 38) (see also Chapter "The Eastern Dead Sea Rift Continental Groundwater Base Level"). It occupies hyper-saline brines typical of a very low electric resistivity. The DS is bounded to the west by north–south directed major faults. The coastal plain between the faults escarpment and the DS consists mainly of Quaternary continental detrital sediments (Fig. 39). Since the area is arid, the main freshwater recharge into the Quaternary aquifer is a lateral flow across the faults scarp from the Judea Group aquifer (Fig. 39), which is replenished in the mountain area some 10–30 km to the west. The DS is hydraulically connected to the adjacent Judea Group and the coastal aquifers and thus the continuous decline of the DS level, affects the parallel decline of the groundwater level in the above-mentioned aquifers (Kafri 1982; Yechieli 1993).

**TDEM measurements**

In order to understand the hydrogeological setup and its related salinity distribution, TDEM soundings were performed along generally W-E traverses between the DS

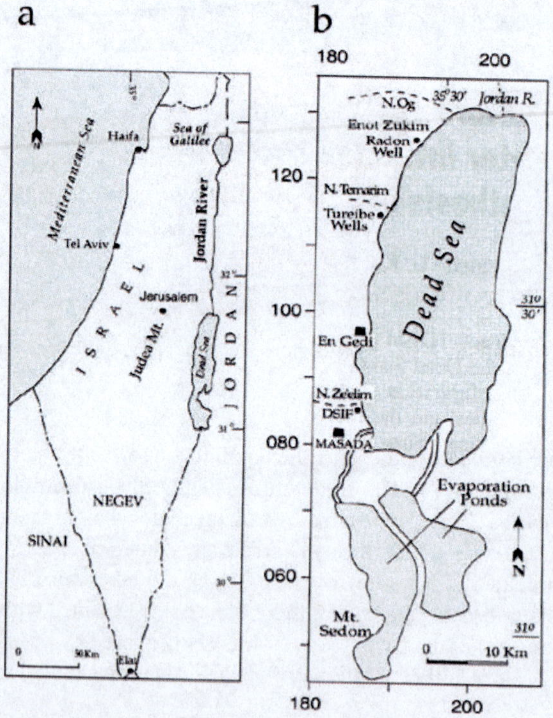


Fig. 38 Location maps: general map (a); location map (b) of TDEM traverses (dashed lines) (after Yecheli et al. 2001)

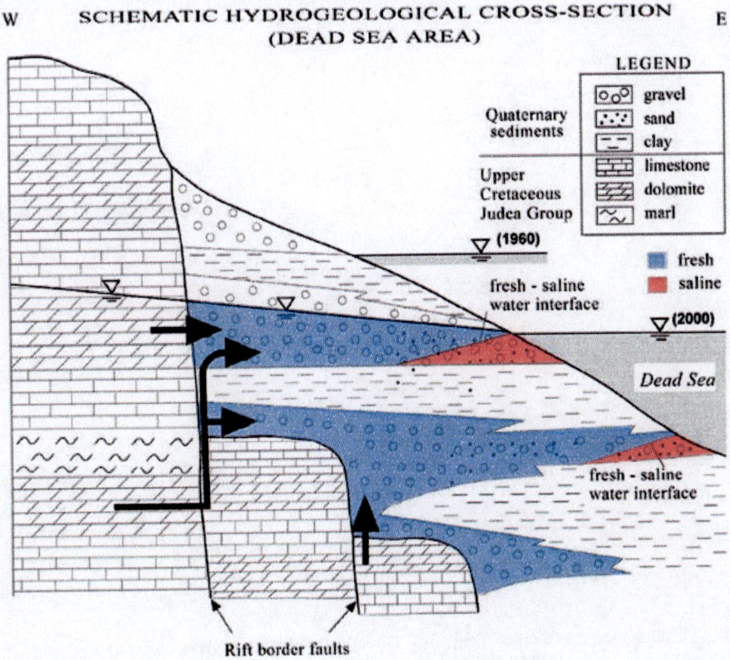


**Fig. 39** Schematic hydrogeological cross-section of the Dead Sea aquifer system (after Yecheli et al. 2001)

shore and the fault scarp (Fig. 38b). The Temarim stream traverse serves, for example, (Fig. 40) enables a spatial interpretation of the salinities in the coastal aquifer as follows:

The obtained configuration of the fresh (>5 ohm-m)-saline (<1 ohm-m) water interface exhibits its westward dipping and the transition (mixing) zone (1–5 ohm-m) that exists in between them (Fig. 40). Two features are recognized, common to other traverses in the region:

1. An unflushed brine body (sounding 23/7), below and above the present water table, and
2. A westward steepening of the interface in the west close to the fault scarp.

The suggested hydrogeological interpretation of the both phenomena is as follows:

1. The DS coastal aquifer in most places is subdivided into sub-aquifers, as detected in some of the boreholes (Yechieli 1993). This setup permits the possibility that the lower fresher water body being separated from the upper more saline one.

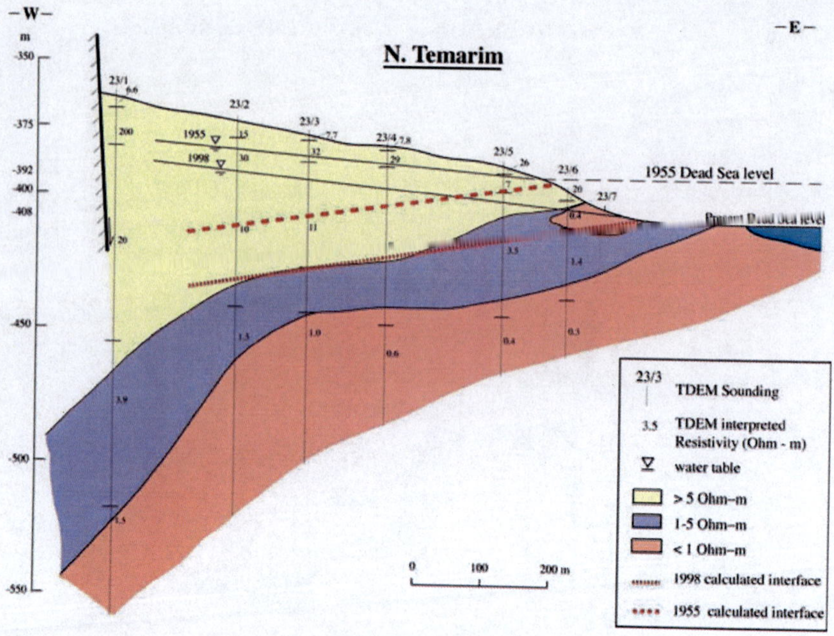


**Fig. 40** TDEM traverse at Nahal Temarim. Location of interface was calculated using Ghyben–Herzberg approximation modified for the Dead Sea conditions (after Kafri et al. 1997)

Moreover, the irregularity may also be the result of the fact that the DS system is not in steady-state conditions, due to the continuous drop of the DS and the adjoining groundwater levels. As a result, parts of the sequence have not yet been completely flushed, although they are already located in the vadose zone far above the fresh–saline water interface, as happens in some other locations in the region.

2. The steepening gradient of the water levels across the border faults, caused by anisotropic permeability in the fault zone and/or by different permeability in both sides of the fault (Yechieli et al. 2001), would demand there a parallel steepening of the freshwater/brine interface, as was indeed recognized here.

## 4.3 *Geophysical Delineation of Sinkhole Development in the Dead Sea Coastal Plain*

Thousands of sinkholes have been developed and are still developing, since the early 1980s, along the Israeli, as well as the Jordanian coastal plains of the Dead Sea (hereafter DS) following the rapid recession of the DS level since the 1960s. The phenomenon and the mechanism of sinkholes formation was studied in detail and discussed in several articles as well as in a separate chapter of this book.

Arkin and Gilat (2000) attributed this phenomenon to the resultant removal of fine sediments and to piping within the young detrital sequence of the DS coastal aquifer. The more commonly accepted mechanism of the sinkholes formation, however, explains the phenomenon by dissolution of a salt layer known at a depth of 20–70 m below surface (i.e., Yechieli et al. 2006; Abelson et al. 2006). Until the beginning of the rapid drop of the DS level, this salt layer was saturated with the concentrated DS brine (hereafter DSB), which was unable to dissolve salt. Above the interface, fresher groundwater or diluted DSB (termed herein DSBD) prevailed. Following the continuous drop of the DS level, accompanied by the drop of the interface, the salt layer was "exposed" to the DSBD, which has the capability to initiate salt dissolution, resulting in the appearance of subsurface cavities, collapse of overlaying sediments and sinkholes formation.

The general mechanism of sinkholes formation all over the world is related to dissolution of soluble rocks leading to appearance of subsurface cavities that collapse and form sinkholes. The dissolved rock types exhibit different dissolution rates, which dictate a strategy of geophysical delineation of zones subjected to potential sinkhole development. In the case of carbonates, typical of relatively slow dissolution rates, geophysical methods usually focus on detecting only the cavities. In the case of salts, which exhibit high dissolution rates, detecting only cavities might be inefficient and the methods are required to delineate and forecast preferable zones or paths of salt dissolution prior to the creation of the cavities. Thus, in the case of the DS coastal plain sinkholes, the geophysical methods were primarily used to delineate the underlying salt layer and to detect the zones, where groundwater salinity is less than that of the DSB. Two other possible mechanisms, namely subsidence and erosion ("piping"), which might also contribute to the sinkhole development, are beyond the scope of this chapter.

The most suitable combination of geophysical techniques to delineate salt layers and to detect the DSBD bodies includes the seismic refraction and high-resolution reflection methods, the TDEM and/or the integrated TDEM and SNMR methods. The seismic refraction method was used to delineate the salt layer and to locate a buried "salt edge," which assumingly represents a dissolution front controlling, at least partly, the sinkhole formation (Ezersky et al. 2017). The use of seismic refraction to delineate the "salt edge" seems somewhat problematic though. The high-resolution seismic method allowed to detect faults, which might serve as preferential hydraulic conduits to DSBD flow toward a contact with the salt layer (Abelson et al. 2003). The TDEM method is known to be feasible for characterization of groundwater salinity (Goldman et al. 1991), but might be more efficient in detecting the DSBD when combined with SNMR (Legchenko et al. 2009).

Due to the complexity of the problem, additional geophysical methods were involved in the study of sinkhole development in the DS coastal plain. Among them are micro-gravimetry (Rybakov et al. 2001), electrical resistance tomography (ERT) (Ezersky 2008; Al-Zoubi et al. 2007), ground-penetrating radar (GPR) (Frumkin et al. 2009, 2011), gravimetry and radar differential interferometry (Closson et al. 2003,

2005; Nof et al. 2013), nano-seismicity and microseismic activity (Wust-Bloch and Joswig 2006; Abelson et al. 2013). The comprehensive review of all these methods and the obtained results is discussed by Ezersky et al. (2017).

Since most of the above geophysical methods, besides seismic refraction/reflection, TDEM and SNMR, contributed only partly or indirectly to the understanding the main mechanism of the Dead Sea coast sinkhole development, the further discussion focuses exclusively on these four methods.

**Seismic refraction**

A typical P-wave velocity versus depth profile of the Nahal Hever South test site is shown in Fig. 41. The cross-section consists of three velocity layers, the lowermost, with a velocity of 2890 m/s has been identified with a salt layer (Wachs et al. 2000). The depth to this layer varies between 20 and 25 m in the western part of the profile, dipping to some 30–35 m in its eastern part. Borehole HS-2, located close to the western segment of profile RP-004, has in fact encountered the salt layer at a depth of 24 m (Wachs et al. 2000). The geophysical and log data of this borehole allowed to correlate the velocity layers with the borehole lithology as follows:

Further seismic refraction surveys all over the DS coastal plain, supported by borehole data, has shown that the P-wave velocity of the salt layer varies between 2900 and 4250 m/s (Ezersky 2006). The good velocity contrast between the target salt layer and the background lithology allowed a reliable delineation of the salt layer along the entire DS coast of Israel (Ezersky et al. 2006). Unfortunately, the second

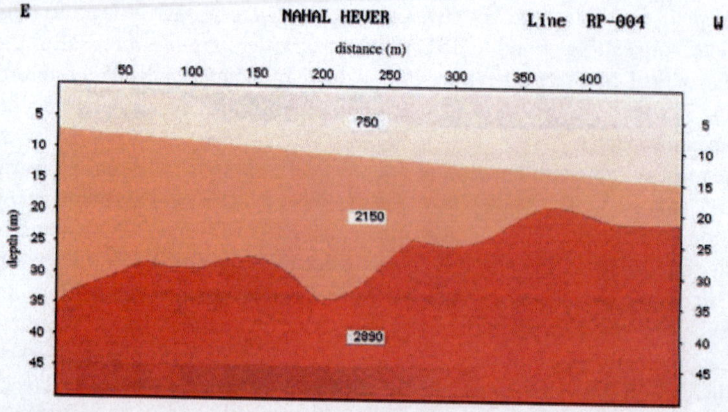


**Fig. 41** Typical velocity (in m/s) versus depth profile along line RP-004 (after Wachs et al. 2000)

important element related to the delineation of the DSBD turned out to be more complicated and difficult for geophysical treatment.

**TDEM**

Countless TDEM surveys, carried out all over the world, showed the great efficiency of the method in characterization of groundwater salinity and in accurate delineation of saline groundwater bodies at different depths, from tens of meters to several kilometers (Fitterman and Stewart 1986; Goldman et al. 1991; Kafri et al. 2008). No wonder that the TDEM method was considered as a principal technique to delineate the DSBD in the DS region. The very first survey in the Nahal Hever South test site has indeed exhibited, as described before, very promising results (Goldman and Ezersky 1999a). Figure 42 shows the quasi-2D TDEM resistivity cross-section obtained for one of the N-S TDEM profiles in the Nahal Hever South test site. The profile crosses the E-W sinkhole lineament between TDEM stations NH5-1 and NH5-2.

Two main resistivity features are recognized in the cross-section, both exhibiting significantly higher resistivities than those of the background ones. The relatively shallow and smaller resistive body is located below station NH5-1, at the depth interval between 30 and 35 m. Since the water table here is at a depth of approximately 20 m, this feature, most likely, testifies to either a lower porosity of the aquifer or to a lower salinity of its hosted groundwater, or a combination of both. Since no information regarding the porosity of the aquifer was available, it was assumed that it represents the DSBD body. According to borehole HS-2, the bottom of the salt layer roughly coincides with the depth to the DSBD, enabling the process of salt dissolution there. The location of this feature just below the sinkhole lineament in the Nahal Hever South test site testifies to the validity of the proposed mechanism (Goldman et al. 2000).

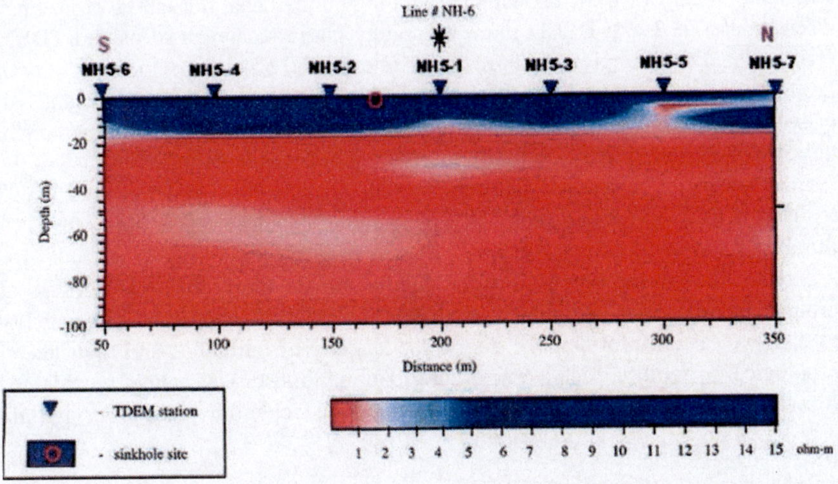


**Fig. 42** An example of a quasi-2D TDEM resistivity cross-section in the Nahal Hever South test site (after Wachs et al. 2000)

Concerning the deep feature at the depths between 50 and 60 m (Fig. 42), the occurrence of clay interlayers within the salt layer and the assumed DSDB might have blocked the process. This could have been overcome by an existing fault that cuts across the clay unit forming a hydraulic conduit, which was indeed detected by the high-resolution seismic method (see below).

The apparent success of the described integrated seismic/TDEM method at the Nahal Hever South test site greatly motivated a wide application of the method along the entire DS coastal plain of Israel. However, it was realized that the ability of TDEM method to characterize extremely saline groundwater is rather limited. In the next survey, at the Nahal Hever North site, despite an abundance of sinkholes on the surface, no increased TDEM resistivities have been found underneath them in the entire site (Goldman and Ezersky 2000). All TDEM resistivities, at depths below the water table, varied within a narrow range between 0.4 and 0.6 ohm-m. Such small resistivity variations could not be attributed to the corresponding salinity variations in the absence of the information regarding porosity or any other factor, which might have affected the measured bulk resistivity. As a result, further application of the TDEM method has been temporarily abandoned until empirical quantitative relationships between the TDEM bulk resistivities and chloride concentration were derived for different lithologies in the DS coastal aquifer (Ezersky and Frumkin 2017). These relationships were based on the correlation of TDEM results and borehole data using statistically significant number of measurements adjacent to 28 boreholes throughout the DS coastal plain. An example of the relationships between the measured bulk TDEM resistivities and the hosted groundwater chlorinity for different lithologies that comprise of aquifer are shown in Fig. 43 (Ezersky and Frumkin 2017). It is evident that the chlorinity is in the range between 80 and 200 g/l whereby even small variations in the bulk resistivities reflect considerable chlorinity changes. Thus, the required quantitative estimation of the chlorinity is still ambiguous. However, knowing the lithology underneath the salt layer, one can discriminate DSB and DSBD more accurately based solely on measured TDEM resistivities. For example, if bulk resistivities exceed 0.55 ohm-m, the chlorinity of groundwater is less than 100 g/l for all sedimentary types of the aquifer, and salt dissolution is thus feasible. If resistivities vary between 0.3 and 0.4 ohm-m, then, in the case of a silty aquifer (DS mud), dissolution is possible, whereas in case of the sandy-gravel aquifer, these resistivities are indicative of DSB, which is not capable to dissolve salt. Resistivities below 0.3 ohm-m, most likely, correspond to the DSB for all types of aquifers.

Thus, knowing the lithology of the aquifer, for the above considered range of groundwater salinities, is crucial for the determination of groundwater's capability to dissolve salts. Unfortunately, due to the scarcity of boreholes and high lateral lithological variability, such a crucial information in most cases cannot be obtained from boreholes. Therefore, an attempt was made to tackle this problem using only

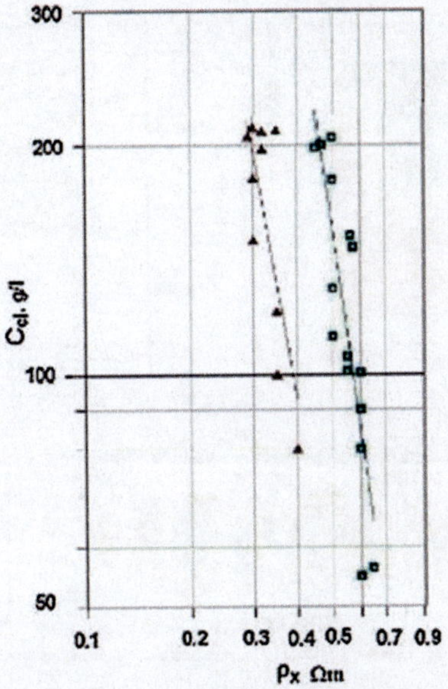


**Fig. 43** Chlorinity versus bulk resistivity relationships for different lithologies of the Dead Sea coastal aquifer: triangles—silty lithology, squares—sandy-gravel lithology (modified after Ezersky and Frumkin 2017)

geophysical methods, namely the combined three-dimensional (3D) SNMR and TDEM methods (Legchenko et al. 2009).

**Combined 3D SNMR-TDEM**

The feasibility study for the combined method was carried out at the Nahal Hever South test site. It included 24 SNMR soundings, each accompanied by a TDEM measurement at exactly the same location (Legchenko et al. 2009). Two important hydrological parameters were extracted from the 3D SNMR interpretation. The first one, which is directly obtained by the SNMR inversion, is water content, and the second is hydraulic conductivity of the aquifer, ($K_{SNMR}$), calculated using inverted SNMR relaxation time ($T_1$) and water content ($w$), using the following empirical relationship (Legchenko 2013): $K_{SNMR} = C_p w T_1{**}2$, with $C_p$ being an empirical constant that can be calibrated to similar hydrogeological parameters obtained from pumping tests.

It should be noted that SNMR is an efficient tool not only for characterizing aquifers but also for locating water-filled voids in the subsurface (Vouillamoz et al. 2003; Boucher et al. 2006).

The results of the 3D SNMR interpretation are shown in Fig. 44. The relaxation time ($T_1$), presented in Fig. 44a, varies between approximately 50 and 400 ms. Higher $T_1$ values correspond to a sandy-gravel formation. The map (Fig. 44b) and the 2D cross-section (Fig. 44c), showing the water content, which was derived from the SNMR inversion, enable to locate and delineate the aquifer. Substituting this

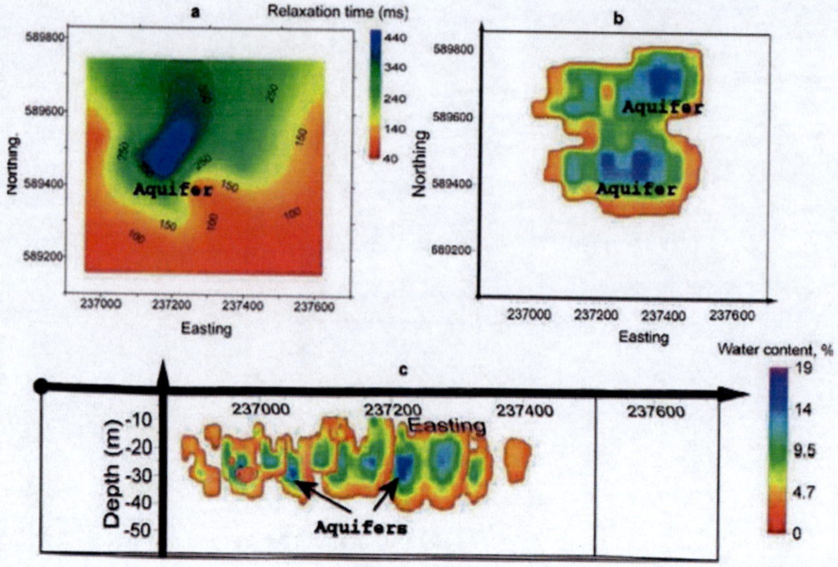


**Fig. 44** 3D-SNMR results in the Nahal Hever site: **a** Map of the relaxation time $T_1$ derived from 3D-SNMR inversion; **b** map of the water content distribution at a depth of 26 m, and **c** variations of the water content within the investigated area (modified after Legchenko et al. 2009)

information along with the measured TDEM bulk resistivities into Fig. 43 allows to estimate the chlorinity of groundwater within the aquifer. The latter results, combined with seismic refraction data regarding the location and depth to the salt layer, allow to delineate the potential zones of sinkholes formation.

As mentioned before, the existence of impermeable clay interlayers within the salt layer and the aquifer required the detection of faults that facilitate the emergence of the DSBD toward the salt layer.

Regarding the above, there are two main mechanisms of the sinkhole formation due to the salt dissolution. The first one, based on the so called "structural model," was described by Abelson et al. (2003). It assumes that sinkholes tend to develop along faults, which serve as preferential hydraulic conduits to emerging confined aggressive water upward to reach salt layer. The second mechanism is based on the so called salt edge model that associates sinkholes with karstification of the salt edge by deep and shallow aggressive groundwater (Ezersky and Frumkin 2013). The latter model is handled using the above described seismic refraction method, whereas the former model is better geophysically treated by means of the high resolution seismic reflection method (HRSR).

#### High-resolution seismic reflection (HRSR)

The results of a high-resolution seismic reflection survey at the Nahal Zeelim site are presented in Fig. 45. The survey included a 340-m-long 2D profile, which was shot in NE-SW direction across the sinkhole line (Ezersky and Frumkin 2013).

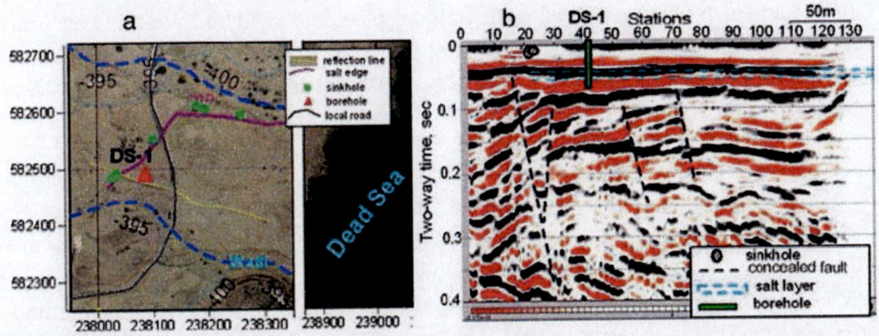


**Fig. 45** **a** Location map of the Nahal Zeelim site. Reflection line (yellow) crosses the salt layer edge and a small sinkhole, which is a part of a sinkhole line. **b** 2-D seismic reflection section through the site in the NW-SE direction (modified after Ezersky and Frumkin 2013)

The edge of a 7–10-m-thick salt layer has been previously delineated by seismic refraction study and verified by borehole DS-1 (Fig. 45a, b). A distortion zone is clearly seen in the vicinity of the intersection of the reflection profile and the sinkhole line (Fig. 45b). The results testify to the presence of faults close to the sinkhole zone. The upper reflector is generally smooth but distorted between stations 10 and 30. Two sinkholes and the salt edge indeed coincide with the fault site.

The following procedure for geophysical evaluation of a sinkhole development hazard in a selected site, based on the above, is presented herein:

1. Seismic refraction survey for the detection of a salt layer in the site, and delineation of its edge.
2. A combined 3D SNMR-TDEM survey for evaluating the chlorinity of the groundwater.
3. A HRSR survey, to detect faults that serve as conduits for emerging DSBD.

## 5 Integrated Geophysical Surveys in the Mediterranean and Dead Sea Coastal Aquifers

Each GE/GEM method has inherent limitations, which reduce its application to specific targets, geoelectric conditions, noise levels and surface conditions. Sometimes, the choice of an appropriate single method might be sufficient to solve the required hydrogeological problem, but in most cases a single method provides only partial information that must be accompanied by independent data. In the past, such data were mainly extracted from nearby boreholes, but the scarcity or absence of boreholes, required to look for an alternative solution. Several decades ago, the idea to jointly use different GE/GM methods for solving specific geological and hydrogeological problems was proposed by Jupp and Vozoff (1975). Initially, the idea was only realized in the so-called joint inversion of different GE/GM data for reducing ambiguity of geophysical interpretation. At present, integrating different

GE/GM and also other geophysical techniques is widely used both for more accurate geophysical interpretation and for solving particularly complicated geological and hydrogeological problems. Two examples of the latter applications are discussed herein.

## 5.1 SNMR-TDEM

It is widely recognized that among all geophysical techniques, the GE/GEM methods are the most efficient ones in solving various hydrogeological problems, particularly those related to delineating and monitoring groundwater bodies and processes. In the case of saline groundwater, in most cases the methods are feasible irrespective of the environment, since saline groundwater provides excellent resistivity contrast practically with any background lithology, whether dry or saturated with freshwater. Unfortunately, it is not always the case with freshwater-saturated aquifers, which in many cases cannot be even delineated by the GE/GEM methods, not mentioning more delicate problems such as hydraulic properties.

On the other hand, the SNMR method proved to be of high efficiency, not only in detecting groundwater of any salinity, but also in quantitative estimation of porosity and transmissivity of aquifers. However, the method suffers from a rather limited exploration depth, and in most cases is unable to detect a fresh/saline water interface. Therofore,by combining the TDEM and SNMR measurements in coastal aquifers, one might expect not only to delineate the freshwater portion of the aquifer, but also to quantitatively estimate the amount of freshwater within the aquifer and, under favorable conditions, the transmissivity of the latter. The feasibility of this integrated technique has been examined in two coastal aquifers of Israel.

**The Mediterranean coastal aquifer**

Two perpendicular profiles, each including four SNMR/TDEM stations, were run in close proximity to existing observation wells at the Nitzanim coastal area (Fig. 46). The comparison of the geophysical and borehole information from all three wells available clearly indicated the limits of applicability of each method. As an example, Fig. 47 shows the results of the comparison.

The SNMR method manages to accurately determine the depth to the water table, while the TDEM method misses it. The opposite is true with regard to the fresh/sea water interface. Due to the limited exploration depth, the SNMR data does not manage to detect the interface, whereas the TDEM method does it with high accuracy. Combining both interpreted models, and taking into account the discussed limits of application of each method, the freshwater portion of the aquifer was found in the depth interval of 10–66 m, as compared to a 12–65 m interval in an adjacent borehole. In the absence of pumping tests, carried out in that well, the alternative way is to quantitatively estimate the water content in the borehole by the SNMR method. Based on statistically significant correlation data gathered in granular coastal aquifers

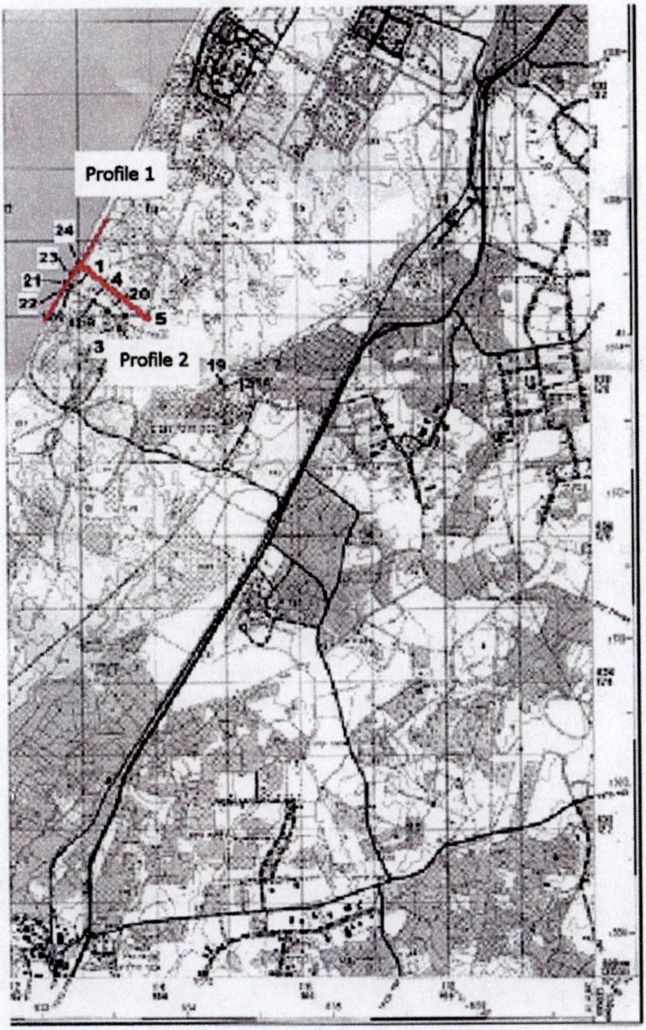


**Fig. 46** NMR/TDEM survey location map in the Nitzanim area

all over the world (Yaramanci 2003), one can use the interpreted SNMR water content to calculate the amount of fresh ground water with great confidence.

Figure 48 compiles the results of the integrated SNMR/TDEM measurements along profile 2 perpendicular to the seashore (Fig. 46). The freshwater portion of the aquifer is clearly indicated in the combined quasi-2D resistivity/water content cross-section. The fairly consistent water content data along the entire profile is evident, varying in a narrow range, roughly between 15 and 20%.

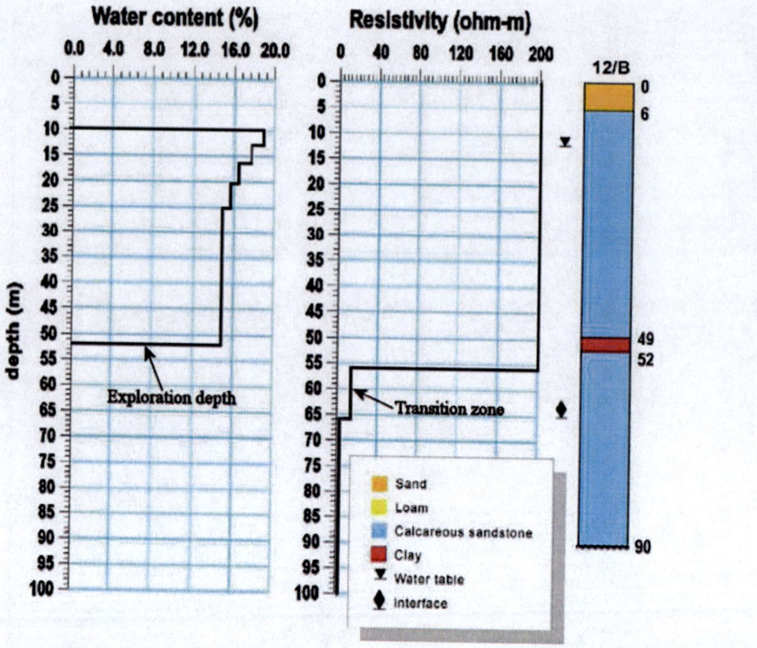


**Fig. 47** SNMR-TDEM interpretation results near observation well 12/B (after Legchenko et al. 1998)

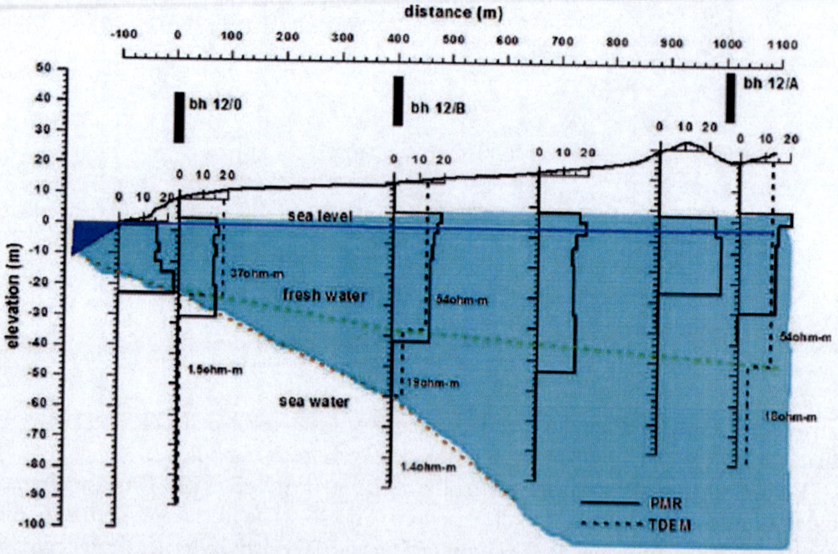


**Fig. 48** Combined SNMR-TDEM columnar cross-section along profile 2. Blue color designates fresh groundwater. PMR = SNMR (after Legchenko et al. 1998)

**The Dead Sea coastal aquifer**

In order to test the proposed integrated SNMR-TDEM technology under a more complicated hydrogeological setup in an arid area, a similar survey has been carried out in the Ein Gedi area (Fig. 49). Due to topographic and technical limitations, it was impossible to align the SNMR-TDEM stations along straight lines. Therefore, the discussed profiles include also projections of the real measurements on the profile lines.

Unfortunately, in the absence of observation wells in the close vicinity of the SNMR-TDEM stations, the calibration measurements were performed under a similar hydrogeological set up of the nearby Tureiba area, adjacent to observation wells $T_2$ and $T_4$ (Fig. 49).

Figure 50 shows, as an example, the comparison between the geophysical results and the well log data from observation well $T_4$. One can see the same phenomenon

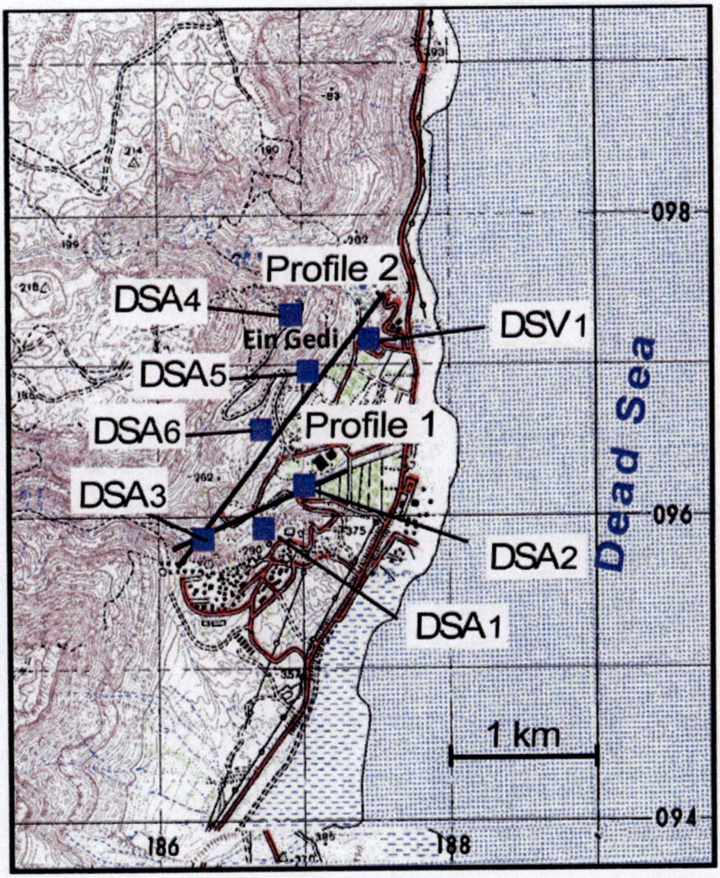


**Fig. 49** SNMR-TDEM survey location map in the Ein Gedi area

that was observed in the Mediterranean coastal aquifer. Due to a rather low resistivity of the vadose zone, that does not provide a sufficient resistivity contrast with the saturated part of the aquifer, the TDEM method failed to detect the water table. However, the latter was accurately detected by the SNMR method (Fig. 50). Unfortunately, the Tureiba wells did not reach in depth the intruded DS brine body, and thus, a quantitative comparison with the TDEM results was not possible. Nevertheless, taking into account the very high reliability of the TDEM method in the detection of fresh/saline water interface, it is reasonable to assume that the resistivity drop slightly below 1 ohm-m at a depth of approximately 60 m (Fig. 50) reflects the upper boundary of the intruded, and somewhat diluted DS brine. Thus, similarly to the Mediterranean coastal aquifer, one can quantitatively estimate the amount of fresh groundwater in the Dead Sea coastal aquifer. Figure 51 demonstrates such an estimate along profile 1 as shown on the location map (Fig. 49).

It should be noted that near the Tureiba $T_4$ observation well, the SNMR decay time, which is closely connected to a transmissivity of the aquifer, accurately fits the location of the pebble layer in the depth interval of 33–40 m (Fig. 50). Unfortunately, such a good fit is not very common and is not statistically supported by the rest of the SNMR measurements.

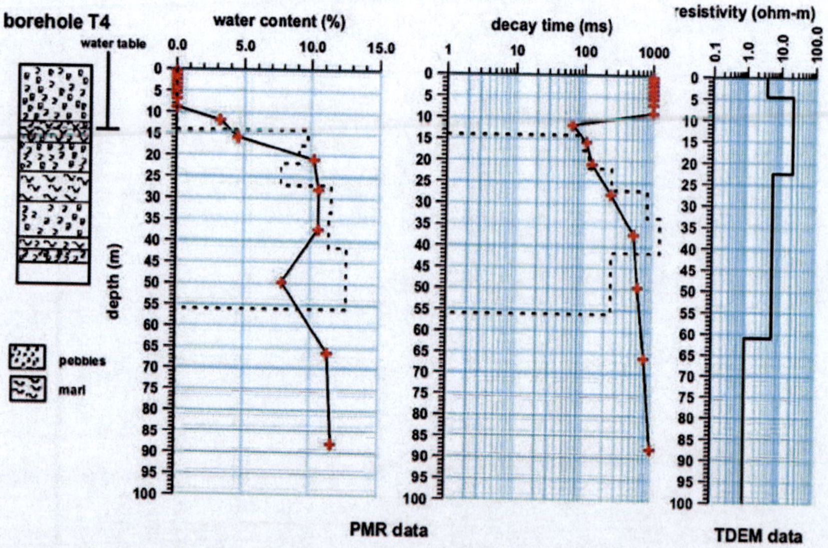


**Fig. 50** SNMR-TDEM calibration measurements near the Tureiba $T_4$ observation well. The dashed and solid lines in the SNMR interpretation represent different inversion algorithms used (after Legchenko et al. 1998)

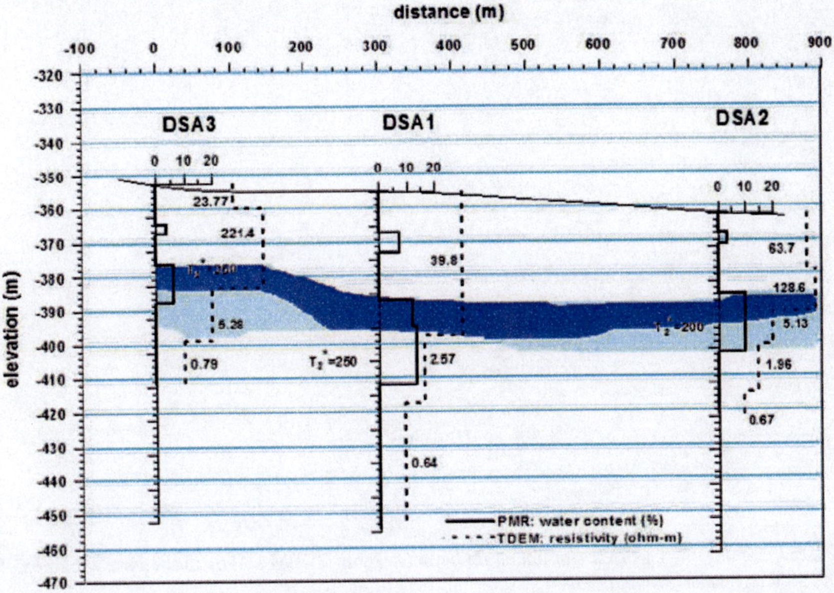


**Fig. 51** SNMR-TDEM columnar cross-sections along profile 1. Dark blue color designates fresh groundwater, and light blue color designates brackish water. The amount of water is estimated by using the water content provided by SNMR. PMR = SNMR (after Legchenko et al. 1998)

## 5.2 *Reflection Seismics—TDEM*

The so-called hydrological reversals representing freshwater bodies below saline ones exist in all coastal multiple aquifer systems of Israel (Mediterranean, Dead Sea and Gulf of Eilat (Aqaba). They were detected and studied by hydrogeological and geophysical methods (e.g., Kafri et al. 1997; Goldman et al. 2003; Kafri and Goldman 2006). In many cases, the TDEM method alone is able to detect these bodies. However, if the thickness of the electrically conductive impermeable layers dividing the aquifer into sub-aquifers is similar or exceeds the thickness of the freshwater-saturated sub-aquifers, the resolution of the method could be insufficient to detect the latter. This phenomenon was observed at some places in the Mediterranean coastal aquifer and was particularly significant in the Gulf of Eilat (Aqaba) coastal aquifer (Goldman et al. 2003). In order to resolve this problem, additional detailed information, regarding the subsurface geological sequence, is essential. Unfortunately, obtaining such information from boreholes is problematic due to both their scarcity and high lateral lithological variability. Integrating TDEM, which is most sensitive to groundwater salinity with high-resolution reflection seismics, which is able to differentiate between the impermeable (clays and loams) and aquiferous (calcareous sandstones) layers was found as a reasonable choice to resolve the problem.

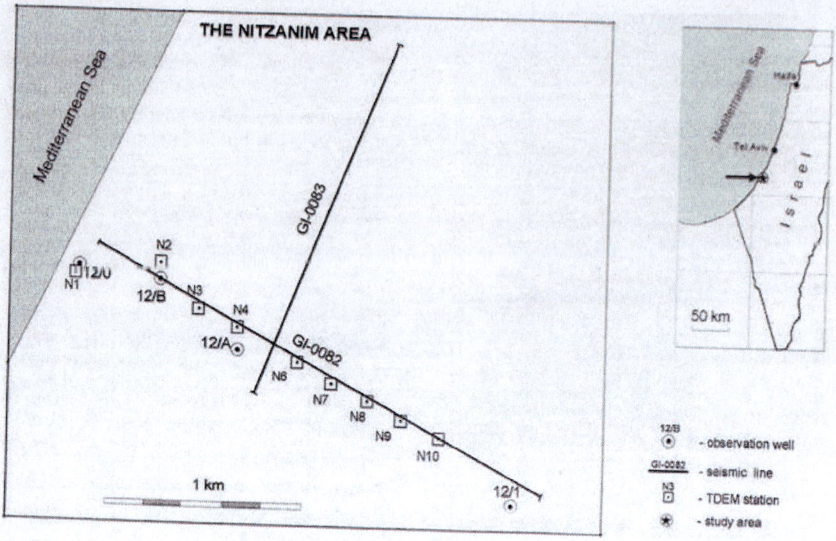


**Fig. 52** Location map of the combined reflection seismic-TDEM survey in the Nitzanim area of the Mediterranean coastal plain

In order to test this approach, the TDEM and reflection seismic methods were jointly applied at the Nitzanim area in the Mediterranean coastal plain of Israel (Shtivelman and Goldman 2000) (Fig. 52). This area has been selected, since previous TDEM measurements testified to the existence of the above-mentioned hydrological reversal and lithological logs and salinity measurements from nearby observation wells were available to assist the interpretation of the geophysical data. Figure 53 shows the schematic W-E hydrogeological cross-section, based on the borehole data across the investigated site. Three of the wells (12/0, 12/A and 12/1) penetrate the entire aquifer down to the Saqiye Group, located at a depth of about 160 m in this area. The water table along the W-E profile (Fig. 52) varies from 2 to 3 m above mean sea level (MSL). The seawater intrusion into the upper sub-aquifers extends to approximately 1 km east of the shoreline.

The high-resolution seismic reflection survey included two seismic lines, GI-0082 and GI-0083 (Fig. 52), which were shot using the conventional *P*-wave technique. The TDEM measurements consisted of nine central loop soundings carried out along seismic line GI-0082 (Fig. 52). The transmitter loop size varied in accordance with the distance to the seashore. The measurements, close to the seashore, were performed using 50 by 50 m transmitter loops, while those located in the eastern part of the profile employed 200 by 200 m transmitter antennae. The corresponding penetration depths varied between roughly 100 m and approximately 300 m, respectively.

An example of a seismic time section for the western part of the reflection line GI-0082 is shown in Fig. 54. The horizontal axis on the section shows the station numbers, and the vertical axis exhibits the two-way travel time in milliseconds.

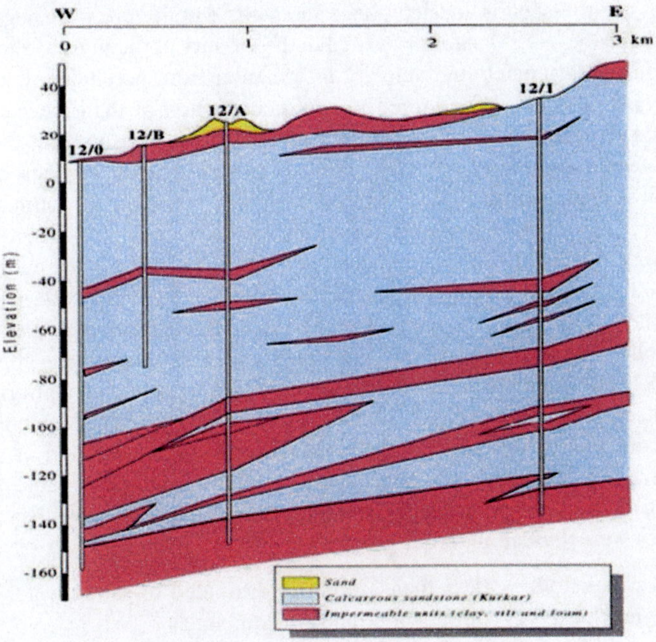


**Fig. 53** W-E geological cross-section of the Nitzanim site

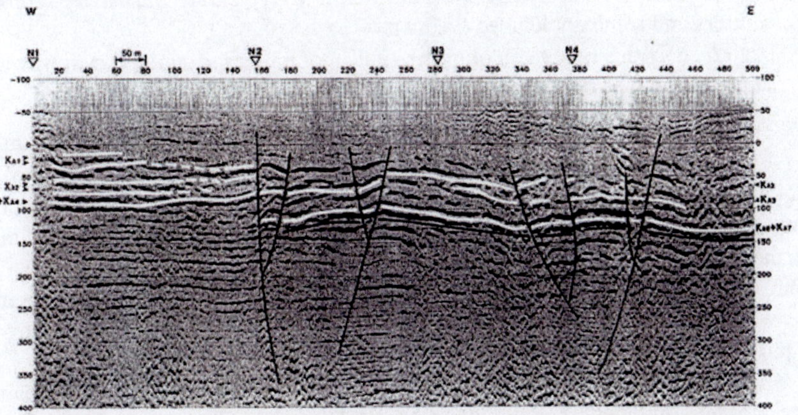


**Fig. 54** Seismic time cross-section for the western part of the reflection line GI-0082 shown in Fig. 52 (modified after Shtivelman and Goldman 2000)

The times for the section located above the MSL datum appear as negative. The boreholes and the TDEM stations located in the vicinity of the line are shown above the section. Unfortunately, no velocity information from boreholes was available in the investigated area. Therefore, no depth conversion of the time sections was carried out. However, a *P*-wave velocity of 2000 m/s, measured on the average in boreholes throughout the coastal plain, was assigned to the relevant calcareous sandstone sequence (Fig. 53). This in turn enabled approximate estimates of the elevations of the reflected horizons, appearing on the sections.

A total of nine TDEM measurements were carried out along the entire seismic line GI-0082 (Fig. 52). Four of the measurements executed in the western part of the seismic line GI-0082 (Fig. 54) were used for the integrated interpretation.

The collected TDEM data were processed and interpreted using the Interpex TEMIX-XL 1-D interpretation software (Interpex 1996), while no a priori information has been used at this stage. The initial model for each sharp boundary inversion was obtained from the smooth Occam inversion (Constable et al. 1987) by reducing the number of layers in the recovered smooth model to a possible minimum. According to the above-described hydrogeological setting of the area, the interpreted resistivities were divided into four groups:

- Very low resistivities (less than 3 ohm-m) attributed to saline water-saturated lithologies (hosted in aquifers or trapped in aquicludes).
- Low resistivities (between 3 and 8 ohm-m), characterizing either brackish water-saturated aquifers or aquicludes (mainly clays).
- Moderate resistivities (between 8 and 15 ohm-m) typical for freshwater-saturated aquifers and some aquicludes (loams).
- High resistivities (above 15 ohm-m), typical of either unsaturated or freshwater-saturated aquifers.

The lateral extent of sea water intrusion into the aquifer was estimated from the TDEM measurements alone, by taking into account that the depth to the base of the aquifer (top Saqiye) is consistent at 160 m below sea level (Fig. 53). The fact is that the saline Saqiye beds, in most known cases, are characterized by low resistivities similar to those of seawater-intruded aquifers. Therefore, if the detected very low resistivity unit was located at depths shallower than 160 m, it was supposed to represent a seawater intrusion, otherwise it represented the Saqiye Group clays underlying the aquifer.

Figure 55 shows the pseudo-2D resistivity cross-section of the western part of the profile, compiled from the interpreted 1-D models. Two moderately resistive layers, which can be identified with freshwater-saturated sub-aquifers were detected beneath station N3. However, the lateral extent of the lower layer, hydrogeologically interpreted as a sub-aquifer, is unclear since it was not recognized at stations N2 and N4. It seems to wedge out laterally toward an aquicludal layer, having a resistivity of approximately 5 ohm-m, somewhere between stations N3 and N4, whereas between N3 and N2 the sub-aquifer probably becomes saturated with saline water. Thus, according to the preliminary TDEM interpretation, which is not based on any

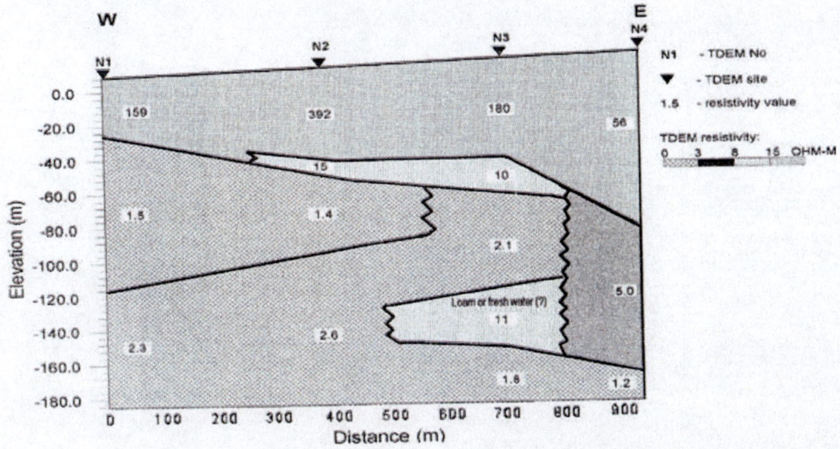


**Fig. 55** Pseudo-2D resistivity cross-section of the Nitzanim site (preliminary interpretation) (modified after Shtivelman and Goldman 2000)

additional independent information, the lower freshwater-saturated sub-aquifer is confined to an area just below station N3.

In order to verify the above preliminary inversion results, a combined interpretation of the TDEM and seismic data was performed. According to the seismic interpretation, the lower calcareous sandstone units, designated as $K_{A6}$ and $K_{A7}$ extend from the vicinity of TDEM station N2 continuously through stations N3 and N4 and farther eastward (Fig. 54). The moderately resistive layer, detected beneath station N3 at approximately the same elevation (−120 m), is most likely identified with the detected seismic horizon saturated with freshwater. Provided that this assumption is correct, the data collected at station N4, where the lower sub-aquifer was not detected by the preliminary TDEM inversion, should be re-interpreted. In order to make the TDEM interpretation consistent with the seismic results, an additional moderately resistive layer has been included in the initial model for the follow up TDEM inversion. The results of this inversion are shown in Fig. 56b. For the sake of convenience, the figure also shows the results of the preliminary inversion (Fig. 56a). The uncertainty in the model parameters is approximately estimated by using the so called linear equivalence analysis (Goldman et al. 1994b). The best-fit models are drawn by solid lines in the resistivity versus depth sections in Fig. 56a, b. Dashed lines in the figures represent alternative (equivalent) models, for those where the misfit error is only slightly greater than that for the best-fit model and whereby the misfit error is still bellow an assumed noise level of the measurements.

As expected, both the resistivity and thickness of the lower moderately resistive layer are poorly resolved, and therefore the inclusion of this layer in the interpreted model is only justified, if it is supported by additional independent information, such as in our case the results of seismic interpretation in this case. Figure 56b shows that the inclusion of this layer in the final combined interpretation decreased

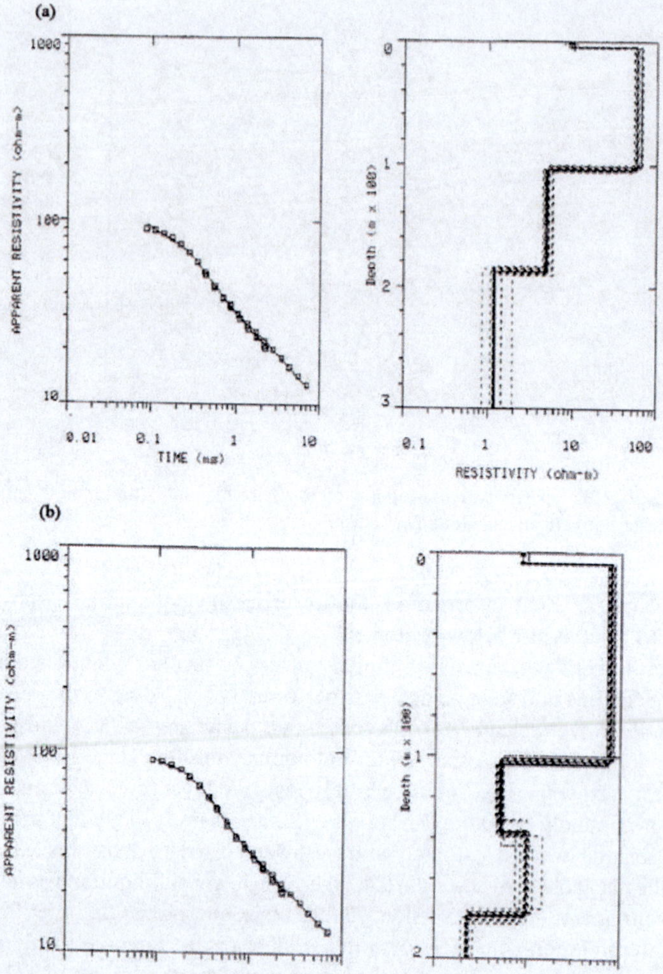


**Fig. 56** TDEM results at station Nitzanim N4. Left panels show the acquired data (open squares) and the best-fit synthetic responses (solid lines). Right panels show the best-fit (solid lines) and equivalent (dashed lines) models. **a** Preliminary interpretation based on the minimum possible number of layers. **b** Final interpretation based on the independent seismic results (modified after Shtivelman and Goldman 2000)

the misfit error of the TDEM inversion from approximately 3%, obtained in the preliminary inversion, to slightly more than 2%. From a practical point of view, such an improvement of the inversion by itself is not that crucial because the estimated total noise level is larger than 3% anyway. Therefore, it can be considered just as a bonus to the main result, namely the consistency of the final TDEM inversion with the independent seismic results. Thus, the hydrogeological interpretation, of the combined TDEM-seismic results, which testifies to the continuation of the relatively resistive freshwater body below station N3 toward station N4, seems to be justified.

The final resistivity cross-section, accompanied by the borehole data, is shown in Fig. 57a. One can see that the location of the lower moderately resistive unit roughly coincides with the lower sub-aquifer in well 12/A, although the coincidence is not perfect, due to the above-mentioned uncertainty in the layer parameters. The resistivity of the aquiclude layer, separating the two sub-aquifers, varies laterally from approximately 5 ohm-m in the east (station N4) to slightly more than 2 ohm-m

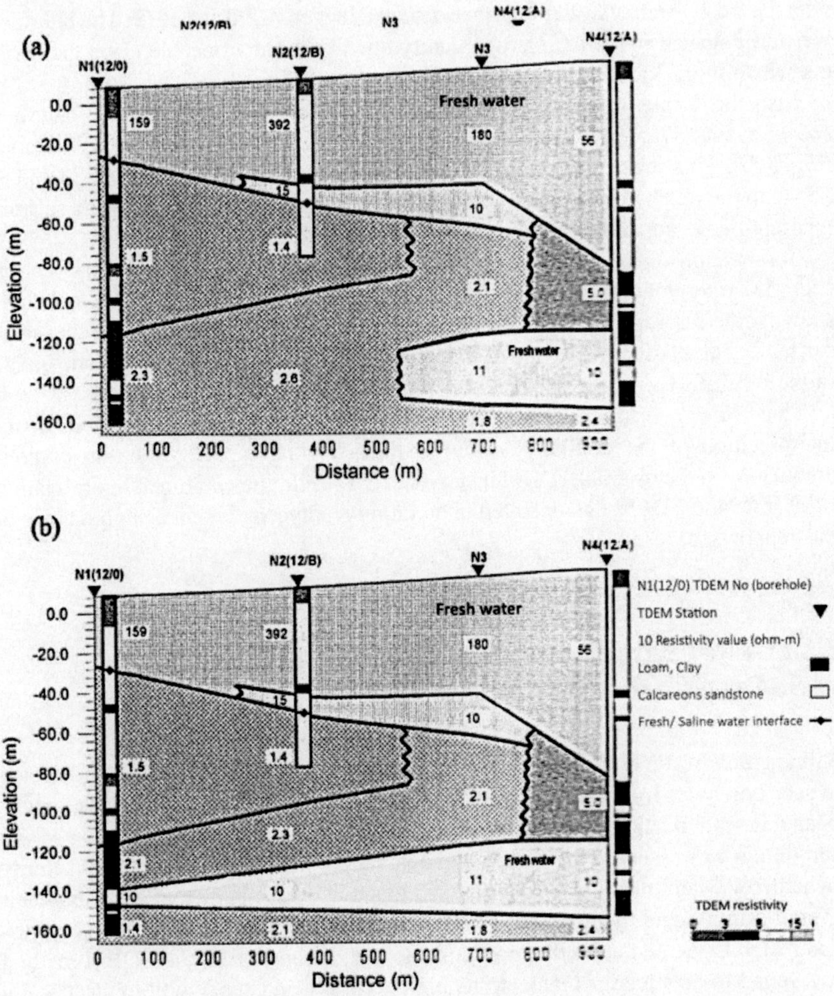


Fig. 57 Pseudo-2D resistivity cross-section of the Nitzanim site (final interpretation) (modified after Shtivelman and Goldman 2000)

in the west (station N3). Such a lateral variation can be reasonably explained by expected higher salinities of water, trapped in the aquiclude in the west, closer to the sea. It also explains, why the lower sub-aquifer was not detected in station N4 during the preliminary interpretation. Most likely, it happened due to the much smaller resistivity contrast between the aquiclude and underlying sub-aquifer at station N4, as compared to station N3.

Another small resistivity contrast within the very low-resistivity unit of approximately 1.5 and 2.3 ohm-m can be most likely identified with a boundary between the sea water-saturated aquifer and the aquiclude. Finally, the boundary between the moderately/highly resistive layers and very low-resistivity unit exactly coincides with the freshwater/seawater interface detected in wells 12/0 and 12/B. The interface was not detected in well 12/A, apparently due to the impermeable clays that exist elsewhere in the vicinity of station N3.

Also, the re-interpretation of TDEM measurements N1 and N2 similarly to the one shown in Fig. 57a, led to the appearance of a thin freshwater-saturated sub-aquifer (Fig. 57b). This sub-aquifer was indeed known at the depth interval of −138 and −154 m in the nearby observation well 12/0. The unit was not detected by the seismic measurements, apparently since it is too thin at this location.

Comparison of the seismic results (Fig. 54) with the geological cross-section (Fig. 53) interpolated using the well log data between boreholes 12/A and 12/B shows some discrepancy typical for regions with sparse borehole data. In this case, seismic sections usually provide a continuous and, therefore, more diagnostic image of the subsurface.

Thus, the consideration of independent seismic results motivated a re-interpretation of the preliminary TDEM data, leading to the resolution of non-uniqueness in the original TDEM inversion. As a result, the combined interpretation of seismic and TDEM data resulted in an improved hydrogeological significance of the geophysical results.

## 6 TDEM Characterization of Groundwater Salinity in Deep Carbonate Aquifers

Saline groundwater bodies hosted in deep, mostly carbonate, aquifers were detected in deep boreholes from all over Israel. The geochemistry of these aquifers was investigated in several studies (e.g., Starinsky 1974; Kafri and Arad 1979; Nativ 1984). In addition, it was essential to spatially delineate these bodies and assess their salinity (whether seawater or brine concentrations) that could have helped in obtaining their overall salinity, and to shed light as to their origin. For this purpose, the geophysical methods seemed to be the most suitable and, among them, the TDEM method was found to be superior thanks to its high accuracy and sensitivity to electrically conductive targets (Goldman et al. 1988, 1991).

The first TDEM measurements using SHOTEM transmitter/receiver configuration were aimed to check the possibility of subsurface seawater intrusion into the carbonate Judea Group Aquifer (hereafter JGA) close to the seashore. The survey was carried out in the Mount Carmel area and, later on, more to the south in the central coastal plain (Goldman et al. 1988). The obtained promising results allowed to tackle a more complicated geophysical and hydrogeological issue, namely the possibility of subsurface seawater intrusion from the Mediterranean Sea to the DSR base level as was suggested by Kafri and Arad (1979). In order to examine this assumption, deep TDEM traverses connecting the Mediterranean Sea and DSR were carried out to provide the resistivity and thus the salinity of the groundwater bodies along the traverses as well as the configuration of the interfaces between the saline bodies and the overlying fresh groundwater system.

A total of four generally W-E directed deep TDEM profiles were run, between the Mediterranean Sea and the northern DSR and between the Mediterranean Sea and the Dead Sea. Also, a deep TDEM traverse was executed in the south along the Arava Valley between the Dead Sea and the Gulf of Eilat (Aqaba). These measurements were performed in order to delineate and characterize the different saline water bodies in depth and thereby to determine their origin.

The results of all the above-mentioned studies are described herein.

## 6.1 *Mount Carmel-Central Coastal Plain Area*

A number of SHOTEM measurements were carried out to detect possible subsurface seawater intrusion into the JGA in the Mount Carmel area in northern Israel close to the Mediterranean Sea. The aquifer there is phreatic and open to the sea. Salinization processes were detected in some of the groundwater exploitation wells in this area. As a result, it was assumed that the aquifer is intruded by seawater (Kafri 1967, 1969; Arad et al. 1975; Kafri and Arad 1979). For calibration purposes, the measurements were performed adjacent to relevant existing water wells (Fig. 58).

The results of the measurements can be summarized, as follows (Goldman et al. 1988):

1. The actual freshwater/seawater interface was not encountered in the discussed wells since it was assumingly below the bottom of the wells. Therefore, in order to be correlated with the relevant SHOTEM results, the depth to the interface was calculated for each well, based on the freshwater head through the Ghyben–Herzberg equation.
2. The K1 TDEM station is adjacent to Nahal Oren 2 well. As shown in Fig. 59, the upper resistivity contrasts coincide with parallel lithological boundaries in the well log. However, the lower TDEM boundary between very high and low resistivities, which is typical to lithological unit saturated with seawater, does not correspond to a lithological boundary. The depth of this boundary is indeed compatible with the depth of the calculated interface at around 200 m.

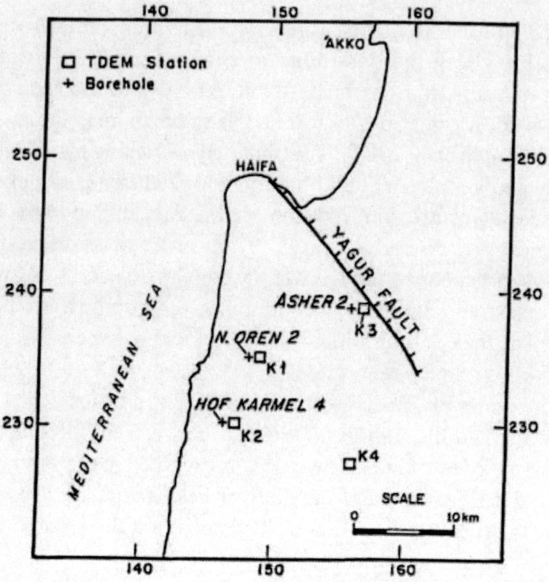


Fig. 58 Location map of the TDEM survey in the Mount Carmel area

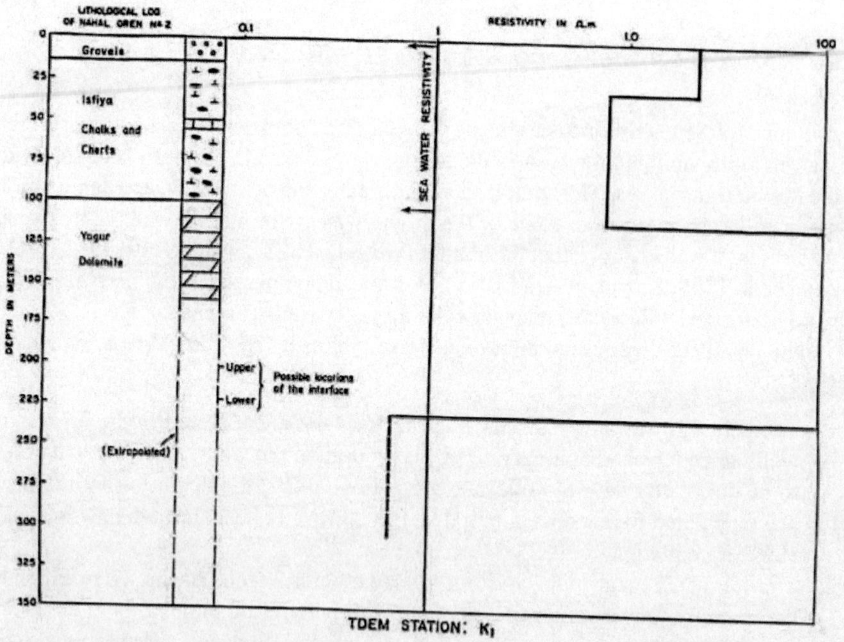


Fig. 59 Correlation of the TDEM results and well log data from the Nahal Oren-2 site (after Goldman et al. 1988)

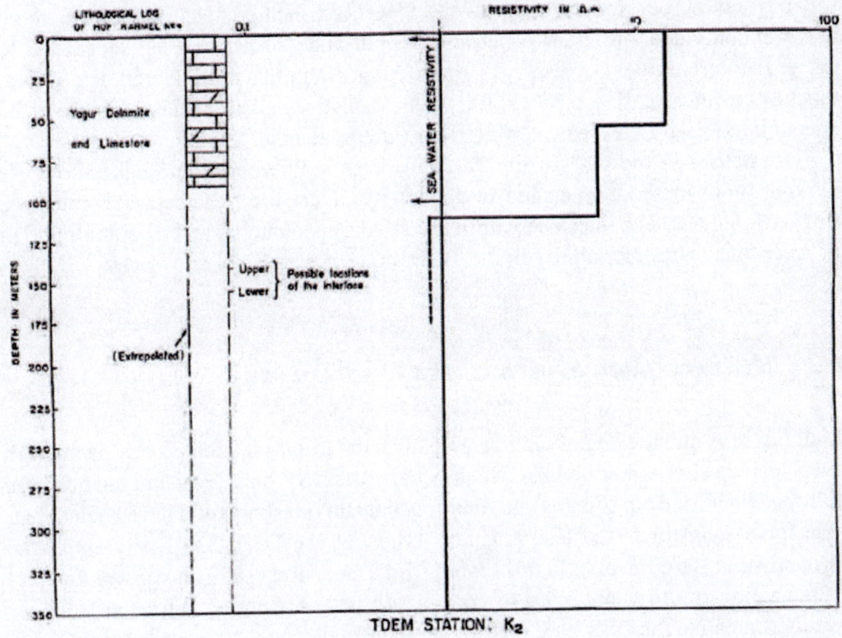


**Fig. 60** Correlation of the TDEM results and well log data from the Hof Carmel-4 site (after Goldman et al. 1988)

3. Similarly, the K2 TDEM station is adjacent to Hof Carmel 4 well. In this case (Fig. 60), the upper resistivity contrast within a homogeneous lithology assumingly represents a boundary between the upper freshwater body and a mixing zone with a saline end member underneath. Here, the depth to the lower resistivity boundary is somewhat lower than that of the calculated fresh/seawater interface (roughly at 140 m), apparently, due to the existence of the interpreted mixing zone.

In the cases of TDEM stations K3 and K4, the penetration depth of the measurements was not sufficient enough to detect the calculated depths to the interface.

In the southern fringes of Mount Carmel and farther south in the central coastal plain of Israel, groundwater salinity was encountered in some of the wells exploiting the JGA as part of the northern Yarkon–Taninim aquifer system. It was agreed in the past that the system is blocked to the sea and thus not subjected to seawater intrusion. Contrary to it, Kafri (1967) and Kafri and Arad (1979) claimed that the encountered salinity in those wells is related to current seawater intrusion. In exploration boreholes, drilled later on in the northern Yarkon–Taninim aquifer system, indeed, groundwater of approximately seawater salinity was encountered at their bottom at a considerable depth (Goldman et al. 2006). Unfortunately, due to the "masking" effect of the overlying rather thick conductive Saqiye Group shale sequence, this target was beyond the penetration depth of the SHOTEM method. The LOTEM method, however, provided sufficient penetration depth and, in some cases, succeeded to detect low resistivities corresponding to the high salinities in the boreholes. Unfortunately, in most cases, the LOTEM method also failed to accurately detect the encountered deep fresh/seawater interface (Goldman et al. 2006).

Fortunately, in the case of the Taninim deep well, where the Saqiye Group is missing, both methods managed to accurately detect the fresh/seawater interface (Figs. 61 and 62). The interface within the JGA is encountered at the depth of 600 m in accordance with the salinity in the Taninim deep well.

## 6.2 Mediterranean Sea-Northern DSR Traverse

Both the JGA and the Sea of Galilee are subjected to salinization processes through deep-seated saline water bodies, which occupy the DSR base level and the northern valleys (Fig. 63). Both sources and their mechanism of salinization have been debatable for a long time (see also Chapter "The Eastern Dead Sea Rift Continental Groundwater Base Level"). In this regard, Kafri and Arad (1979) suggested that part of the salinization is connected to current subsurface seawater intrusion from the Mediterranean to the DSR base levels due to the elevation difference between them.

In order to detect those deep-seated saline water bodies, the geoelectromagnetic (GEM) methods were employed. The objective of the initiated survey was to obtain the spatial distribution of the deep-seated saline water bodies as well as the configuration of the interface between them and the overlying fresher ones. This in turn could help to clarify the source and mechanism of groundwater salinization in the region. The results of the study were detailed by Kafri et al. (2007) as follows:

A total of 28 central loop SHOTEM soundings were executed between the Mediterranean Sea and the Sea of Galilee along two traverses (Fig. 63) using mostly 500 by 500-m transmitter loops. The maximum penetration depth of the SHOTEM soundings varied between 500 and 800 m. LOTEM measurements, to a depth interval between 300 and 3000 m in the same places, were conducted using 16 transmitters (Tx) and 29 receiver (Rx)-stations with Tx-Rx separations between 3 and 4 km. Calibration measurements were also carried out adjacent to existing deep water wells.

Based on the above-mentioned working hypothesis (Kafri and Arad 1979), TEM, both LOTEM and SHOTEM measurements were conducted between the Mediterranean and the DSR, primarily at those locations, where deep-seated saline water bodies were expected to exist. The alignment of the measurement sites and the resultant traverses (designated A-A′ and B-B′ in Fig. 63) were supposed to follow expected intruded sea water flow paths from the Mediterranean to the DSR. The target of the survey was the relevant deep electric conductor that reflects the saline water bodies, occupied within the JGA which could have been obtained mainly by the LOTEM measurements. The bulk resistivities of the shallower, mostly younger graben fill

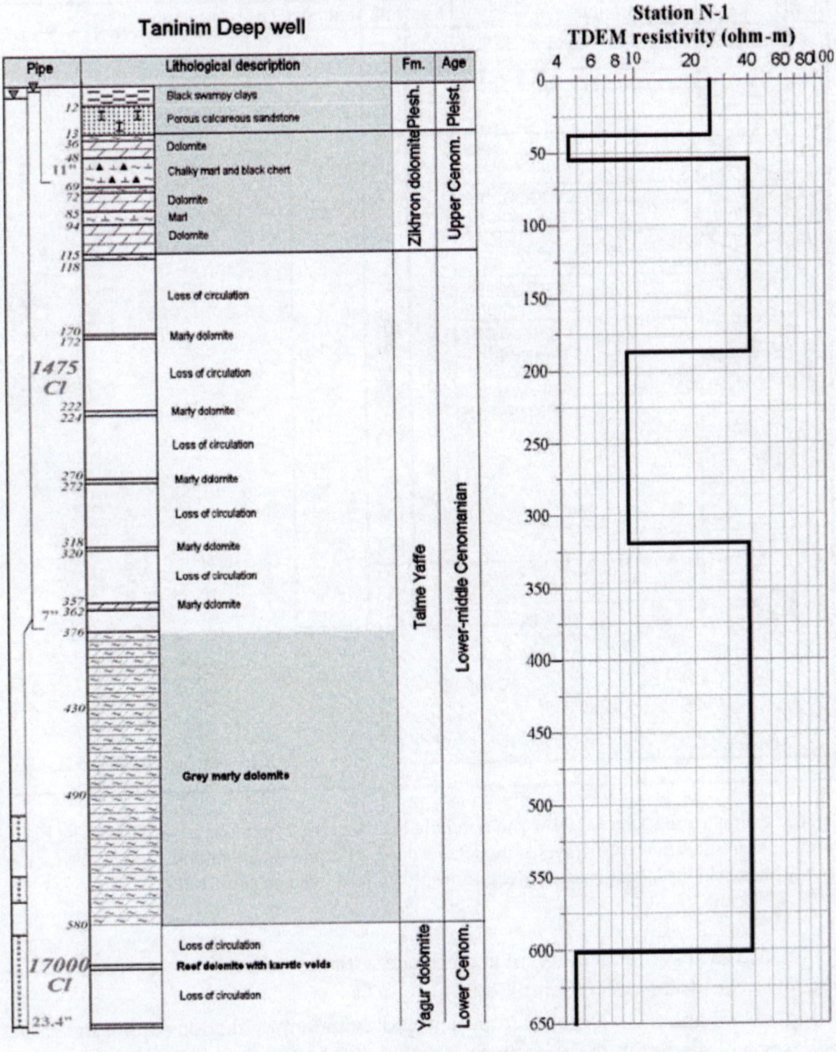


**Fig. 61** Comparison of the deep SHOTEM (Cycle-5 M) and borehole results at the Taninim deep well site (after Goldman et al. 2006)

sediments, that overly the deep-seated JGA, were obtained by the SHOTEM measurements as also described by Kafri et al. (2007). However, in order to remain focused only on the deep target, the shallow data appear on the TDEM cross-sections in blank (Figs. 65 and 66).

Figure 64 exhibits the columnar resistivity cross-sections obtained by the mainly LOTEM measurements along traverses A-A′ and B-B′. Both exhibit deep-seated conductors with resistivities below 10 ohm-m, which were interpreted as groundwater

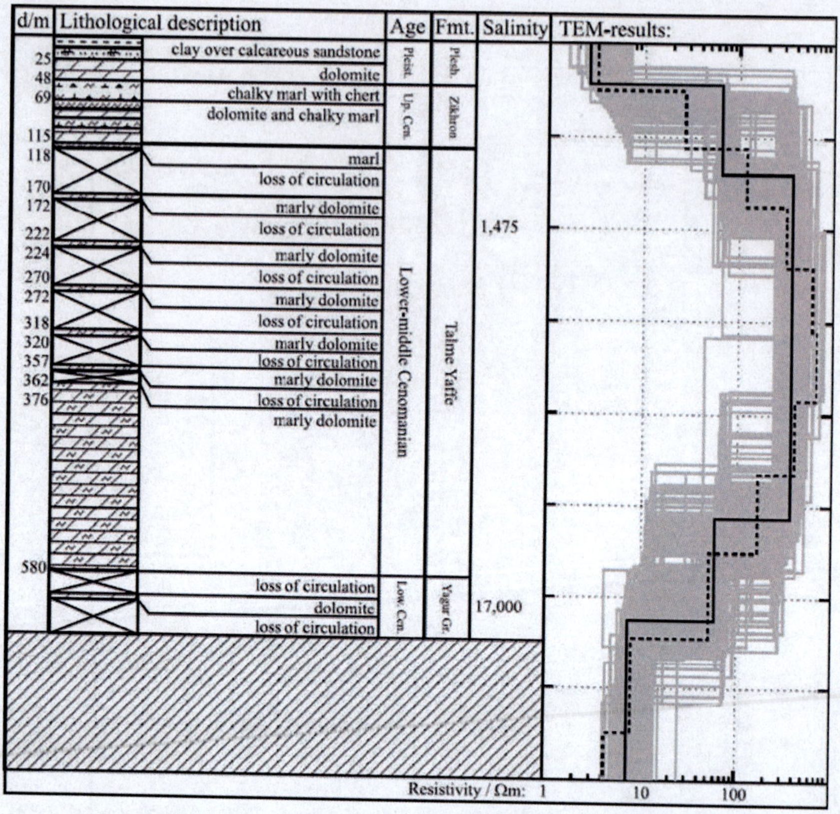


**Fig. 62** Comparison of the LOTEM and borehole results at the Taninim deep well site. The black solid line represents the best-fit layered model, the dashed line shows the smooth model. The family of gray lines exhibits all equivalent inversion results (after Goldman et al. 2006)

of brackish to seawater salinity, in accordance with the calibration measurements in adjacent deep water wells (Scholl 2005).

Figures 65 and 66 exhibits a simplified geological cross-sections along the same traverses, coupled with the superimposed water table and the assumed fresh/saline groundwater interface representing the top of the detected deep conductor.

**The Afeq-Bet Shean traverse A-A′ (Fig. 65)**

The fresh-saline water interface along this traverse continuously declines eastward within, above and below the JGA, crossing different geological structures all the way from the Mediterranean Sea to the DSR. The decline of the interface follows that of the freshwater table, both below sea level along most of the eastward flow system. Close to the DSR margin, near the Bet Shean site, the deep-seated saline waters emerge to shallower levels or even to the surface through the fault system by the driving head of the freshwater mechanism, as described by Goldshmidt et al.

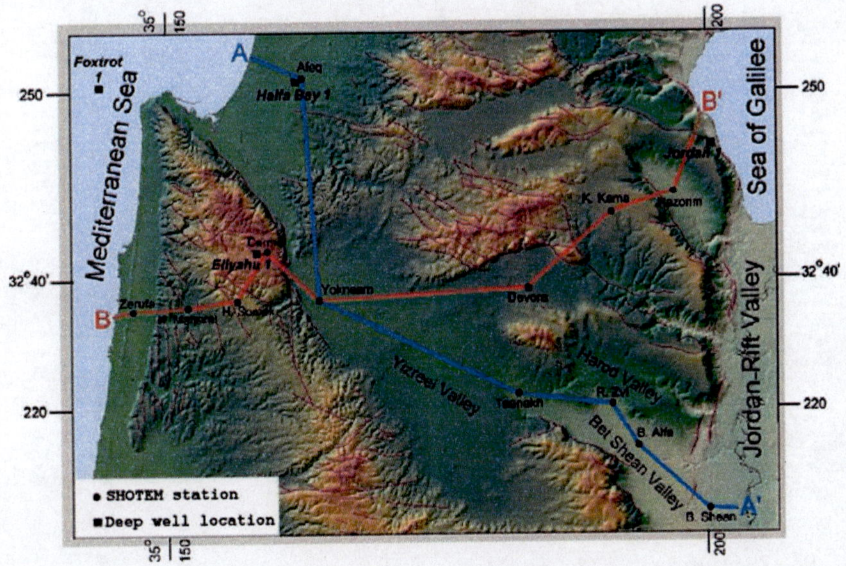


**Fig. 63** Location map of the deep TDEM survey in northern Israel. Lines A-A′ and B-B′ designate the geophysical traverses (modified after Kafri et al. 2007)

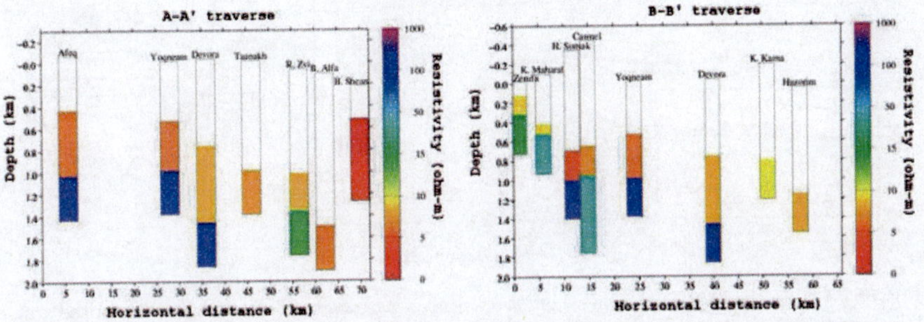


**Fig. 64** Columnar resistivity cross-sections obtained by the TEM (mainly by LOTEM) measurements along traverses A-A′ and B-B′. The shallower resistivity sequence obtained by SHOTEM appears in blank (modified after Kafri et al. 2007)

(1967) and Kafri and Arad (1979). The steepness of the interface between the Bet Alfa and Bet Shean sites (Fig. 65) is controlled, as expected, by the very steep fresh groundwater flow gradient between those sites.

### The Zerufa–Hazorim traverse (B-B′) (Fig. 66)

This traverse is running from the Mount Carmel coastal plain across the uplifted Mount Carmel block, along the northern valleys, crossing a structurally more elevated area close to the Sea of Galilee. The obtained configuration of the interface is as follows:

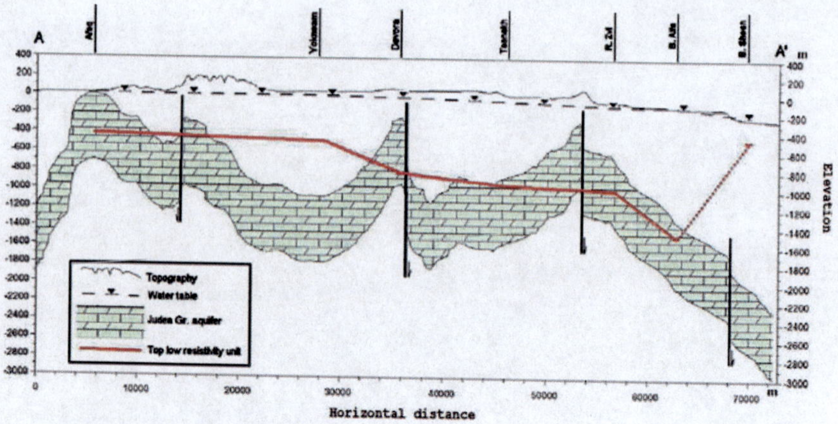


**Fig. 65** A simplified hydrogeological cross-section along traverse A-A′ with superimposed water table and top of the deep conductor detected by TEM and identified with fresh/saline groundwater interface (modified after Kafri et al. 2007)

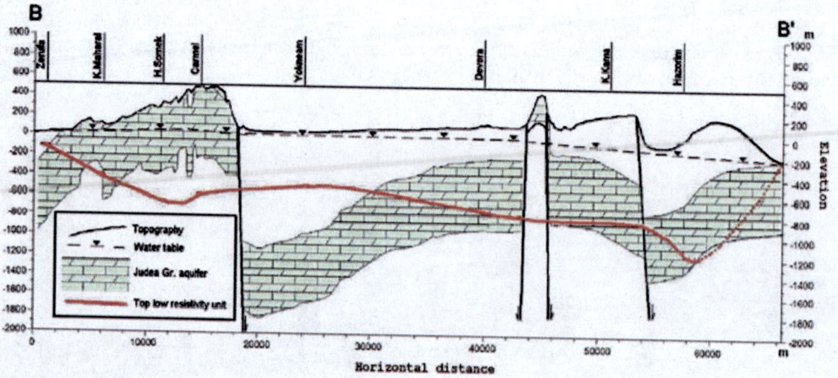


**Fig. 66** A simplified hydrogeological cross-section along traverse B-B′ with superimposed water table and top of the deep conductor detected by TEM and identified with fresh/saline groundwater interface (modified after Kafri et al. 2007)

Groundwater level in the Mount Carmel block declines from the divide to the west and as a result, the interface declines eastward in accordance with the Ghyben–Herzberg approximation. From the divide eastward and farther to the Yoqneam site, the interface starts, as expected, to rise due to the drop of the groundwater level east of the divide. Farther east, groundwater levels mildly decline to slightly below sea level, accompanied by a mild decline of the interface. Close to the DSR, where the groundwater level gradient steepens, a steeper decline of the interface takes place accordingly. Due to the absence of TDEM measurement and hydrological information east of the Hazorim site, the extrapolated interface is shown as a dashed line in Fig. 66. The obtained configuration provides a picture similar to the one obtained in the Bet Shean site in traverse A-A′.

All the above apparently support the proposed idea of subsurface seawater intrusion from the Mediterranean to the DSR. In addition, a hydrological model, carried out is also in accordance with the above described results (Kafri et al. 2007).

## 6.3 *Mediterranean—Dead Sea Traverse*

The subsurface of the Be'er Sheva valley between the Mediterranean Sea and the DSR occupies a relatively shallow and rather fresh groundwater system, which is underlain by deeper saline water bodies The different deep-seated saline sources were discussed and interpreted by Starinsky (1974), Nativ (1984) and Fleischer et al. (1997). Two basic types of brines were proposed as follows:

1. The C-type brine with rNa/rCl ratio similar to the marine one (0.82–0.86), found in the western part of the study area and
2. The R-rift type of Ca-chloride brine with rNa/rCl ratios below the marine ones found mostly in the DSR.

Stanislavsky and Gvirtzman (1999), demonstrated both the distribution of the above brine underneath the fresh groundwater system. They showed that the R-type brine is being density driven from the DSR westward to underneath the C-type brine. According to their model, the fresh groundwater system close to the DSR directly overlies the westward intruding Dead Sea brine (hereafter DSB) of the R-type.

Regarding geoelectric studies, shallow TDEM measurements managed to detect the interface between a fresher water body and the DSB body in the Dead Sea coastal plain aquifer (Kafri et al. 1997).

A deeper SHOTEM survey, aimed to the JGA, adjacent to the Dead Sea coastal plain (Levi et al. 2008), exhibited a different sequential salinity configuration from the one described by Stanislavsky and Gvirtzman (1999). The upper, high-resistivity freshwater body was found to be underlain below the theoretical depth of the freshwater-DSB interface, by a water body with resistivities higher than the expected ones for the latter. This could hint to the existence of an additional intermediate normal intruded seawater body from the west, between the upper freshwater body and the lower DSB one, as proposed for northern Israel (see Sect. 6.2).

In order to check this hypothesis, a deep TDEM survey from the Mediterranean Sea to the DS was executed, combined with geochemical considerations. The results of this study (Kafri et al. 2014) suggest that the hydrogeological setup is basically similar to the one discussed one in the northern valleys. It is assumed, contrary to previous assumptions, that the JGA is not blocked to the Mediterranean Sea, enabling seawater intrusion into it. However, unlike the case in the north, the DSR contains here a concentrated brine.

In order to understand the type and genesis of the different groundwater bodies in the study area and to delineate their spatial distribution, two roughly WNW-ESE directed TDEM traverses were executed between the Dead Sea and the Mediterranean coastal plain (Figs. 67 and 68).

The traverses consist of 30 deep TDEM measurements located as close as possible to the existing wells for calibration (Fig. 68). Unfortunately, the TDEM traverses could not be reaching the west to the proximity of the Mediterranean shoreline since the shallow subsurface of the western coastal plain consists of a thick Neogene shale sequence of a very low resistivity which electrically masks the underlying

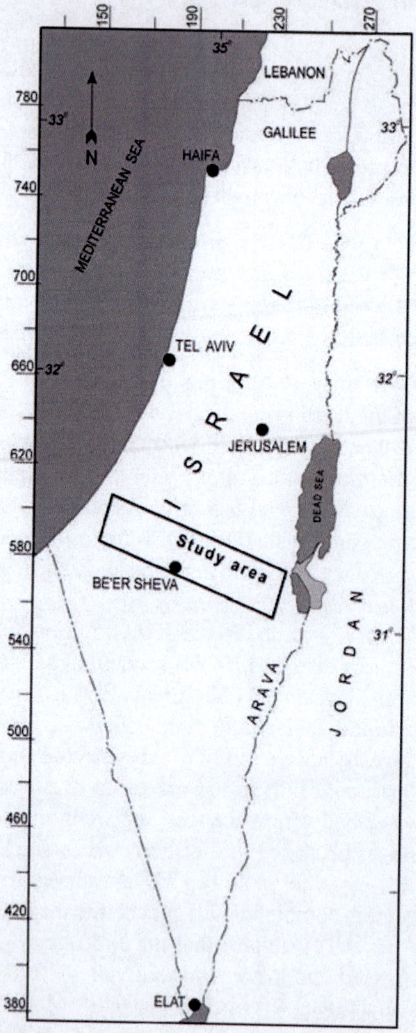


Fig. 67 Location map of the study area

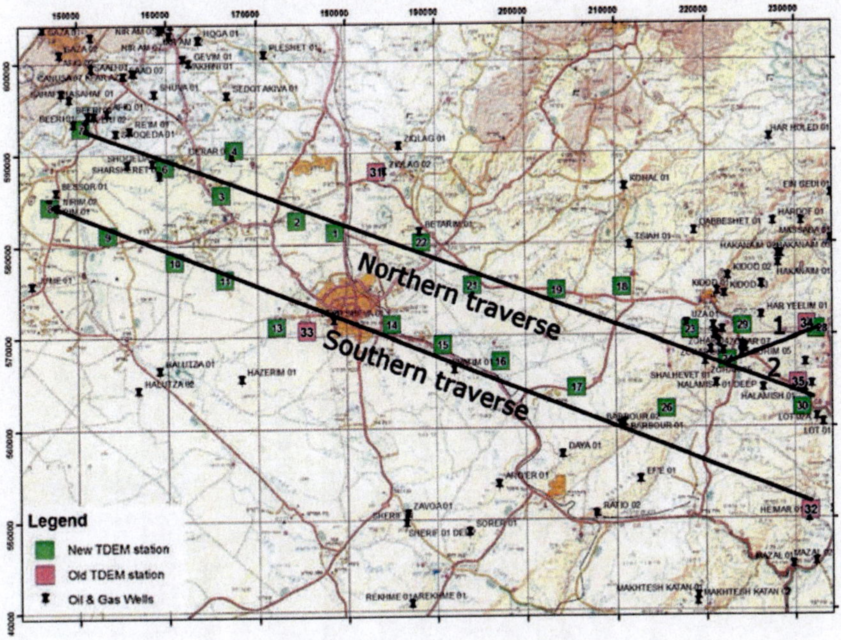


**Fig. 68** Location map of the TDEM traverses (after Kafri et al. 2014)

depositional sequence. Hence, measurements were carried out only at sites, where the thickness of the shale sequence does not exceed some 250 m.

To enable a reasonable interpretation a known resistivity–salinity relationship was employed based on the following:

1. In the Mediterranean coastal aquifer with porosities between 0.2 and 0.4, the bulk resistivity of the seawater-intruded portion is known to be between 1 and 3 ohm-m with the mean value of approximately 1.7 ohm-m (Goldman et al. 1991, 2003: Kafri and Goldman 2006).
2. The Dead Sea coastal aquifer that is intruded by the concentrated DSB exhibits resistivities well below 1 ohm-m (Kafri et al. 1997).
3. Seawater-intruded carbonate aquifers in Israel (see also Sect. 7) are characterized by higher resistivities as compared to the granular aquifers (approximately, 5 ohm-m) due to their lower porosities (Goldman and Kafri 2004; Kafri and Goldman 2005).

The results of the TDEM campaign are shown on three columnar resistivity cross-sections along the measured traverses (Figs. 69, 70 and 71). These include the TDEM resistivity models, the main geological units, the water table and the groundwater salinities in g/l TDS (in blue when rNa/rCl ratio matches the marine one and in red when it is considerably low). Also, the minimum calculated possible depth to the assumed deep-seated brines, which are located beyond the maximum exploration depth of the TDEM measurements, is designated by a dashed line. Bulk resistivities

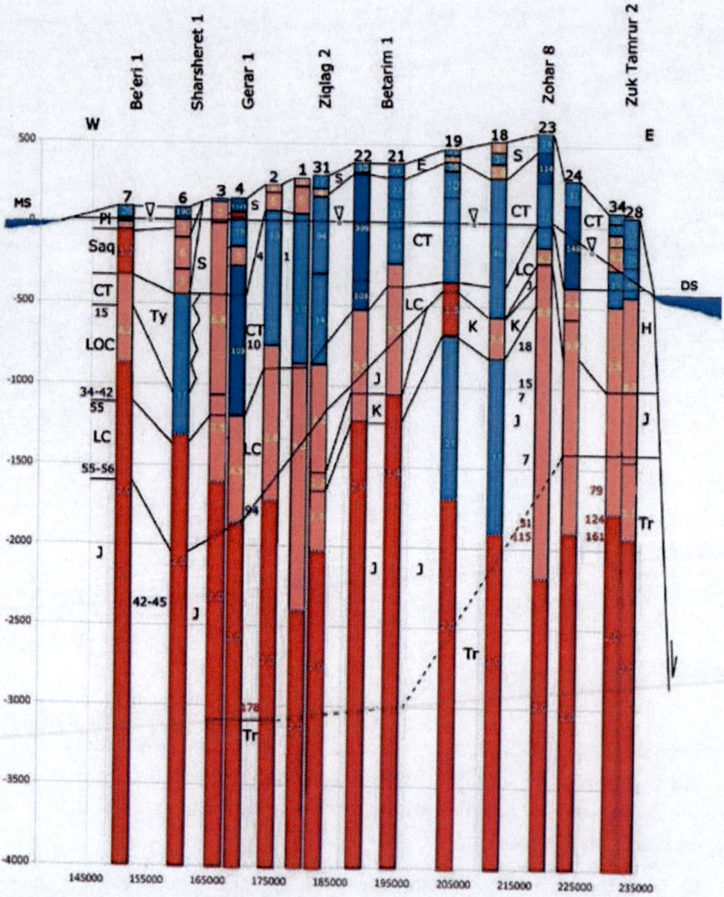


**Fig. 69** Northern-1 TDEM columnar cross-section (after Kafri et al. 2014)

exceeding 10 ohm-m, represent fresh to slightly brackish water and appear on the cross-sections in blue color. Those between 3 and 10 ohm-m, representing roughly seawater concentration, appear on the cross-sections in pink color. A diluted deep DSB, evidenced by hydrochemical information from below a depth of 2 km, appears in red color on the cross-sections. For the assumed 0.1 porosity, a roughly calculated 2 ohm-m bulk resistivity was obtained for such brines and at some sites, as was indeed revealed by the TDEM method in depth. At all other sites, where these brines were assumed to be found below the maximum exploration depth of the TDEM

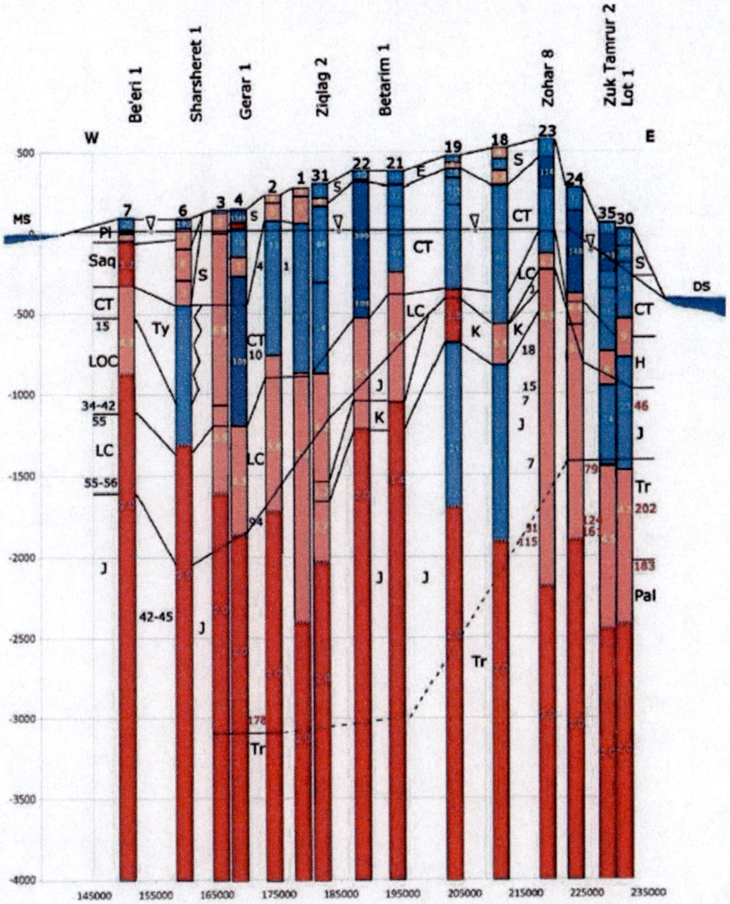


Fig. 70 Northern-2 TDEM columnar cross-section (after Kafri et al. 2014)

measurements, the 2 ohm-m resistivity was designated to a fictitious deep layer located at a minimum theoretically possible depth.

**Northern cross-sections 1–2 (Figs. 69 and 70)**

These two cross-sections are basically the same except that they end up at different locations in the easternmost part, close to the Dead Sea. Their proposed interpretation if as follows:

The uppermost high-resistivity fresh to somewhat brackish water body extends along the entire cross-sections from west to east. The depth of its base is assumingly controlled by the equipotential groundwater table at 10–20 m asl along most of the cross-section, except that it drops steeply to the DSR in the east. A relatively conductive unit (1.8–3.4 ohm-m) is recognized in measurements 18 and 19, "interbedded" within the uppermost high-resistivity layer most likely correlates well with the known

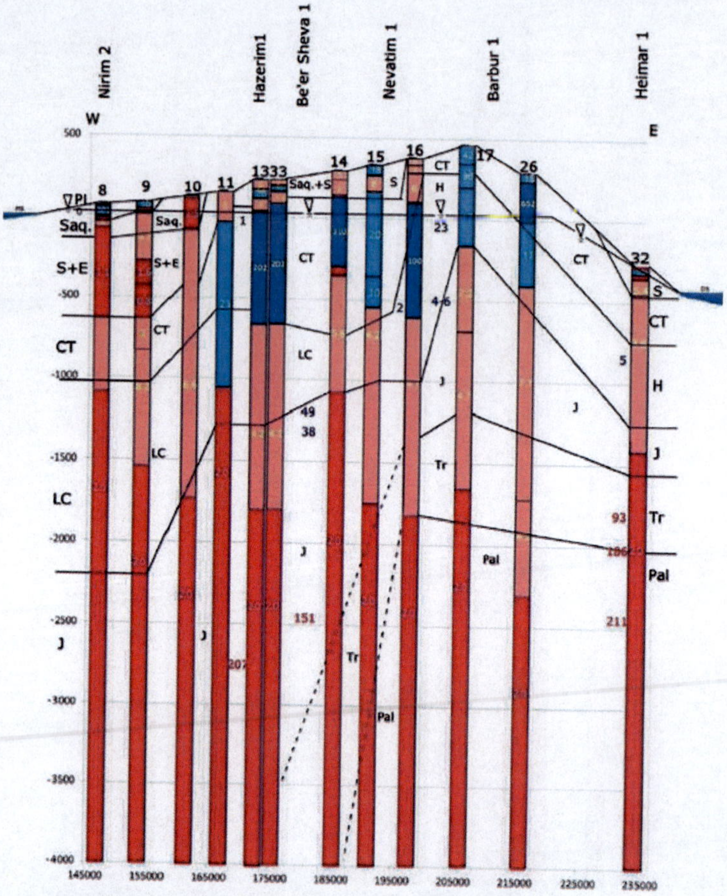


Fig. 71 Southern TDEM columnar cross-section (after Kafri et al. 2014)

conductive Jurassic Kidod shale Formation (designated K on the cross-sections) and known at the same depth interval. The electrical conductivity of this unit stems from its considerable amount of pyrite. Toward the Dead Sea in the east and the Mediterranean Sea in the west, the thickness of the uppermost high-resistivity body is being reduced considerably as expected toward both base levels on top of the declining interfaces underneath. In the west, its thickness is also reduced due to the existence of the overlying low-resistivity Neogene shale sequence.

The intermediate moderate resistivity layer, beneath the high-resistivity one, extends along almost the entire northern cross-section except for the more structurally elevated area in the north (measurements 18 and 19), which is outside the path of the assumed intruding seawater. Based on the hydrochemical interpretation, this layer represents a triple component mixture of deep-seated brines, intruding seawater and freshwater from above.

The deep and very low-resistivity brine body (≈2 ohm-m) was not detected in this cross-section except for two measurements (21 and 22). Its upper boundary matches the low-resistive Kidod Formation, whereby it assumingly masks the underlying sequence and, as a result, as an artifact, the brine body is artificially detected at a rather shallow level. Thus, the existence at depth of the brines in the west and east is detected, whereas in the middle of the cross-section it is mainly deduced from boreholes hydrochemical information.

Figure 72 represents a simplified northern cross-section that exhibits the main resistivity units.

**Southern cross-section (Fig. 71)**
The southern cross-section is located closer to the center of the Be'er Sheva Valley and thereby is less affected by the elevated higher groundwater levels to the north of it. It is thus most likely located entirely along the expected path of the intruding seawater.

Here too, the uppermost high-resistivity freshwater body is hosted in the Judea Group (CT) and the lower Cretaceous (LC, H) aquifers, extending along the entire

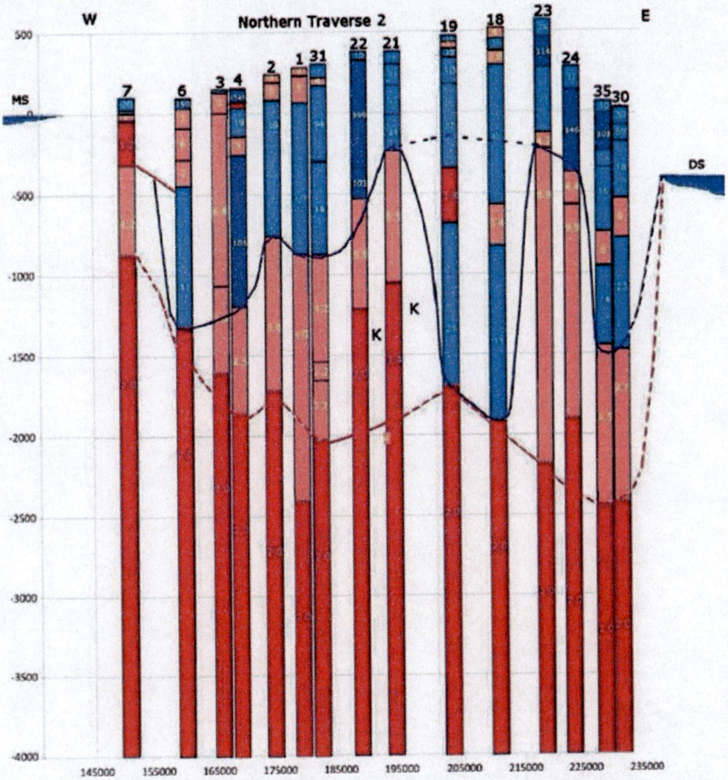


**Fig. 72** Simplified northern-2 TDEM columnar cross-section (after Kafri et al. 2014)

cross-section. Here too the depth of its lower boundary is controlled by the equipotential water table, as noticed in the northern cross-section. Similarly, its thickness is being reduced toward the Mediterranean and the DSR base levels as expected.

The underlying intermediate moderate-resistivity body also extends here along the entire cross-section.

The very low-resistivity unit (2 ohm-m) underneath the above, which assumingly represents the deep-seated brines, was detected here in TDEM measurements 11, 5, 17 and 32 whereby its uppermost theoretical possible depth was calculated for the rest of the measurements. Similarly to the northern cross-section, the upper boundary of the unit roughly exhibits the presumed inclined interfaces the eastern and western base levels (Fig. 73) except of measurements 11 and 14, where the anomalously shallow appearance of this boundary seems to be affected by the rather shallow Kidod Formation at this site and thus, the top of the very low-resistivity brine unit is expected to be, in fact, at a deeper level in accordance with the rest of the cross-section.

Figure 73 represents a simplified southern cross-section that exhibits the main resistivity units.

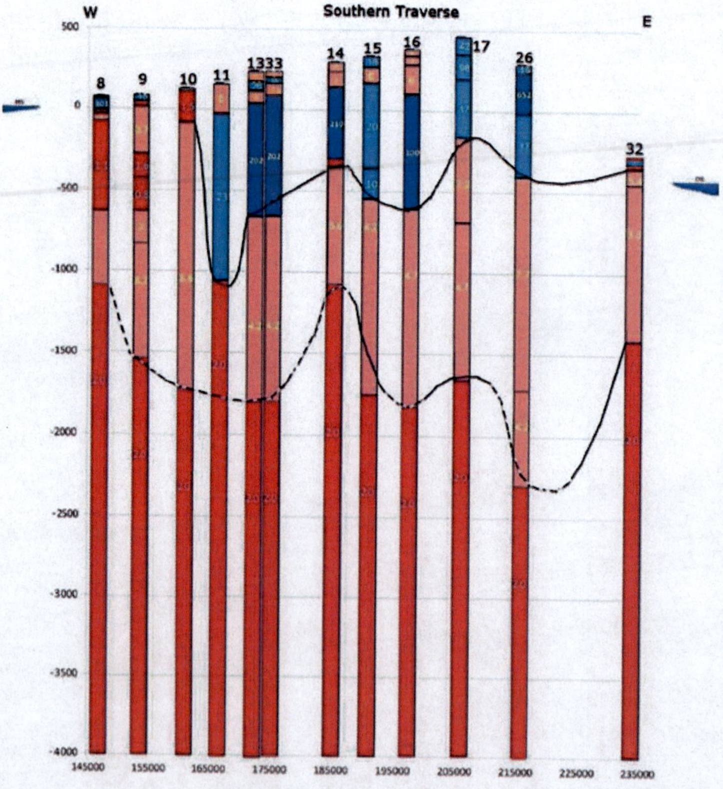


**Fig. 73** Simplified southern TDEM columnar cross-section (after Kafri et al. 2014)

Based on the TDEM results, the combined geochemical considerations and the employed hydrological FeFlow model, the discussed groundwater system includes the following resistivity/salinity units from top to bottom:

1. An upper high-resistivity fresh to brackish water unit, which flows to the western Mediterranean and eastern DSR base levels, occupied by the Judea Group and the Lower Cretaceous aquifers.
2. An intermediate moderate resistivity unit, interpreted also geochemically as a mixing zone of three components, namely deep-seated brines, seawater intrusion from the west and freshwater.
3. A lowermost, very low-resistivity body representing the C- and R-type brines, whereby the latter are density driven from the DSR westward.

## 6.4 Dead Sea—Gulf of Eilat Traverse

The Arava Rift Valley extends between the Gulf of Eilat (Aqaba) in the south and the Dead Sea in the north (Fig. 74). The main problem there is related to the salinization processes of the groundwater systems caused through different mechanisms as well as from deep-seated saline sources *r* as proposed (see below) in numerous studies. In most cases, the existing boreholes or geophysical soundings were not deep enough to penetrate or detect those deep saline sources. A deep TDEM survey was thus carried out along the entire Arava Valley, targeted to detect the deep saline sources, and to spatially delineate their extension and to reveal the genetic relationship between them. It also aimed to delineate the different base levels and paleo-base levels in the south and north. The results of the survey were described by Kafri et al. (2008).

The hydrogeology and groundwater chemistry of the Arava Rift Valley and its western margins were discussed in numerous studies (i.e., Bartov and Bein 1994; Bein et al. 1995, 1998; Fleisher et al. 1996, 1997; Guttman et al. 1999; Issar et al. 1972; Kroitoru et al. 1981; Kronfeld et al. 1992; Magal and Yechieli 2002; Rosenthal et al. 1992). A concise summary based on previous studies as well as on some of the present observations is given below:

The regional mountain aquifers include the Paleozoic to early Cretaceous Kurnub Group (or Nubian sandstone) aquifer (see also Chapter "The Nubian Sandstone Aquifer in the Sinai Peninsula and the Negev Desert") and the late Cretaceous Judea Group aquifer. The graben fill aquifer, mainly within the Rift, includes the Neogene to Recent Hazeva Formation, the Arava conglomerate and the younger graben fill.

The general groundwater flow is from both western and eastern mountain aquifers into the graben fill aquifer. A hydrological divide within the Rift Valley (see below) dictates the separation and diversion of the flows to the Gulf of Eilat base level in the south and to the Dead Sea base level in the north.

Regarding—salinity, the Kurnub and the JG aquifers contain only fresh to brackish waters along the entire Arava Valley and its margins. Within the graben fill aquifer, freshwater is found close to the divide and to the outlets of the large streams that drain from the west the northern part of the valley. Elsewhere, fresh to brackish waters

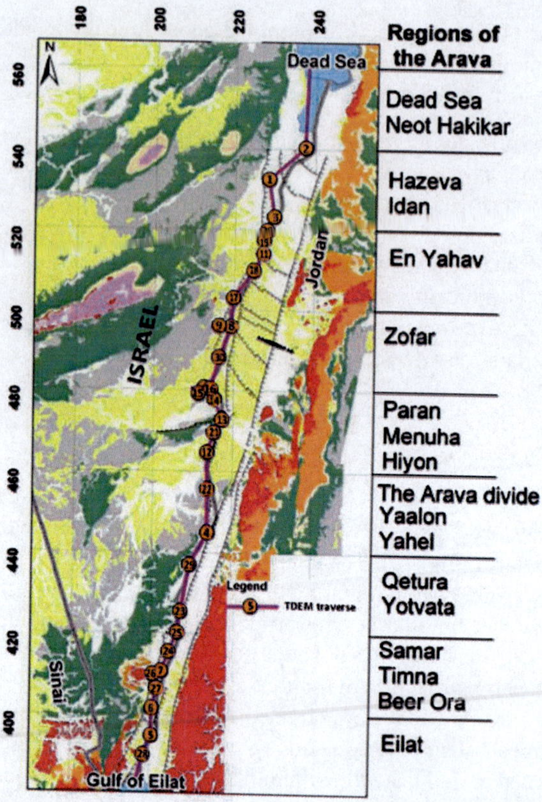


**Fig. 74** TDEM locations superimposed on the geologic map of the study area (after Kafri et al. 2008)

are abundant. Groundwater with seawater salinity is encountered in the immediate vicinity of the Gulf of Eilat, whereas concentrated brines, with up to 202 gr/l Cl, were detected in the Arava 1 deep well close to the Dead Sea. Superficial to shallow brines were encountered in the so-called Sabkha sediments at Yotvata (up to 100 gr/l Cl) and Evrona. The latter were formed by emergence of brackish waters from depth and their evaporation close to the surface (Michaeli and Galai 1963; Michaeli 1971; Kafri 1977).

Several shallow TDEM surveys, up to a depth of a few hundred meters, were executed in the past along the Arava Valley, the results of which are described by Bartov et al. (1998), Goldman and Ezersky (1999b), and Kafri et al. (2001). Geoelectromagnetic surveys conducted in the framework of the Dead-Sea-Rift-Transec (DESERT) project provided additional information regarding both shallow (TDEM) and deep (magnetotelluric) structures in the eastern margins of the central Arava valley (Ritter et al. 2003; Koch et al. 2004; Maercklin et al. 2005). In general, the shallow measurements revealed in both the southern and northern Arava, rather shallow, low-resistivity body interpreted as saline groundwater bodies of limited extension. In the north, in the vicinity of the Dead Sea those measurements detected a very low-resistivity unit (below 1 ohm-m) at a depth of several hundred meters. The magnetotelluric measurements in the eastern Arava valley revealed a considerable drop of resistivity to roughly 1–2 ohm-m at a depth of approximately 1.2 km.

The working hypothesis of Kafri et al. (2008) assumed that the regional base levels in the north and south served as sources of salinization in the past and at present. It was thus most important to reveal the salinization as well as flushing mechanisms of the system during the different time intervals of the system. Those mechanisms were suggested below for the conceptual model of the Arava Valley (Fig. 75).

The present base levels are at the elevations of 0 m and at 430 m bsl in the Gulf of Eilat and the Dead Sea, respectively. Based on their respective dated deposits, the Pliocene marine base level in the northern and southern Arava valley is assumed to be at 60–70 m asl, and the Quaternary Lisan lake base level, which existed only in the northern Arava valley at 180 m bsl. The marine base level, at the peak of the last glacial, all over the world and thus also in the Gulf of Eilat is at 120 m bsl. Current water table elevations are monitored and known. In the case of paleo-water tables, these were reconstructed, taking in account the location of the paleo base levels and the accepted assumed groundwater flow gradients, similar to the present day ones. The known saline water densities that occupy or were occupied in the different base levels are: 1.029 for normal seawater, 1.115 for the Lisan lake brine and 1.23 for the present Dead Sea brine. The configurations of the interfaces for each paleo-stage

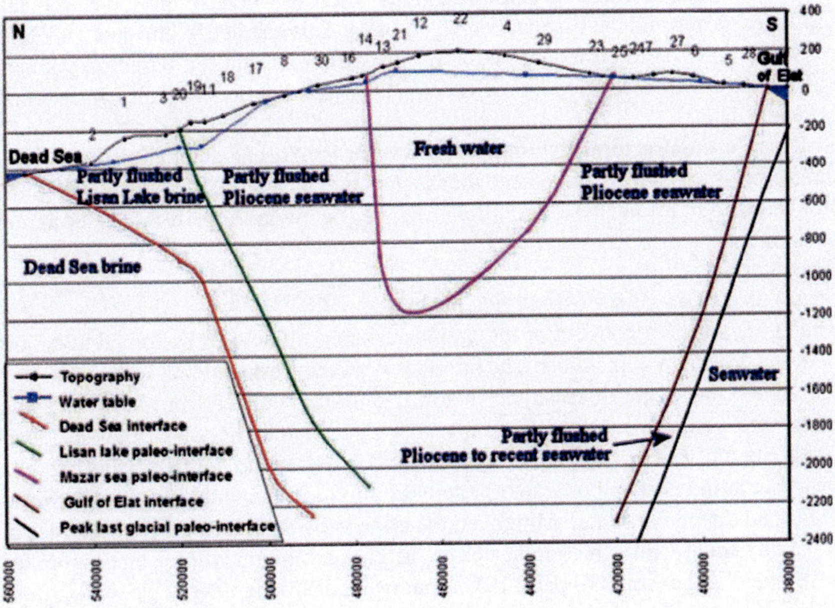


**Fig. 75** A N-S cross-section of the conceptual model (after Kafri et al. 2008)

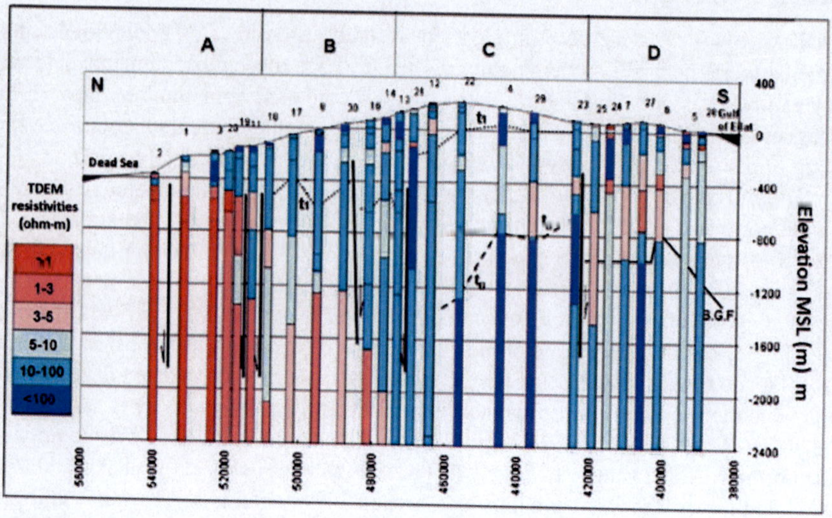


**Fig. 76** A transversal columnar TDEM resistivity cross-section between the Gulf of Eilat and the Dead Sea. tB—top basement, tJ—top Judea Group, B.F.G—base graben fill (after Kafri et al. 2008)

were calculated using the Ghyben–Herzberg approximation. Based on the results of the above mentioned deep TDEM survey along the Arava valley, a cross-section between the Gulf of Eilat and the Dead Sea was constructed (Fig. 76). The cross-section exhibits, in addition to the resistivity intervals, the/also the top Basement, top Judea Group and the base graben fill aquifers. This enabled to attribute a specific resistivity interval to either a granular or to a carbonate aquifer, based on porosity considerations as discussed below (Sect. 7). The correlation was also insisted on borehole data.

Based on the structural, hydrological and the resistivity characteristics, the longitudinal cross-section was divided into four segments or basins (A, B, C and D in Fig. 76) from north to south and the resistivities, discussed below, are already translated herein to salinities:

### Segment A (The Hazeva-Dead Sea basin)

The entire segment consists of the granular graben fill aquifer bottomed at depths between 2500 and 4000 m bsl. The interface between fresh and brackish waters and the underlying, brine of resistivity below 1 ohm-m, declines from north to south as encountered at measurement points (mp) 2, 1, 3, 20. The top of the very low-resistivity unit reaches a depth of at least 1400 m bsl. In mp 3, the brine is overlain by a 2 ohm-m resistive body, interpreted as a mixing zone with fresher water or as unflushed trapped seawater. Moreover, the brine body exhibits a gradual increase of resistivity southward, which indicates an increasing dilution with distance from the base level. Farther south (mp 11, 19), a southward declining interface between fresh to brackish waters and underlying water of approximately seawater concentration, is encountered.

**Segment B (The En Yahav-Zofar basin)**

This segment is aligned at the margins of the Rift where by the soundings penetrated only the mountain aquifers. The Judea Group carbonate aquifer contains fresh to brackish waters in its upper part, underlain in places (mp 17 and 8) by water of seawater concentration. The upper Kurnub Group granular aquifer below it contains waters of seawater concentration.

**Segment C (The Menuha-Qetura basin)**

This segment partly coincides with the central structural divide of the Arava Valley, where the Rift narrows and the graben fill aquifer, overlying the Judea Group aquifer, is very thin. The top of the crystalline basement here (mp 22, 4, 9 and 23) is relatively shallow, between 600 and 1200 m bsl, as estimated by downward stratigraphic extrapolation and through geoelectromagnetic measurements (Kafri et al. 2001; Ritter et al. 2003).

In its northern part, the segment is characterized by the absence of low-resistivity saline water along the entire penetrated sequence, supported by the occurrence of only freshwater encountered in the Yaalon 7 borehole adjacent to mp 4. Farther south (mp 23, 29), high (freshwaters) resistivities are encountered, also in complete agreement with adjacent boreholes both within the Judea and the Kurnub Group aquifers. However, contrary to the northern part of the segment, here those high resistivities are underlain by lower resistivities (brackish water) within the Kurnub Group aquifer.

**Segment D (The Yotvata-Eilat basin)**

This segment is aligned in the graben fill aquifer that thickens southward. Thin and low-resistivity bodies are encountered up to the depth of 100 m. These are interpreted herein as saline water bodies, related to existing or ancient sabkhas or to alternatively to unflushed clay layers (Kafri et al. 2001).

The upper high-resistivity (freshwater) body is not thick, and is bottomed by an interface declining from 100 to 600 m bsl in a south–north direction, and the underlying water body exhibits brackish to seawater concentrations. Moreover, the current seawater intrusion here from the south is limited to the close vicinity of the Gulf of Eilat, and it only occupies the uppermost graben fill aquifer (Bein et al. 1998). The underlying high resistivities are interpreted herein as the crystalline basement since no freshwater is expected at such depths below the saline ones.

No deep, very low-resistive brines have been encountered in the entire segment despite the fact that saline water in the nearby Timna site were interpreted as originating from deep-seated brines. Based on the above, an alternative interpretation suggests herein that the latter are density-driven brines downward, from the nearby existing or ancient surficial hallow sabkhas.

It is, thus, concluded that the overall salinity distribution configuration along the Arava Valley, as deduced from the longitudinal TDEM cross-section (Fig. 76) is basically in agreement with the one shown in the conceptual model (Fig. 75), Kafri et al. (2008).

# 7 Evaluating Porosity/Storativity of Saline Ground Water-Saturated Aquifers Using TDEM Method

## 7.1 General

Contrary to groundwater salinity and geometry of hydrogeological targets, the determination of hydraulic properties of porous media, using conventional geophysical methods, is highly inaccurate and ambiguous. This is caused by an inherent non-uniqueness of the geophysical interpretation accompanied by numerous uncertainties in mostly empirical relationships between hydraulic and geophysical parameters (Archie 1942; Winsauer et al. 1952; Glover 2016). Therefore, as mentioned before, during the last two decades major efforts were made in the development of novel geophysical methods (e.g., borehole tomography, SNMR, seismoelectric, electro-seismic methods, etc.) as well as in the adaptation of some conventional geophysical techniques (e.g., induced polarization (IP) and self-potential (SP)), which are based on studying parameters more closely related to porosity, hydraulic conductivity and transmissivity of the aquifers than traditional electric resistivities or seismic velocities (e.g., Garambois and Dietrich 2001; Rubin and Hubbard 2005; Linde et al. 2006; Hördt et al. 2007). Unfortunately, most of these studies are still experimental and some are only performed on samples in labs or using synthetic modeling (Binley et al. 2005). In fact, among surface geophysical methods, only SNMR has an extensive record of the successful field applications for direct detection of the water content and, to some lesser extent, the transmissivity of aquifers (Yaramanci 2003). However, taking into account the very limited penetration depth and high sensitivity to ambient EM noise, the application of SNMR in Israel seems to be questionable.

An alternative approach to obtain the aquifer's porosity or storativity was to employ the TDEM method to aquifers that occupy saline water of a known salinity (Kafri and Goldman 2005). The approach is based on the use of conventional GEM methods and particularly recommends to use the TDEM method, which allows to overcome the above mentioned limitations of the novel techniques.

Similar to previous conventional studies, the theoretical basis of the described approach is being constituted by the empirical Archie's equation (Archie 1942):

$$\rho_w/\rho = \alpha\Phi^m \tag{1}$$

where $\rho_w$ is the resistivity of water within the pore space, $\rho$ is the bulk resistivity of the rock, $\Phi$ is porosity of the rock that approximately represents the volume of water filling the pore space, $\alpha$ and $m$ are material-dependent empirical parameters, which are introduced to force the equation to fit the behavior of a rock in question.

Since $m$ increases with cementation, Archie (1942) named it *cementation index*. Its value varies from 1.3 for unconsolidated sands to 1.8–2 for consolidated sandstones.

The most widely used simplified form of the Archie's equation for sandstone aquifers is obtained from (1) by letting $\alpha = 1$ and $m = 2$. Thus,

$$\rho_w/\rho = \Phi^2 = 1/F, \tag{2}$$

where $F$ is the so-called formation factor, related to the porosity for $m = 2$ and $F = 1/\Phi^2$.

Even in such an oversimplified case, the required porosity cannot be determined from Eq. (2) without a priori knowledge of the water resistivity, $\rho_w$. Moreover, the bulk resistivity of rocks, $\rho$, which is the only parameter provided by surface GE and GEM measurements, in many cases is ambiguously determined due to the so-called equivalence problem (e.g., Kafri and Goldman 2005).

The resistivity of the pore water, $\rho_w$ is customarily estimated using borehole salinity measurements. Also, an ambiguity in geophysical interpretation is mainly based on using independent borehole information. Unfortunately, in most cases, the number of existing boreholes is insufficient to provide accurate estimates of the both parameters Therefore, most of the attempts to apply surface GE and GEM methods to determine hydrological parameters in general and porosity in particular suffer from one common drawback: They are apparently "successful" in close proximity to boreholes, but could be rather unreliable far away from them. Moreover, even in the proximity of boreholes, the required a priori information provided by well logs is inevitably scale dependent and, thus, does not necessarily coincide with the required bulk geoelectric parameters, as will be discussed below. The scale dependency phenomenon, which is less pronounced in the case of granular clastic aquifers, becomes crucial for fractured, cavernous, karstic or crystalline aquifers.

In order to significantly reduce ambiguities of both hydrogeological and geoelectrical interpretations in case of coastal aquifers, particularly, in view of scarce amount or even a total absence of boreholes, Goldman and Kafri (2004) suggested to use properly selected GE and/or GEM techniques for detecting the fresh/saline groundwater interface, followed by application of the Archie's equation only to the seawater-intruded portions of the aquifer.

By using this approach, one can exactly determine one of the parameters in the Archie's equation and significantly improve the accuracy of the other. Indeed, the pore water resistivity, $\rho_w$ is exactly known for the seawater salinity (i.e., Repsold 1989) and the use of the TDEM method allows to significantly reduce the ambiguity in determining the bulk resistivity, $\rho$, of the seawater-intruded portions of the aquifer (discussed below). Then, by applying Eqs. (1) or (2), one can more accurately calculate the porosity of the saline water-saturated portion of the aquifer. Since hydraulic properties of the aquifer should not depend on the location of the fresh/saline groundwater interface, one might expect that the porosity determined below the interface adequately represents the entire aquifer or sub-aquifer, in which the interface was detected.

It is important to emphasize that the approach in question is particularly feasible in complicated groundwater systems including partial saturation or clay conductivity effect. Such systems are described by more general equations, connecting fluid and electrical properties, such as the empirical Schlumberger equation or the theoretical Clausius–Mossotti equation, the application of which require additional a priori information (Wempe 2000). In brine-saturated aquifers, however, where the clay conductivity effect becomes insignificant (Hoyer and Spann 1975), the amount of the required independent information can be essentially reduced.

The proposed methodology is particularly feasible in coastal aquifers, where $\rho_w$ of the intruding seawater is well known (Repsold 1989) and where the normally wide extend of the intrusion allows the gathering sufficient amount of GE/GEM data to be statistically representative of the entire aquifer.

The main requirement for a GE/GEM method is the ability to accurately detect the fresh/saline interface and to determine the bulk resistivity of the aquifer below the interface. Experience shows that, in most cases, the TDEM method could be a right choice for this purpose (Goldman et al. 1991). However, the application of TDEM by itself is sometimes limited, due to technical reasons, such as a high level of ambient EM noise (in urban areas) or laying down transmitter loops, in a rough topography. In such cases, the 2D resistivity imaging method seems also to be feasible for this purpose. In addition, the latter can be used either instead of, or along with TDEM method in areas with expected sharp lateral variations of the subsurface resistivity.

The most important feature, which can significantly affect the ability of a geophysical method to quantitatively estimate the hydraulic properties of the aquifer, is an inherent non-uniqueness of the geophysical interpretation. The following example demonstrates how crucial is this feature for the accuracy in estimating porosity of the coastal aquifer. Figure 77 shows TDEM data collected by the Geonics EM-67 system (open squares at the left-hand side) along with a family of the equivalent interpreted models (at the right-hand side of the figure). The model, drawn by a solid line, represents the so-called best-fit model, whereas the rest of the models, shown by dashed lines, exhibit the equivalent models. At the left-hand side, the solid line represents a family of synthetic responses for both the best-fit and the equivalent

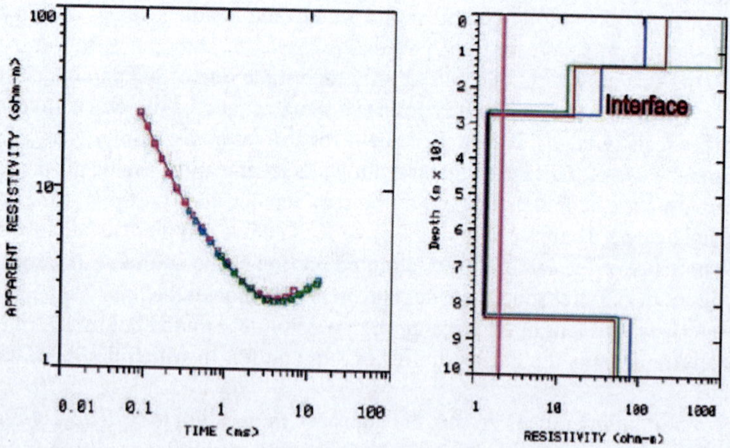


**Fig. 77** TDEM data collected at station HS557 in the coastal aquifer, and the interpretation, including linear equivalence analysis by means of TEMIX-XL interpretation software (modified after Kafri and Goldman 2005)

models. The difference between the responses is so small that they all appear as a single solid line. In this specific example, the best-fit model provides a misfit error of approximately 1.9%, while misfits for all the equivalent models do not exceed 2%. Thus, the maximum difference between all the synthetic responses does not exceed 0.1%, which is considerably less than the overall noise level of the measurement. Thus, in practical terms all the interpreted models are equally consistent with the data and cannot be distinguished without a priori information.

In this example, both the data interpretation and the equivalence analysis have been carried out using TEMIX-XL interpretation software (Interpex 1996). It should be noted that the software does not provide a full (nonlinear) equivalence analysis, but only the one, which is based on the use of a parameter resolution matrix (Goldman et al. 1994b). Nevertheless, for illustration purposes, it is enough to consider the partial equivalence analysis since in reality the variations of some model parameters could be even wider than those shown in Fig. 1. Some further details regarding the difference between linear and nonlinear equivalence problems can be found in Goldman et al. (1994b).

The equivalence analysis of the TDEM measurement (Fig. 77) exhibits the following resistivity sequences:

It should be noted that since the estimated depth equivalence range (Fig. 77) is considerably narrower than that of the resistivities, the depth values in the above table are taken for the best-fit model only.

## 7.2 Porosity Calculations

The above example (Fig. 77) is used to describe the methodology for calculating porosity of saline water-intruded aquifers.

Based on the fact that bulk resistivity values below 2 ohm-m are solely indicative of seawater intrusion into the Mediterranean coastal aquifer (Goldman et al. 1991), the location of freshwater/seawater interface was determined at the depth of approximately 26 m. The high-resistivity sequence from the surface to the depth of 14 m represents both the unsaturated zone and the underlying freshwater-saturated aquifer. The thickness of the unsaturated zone at this site is only 2 m, and therefore, it is not taken in the further considerations. In fact, no resistivity drop is detected at the depth of the water table, apparently due to the limited resolution of the TDEM method. The resistivity values within the intermediate zone between 14 and 26 m are typical for either an aquiclude, an aquitard or an aquifer saturated with brackish waters, with a variable salinity.

The very low-resistivity geoelectric layer below 26 m depth is undoubtedly identified with seawater intrusion. By applying Eq. (2) to this layer, one can accurately calculate the porosity of the seawater-intruded portion of the aquifer by using solely the TDEM data without any additional borehole information. Indeed, the salinity of the Mediterranean water is known, and the calculated fluid resistivity ($\rho_w$), according to Repsold (1989) is 0.18 ohm-m. Similar values were also obtained from measured boreholes conductivity logs (Friedman 2003, personal communication). Thus, the porosity of the seawater-intruded part of the aquifer, as calculated using Eq. (2), is $\Phi = 0.36$.

Similar porosity calculations for the shallow-resistivity sequence led to a great uncertainty due to the wide equivalence range of resistivities ($\rho$). Even when fixing $\rho_w = 10$ ohm-m, which is typical for freshwater-saturated aquifers in the area, the calculated porosities are found to vary in a wide range between 0.13 and 0.35, due to the geophysical non-uniqueness only.

The uncertainties could become even greater in the intermediate resistivity sequence, where water salinity varies in a wide range and thus, accordingly leading to a wide range of possible $\rho_w$ values. Since the resistivity, in this sequence, also $r$ varies due to equivalence (Fig. 77), the resulting uncertainty in the calculated porosity might become significant.

Since it is reasonable to assume that the porosity of the aquifer does not significantly depend on the location of the fresh/saline water interface, the suggested approach can be also used to resolve non-uniqueness in the upper resistivity sequence. Indeed, by applying Eq. (2) for $\Phi = 0.36$ and $\rho_w = 10$ ohm-m, one obtains $\rho =$ 77 ohm-m, which is fairly similar to the lowest resistivity value in the equivalence range of this sequence.

Some additional semi-quantitative information, regarding the groundwater salinity, can be also obtained for the intermediate resistivity zone above the interface. Indeed, in order to obtain a porosity value of 0.36, $\rho_w$ must vary in a certain range, corresponding to the equivalence range of $\rho$ in this zone. Applying the Archie's Eq. (2), and considering the possible resistivity range for the intermediate layer (Fig. 77 and the above Table), one can obtain $\rho_w$ values varying between 1.4 and 3.5 ohm-m. This resistivity range roughly corresponds to salinities varying between 300 and 3000 mg Cl/l (Repsold 1989).

The applicability of the proposed methodology has been tested in different coastal aquifers of Israel by comparing TDEM results with borehole data. The study focused mainly on the Mediterranean coastal aquifer, where the greatest amount of both geophysical and borehole data is available. In addition, the analysis also included coastal aquifers of the Gulf of Eilat and the Dead Sea, as well as the regional carbonate Judea Group aquifer.

**The porosity of different types of aquifers**

The total porosity of the aquifer includes both, the connected pores (named also as the aquifer's storativity) and the unconnected pores. The total porosity is determined usually in a laboratory from core samples, whereas storativity is obtained from hydrological pumping interference tests, which are rarely available in most cases.

Two main types of existing aquifers are granular (Fig. 78a) and hard rock aquifers, where the porosity is mostly gained by solution (Fig. 78b) or fracturing (Fig. 78c).

The porosity of granular aquifers is mostly inter-granular and the aquifers are usually moderately heterogeneous. Therefore, a porosity value obtained from a core sample approximately represents larger volumes of the aquifer. Contrary to the above, in carbonate aquifers, most of the hydraulic conductivity, as well as storativity, is

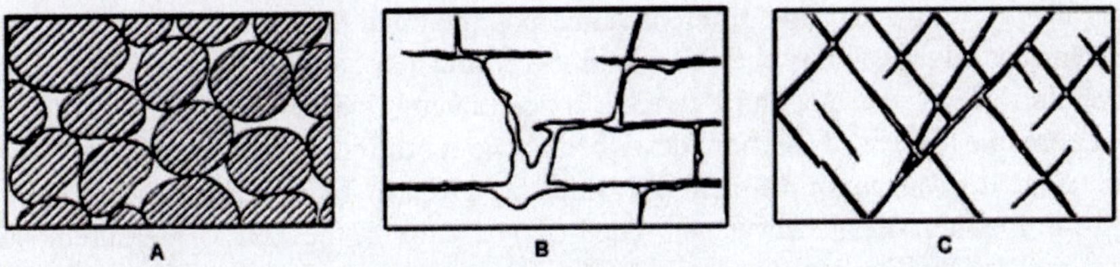


**Fig. 78** Diagram showing different types of rock interstices and the relation of rock texture to porosity. **a** Well-sorted sedimentary deposit having high porosity; **b** rock rendered porous by solution; **c** rock rendered porous by fracturing (after Meinzer 1942)

related to open fissures and solution channels, and the aquifers are often extremely heterogeneous. In many cases, an aquifer might have a double or even triple porosity system, which includes both the initial matrix porosity and the secondary fissured and solution porosity. As a result, the hydraulic parameters obtained in carbonate aquifers are usually scale dependent and often show preferable spatial parameters with increasing test radii, from a core sample to a pumping test and toward a regional study (Rover and Cherkauer 1995; Zuber and Motyka 1998; White 2002).

It is thus essential to know, particularly for carbonate aquifers, whether a given porosity value represents a small or a large aquifer volume. Furthermore, when a resistivity value is used for calculating porosity, it should be considered whether it is taken from a borehole log representing a small radius and thus lower volumes, or from TDEM measurements that integrate larger aquifer volumes and thus representing assumingly higher porosities.

Porosity and storativity, both measured and calculated values for different granular and carbonate aquifers of Israel are discussed herein, and are summarized in Tables 1–2.

Storativities in coastal aquifers, based on pumping tests or models, vary between 0.25 and 0.4 representing only the aquiferous portion of the sequence. It is assumed that the storativities in this portion of the aquifer are not quite similar to those of the total porosity of the sequence. The situation, however, is different within the clayey aquicludal interlayers, which might possess a high non-effective porosity that contains trapped and unflushed fluids.

The storativities of carbonate aquifers mostly vary between 0.03 and 0.08, since the initial sedimentary (matrix) porosity is small, and most of the storativity is related to secondary porosity gained by fracturing and solution. In places, like Mount Carmel, where a reef facies is known in both surface and subsurface, the initial porosity and as a result the storativity are high, attaining a value of 0.15. In the case of the discussed carbonate aquifers one should consider, in addition, the large aquifer volume with a small initial non-effective porosity.

**Calculated versus actual resistivities in Israeli aquifers**

Using the simplified Archie's Eq. (2), one can calculate the expected bulk resistivities of the different aquifers, and based on their related porosities and hosted saline waters compare them with the actually measured resistivities by the TDEM method.

The $\rho_w$ values, for the different saline potential end members, can be obtained from a resistivity/salinity chart (Repsold 1989), and they are 0.18 and 0.04 ohm-m for Mediterranean seawater and Dead Sea concentrated brine, respectively. The ranges of porosities ($\Phi$) used are those described above for different aquifers. The resultant calculated $\rho$ values for different porosities are given in Table 2. It can be seen that in the Mediterranean seawater-intruded coastal aquifer, the TDEM measurements yielded resistivities between 1 and 2 ohm-m, which are similar to the ones shown in Table 2 for the porosity values greater than 0.3. One can note the difference between the lower and the upper parts of the Mediterranean coastal aquifer with regard to the expected and actually measured bulk resistivities. The reason is that in the lower sub-aquifers, the ratio between the cumulative thickness of the clayey interlayers and that of the aquiferous ones is higher, as compared to the situation in the upper sub-aquifers. Therefore, in some cases, where freshwaters were sampled in the aquiferous parts there, TDEM bulk resistivities were below the expected values. It is reasonable to assume that in these cases, the TDEM measurements integrate both aquiferous and clayey layers, which apparently host trapped and unflushed saline water. This

**Table 1** Storativity and porosity of granular and karstic aquifers

| Aquifer | Location/basin | Storativity | References |
|---|---|---|---|
| Coastal (granular) | Shiqma | 0.3–0.4 | Sellinger and Kapuler (1972) |
| | Nizzanim | 0.3–0.4 | Sellinger and Kapuler (1973) |
| | Western Galilee | 0.25–0.35 | Sellinger and Levin (1977) |
| | General | 0.25 | Mercado (2000) |
| | Dead Sea | 0.012 | Hirsch (1975) |
| Judea Group (carbonate) | Naaman | 0.03–0.08 | Michaeli et al. (1973) |
| | Naaman | 0.06 | Mercado and Ben Zvi (1983) |
| | Kabri | 0.04 | Shachnai and Goldshtoff (1978) |
| | Yarkon–Taninim | 0.05 | Baida et al. (1978) |
| | Bet Shean | 0.03–0.07 | Goldshtoff and Shaliv (1979) |
| | Mt. Carmel Na'aman | 0.03 | Bar Yosef et al. (1997) |
| | Atlit 1 (Carmel) | 0.013 | Mercado and Ben Zvi (1983) |
| | N. Oren Carmel) | 0.15 | Mercado and Ben Zvi (1983) |
| | | Porosity | |
| | Mt. Carmel reefs | Up to 0.15 | Bein (1971) |
| | Ashqelon 3 | 0.15–0.20 | Bein (1971) |

**Table 2** Bulk resistivities calculated using Archie's equation for different porosities and water resistivities and actual resistivities measured by TDEM in different aquifers

| Aquifer and saline water body | $\Phi$ | $\rho_w$ (ohm-m) | $\rho$ (ohm-m) | |
|---|---|---|---|---|
| | | | Calculated | TDEM |
| Coastal, granular aquifer saturated with normal seawater | 0.2 | 0.18 | 3.6 | Between 1 and 2 |
| | 0.3 | 0.18 | 1.6 | |
| | 0.4 | 0.18 | 0.91 | |
| Coastal, granular aquifer saturated with Dead Sea concentrated brine | 0.2 | 0.04 | 0.81 | Between 0.2 and 0.8 |
| | 0.3 | 0.04 | 0.36 | |
| | 0.4 | 0.04 | 0.20 | |
| Judea Group carbonate aquifer saturated with normal seawater | 0.03 | 0.18 | 200 | Between 4 and 8 |
| | 0.05 | 0.18 | 72 | |
| | 0.07 | 0.18 | 37 | |
| | 0.08 | 0.18 | 28 | |
| | 0.15 | 0.18 | 8 | |
| | 0.20 | 0.18 | 4.5 | |
| Judea Group carbonate aquifer saturated with Dead Sea concentrated brine | 0.05 | 0.04 | 16 | Between 5 and 10 |
| | 0.15 | 0.04 | 1.8 | |

In the calculations for granular aquifers, parameter $\alpha$ was chosen to provide a well-known empirical relation $F = 0.81/\Phi^2$ normally used for sands

results in the lowering of resistivities of the entire sequence. As described before, quite a similar phenomenon was encountered in the Gulf of Eilat coastal aquifer.

In the Dead Sea coastal aquifer, where the TDEM resistivities, well below 1 ohm-m, were measured for the density driven intrusion of the Dead Sea brine into the aquifer, similar values were calculated using Archie's equation (Table 2).

The situation is more complicated for the seawater-intruded Judea Group carbonate aquifer. The calculated resistivities for relatively low storativities (0.03–0.08) are significantly higher (between 28 and 200 ohm-m) than those measured by the TDEM method. Only for higher storativities (0.15–0.20), the calculated resistivities become closer to actual TDEM, or well log resistivities.

Deep TDEM measurements near Nahal Oren in the Mount Carmel area, where high storativities were previously obtained from pumping tests in boreholes, yielded bulk resistivities around 1 ohm-m, while elsewhere, these values for the saline water-saturated parts of the Judea Group aquifer varied approximately between 5 and 10 ohm-m (Goldman and Kafri 2002), assumingly due to lower storativities. It should be noted that in a few wildcats (Gaash 2, Netania 1 and Cesarea 3), carbonate sequences saturated with waters of normal seawater salinity (18,000–20,000 mg Cl/l) exhibited well log resistivities between 4 and 6 ohm-m, similar to those obtained by the deep TDEM survey (Kafri et al. 2007).

## References


Abelson M, Gidon B, Shtivelman V, Wachs D, Raz E, Crouvi O, Kurzon I, Yechieli Y (2003) Collapse sinkholes and radar interferometry reveal neotectonics concealed within the Dead Sea basin. Geophys Res Lett 30(10):1545

Abelson M, Yechieli Y, Crouvi O, Baer G, Wacks D, Bein A, Shtivelman V (2006) Evolution of the Dead Sea sinkholes. In: Enzel Y, Agnon A, Stein M (eds) Geol SocAm SpecPap 401:241–253

Abelson M, Yechieli Y, Aksinenko T, Pinski V, Baer G (2013) Microseismic monitoring of concealed sinkholes activity at Mineral Beach, Dead Sea, Israel. Geol Surv Isr Rep GSI/29/2013

Albouy Y, Andrieux P, Rakotondrasoa G, Ritz M, Descloitres M, Join JL, Rasolomanana E (2001) Mapping coastal aquifers by joint inversion of DC and TEM soundings—three case histories. Ground Water 39:87–97

Al-Zoubi A, Abueladas AR, Camerlynck C, Al-Ruzouq R, Al-Rawashdeh S, Ezersky M, Ali W (2007) Use of 2D multi electrodes resistivity imagining for sinkholes hazard assessment along the eastern part of the Dead Sea, Jordan. Am J Environ Sci 3(4):229–233

Arad A, Kafri U, Fleisher E (1975) The Na'aman springs, northern Israel: salination mechanism of an irregular freshwater-seawater interface. J Hydrol 25:81–104

Archie GW (1942) The electrical resistivity log as an aid in determining some reservoir characteristics. Trans Am Inst Min Met Eng 146:54–62

Arkin Y, Gilat A (2000) Dead Sea sinkholes—an ever-developing hazard. Environ Geol 39(7):711–722

Baida U, Goldstoff Y, Kidron Y (1978) Numerical model of the Cenomanian aquifer in the southern part of the Yarkon-Taninim basin (Beer Sheva Basin). Tahal Rep 01/78/08 (in Hebrew)

Bartov Y, Bein A (1994) The geology and hydrogeology of the central Arava between Yotveta and Paran. Geol Surv Isr Rep GSI/5/94 (in Hebrew)

Bartov Y, Goldman M, Calbo R, Ronen A (1998) TDEM survey in Northern Arava. Combined Geol Surv Isr-Geophys Inst Isr Reps TR-GSI/2/98-L11-188 95 (in Hebrew)

Bar Yosef Y, Michaeli A, Wolman S (1997) Hydrological planning of exploitation from the Naaman basin. Summarizing report. NRD Rep NR/221/97 (in Hebrew)

Bein A, Yechieli Y, Halicz L (1995) The geochemistry of groundwater in the southern Arava. Geol Surv Isr Rep GSI/43/95 (in Hebrew)

Bein A (1971) Rudistid reef complexes (Albian to Cenomanian) in the Carmel and the Coastal Plain Israel Inst Petr Res Geophys Rep 1051 and Geol Surv Israel Rep OD/2/71

Bein A, Yechieli Y, Ben Shabat J (1998) Hydrogeological regional model of the southern Arava. Geol Surv Isr Rep GSI/15/98 (in Hebrew)

Binley A, Slater LD, Fukes M, Cassiani G (2005) Relationship between spectral induced polarization and hydraulic properties of saturated and unsaturated sandstone. Water Resour Res 41:W12417

Boucher M, Girard JF, Legchenko A, Baltassat JM, Dörfliger N, Chalikakis K (2006) Using magnetic resonance soundings to locate a water-filled karst conduit. J Hydrol 330:413–421

Closson D, Karaki NA, Hussein MJ, Al-Fugha H, Ozer A (2003) Space-borne radar interferometric mapping of precursory deformations of a dyke collapse, Dead Sea area, Jordan. Int J Remote Sensing 24:843–849

Closson D, Karaki NA, Klinger Y, Hussein MJ (2005) Subsidence and sinkhole hazards assessment in the southern Dead Sea area, Jordan. Pure Appl Geophys 162:221–248

Constable S, Parker R, Constable C (1987) Occam's inversion: a practical algorithm for generating smooth models from electromagnetic sounding data. Geophysics 52(3):289–300

Ezersky M (2006) The Geophysical properties of the Dead Sea salt applied to the sinkhole problem. J Appl Geophys 58(1):45–58

Ezersky M (2008) Geoelectric structure of the Ein Gedi sinkhole occurrence site at the Dead Sea shore in Israel. J Appl Geophys 64:56–69

Ezersky M, Frumkin A (2013) Faults–dissolution front relations and the DS sinkholes problem. Geomorphology 201:35–44

Ezersky M, Frumkin A (2017) Evaluation and mapping of Dead Sea coastal aquifers salinity using Transient Electromagnetic (TEM) resistivity measurements. C R Geosci 349:1–11

Ezersky M, Legchenko A, Camerlynck C (2006) Part III. TEM mapping of shallow aquifers and study of their link with sinkhole development in the Dead Sea coastal area of Israel. NATO Programme Science for Peace, Project SfP 981128. Rep 232/362/08 Geophys Inst Isr- Paris 6 University-IRD, 122 p

Ezersky M, Legchenko A, Eppelbaum L, Al-Zoubi A (2017) Overview of the geophysical studies in the Dead Sea coastal area related to evaporite karst and recent sinkhole development. Int J Speleol 46(2):277–302

Fitterman DV, Labson VF (2005) Electromagnetic induction methods for environmental problems. Near Surf Geophys 301–356

Fitterman DV, Stewart MT (1986) Transient electromagnetic soundings for groundwater. Geophysics 53:118–128

Fleisher E, Fleischer L, Frieslander U (1996) Groundwater in the Arava, Yaalon-Yotveta region. Geophys Inst Isr Rep 820/56/96 (in Hebrew)

Fleisher E, Fleischer L, Frieslander U (1997) The hydrogeology of the Idan-En Yahav region. Geophys Inst Isr Rep 820/36/96 (in Hebrew)

Frumkin A, Kofman L, Ezersky M (2009) Improvement of the reliability of subsurface void detection, including sinkhole development, at the Dead Sea shore area by means of Ground Penetration Radar (GPR). Hebrew University—TECHNION Research and Development Foundation LTD—Geophys Institute Isr Rep MNI-ES-36-2008, 145 p

Frumkin A, Ezersky M, Al-Zoubi A, Akkawi E, Abueladas AR (2011) The Dead Sea hazard: geophysical assessment of salt dissolution and collapse. Geomorphology 134(1–2):102–117

Garambois S, Dietrich M (2001) Seismoelectric wave conversions in porous media: field measurements and transfer function analysis. Geophysics 66:1417–1430

Ginzburg A, Levanon A (1976) Determination of a salt-water interface by electric resistivity depth soundings. Hydrol Sci Bull 21(4):561–568

Glover PWJ (2016) Archie's law—a reappraisal. Solid Earth 7:1157–1169

Goldman M, Ezersky M (1999a) Geoelectrical measurements at the Nahal Hever site. Geophys Inst Isr Rep 841/92/98 (in Hebrew)

Goldman M, Ezersky M (1999b) TDEM survey in the vicinity of water reservoirs in Northern Arava. Geophys Inst Isr Rep 822/90/98 (in Hebrew)

Goldman M, Ezersky M (2000) TDEM measurements in the sinkhole development area at the Nahal Hever North site. Geophys Inst Isr Rep 824/87/99

Goldman M, Kafri U (2004) The use of the time domain electromagnetic (TDEM) method to evaluate porosity of saline water saturated aquifers. In: 18th salt water intrusion meeting (SWIM) selected papers, Cartagena. Publicaciones del Instituto Geologico Y Minero de Espana Hydrog Y aguas subterraneas, vol 15, pp 327–340

Goldman M, Arad A, Kafri U, Gilad D, Melloul A (1988) Detection of freshwater/seawater interface by the time domain electromagnetic (TDEM) method in Israel. Natuurwet Tijdschr 70:329–344

Goldman M, Gilad D, Ronen A, Melloul A (1991) Mapping of seawater intrusion into the coastal aquifer of Israel by the time domain electromagnetic (TDEM) method. Geoexploration 28:153–174

Goldman M, Rabinovich B, Rabinovich M, Gilad D, Gev I, Schirov M (1994a) Application of the integrated NMR-TDEM method in groundwater exploration in Israel. J Appl Geophys 31:27–52

Goldman M, Du Plooy A, Eckard M (1994b) On reducing ambiguity in the interpretation of transient electromagnetic sounding data. Geophys Prospect 42:3–25

Goldman M, Shtivelman V, Wachs D, Yechieli Y (2000) Shallow integrated geophysical study in the sinkhole development area at the Dead Sea coast, Israel. 6th EEGS Meeting, Bochum, Germany, Expanded Abstracts, CH04

Goldman M, Kafri U, Yechieli Y (2003) Application of the TDEM method for studying groundwater salinity in different coastal aquifers in Israel. In: Lopez-Geta JA, de Dios Gomes J, de la Orden

JA, Ramos G, Rodriguez L (eds) Coastal aquifers intrusion technology: mediterranean countries. The Institute of Geology and Mineralogy of Spain (IGME), pp 45–56

Goldman M, Helwig SL, Kafri U, Scholl C, Tezkan B (2006) Integrated electromagnetic geophysical technology applied to the detection of deep fresh/saline interface in northern and central Israel. German-Israeli Cooperation in Water Technology, Joint BMBF-MOST project WT 00103/1770, 252p

Goldman M, Levi E, Tezkan B, Yogeshwar P (2011) The 2D coastal effect on marine time domain electromagnetic measurements using broadside dBz/dt of an electrical transmitter dipole. Geophysics 76:F101–F109

Goldshmidt MJ, Arad A, Neev D (1967) The mechanism of the saline springs in the Lake Tiberias depression. Geol Surv Isr Bull 45:1–19

Guttman J, Burg A, Lifshitz A, Lumelsky S, Zukerman H (1999) Flow model for the evaluation of the water potential of the Arava aquifer. Water Planning for Israel, Tahal (in Hebrew)

Haroon A, Lippert K, Mogilatov V, Tezkan B (2018) First application of the marine differential electric dipole for groundwater investigations: a case study from Bat Yam, Israel. Geophysics 83(2):B59–B76

Hirsch H (1975) Possibility for exploitation of freshwater in the Ein Tureibe-Ein Ghuweir region.Tahal Rep 01/75/67

Hördt A, Blaschek R, Kemna A, Zisser N (2007) Hydraulic conductivity estimation from induced polarization data at the field scale—the Krauthausen case history. J Appl Geophys 62:33–46

Hoyer WA, Spann MM (1975) Comments on obtaining accurate electrical properties of cores. Trans SPWLA XVI

Hurwitz S, Goldman M, Ezersky M, Gvirtzman H (1999) Geophysical (TDEM) delineation of a shallow brine beneath a freshwater lake, the Sea of Galilee, Israel. Water Resour Res 35:3631–3638

Interpex (1996) TEMIX-XL user's manual. Interpex Ltd., Golden, Colorado

Issar A, Bein A, Michaeli A (1972) On the ancient water of the Upper Nubian Sandstone aquifer in central Sinai and southern Israel. J Hydrol 17:353–374

Jupp DLB, Vozoff K (1975) Stable iterative methods for the inversion of geophysical data. Geophys J Int 42(3):957–976

Kafri U (1967) Facies changes in the southwestern Carmel (Israel) and their influence on groundwater regime. Isr J Earth Sci 16:206–214

Kafri U (1969) Geology and groundwater of the Cenomanian formations in Galilee, west of the watershed. Ph.D. thesis, Hebrew Univ., Jerusalem, 188 p (in Hebrew)

Kafri U (1977) Disposal of waste effluents formed during the proposed production of titanium dioxide in Timna. Isr Geol Surv Rep Hydro/2/77, 11 p

Kafri U (1982) Relationship between the Dead Sea and groundwater levels in the western Dead Sea catchment area. Isr Geol Surv Curr Res 1981:90–94

Kafri U, Arad A (1979) Current subsurface intrusion of Mediterranean seawater. A possible source of groundwater salinity in the Rift Valley system, Israel. J Hydrol 44:267–287

Kafri U, Goldman M (2005) The use of the time domain electromagnetic (TDEM) method to delineate saline groundwater in granular and carbonate aquifers and to evaluate their porosity. J Appl Geophys 57:167–178

Kafri U, Goldman M (2006) Are the lower sub-aquifers of the Mediterranean coastal aquifer of Israel blocked to seawater intrusion? Results of a TDEM (time domain electromagnetic) study. Isr J Earth Sci 55:55–68

Kafri U, Goldman M, Lang B (1997) Detection of subsurface brines, freshwater bodies and the interface configuration in-between by the time domain electromagnetic method in the Dead Sea Rift, Israel. Environ Geol 31:42–49

Kafri U, Goldman M, Gendler M (2001) TDEM survey for characterization of groundwater salinity in the Southern Arava and the coast of the Bay of Elat. Geol Surv Isr Rep TR-GSI/3/2001 and Geophys Inst Isr Rep 906/94/00

Kafri U, Goldman M, Lyakhovsky V, Scholl C, Helwig S, Tezkan B (2007) The configuration of the fresh-saline groundwater interface within the regional Judea Group carbonate aquifer in northern Israel between the Mediterranean and the Dead Sea base levels as delineated by deep geoelectromagnetic soundings. J Hydrol 344:123–134

Kafri U, Goldman M, Levi E (2008) The relationship between saline groundwater within the Arava Rift Valley in Israel and the present and ancient base levels as detected by deep geoelectromagnetic soundings. Environ Geol 54(7):1435–1445

Kafri U, Goldman M, Levi E, Wollman S (2014) Detection of saline groundwater bodies between the Dead Sea and the Mediterranean Sea, Israel, using the TDEM method and hydrochemical parameters. Environ Process 1:21–41

Kaufman AA, Keller GV (1983) Frequency and transient sounding: methods in geochemistry and geophysics. Elsevier, Amsterdam

Kelly WE, Mareš S (1993) Applied geophysics in hydrogeological and engineereing practice. Elsevier, Amsterdam, 290 p

Kemna A, Binley A, Cassiani G, Niederleithinger E, Revil A, Slater L, Williams LKH, Orozco AF, Haegel FH, Hördt A, Kruschwitz S, Leroux V, Titov K, Zimmermann E (2012) An overview of the spectral induced polarization method for near-surface applications. Near Surf Geophys 10:453–468

Koch O, Helwig SL, Meqbel N (2004) Vertical near surface conductivity anomaly detected at the Dead-Sea-Transform. In: Proceedings of the 17th workshop on electromagnetic induction in the earth, Hyderabad, India. Available at http://www.emindia2004.org: 1/6–6/6

Kolton Y (1988) Examination of the connection between seawater in the Pleistocene aquifer in the Mediterranean shelf of central Israel. Tahal, Water Planning for Israel. Rep 01/88/31 (in Hebrew)

Kroitoru L, Kronfeld J, Ginzburg A (1981) The geohydrology of the Gerofit-Ya'alon area. Isr J Earth Sci 30:24–30

Kronfeld J, Weinberger G, Yaniv A, Zafrir J, Vulcan U, Agami M, Rosenthal E (1992) Uranium isotope disequilibrium studies and the geohydrology of the Arava rift valley, Israel. Nucl Geophys 6:535–545

Legchenko A (2013) Magnetic resonance imaging for groundwater. ISTE Ltd., Wiley, Great Britain, 158 p

Legchenko A, Baltassat J-M, Beauce A, Bernard J (2002) Nuclear magnetic resonance as a geophysical tool for hydrogeologists. J Appl Geophys 50(1):21–46

Legchenko A, Ezersky M, Kamerlynck C, Al-Zoubi A, Chalikakis K (2009) Joint use of TEM and MRS method in complex geological setting. Comput Rendus Geosci 341(10–11):908–917

Legchenko A, Valla P (2002) A review of the basic principles for proton magnetic resonance sounding measurements. J Appl Geophys 50(1):3–19

Levi E, Goldman M (2017) Application of the marine Ex-Bz transient system for delineating near shore resistive targets. J Appl Geophys 144:57–68

Levi E, Goldman M, Hadad A, Gvirtzman H (2008) Spatial delineation of groundwater salinity using deep TDEM geophysical measurements: a feasibility study. Water Resour Res 44:W12404

Levi E, Goldman M, Tibor G, Herut B (2018) Delineation of subsea freshwater extension by marine geoelectromagnetic soundings (SE Mediterranean Sea). Water Resour Manage 32:3765–3779

Linde N, Binley A, Tryggvason A, Pedersen LB, Revil A (2006) Improved hydrogeophysical characterization using joint inversion of cross-hole electrical resistance and ground penetrating radar travel time data. Water Resour Res 42:W12404

Linde N, Tryggvason A, Peterson JE, Hubbard SS (2008) Joint inversion of crosshole radar and seismic traveltimes acquired at the South Oyster Bacterial Transport Site. Geophysics 73(4):G29–G37

Lippert K (2015) Detection of a submarine aquifer offshore of Israel using LOTEM. Ph.D. thesis, University of Cologne (in German)

Loke MH (2004) Tutorial: 2-D and 3-D electrical imaging surveys. Geotomosoft.com

Maercklin N, Bedrosian PA, Haberland C, Ritter O, Ryberg T, Weber M, Weckmann U (2005) Characterizing a large shear-zone with seismic and magneotelluric methods: the case of the Dead Sea Transform. Geophys Res Lett 32

Magal E, Yechieli Y (2002) A literature survey of the Yotveta sabkha. Geol Surv Isr Rep GSI/9/2002 (in Hebrew)

Meinzer OE (1942) Hydrology. McGraw-Hill Book Co., New York

Mercado A (2000) Hydrological examination of sea water supply to desalination plants by wells in the southern coastal plain. Progress report no 1 Geol Surv Israel

Mercado A, Ben Zvi A (1983) A preliminary hydrogeological survey in chosen areas for siting of subsurface nuclear power plants. Tahal Rep 04/83/11 (in Hebrew)

Michaeli A (1971) A survey of brackish water salinity. Southern Arava. Tahal Water Planning for Israel (in Hebrew)

Michaeli A, Galai A (1963) A hydrological study in the Yotveta region. Tahal Water Planning for Israel P M 359 (in Hebrew)

Michaeli A, Shaharabani M, Blank D (1973) Development and calibration of groundwater flow model in the Na'aman basin. Tahal Rep 01/73/29 (in Hebrew)

Nativ R (1984) The water potential of the deep Jurassic to Paleozoic aquifers in the Negev, Israel. Ph.D. thesis, Ben Gurion Univ Beer Sheva, 311 p (in Hebrew)

Nof RN, Baer G, Ziv A, Raz E, Atzori S, Salvi S (2013) Sinkhole precursors along the Dead Sea, Irsael, revealed by SAR interferometry. Geology 41:1019–1022

Repsold H (1989) Well logging in groundwater development. Int Assoc Hydrogeol 9:136

Ritter O, Ryberg T, Weckmann U, Hoffmann-Rothe A, Abueladas A, Garfunkel Z (2003) Geophysical images of the Dead Sea Transform in Jordan reveal an impermeable barrier for fluid flow. Geophys Res Lett 30

Rosenthal E, Adar E, Issar A, Batelaan O (1992) Definition of groundwater from patterns by environmental traces in the multiple aquifer system of the southern Arava valley, Israel. J Hydrol 117:339–368

Rover CW, Cherkauer DS (1995) Scale dependency of hydraulic conductivity measurements. Ground Water 33:769–780

Rubin Y, Hubbard SS (2005) Hydrogeophysics, water science and technology library. Springer, Berlin, 527 p

Rybakov M, Goldshmidt V, Fleischer L, Rotstein Y (2001) Cave detection and 4-d monitoring: a microgravity case history near the Dead Sea. Lead Edge 20(8):896–900

Scholl C (2005) The influence of multidimensional structures on the interpretation of LOTEM data with one-dimensional models and the application to data from Israel. Ph.D. dissertation, University of Cologne, Institute for Geophysics and Meteorology

Sellinger A, Kapuler I (1972) The Shiqma coastal collector; proposal for hydrological planning of the exploitation system. Tahal Rep HR/72/118 (in Hebrew)

Sellinger A, Kapuler I (1973) The Nitzanim coastal collector; proposal for hydrological planning of the exploitation system. Tahal Rep HR/73/01 (in Hebrew)

Sellinger A, Levin M (1977) Hydrology of the Pleistocene aquifer in the Western Galilee. Tahal Rep 01/77/18 (in Hebrew)

Shachnai E, Goldshtoff Y (1978) The upper Cenomanian-Turonian aquifer of the Qeren-Kabri basin. Calibration of digital model to simulate groundwater flow. Tahal Rep 01/77/42

Shalem Y, Weinstein Y, Levi E, Herut B, Goldman M, Yechieli Y (2014) The extent of aquifer salinization next to an estuarine river: an example from the eastern Mediterranean. Hydrogeol J 23:69–79

Shtivelman V, Goldman M (2000) Integration of shallow reflection seismics and TDEM for detailed study of the coastal aquifer in the Nitzanim area of Israel. J Appl Geophys 44:197–215

Stanislavsky E, Gvirtzman H (1999) Basin-scale migration of continental-rift brines: paleohydrologic modeling of the Dead Sea basin. Geology 27:791–794

Starinsky A (1974) Relationship between Ca-chloride brines and sedimentary rocks in Israel. Ph.D. thesis, Hebrew Univ Jerusalem, 176 p (in Hebrew)

Stiller M (1994) The chloride content in pore water of Lake Kinneret sediments. Isr J Earth Sci 43:179–185

Strack KM (1992) Exploration with deep transient electromagnetics, methods in geochemistry and geophysics, vol 30. Elsevier, Amsterdam

Swarzenski PW, Burnett WC, Greenwood WJ, Herut B, Peterson R, Dimova N, Shalem Y, Yechieli Y, Weinstein Y (2006) Combined time-series resistivity and geochemical tracer techniques to examine submarine groundwater discharge at Dor Beach. Geophys Res Lett 33:L24405. https://doi.org/10.1029/2006GL028282

Thompson A, Gist G (1993) Geophysical applications of electrokinetic conversion. Lead Edge 12:1169–1173

Vanyan LL (1967) Electromagnetic depth soundings. Springer, Berlin, 316 p

Vouillamoz JM, Legchenko A, Albouy Y, Bakalowicz M, Baltassat JM, Al-Fares W (2003) Localization of saturated karst aquifer with magnetic resonance sounding and resistivity imagery. Ground Water 41:578–587

Wachs D, Yechieli Y, Shtivelman V, Itamar A, Baer G, Goldman M, Raz E, Rybakov M, Shatner U (2000) Sinkholes development along the Dead Sea coasts. Geol Surv Isr Rep GSI/41/2000

Wempe WL (2000) Predicting flow properties using geophysical data: improving aquifer characterization. Ph.D. thesis, Stanford University

White WB (2002) Karst hydrology: recent developments and open questions. J Eng Geol 65:85–105

Winsauer WO, Shearing HM Jr, Masson PH, Williams M (1952) Resistivity of brine saturated sands in relation to pore geometry. Am Assoc Pet Geol Bull 36:253–277

Wust-Bloch GH, Joswig M (2006) Pre-collapse identification of sinkholes in unconsolidated media at Dead Sea area by 'nanoseismic monitoring' (graphical jackknife location of weak sources by few, low-SNR records). Geophys J Int 167:1220–1232

Yang CH, Tong LT, Huang CF (1999) Combined application of dc and TEM to sea-water intrusion mapping. Geophysics 64:417–425

Yaramanci U (2003) Surface Nuclear Magnetic Resonance for direct assessment of groundwater. Bollettino di Geofisica Teorica et Applicata 44:237–252

Yechieli Y (1993) The effects of water level changes in closed lakes (Dead Sea) on the surrounding groundwater and country rocks. Ph.D. thesis, Weizmann Institute of Science, 197 p

Yechieli Y, Abelson M, Bein A, Crouvi O, Shtivelman V (2006) Sinkhole "swarms" along the Dead Sea coast: reflection of disturbance of lake and adjacent groundwater systems. Geol Soc Am Bull 118:1075–1087

Yechieli Y, Kafri U, Goldman M, Voss CI (2001) Factors controlling the configuration of the fresh-saline water interface in the Dead Sea coastal aquifers: synthesis of TDEM surveys and numerical groundwater modeling. Hydrogeol J 9:367–377

Zuber A, Motyka J (1998) Hydraulic parameters and solute velocities in triple-porosity karstic-fissured-porous carbonate aquifers: case studies in southern Poland. J Env Geol 34:243–250

# Numerical Modeling of Groundwater in Israel

**Amir Paster and Nitzan Matan**

## 1 Introduction

Groundwater modeling is aimed to simulate the flow and transport in the aquifer. By definition, models are a rough description of the actual intricate phenomena that take place in the aquifer. In many cases, models are used to anticipate the effects of hydrological changes (such as pumping) on the water quantity and quality, which have significant implication upon management. In some simple problems, analytical solutions for flow and/or transport exist. However, in most real-life problems, particularly those with irregular boundaries, complex boundary, non-trivial and initial conditions, or heterogeneities, analytical solutions are not suitable (Bouwer 1978). Historically, physical (e.g., Hele-Shaw cells, Hele-Shaw 1898) and electric-analog models were used to simulate flow in aquifers. With the development of computing power and software tools, the solution of the mathematical equations that describe flow became increasingly practical (Shechter and Schwarz 1970). Due to these developments, it is nowadays the common practice to model groundwater by numerical codes. One of the advantages of using numerical models, with respect to physical models, is the ease of changing parameters such as size and material properties. Still, physical laboratory models and field experiments are utilized in some cases to support and validate the numerical models (e.g., Turkeltaub et al. 2015).

Specifically, in semi-arid regions like Israel, sustainable management of groundwater is relatively complex. Due to this complexity, there is a clear need to use models as tools for providing the predictions needed for decision makers. This need was well understood since the early years of the state of Israel (Schwarz et al. 2016).

A. Paster (✉) · N. Matan
School of Mechanical Engineering, Tel Aviv University, Ramat Aviv, Tel Aviv, Israel
e-mail: amirp@breezometer.com

N. Matan
e-mail: nitzanmatan@gmail.com

BreezoMeter, Haifa, Israel



U. Kafri and Y. Yechieli (eds.), *The Many Facets of Israel's Hydrogeology*,
Springer Hydrogeology, https://doi.org/10.1007/978-3-030-51148-7_17

Thus, significant efforts had been put into development of models in academic institutions (mainly in The Hebrew University, Ben Gurion University, Technion, and Tel Aviv University) and in government bodies (e.g., Mekorot, Tahal, and the Geological Survey of Israel). This chapter aims to describe the results of these efforts. There are numerous works that focus on groundwater modeling in Israel. A significant fraction of these works are reports that were intended for governmental use and were written in Hebrew (see e.g., the references in Paster 2010 and Kessler and Kafri 2007). We deliberately turn our attention in this chapter to the works that have international relevance and were published in peer-reviewed journals in English. Also, we only consider works that are mainly focused on groundwater. Therefore, studies of surface hydrology, runoff, and recharge estimation are not necessarily mentioned herein.

Roughly speaking, we can divide the studies surveyed in this chapter into two groups: (1) studies that involved development of a new model, and (2) studies that utilized an existing model or models. Some examples to such existing models include the popular codes MODFLOW (Harbaugh et al. 2000), MT3D (Zheng and Wang 1999) FEFLOW (Diersch 2005), SUTRA (Voss 1984), and Hydrus (Šimůnek et al. 2005). Examples of numerical models that were developed by Israeli researchers include (a) finite element models for constant density flow (Berger 1999; Shavit and Furman 2001); (b) finite element models for density-driven flow (Sorek et al. 2001; Zhou et al. 2005); and (c) cell models for the Kineret watershed (Assouline 1993; Kessler 2012; Rimmer and Salingar 2006) and the Na'aman aquifer (Kessler and Kafri 2007). A complete list of the numerical codes mentioned in this chapter is provided in Table 1.

The scope of this chapter is therefore (1) to review the studies that involved development of groundwater models by Israeli hydrologists; (2) to review the studies that applied existing numerical models to solve flow and transport in aquifers in Israel; and (3) to describe the present trends in groundwater modeling in Israel. We divided the studies surveyed in this chapter into groups according to their main theme. The groups are (1) regional aquifer management (Sect. 3); (2) groundwater salinization (Sect. 4); (3) contaminant fate and transport (Sect. 5); and (4) vadoze zone (Sect. 6).

As far as we know, this is the first review dedicated to groundwater modeling in Israel. We tried our best to cover all the relevant studies in this chapter. However, some may have been omitted, and we apologize in advance if this is the case.

## 2 Survey of Models

In this section, we describe the development and application of numerical models by Israeli researchers in academic institutions and elsewhere. This review will not be complete without mentioning the seminal books of Jacob Bear (Bear 1972) and Gedeon Dagan (Dagan 1989). Bear (1972) is a classic textbook on flow and transport on porous media focusing on homogeneous media and analytical solutions. Dagan (1989) is a classical textbook on flow and transport in heterogeneous formations. The derivations in Dagan (1989) relate to the pore-scale, laboratory-scale, and field-scale

**Table 1** Numerical codes for modeling flow and transport in groundwater which are mentioned in this chapter

| Model name | Features | References |
|---|---|---|
| AKCell | Solving mass balance for a cell model of an aquifer | Kessler and Kafri (2007) |
| FEAS | Solving variable density flow and transport using a finite elements approach | Zhou et al. (2005) |
| FEFLOW | Solving variable density flow, mass, and heat transport in two- and three-dimensional for variably saturated media | Diersch (2005, 2013) |
| Hydrus 1D | Solving flow, heat, and solute transport in one-dimensional variably saturated media | Šimůnek et al. (2005) |
| Hydrus 2D/3D | Solving flow, heat, and solute transport in two- and three-dimensional variably saturated media using a finite element approach | Šimůnek et al. (2016) |
| JHU2D | Solving a transient two-dimensional flow and heat transport using a finite element approach | Garven and Freeze (1984b) |
| MEL2DSLT | Solving an aerial model of vertically averaged flow and transport | Sorek et al. (2001) |
| MODFLOW | Solving a two- and three-dimensional groundwater flow by a finite difference method | Harbaugh et al. (2000) |
| MT3DMS | Solving three-dimensional multispecies transport (advection, dispersion, and chemical reactions) of contaminants in groundwater. The flow is derived from MODFLOW results | Zheng and Wang (1999) |
| OILGEN | Solving a steady-state two-dimensional flow and heat transfer using a finite element method | Garven (1989) |
| SUTRA | Solving a two-dimensional variable density flow and solute or energy transport in variably saturated media using a finite element/a finite difference approach | Voss (1984) |
| UNSATCHEM | Solving one-dimensional water flow, heat transport, carbon dioxide transport, and solute transport with major ion equilibrium and kinetic chemistry in variably saturated media | Šimůnek et al. (1996) |

The codes are ordered by alphabetical order

phenomena, and include many analytical solutions based on the small perturbation approach. Jacob Bear is an author of few additional important books on hydrology including Bear (1979) and Bear and Bachmat (1990). Bear (1979) is focused more on modeling of flow and solving the inverse problem. Bear and Bachmat (1990) focus on modeling transport phenomena in porous media.

In the rest of this section, we present studies which are more particular and relate to a specific problem in groundwater modeling and management. The studies are presented in the following sub-sections, based on the relevant topic of the model.

## 3 Regional Aquifers Management

There are many models that were developed in the past to describe flow at a regional scale in aquifers in Israel. Many of those models were conducted by TAHAL, Mekorot, and GSI (e.g., references in Sheffer et al. 2010). The models were mainly intended for governmental use and were published as technical reports in Hebrew. Herein, we briefly mention some of these works and mainly focus on studies published in scientific journals.

One of the earlier works on groundwater management is the study of Schwarz (1976). In this study, a general GW management problem was considered. The problem is solved by a linear programming (LP) approach. LP was developed in the 1940s for solving optimization problems (see references in Schwarz 1976). In the work of Schwarz (1976), the decision parameters are pumping rates and artificial recharge rates, i.e., only water quantity is taken into account. The flow is solved by a 2D finite difference model of the aquifer and some constraints, such as minimal/maximal GW level and minimal/maximal pumping rates are imposed.

The Coastal Aquifer was modeled in the 1970s by Mercado (1976) using a single-cell model in the order to study the pollution of the aquifer by nitrate and chloride. The model indicated that the main factors that affect the transport of these pollutants are the specific sources at the surface, the thickness of the unsaturated zone, and the processes that can modify the nitrate concentration. Most other studies dealt with the inter-relation of the aquifer and the sea, examining the effect of seawater intrusion (e.g., Amir et al. 2013; Prieto et al. 2001; Yechieli et al. 2010) and Submarine groundwater discharge (Amir et al. 2013; Prieto et al. 2001). The connection of the aquifer and the sea and the issue of submarine groundwater discharge was modeled by Prieto et al. (2001). The possible effect of future sea level rise was examined using FEFLOW code, The results showed showing that due to the moderate topography of the Israeli coast, the inland extent will not be that significant (Yechieli et al. 2010), unlike other locations (e.g., Bangladesh) where the topography is lower and flatter.

Models of the flow in the Yarkon-Taninim aquifer were presented in multiple studies, including Guttman and Zukerman (1995), Berger (1999), Dafny et al. (2010), and Paster (2010). These studies model the entire aquifer. Guttman and Zukerman (1995) modeled the aquifer by a 2D (aerial) MODFLOW model and calibrated it manually. The model was later updated by Zukerman and Shachnai (1999). A different numerical approach was proposed by Berger (1999) who developed a finite elements model for the aquifer. He solved the inverse problem following the approach presented by Sun and Yeh (1992). The inverse problem solved by Berger (1999) is the problem of identifying the values of transmissivity and storativity in all the model cells. Dafny et al. (2010) modeled the aquifer using FEFLOW. The model was three-dimensional and assumed constant density. Dafny et al. (2010) provides a summary of previous numerical models for the Yarkon-Taninim basin (all in Hebrew).

Studies of specific, local, Karst systems in the Yarkon-Taninim aquifer were conducted as well. Weiss and Gvirtzman (2007), Peleg and Gvirtzman (2010), and

Peleg et al. (2012) studied such Karst systems and used FEFLOW to model the saturated zone. The study of Sheffer et al. (2010) also focused on the same sites and used FEFLOW as well. It differs from the prior works by adding two additional aspects: (1) modeling the unsaturated zone within the FEFLOW model, and (2) modeling the recharge via the daily recharge assessment model (DREAM).

Rimmer and Salingar (2006) modeled the Mt. Hermon GW Basin by a cell model. This approach is based on the assumption that the aquifer is a set of linear reservoirs. The term linear here refers to the assumption used in such models where the flow into and out of the reservoir depends linearly on the level in that reservoir. Kessler and Kafri (2007) cited many other studies that used cell models to simulate groundwater in Israel and elsewhere. The work of Rimmer and Salingar (2006) was extended to include a sensitivity test by Hartmann et al. (2013). In contrast, Asmael et al. (2015) also modeled the Hermon GW basin using a FEFLOW model.

The lake Kinneret watershed has received significant attention in terms of modeling, e.g., Rimmer and Gal (2003) and Rimmer et al. (1999), Assouline (1993), Abbo et al. (2003) who focus on the relationship of groundwater and the lake. The review by Rimmer (Rimmer and Givati 2014) describes these works as well as others. Additional works that focus more on the saline springs are mentioned in the next section (Sect. 4).

Other smaller aquifers were modeled as well. The Golan basalt aquifer was modeled by Dafny et al. (2003), using MODLFOW. The Arava valley aquifer was modeled by Adar et al. (1992), who modeled the flow and transport of dissolved species in the aquifer using mixing cells model. They calibrated the model by the Wolf algorithm approach (Adar et al. 1988). The Na'aman groundwater basin was modeled by Kessler and Kafri (2007), who used a cell model approach. The East Judea aquifer, i.e., the carbonate aquifer beneath the Judea Desert, was modeled by Ben-Itzhak and Gvirtzman (2005). In this work, steady-state 3D numerical groundwater flow simulations were conducted using the MODFLOW code. The results of the model led to multiple insights on the properties of the aquifer, e.g., a 'river-like' flow pattern in the phreatic aquifer, where flow is strongly effected by geological folding. In the work of Yechieli et al. (2007), the connection between geological flexure and reduction in hydraulic conductivity was studied. The Judea Group Aquifer was taken as a case study. Using a FEFLOW 2D model for flow, this study suggested that the increased compressional stress in the flexure reduces the porosity and consequently the hydraulic conductivity. The work by Yechieli et al. (2009) studied the effect of changes in the Dead Sea level on the groundwater in the Judea Group aquifer. They considered a cross section between the Mediterranean Sea and the Dead Sea and used two approaches: (1) an analytical 1D steady-state flow model, and (2) a numerical 2D model in FEFLOW, taking into account flow and brine transport. The model was run to predict 1000 years from present. The results showed that the 1D model gives a good approximation of the 2D model, and that the decline in the Dead Sea level mainly affects the region close to the sea, where the transition zone moves (see Fig. 9 in Chapter "Paleohydrogeology of Israel").

## 4 Groundwater Salinization

Salinization has been a major point of concern in GW management in Israel in the 1960s and the 1970s, especially in the context of seawater intrusion in the coastal aquifer and the saline springs of lake Kinneret (Sea of Galilee). Over the years, multiple studies have tackled the topic of groundwater salinization via models. Two main modeling approaches have been used for describing the relationships between freshwater and saltwater in aquifers: (1) a sharp interface approach and (2) transport models, taking into account the transition zone between freshwater and saline water bodies.

Salinization processes can also be modeled on a regional scale via cell models. In such models, the concentration in each single cell is a function of time. In other words, the spatial distribution of salt within the cell is neglected (Adar et al. 1992; Kessler and Kafri 2007; Mercado 1976; Rimmer and Gal 2003; Rimmer et al. 1999). Hence, cell models do not resolve transport at scales below the cell size.

### *4.1 Sharp Interface Models*

Sharp interface models assume that the freshwater and saltwater in the aquifer are stably layered and that the transition between the two water bodies is abrupt, i.e., there is a sharp interface between them (Bear 1972). The position of the interface is determined by the continuity of pressure at the interface.

One of the seminal works using this approach is the study of Bear and Dagan (1964) who developed analytical solutions for the steady-state position of the sharp interface in the case of seawater intrusion in a costal aquifer. They assumed a homogenous aquifer and used the hodograph method to solve the problem. They considered a few cases, including a point sink and a drain. This work was extended to the case of a transient interface by Dagan and Bear (1968) who developed the analytical solutions by using the small perturbation assumption. In addition, they validated the results by sandbox experiments.

A short time later, Schmorak and Mercado (1969) published their seminal work on field experiments of pumping freshwater above a saline water body. The results of the experiment were compared to the results of the analytical model of Dagan and Bear (1968), and good match was observed.

Paster et al. (2006) studied the saline water body in the deep parts of the Yarkon-Taninim aquifer at its northwestern boundary. They used the Ghyben-Herzberg formula to analyze water levels at freshwater and saltwater monitoring wells and deduced the position of the interface in two separate locations. They found a clear trend of interface rise over the considered period (1984–2005, see Fig. 1).

They hypothesized that the rise of the interface is due to the increased exploitation of freshwater from the aquifer. This increase led to a decline in the freshwater head, which leads to a respective decline in saltwater head, which eventually results in

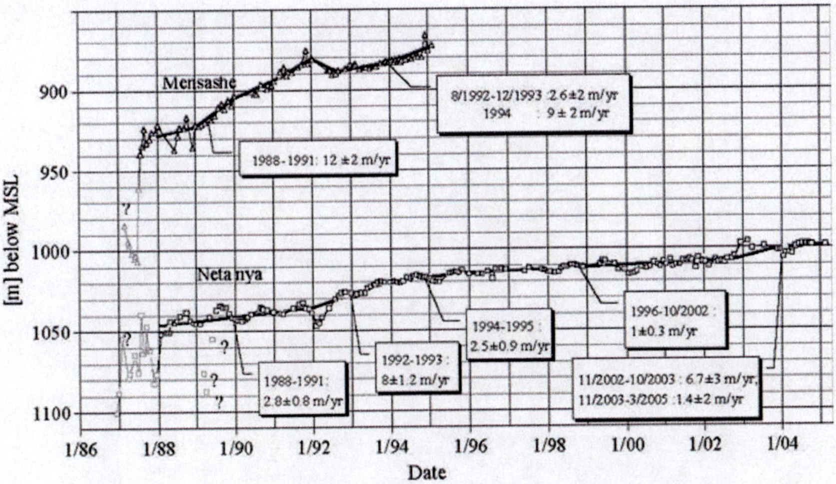


**Fig. 1** Calculated interface elevation and approximate trends in two locations in the Yarkon-Taninim aquifer (used by permission of Elsevier from Paster et al. 2006, Fig. 8)

increased seawater intrusion. In a later study, Paster (2010) used a sharp interface approach to solve the flow in the entire freshwater body in the aquifer (see Sect. 4.2).

## 4.2 *Models Taking into Account Transport*

In many cases, the mixing of saltwater and freshwater must be taken into account. In such cases, using sharp interface models is not proper, and a different approach is required. This modeling approach is based on solving the equation of solute transport for the salt as well as the equation of water flow. Multiple studies have used this kind of modeling approach, and some of them are described in the following. In this section, we describe first the works that included a numerical model development, and later we describe the works that mainly used existing numerical codes to study specific sites or aquifers.

**Development of Models**

Shavit and Furman (2001) studied the deep salinity sources in the Coastal aquifer in the Be'er Tuvia region where the chloride concentration rose significantly since the 1960s and reached values of 400–700 mg/L in the 1990s. They developed a two-dimensional (Dupuit) model, solving the water flow and solute transport equations by a finite-element method. They used this model to solve the inverse problem of identifying the salinity sources in the region. This solution includes the water and the salt mass flow rates at every node as a function of time (Fig. 2). The model results agreed reasonably well with the geochemical study of Vengosh et al. (1999).

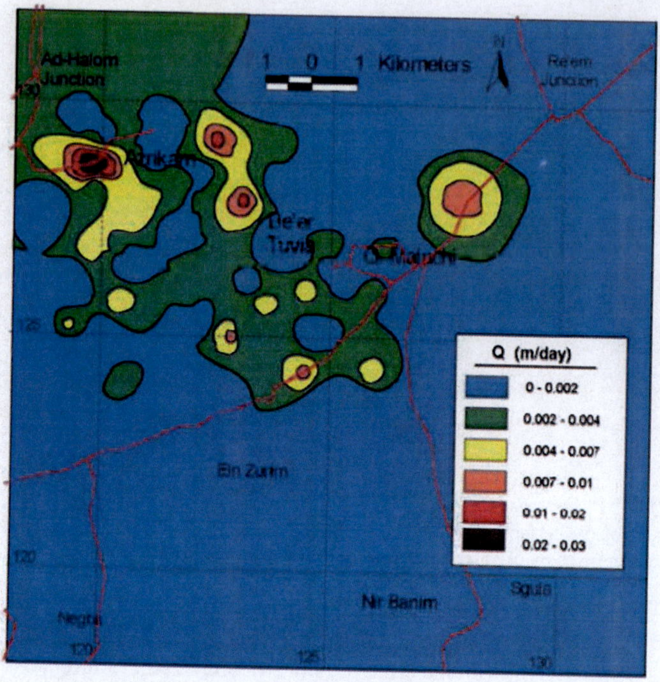


**Fig. 2** Model reconstruction of the average water source $Q$ in the Be'er Tuvia region, 1970–1990 (used by permission of Elsevier, from Shavit and Furman 2001)

Zhou et al. (2005) considered the problem of density-dependent flow and transport, where the equations are coupled. They developed a finite element code for both 2D and 3D cases and incorporated a particle tracking algorithm to model solute transport (Gordon et al. 2000). They verified the code in a number of benchmark cases (Zhou 1999). Zhou et al. (2005) illustrated their code by solving the case of transient upconing-decay process below a pumping well (Fig. 3). They showed that the process is sensitive to both transverse and longitudinal dispersivity, and that their code is practically free of numerical dispersion.

Another code for modeling density-dependent flow and transport was developed by Sorek et al. (2001). Their approach is based on averaging the flow and transport equation over the vertical coordinate and approximation of some of the averaged terms. The resulting equations are two-dimensional, and the variables depend on horizontal coordinates and time. The solution is obtained via a modified Eulerian–Lagrangian scheme (Sorek et al. 1999). The approach was applied to a 2D and 3D test cases, including one in the Khan Yunis area of the Gaza Strip (Yakirevich et al. 1998) where overexploitation of the coastal aquifer led to severe salinization.

A different approach for modeling density-driven flow in groundwater was presented by Paster and Dagan (2007), who developed a boundary layer (BL) approach for describing the mixing zone between freshwater and saltwater for the

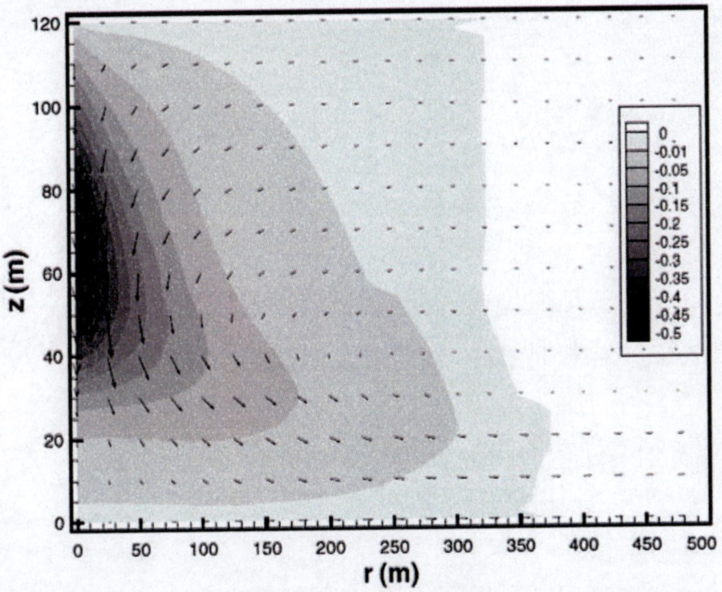


**Fig. 3** Results of the transient well upconing simulation by the FEAS model (Zhou et al. 2005). The contours represent the vertical velocity (m/d), and the arrows represent the velocity vector field. The well was shut off at $t = 3.95$ years, and the results are shown at $t = 5$ years. The circulation observed here is the result of the decay of the transition zone between freshwater and saltwater, following shut off

case of 2D seawater intrusion. In this approach, first a sharp interface solution is obtained, and then, a transition zone is approximately described via a BL approach (e.g., Rubin 1983). The approach was generalized to two additional cases where the sharp interface has analytical solutions: (1) axi-symmetric upconing (Paster and Dagan 2008b), and (2) a pumping well in a coastal aquifer (Paster and Dagan 2008a). An extension to a general aquifer, where the sharp interface solution is obtained by a numerical model, was described in Paster (2010). The approach was illustrated for the saltwater body in the Yarkon-Taninim aquifer (Fig. 4). In this work, the apparent transverse dispersivity of the aquifer was calibrated by the model and found to be $\alpha_T = 4$ (cm), a surprisingly low value [e.g., Zech et al. 2019 for comparison].

The study of Mandell et al. (2003) aimed to identify the salinity source in the Kefar Uria area of Yarkon-Taninim aquifer. In this work, flow is solved analytically, and the salt transport is solved by a particle tracking model. Their model neglected dispersion and took into account only the advective term of solute transport. The results of the model led to the identification of the salinity source, following the work of Avisar et al. (2003).

Optimization of the net value of the pumped water from the aquifer was studied by Shechter and Schwarz (1970) and Gordon et al. (2000). Shechter and Schwarz (1970) studied the case of a coastal collector. Their goal was to find the optimal pumping

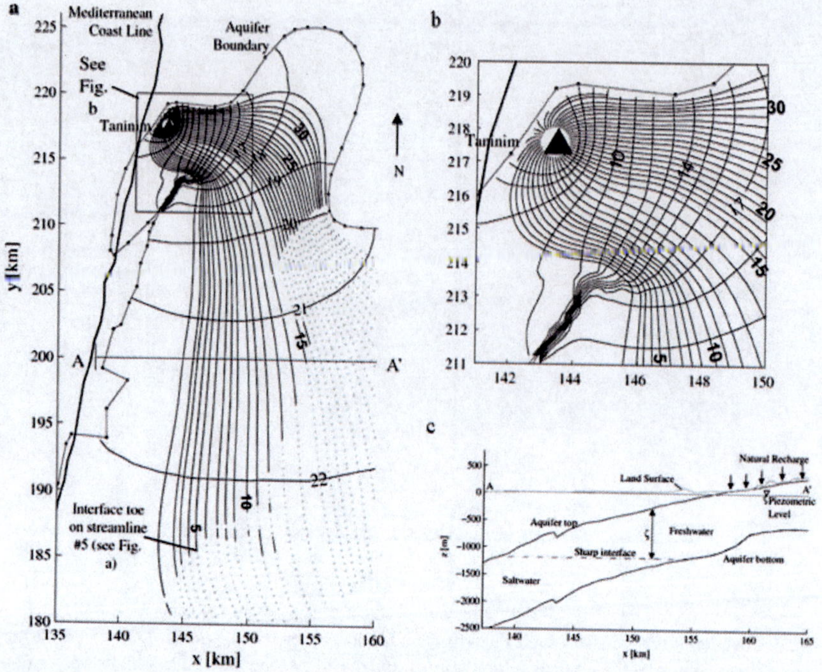


**Fig. 4** Sharp-interface flow solution in the northern part of Yarkon-Taninim aquifer. **a** Overview, with head contours and streamlines depicted. The streamlines are dotted in the freshwater-only zone. **b** An enlarged view of the boxed area. **c** A typical W-E cross section of the northern part of the aquifer (A–A′). The aquifer is confined except at the eastern part. The saltwater body fills the aquifer in the western part of the cross section (modified after Paster 2010)

rates of the collector. At first, they modeled the seawater intrusion and the dynamics of the freshwater–saltwater interface using Hele-Shaw cell experiment results. Then, they used the results to find an optimal solution for the management problem. One important conclusion they obtained was that the pumping rate should be higher at first and then should be reduced after a few years. The first pumping stage is transient and uses a one-time reserve, while the second pumping stage maintains a steady state (Fig. 5).

Gordon et al. (2000) studied the management optimization problem of a regional aquifer, when the aquifer is subject to salinization by relatively low concentrations. They developed a 2D finite elements model for flow and transport in the aquifer. Then, they used the model results to search for an optimal solution for the management problem. The optimization algorithm was based on a bundle-trust method. One of the main conclusions of this work was that the optimization problem is non-convex,

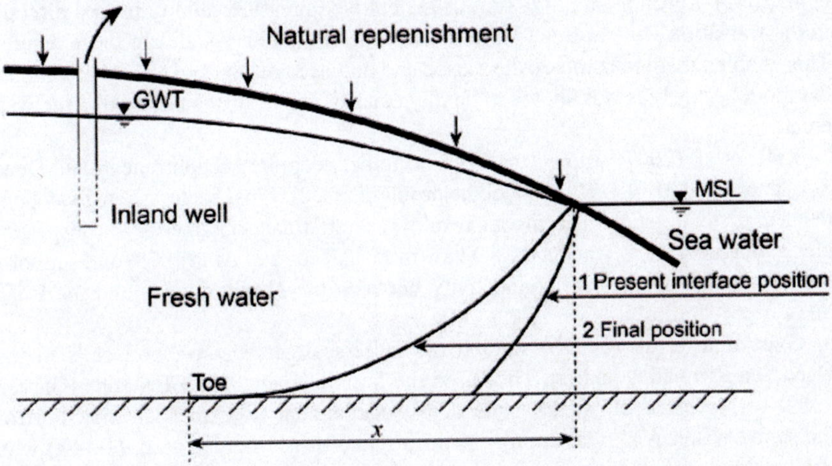


**Fig. 5** A cross section of a coastal aquifer, showing the response of the interface to a two-stage pumping by a coastal collector (following Shechter and Schwarz 1970)

i.e., the search for an optimal solution may lead to local minima and not to the global one.

**Use of Existing Models**

Multiple works utilized existing models, mainly MODFLOW, MT3DMS, SUTRA, and FEFLOW, to solve salinity transport in the aquifer. Abbo et al. (2003) studied the hydrology of saline onshore and offshore springs in lake Kinneret. They used the MODFLOW and MT3DMS codes to describe the density-dependent flow and transport in the Judea group aquifer. After calibration of the model, they were able to match the measurements of salinity and head levels. They were able to identify, using their model, that some of the springs are dominated by the lake and some by the aquifer.

Yakirevich et al. (1998) studied the seawater intrusion in the Khan Yunis area of Gaza Strip by a SUTRA model. The model domain was two-dimensional vertical cross section, and solved for transient density-dependent flow and transport. Both the saturated and unsaturated zone were taken into account in the model. The model predicted significant decline in groundwater levels and increased seawater intrusion in the aquifer due to overexploitation by pumping.

The FEFLOW model is recently becoming popular in modeling density-dependent flow and transport. Shalev et al. (2009) and Levanon et al. (2016, 2017) studied the response to tidal fluctuations of the transition zone between the freshwater and saltwater in the coastal aquifer. Shalev et al. (2009) used a 3D FEFLOW model to study the salinity profile in a well that monitors the transition zone. They found that the measured fluctuations of the transition zone are biased by the well. The magnitude of the fluctuations of the transition zone inside the well was found to be an order of magnitude larger than the magnitude in the aquifer (see also Levanon et al. 2013). Levanon et al. (2016, 2017) deployed pressure/salinity probes directly in the transition zone at a field site in the coastal aquifer to validate these results. This enabled them to compare the biased and unbiased results and to further validate the model. This bias was shown to be the result of the vertical shortcut that the well creates.

Kafri et al. (2007) studied the deep brine between the Mediterranean and Dead Sea. They used FEFLOW to model the problem as a 2D cross section, density dependent, and steady state. The model results resemble data obtained by a deep geo-electromagnetic survey and a Hele-Shaw model (Kafri and Arad 1979) and supports the hypothesis of physical connectivity between the Mediterranean and the Dead Sea.

Gvirtzman et al. (1997b) studied the large-scale free convection cells in the Dead Sea Rift valley aquifer. They used the finite elements OILGEN code (Garven 1989) to solve the 2D steady-state density-dependent flow and heat transport of the saline water. A similar approach was applied by Gvirtzman et al. (1997a) who studied the deep brine in the vicinity of Lake Kinneret. Another finite elements code, JHU2D (Garven and Freeze 1984a), was used by Stanislavsky and Gvirtzman (1999), Gvirtzman and Stanislavsky (2000), and Hurwitz et al. (2000) for similar problems.

The regional model of Zukerman and Shachnai (1999) (see also Sect. 3) took into account the flow of saltwater into the Yarkon-Taninim aquifer. Their model neglected the density variation between saltwater and freshwater, i.e., the density is uniform in the entire aquifer. Thus, the seawater flow was treated via a boundary condition of head dependent flux. Their model yielded rough estimates of the variable rate of recharge of the aquifer by seawater.

## 5 Contaminant Fate and Transport

While groundwater salinization by natural saline end members (seawater, brine) has received significant attention by modelers, other anthropogenic contaminants had also been the subject of multiple studies involving models, especially in the recent decade.

Ezra et al. (2006) studied solute transport through a single fracture in a saturated porous matrix using the analytical solution of Tang et al. (1981). The study was aimed at understanding the transport of chemical products in the fractured chalk aquitard underlying the industrial complex of Ramat Hovav, Negev.

The flow and transport at the vicinity of Shafdan site were studied by Abbo and Gev (2008) and Rona et al. (2014). The effluent water of Shafdan treatment plant are treated by SAT (Soil Aquifer Treatment). They are recharged into the aquifer via shallow infiltration ponds, and are pumped by pumping wells, transported, and supplied to farmers in northern Negev. The Shafdan water contain a higher salinity then that of the background groundwater, and also micro-pollutants. Hence, it is desired that the Shafdan water plume will be abstracted by the Shafdan pumping wells and will not spread toward other freshwater pumping wells. Abbo and Gev

(2008) used MODFLOW and MT3DMS to model the 3D flow of water and transport of chloride. Rona et al. (2014) used a 3D FEFLOW model and included the unsaturated zone in the model. In their transport model, they included both chloride and carbamazepine. The use of two tracers instead of a single tracer, over longer period of time, proved to be valuable in the calibration process.

Dvory et al. (2018) also considered carbamazepine in their transport model. They studied the contamination of a Karst aquifer by leaking sewage pipeline in the Beit Zayit area using MODFLOW and MT3D. Levy et al. (2017) studied the nitrate transport from land surface to wells' perforated casings in the coastal aquifer under a cultivated agricultural land. They used Hydrus 1D for water flow and nitrate transport and calibrated the model using field data in Sharon area. Ganot et al. (2018) studied the geochemical processes along the variably-saturated zone under the Menashe infiltration ponds during SAT with reverse osmosis (RO) desalinated seawater. One of the goals of their work was to check whether SAT can enrich the water with $Mg^{2+}$ sufficiently to avoid post-treatment. The flow model in this work is similar to the one in Ganot et al. (2017). Reactive transport of ions (e.g., Ca, Mg) was modeled by UNSATCHEM. This is a variably saturated reactive transport model by Šimůnek et al. (1996) taking into account cation exchange, calcite precipitation, and dissolution. The results support the claim that enrichment of RO water by SAT is a potentially sustainable practice at calcareous sandy aquifers.

Moreno and Paster (2018) studied the transport of contaminants from an industrial site in the western part of the coastal aquifer. The complexity of modeling transport in this site stems from the significant heterogeneity of the aquifer and the limited knowledge on the geological structure. To tackle this problem, they first derived the geostatistics of the geological facies from drilling logs. Then, they used this information to construct 1300 aquifer structures that were conditioned on the boreholes. Next, they used a flow and transport model (MODFLOW and MT3DMS) to solve the forward problem, and compared the heads and concentrations with the measured values. The aquifer realizations that had reasonable fit were assigned as valid. These valid realizations were used to predict the efficiency of a planned pump and treat scheme. The results describe the statistical distribution of the predicted efficiency, and not just a single expected value (see also Moreno and Paster 2018).

Anomalous transport (e.g., Fig. 6) has received attention in multiple works (Berkowitz et al. 2006; Berkowitz and Scher 1997, 1998). The main focus in these works is the development of a mathematical model to quantitatively account for anomalous transport. The formulation adopted is the continuous-time random walk (CTRW) that has been demonstrated to provide general and effective means to quantify non-Fickian transport (Kosakowski et al. 2001).

## 6 Vadoze Zone

The vadoze zone (unsaturated zone) is the part of the subsurface above the groundwater. The processes of flow and transport in the vadoze zone are important with

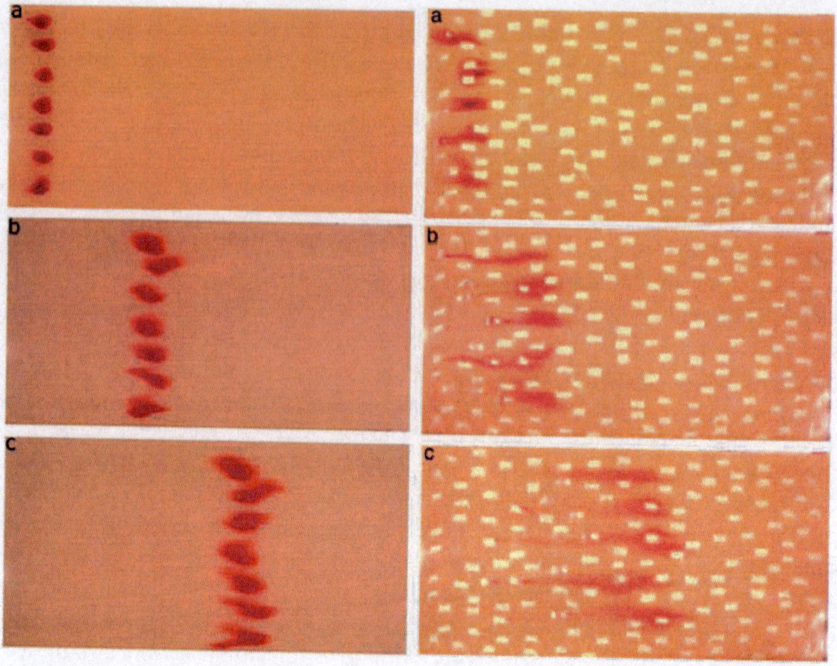


**Fig. 6** Transport in homogeneous versus heterogeneous porous media. Left panel: Photographs of a homogeneous, saturated sand pack with seven dye tracer point injections being transported, under constant flow of 53 mL/min, from left to right. Times at **a** $t = 20$, **b** $t = 105$, **c** $t = 172$, and **d** $t = 255$ min after injection. Internal dimensions of the flow cell are 86 cm (length), 45 cm (height), and 10 cm (width). Right panel: Photographs of a randomly heterogeneous, saturated sand pack with five dye tracer point injections being transported, under constant flow of 65 mL/min, from left to right at **a** $t = 16$ min, **b** $t = 49$ min, and **c** $t = 115$ min after injection (modified after Levy and Berkowitz 2003)

respect to the infiltration of rain water, pollution by anthropogenic sources, geochemical processes, agriculture, etc. One of the seminal works on the subject is the study by Mualem (1976) who developed an empirical formulation for the hydraulic conductivity of the unsaturated porous media. This work, complemented by the study of Van Genuchten (1980), forms the celebrated Mualem–Van Genuchten (VG) model. The VG model is used extensively in numerical studies of unsaturated zone.

In the following, we describe studies that modeled the vadoze zone and the groundwater. We deliberately ignore the large body of works that modeled only the vadoze zone and neglect the groundwater.

Sheffer et al. (2010) studied the recharge of the Yarkon-Taninim aquifer by precipitation. Their model took as input the daily precipitation over the aquifer's outcrops from hundreds of rain stations, and pan evaporation data from four stations. The data was interpolated into each triangular element in the model mesh. Then, the water balance in the vadose zone was described by considering the vadoze zone as a tank (or cell) with inputs and outputs (Fig. 7). The model took into account the soil and rock type. The innovative approach of Sheffer et al. (2010) led to a high resolution estimation of the recharge into the aquifer. It also allows to evaluate the controls on recharge apart from annual precipitation, e.g., the effect of rainy season length.

Dvory et al. (2018) considered flow and transport in a Karst aquifer, characterized by highly conductive karst/fractures (conduits) and low permeability matrix (see also Sect. 3 in Chapter "The Yarkon-Taninim Basin—An Example of a Major Carbonate Aquifers in Israel"). They used a dual permeability model in both unsaturated and saturated zones. The flow in the unsaturated zone was modeled as a one-dimensional (vertical) flow, and it is coupled with the 3D flow in the aquifer. This innovative concept was developed by Kuznetsov et al. (2012).

Another approach for estimating groundwater recharge was described by Kurtzman and Scanlon (2011) who used chloride concentrations to estimate recharge. Turkeltaub et al. (2015) studied the relationship between precipitation and groundwater recharge in a dune area above the coastal aquifer. They used Hydrus 1D and calibrated the model using water saturation data from the field site. The data was

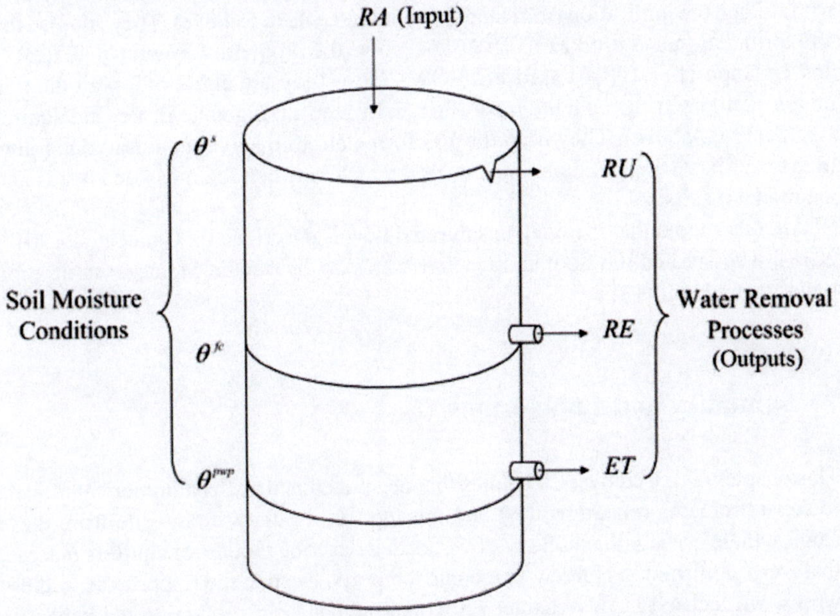


**Fig. 7** Conceptual model of the recharge estimation model. A tank with one inlet for precipitation (RA) and three outputs. The bottom outlet for evapotranspiration (ET) produces output only when the water level in the tank is above the permanent wilting point, $\theta^{pww}$. The second outlet for recharge into the aquifer (RE) produces output only when the water level in the tank is above field capacity, $\theta^{fc}$. The top outlet produces runoff (RU) when the water content in the tank reaches saturation (porosity, $\theta^{s}$) (modified after Sheffer et al. 2010)

collected by an innovative vadoze zone monitoring system installed in slanted boreholes (Rimon et al. 2007). This study and similar works are described in detail in Chapter "Unsaturated-Zone Hydrology in Israel: A Subjective Review". Ben Neriah and Paster (2018) used the T2VOC module of the TOUGH2 numerical model (Falta et al. 1995) to study the effect of the pulse duration on groundwater remediation efficiency in pulsed air sparging. They modeled a laboratory-scale tank with radial symmetry. The numerical model was calibrated using results from a series of laboratory experiments. Using the numerical model, the optimal pulse duration was found, and a methodology for estimating the optimal pulse duration, using groundwater pressure measurements, was developed and tested.

A few works already mentioned in the previous sections (Levanon et al. 2016, 2017; Levy et al. 2017; Rona et al. 2014, 2018) took into account flow and/or transport in the unsaturated zone. We briefly mention these works in the following.

Levy et al. (2017) modeled water flow and nitrogen transport at agricultural lands in both unsaturated and saturated zones. In the unsaturated zone, they used Hydrus 1D, and in the saturated zone, they used MODFLOW and MT3DMS. The models were calibrated with field data obtained by direct push sampling.

Flow and transport in the unsaturated zone were modeled by Rona et al. (2018) who studied the infiltration from Shafdan ponds (see Sect. 5 above). They adopted the VG formulation and used FEFLOW 3D to solve the problem. Levanon et al. (2017) and Levanon et al. (2016) used FEFLOW 2D to study the effect of a periodic tide on the capillary fringe and the freshwater–saltwater mixing zone in a costal aquifer (see Sect. 4.2 above). They used the VG formulation to solve the unsaturated flow in the capillary fringe and obtained a good match between the measured heads and the model results.

Another model that included unsaturated flow is described by Ganot et al. (2017, 2018), who studied the flow and reactive transport below the Menashe infiltration ponds using Hydrus 1D.

## 7 Summary and Conclusions

This chapter reviewed over 70 studies that developed and utilized numerical models to solve problems related to flow and transport in groundwater. Aside from these studies, there is a significant body of works that describe models of aquifers in Israel that were published in Hebrew as technical reports. Despite the importance of these works, we focused in this chapter on works published in peer-reviewed papers in English, which are accessible to a larger audience. As can be seen in Fig. 8, the number of studies published per decade was rather constant during 1960s, 1970s, and 1980s. There is a clear trend of rise during the 1990s and the following decade. This may be attributed to the increasing availability of computers and numerical models. In the present decade, we observe some small decline in the number of publications, but this is not necessarily pointing to a shift in the trend.

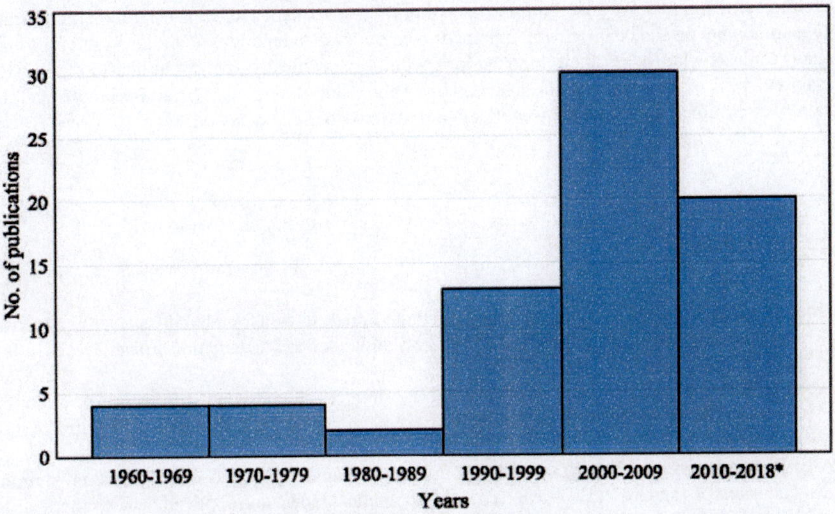


**Fig. 8** Number of studies per decade published in peer-reviewed papers and in books, involving groundwater modeling in Israel. Note that the last bar (*) refers to the present decade, in which one may expect few additional (about 4, by extrapolation) publications yet to appear

About half (34) of the studies that were reviewed used existing ("shelf") models to simulate flow and transport in aquifers. 12 out of 34 studies used the FEFLOW code, making it the most popular software among these studies. Few (6) other works used MODFLOW, with or without the MT3DMS package for transport, and few more (3) used HYDRUS.

About another half (36) of the studies that were reviewed in this chapter involved development of new models for flow and transport. These studies span over multitude of subjects, from optimization problem solving to water flow modeling, tracer transport, and variable density flow. The numerical models used include cell models, finite difference, and finite element methods. Few of the works include sharp interface modeling and boundary layer approach for modeling transition zone between freshwater and saltwater.

The models reviewed in this chapter had a significant impact on the development and utilization of the groundwater resources in Israel. Judging by the number of citations of the works, they also had significant contribution at the larger scale, i.e., for the benefit of the global hydrological community. This is especially true when considering the studies where new methods were developed.

Issues likes global warming, population growth, and diminishing quality of groundwater are expected to make the management of groundwater in Israel even more challenging in the future. Taking into account the complexity of the problems involved, it is reasonable to assume that models will continue to be an important tool for increasing our understanding of the flow and transport in the aquifers. This improved understanding is essential to reliable decision making.

**Acknowledgements** We are grateful to the editors, Yossi Yechieli and Uri Kafri, for their constructive comments on the chapter, which helped improve it considerably. We are also grateful to the many authors who shared their contributions with us. Specifically, we would like to thank Alex Yakirevich, Alex Furman, Gedeon Dagan, Hila Abbo, Haim Gvirtzman, Brian Berkovitz, and the late Alon Rimmer, who sent us his works just few weeks before he passed away.

## References

Abbo H, Gev I (2008) Numerical model as a predictive analysis tool for rehabilitation and conservation of the Israeli coastal aquifer: example of the Shafdan sewage reclamation project. Desalination 226(1–3):47–55

Abbo H, Shavit U, Markel D, Rimmer A (2003) A numerical study on the influence of fractured regions on lake/groundwater interaction; the Lake Kinneret (Sea of Galilee) case. J Hydrol 283(1–4):225–243

Adar E, Neuman S, Woolhiser D (1988) Estimation of spatial recharge distribution using environmental isotopes and hydrochemical data, I. Mathematical model and application to synthetic data. J Hydrol 97(3):251–277. https://doi.org/10.1016/0022-1694(88)90119-9

Adar E, Rosenthal E, Issar A, Batelaan O (1992) Quantitative assessment of the flow pattern in the southern Arava Valley (Israel) by environmental tracers and a mixing cell model. J Hydrol 136(1–4):333–352

Amir N, Kafri U, Herut B, Shalev E (2013) Numerical simulation of submarine groundwater flow in the coastal aquifer at the Palmahim Area, the Mediterranean Coast of Israel. Water Resour Manage 27(11):4005–4020. https://doi.org/10.1007/s11269-013-0392-2

Asmael NM, Dupuy A, Huneau F, Hamid S, Coustumer PL (2015) Groundwater modeling as an alternative approach to limited data in the northeastern part of Mt. Hermon (Syria), to develop a preliminary water budget. Water 7(7):3978–3996

Assouline S (1993) Estimation of lake hydrologic budget terms using the simultaneous solution of water, heat, and salt balances and a Kalman filtering approach: application to lake Kinneret. Water Resour Res 29(9):3041–3048

Avisar D, Rosenthal E, Flexer A, Shulman H, Ben-Avraham Z, Guttman J (2003) Salinity sources of Kefar Uriya wells in the Judea Group aquifer of Israel. Part 1—conceptual hydrogeological model. J Hydrol 270(1–2):27–38

Bear J (1972) Dynamics of fluids in porous media, vol 1. American Elsevier Publishing Company

Bear J (1979) Hydraulics of groundwater. McGraw-Hill series in Water resources and environmental engineering. New York, 463 pp

Bear J, Bachmat Y (1990) Introduction to modeling of transport phenomena in porous media. Kluwer Academic, Dordrecht

Bear J, Dagan G (1964) Some exact solutions of interface problems by means of the hodograph method. J Geophys Res 69(8):1563–1572

Ben-Itzhak LL, Gvirtzman H (2005) Groundwater flow along and across structural folding: an example from the Judean Desert, Israel. J Hydrol 312(1–4):51–69

Ben Neriah A, Paster A (2018) Air sparging: enhancing the volatilization of organic compounds through selection of optimized pulsation frequency. Paper presented at the EGU General Assembly

Berger D (1999) Hydrological model for the Yarqon-Taninim aquifer. Mekorot Ltd., pp 1–47 (in Hebrew)

Berkowitz B, Scher H (1997) Anomalous transport in random fracture networks. Phys Rev Lett 79(20):4038–4041

Berkowitz B, Scher H (1998) Theory of anomalous chemical transport in random fracture networks. Phys Rev E 57(5):5858–5869

Berkowitz B, Cortis A, Dentz M, Scher H (2006) Modeling non-Fickian transport in geological formations as a continuous time random walk. Rev Geophys 44(2):1–49

Bouwer H (1978) Groundwater hydrology. McGraw-Hill Book, New York

Dafny E, Gvirtzman H, Burg A, Fleischer L (2003) The hydrogeology of the Golan basalt aquifer, Israel. Isr J Earth Sci 52:139–153

Dafny E, Burg A, Gvirtzman H (2010) Effects of Karst and geological structure on groundwater flow: the case of Yarqon-Taninim aquifer, Israel. J Hydrol 389(3):260–275. https://doi.org/10.1016/j.jhydrol.2010.05.038

Dagan G (1989) Flow and transport in porous formations. Springer Science and Business Media, Berlin

Dagan G, Bear J (1968) Solving the problem of local interface upconing in a coastal aquifer by the method of small perturbations. J Hydraul Res 6(1):15–44

Diersch H (2005) WASY Software FEFLOW (R)—finite element subsurface flow and transport simulation system: reference manual. WASY GmbH Inst. WASY GmbH, pp 1–292

Diersch H (2013) FEFLOW: finite element modeling of flow, mass and heat transport in porous and fractured media. Springer Science and Business Media, Berlin

Dvory NZ, Kuznetsov M, Livshitz Y, Gasser G, Pankratov I, Lev O, Yakirevich A (2018) Modeling sewage leakage and transport in carbonate aquifer using carbamazepine as an indicator. Water Res 128:157–170

Ezra S, Feinstein S, Yakirevich A, Adar E, Bilkis I (2006) Retardation of organo-bromides in a fractured chalk aquitard. J Contam Hydrol 86(3–4):195–214

Falta RW, Pruess K, Finsterle S, Battistelli A (1995) T2VOC user's guide. Lawrence Berkeley Laboratory, pp 1–168

Ganot Y, Holtzman R, Weisbrod N, Nitzan I, Katz Y, Kurtzman D (2017) Monitoring and modeling infiltration-recharge dynamics of managed aquifer recharge with desalinated seawater. Hydrol Earth Syst Sci 21(9):4479–4493

Ganot Y, Holtzman R, Weisbrod N, Russak A, Katz Y, Kurtzman D (2018) Geochemical processes during managed aquifer recharge with desalinated seawater. Water Resour Res 54(2):978–994

Garven G (1989) A hydrogeologic model for the formation of the giant oil sands deposits of the Western Canada sedimentary basin. Am J Sci 289(2):105–166

Garven G, Freeze RA (1984a) Theoretical analysis of the role of groundwater flow in the genesis of stratabound ore deposits; 1. Mathematical and numerical model. Am J Sci 284(10):1085–1124

Garven G, Freeze RA (1984b) Theoretical analysis of the role of groundwater flow in the genesis of stratabound ore deposits; 2. Quantitative results. Am J Sci 284(10):1125–1174

Gordon E, Shamir U, Bensabat J (2000) Optimal management of a regional aquifer under salinization conditions. Water Resour Res 36(11):3193–3203

Guttman J, Zukerman H (1995) Yarqon-Taninim-Be'er Sheva basin, setting and calibration of flow and salinity model. Rep. Mekorot Co., Israel (in Hebrew)

Gvirtzman H, Garven G, Gvirtzman G (1997a) Hydrogeological modeling of the saline hot springs at the Sea of Galilee, Israel. Water Resour Res 33(5):913–926

Gvirtzman H, Stanislavsky E (2000) Palaeohydrology of hydrocarbon maturation, migration and accumulation in the Dead Sea Rift. Basin Res 12(1):79–93

Gvirtzman H, Garven G, Gvirtzman G (1997b) Thermal anomalies associated with forced and free ground-water convection in the Dead Sea rift valley. Geol Soc Am Bull 109(9):1167–1176

Harbaugh AW, Banta ER, Hill MC, McDonald MG (2000) MODFLOW-2000, The U. S. Geological Survey modular ground-water model-user guide to modularization concepts and the ground-water flow process. Open-file report. U. S. Geological Survey, pp 1–134

Hartmann A, Wagener T, Rimmer A, Lange J, Brielmann H, Weiler M (2013) Testing the realism of model structures to identify karst system processes using water quality and quantity signatures. Water Resour Res 49(6):3345–3358

Hele-Shaw HS (1898) The flow of water. Nature 58:34–36. https://doi.org/10.1038/058034a0

Hurwitz S, Stanislavsky E, Lyakhovsky V, Gvirtzman H (2000) Transient groundwater-lake interactions in a continental rift: Sea of Galilee, Israel. Geol Soc Am Bull 112(11):1694–1702

Kafri U, Arad A (1979) Current subsurface intrusion of Mediterranean seawater—a possible source of groundwater salinity in the Rift Valley system, Israel. J Hydrol 44(3):267–287. https://doi.org/10.1016/0022-1694(79)90135-5

Kafri U, Goldman M, Lyakhovsky V, Scholl C, Helwig S, Tezkan B (2007) The configuration of the fresh–saline groundwater interface within the regional Judea Group carbonate aquifer in northern Israel between the Mediterranean and the Dead Sea base levels as delineated by deep geoelectromagnetic soundings. J Hydrol 344(1–2):123–134

Kessler A (2012) Integrating a cell model in the analysis of the salt flow to the Sea of Galilee. Paper presented at the Annual conference of the Israel Water Resources Association

Kessler A, Kafri U (2007) Application of a cell model for operational management of the Na'aman groundwater basin, Israel. Isr J Earth Sci 56:29–46

Kosakowski G, Berkowitz B, Scher H (2001) Analysis of field observations of tracer transport in a fractured till. J Contam Hydrol 47(1):29–51. https://doi.org/10.1016/S0169-7722(00)00140-6

Kurtzman D, Scanlon BR (2011) Groundwater recharge through Vertisols: irrigated cropland vs. natural land, Israel. Vadose Zone J 10(2):662–674

Kuznetsov M, Yakirevich A, Pachepsky YA, Sorek S, Weisbrod N (2012) Quasi 3D modeling of water flow in vadose zone and groundwater. J Hydrol 450–451:140–149. https://doi.org/10.1016/j.jhydrol.2012.05.025

Levanon E, Yechieli Y, Shalev E, Friedman V, Gvirtzman H (2013) Reliable monitoring of the transition zone between fresh and saline waters in coastal aquifers. Groundwater Monit Rem 33(3):101–110

Levanon E, Shalev E, Yechieli Y, Gvirtzman H (2016) Fluctuations of fresh-saline water interface and of water table induced by sea tides in unconfined aquifers. Adv Water Resour 96:34–42

Levanon E, Yechieli Y, Gvirtzman H, Shalev E (2017) Tide-induced fluctuations of salinity and groundwater level in unconfined aquifers—field measurements and numerical model. J Hydrol 551:665–675

Levy M, Berkovitz B (2003) Measurement and analysis of non-Fickian dispersion in heterogeneous porous media. J Contam Hydrol 64:203–226

Levy Y, Shapira RH, Chefetz B, Kurtzman D (2017) Modeling nitrate from land surface to wells' perforations under agricultural land: success, failure, and future scenarios in a Mediterranean case study. Hydrol Earth Syst Sci 21(7):3811

Mandell AH, Zeitoun DG, Dagan G (2003) Salinity sources of Kefar Uria wells in the Judea group aquifer of Israel. Part 2—quantitative identification model. J Hydrol 270(1):39–48. https://doi.org/10.1016/S0022-1694(02)00217-2

Mercado A (1976) Nitrate and chloride pollution of aquifers: a regional study with the aid of a single-cell model. Water Resour Res 12(4):731–747

Moreno Z, Paster A (2018) Prediction of pollutant remediation in a heterogeneous aquifer in Israel: reducing uncertainty by incorporating lithological, head and concentration data. J Hydrol 564:651–666. https://doi.org/10.1016/j.jhydrol.2018.07.012

Mualem Y (1976) A new model for predicting the hydraulic conductivity of unsaturated porous media. Water Resour Res 12(3):513–522

Paster A (2010) Mixing between fresh and salt waters at aquifer regional scale and identification of transverse dispersivity. J Hydrol 380(1–2):36–44

Paster A, Dagan G (2007) Mixing at the interface between two fluids in porous media: a boundary-layer solution. J Fluid Mech 584:455–472

Paster A, Dagan G (2008a) Mixing at the interface between fresh and salt waters in 3D steady flow with application to a pumping well in a coastal aquifer. Adv Water Resour 31(12):1565–1577. https://doi.org/10.1016/j.advwatres.2008.06.008

Paster A, Dagan G (2008b) Mixing at the interface between two fluids in aquifer well upconing steady flow. Water Resour Res 44(5):1–11

Paster A, Dagan G, Guttman J (2006) The salt-water body in the Northern part of Yarkon-Taninim aquifer: field data analysis, conceptual model and prediction. J Hydrol 323(1–4):154–167

Peleg N, Gvirtzman H (2010) Groundwater flow modeling of two-levels perched karstic leaking aquifers as a tool for estimating recharge and hydraulic parameters. J Hydrol 388(1–2):13–27

Peleg N, Morin E, Gvirtzman H, Enzel Y (2012) Rainfall, spring discharge and past human occupancy in the Eastern Mediterranean. Clim Change 112(3–4):769–789

Prieto C, Destouni G, Schwartz J (2001) Seawater intrusion in coastal aquifers: effects of seasonal variations in extraction and recharge rates. Paper presented at the First international conference on saltwater intrusion and coastal aquifers—monitoring, modeling, and management, Essaouira, Morocco

Rimmer A, Gal G (2003) Estimating the saline springs component in the solute and water balance of lake Kinneret, Israel. J Hydrol 284(1–4):228–243

Rimmer A, Givati A (2014) Hydrology. In: Lake Kinneret—ecology and management. Aquatic ecology series, vol 6. Springer, Berlin, pp 97–111

Rimmer A, Salingar Y (2006) Modelling precipitation-streamflow processes in karst basin: the case of the Jordan River sources, Israel. J Hydrol 331(3–4):524–542

Rimmer A, Hurwitz S, Gvirtzman H (1999) Spatial and temporal characteristics of saline springs: Sea of Galilee, Israel. Ground Water 37(5):663–673

Rimon Y, Dahan O, Nativ R, Geyer S (2007) Water percolation through the deep vadose zone and groundwater recharge: preliminary results based on a new vadose zone monitoring system. Water Resour Res 43(5):1–12

Rona M, Gasser G, Negev I, Pankratov I, Elhanany S, Lev O, Gvirtzman H (2014) A 3-D hydrologic transport model of a water recharge system using carbamazepine and chloride as tracers. Water Resour Res 50(5):4220–4241

Rona M, Lev O, Gvirtzman H (2018) Optimal remediation scheme for a wastewater recharge site: contaminants fate and transport model. Ground Water 56(6):871–880

Rubin H (1983) On the application of the boundary layer approximation for the simulation of density stratified flows in aquifers. Adv Water Resour 6(2):96–105. https://doi.org/10.1016/0309-1708(83)90046-5

Schmorak S, Mercado A (1969) Upconing of fresh water—sea water interface below pumping wells, field study. Water Resour Res 5(6):1290–1311

Schwarz J (1976) Linear models for groundwater management. J Hydrol 28(2–4):377–392

Schwarz J, Bear J, Dagan G (2016) Groundwater development in Israel. Ground Water 54(1):143–148

Shalev E, Lazar A, Wollman S, Kington S, Yechieli Y, Gvirtzman H (2009) Biased monitoring of fresh water-salt water mixing zone in coastal aquifers. Ground Water 47(1):49–56

Shavit U, Furman A (2001) The location of deep salinity sources in the Israeli coastal aquifer. J Hydrol 250(1–4):63–77

Shechter M, Schwarz J (1970) Optimal planning of a coastal collector. Water Resour Res 6(4):1017–1024

Sheffer N, Dafny E, Gvirtzman H, Navon S, Frumkin A, Morin E (2010) Hydrometeorological daily recharge assessment model (DREAM) for the Western Mountain Aquifer, Israel: model application and effects of temporal patterns. Water Resour Res 46:W05510. https://doi.org/10.1029/2008WR007607

Šimůnek J, Suarez DL, Sejna M (1996) The UNSATCHEM software package for simulating the one-dimensional variably saturated water flow, heat transport, carbon dioxide production and transport, and multicomponent solute transport with major ion equilibrium and kinetic chemistry. Version 2.0. Report 141. US Salinity Laboratory, pp 1–197

Šimůnek J, Van Genuchten MT, Sejna M (2005) The HYDRUS-1D software package for simulating the one-dimensional movement of water, heat, and multiple solutes in variably-saturated media. University of California-Riverside Res Rep 3, pp 1–240

Šimůnek J, Van Genuchten MT, Šejna M (2016) Recent developments and applications of the HYDRUS computer software packages. Vadose Zone J 15(7):1–25

Sorek S, Borisov V, Yakirevich A (1999) Modified Eulerian Lagrangian method for density dependent miscible transport. In: Seawater intrusion in coastal aquifers—concepts, methods and practices. Springer, Berlin, pp 363–398

Sorek S, Borisov V, Yakirevich A (2001) A two-dimensional areal model for density dependent flow regime. Transp Porous Media 43(1):87–105

Stanislavsky E, Gvirtzman H (1999) Basin-scale migration of continental-rift brines: paleohydrologic modeling of the Dead Sea basin. Geology 27(9):791–794

Sun NZ, Yeh WWG (1992) A stochastic inverse solution for transient groundwater flow: parameter identification and reliability analysis. Water Resour Res 28(12):3269–3280

Tang D, Frind E, Sudicky FA (1981) Contaminant transport in fractured porous media: analytical solution for a single fracture. Water Resour Res 17(3):555–564

Turkeltaub T, Kurtzman D, Bel G, Dahan O (2015) Examination of groundwater recharge with a calibrated/validated flow model of the deep vadose zone. J Hydrol 522:618–627

Van Genuchten MT (1980) A closed-form equation for predicting the hydraulic conductivity of unsaturated soils 1. Soil Sci Soc Am J 44(5):892–898

Vengosh A, Spivack AJ, Artzi Y, Ayalon A (1999) Geochemical and boron, strontium, and oxygen isotopic constraints on the origin of the salinity in groundwater from the Mediterranean coast of Israel. Water Resour Res 35(6):1877–1894

Voss CI (1984) A finite-element simulation model for saturated-unsaturated, fluid-density-dependent ground-water flow with energy transport or chemically-reactive single-species solute transport. US Geological Survey, pp 1–407

Weiss M, Gvirtzman H (2007) Estimating ground water recharge using flow models of perched karstic aquifers. Ground Water 45(6):761–773

Yakirevich A, Melloul A, Sorek S, Shaath S, Borisov V (1998) Simulation of seawater intrusion into the Khan Yunis area of the Gaza Strip coastal aquifer. Hydrogeol J 6(4):549–559

Yechieli Y, Kafri U, Wollman S, Lyakhovsky V, Weinberger R (2007) On the relation between steep monoclinal flexure zones and steep hydraulic gradients. Ground Water 45(5):616–626

Yechieli Y, Kafri U, Wollman S, Shalev E, Lyakhovsky V (2009) The effect of base level changes and geological structures on the location of the groundwater divide, as exhibited in the hydrological system between the Dead Sea and the Mediterranean Sea. J Hydrol 378:218–229

Yechieli Y, Shalev E, Wollman S, Kiro Y, Kafri U (2010) Response of the Mediterranean and Dead Sea coastal aquifers to sea level variations. Water Resour Res 46(12):1–11

Zech A, Attinger S, Bellin A, Cvetkovic V, Dietrich P, Fiori A, Teutsch G and Dagan G (2019) A critical analysis of transverse dispersivity field data. Groundwater, 57: 632-639. https://doi.org/10.1111/gwat.12838

Zheng C, Wang PP (1999) MT3DMS: a modular three-dimensional multispecies transport model for simulation of advection, dispersion, and chemical reactions of contaminants in groundwater systems; documentation and user's guide. United States Army Corps of Engineers, pp 1–219

Zhou Q (1999) Modeling seawater intrusion in coastal aquifers. PhD thesis, Technion-Israel Institute of Technology

Zhou Q, Bear J, Bensabat J (2005) Saltwater upconing and decay beneath a well pumping above an interface zone. Transp Porous Media 61(3):337–363

Zukerman H, Shachnai E (1999) Yarkon-Taninim-Beer Sheva basin, flow model update. TAHAL Rep 6759-00.133

# Hydrogeology in Archeological Perspective: What Did People in Ancient Times Know About Hydrogeology?

**Yoseph Yechieli, Ronny Reich, Ehud Galili, Tsvika Tsuk, Uzi Dahari, Gideon Avni, and Dorit Sivan**

## 1 Introduction

The term "hydrogeology" used herein relates to the study of groundwater systems and underground flow mechanisms and is different from the term "hydrology" that includes also surface water flow.

This chapter deals solely with the archeology of groundwater systems. In many parts of the world, water is a scarce commodity and people's survival depends on water availability. Since in many locations, there is no surface water source (e.g.,

Y. Yechieli (✉)
Geological Survey of Israel, Jerusalem, Israel
e-mail: yechieli@gsi.gov.il

R. Reich · E. Galili
The Zinman Institute of Archaeology, University of Haifa, 3498838 Haifa, Israel
e-mail: ronny@research.haifa.ac.il

E. Galili
e-mail: galilish@netvision.net.il

T. Tsuk
Department of Archaeology, Israeli Nature and Park Authority, 3 Am Veolamo Str., 95463 Jerusalem, Israel
e-mail: tsuk@npa.org.il

U. Dahari · G. Avni
Israel Antiquities Authority, P.O. Box 586, 91004 Jerusalem, Israel
e-mail: dahari.uzi@gmail.com

G. Avni
e-mail: gideon@israntique.org.il

D. Sivan
Maritime Civilizations Department, L. Charney School of Marine Studies, University of Haifa, 3498838 Haifa, Israel
e-mail: dsivan@research.haifa.ac.il


U. Kafri and Y. Yechieli (eds.), *The Many Facets of Israel's Hydrogeology*, Springer Hydrogeology, https://doi.org/10.1007/978-3-030-51148-7_18

lakes, rivers), there was a need throughout the history to find subsurface water. The subterranean water sources could be relatively easy to find in the case of springs, which issue at the surface. However, since springs are not common everywhere, people in ancient times had to look for a subsurface source, namely, groundwater. However, finding groundwater is much more difficult, even today, since it requires a basic understanding in geological and hydrogeological disciplines and knowledge of the specific conditions in the nearby areas. Thus, it is not yet clear what hydrological knowledge humans possessed and what was their degree of success in finding groundwater in ancient times.

Since a reliable, permanent source of drinking water is the most basic requirement for human physical survival, all temporary encampments or sedentary settlements have been located near or within a short walk of perennial water sources such as springs, rivers, and freshwater lakes. In recent decades, throughout systematic excavations of early Neolithic sites in the Levant (Table 1), a series of discoveries have exposed several wells dated to 9500–8000 BP (before present). This is an additional and unexpected aspect of human activity resulting from the Neolithic revolution. Beside active steps in plant cultivation, animal domestication (Bar-Yosef and Belfer-Cohen 1989; Bar-Yosef and Meadow 1995), and the redesigning of domestic and public spaces, securing sustainable water sources was an important part of the

**Table 1** Relevant prehistoric–historic periods, based on the new encyclopedia of archeological excavations in the Holy Land (modified after Stern et al. 1993)

| Period | Culture | Approximate dates (BCE) |
|---|---|---|
| Pre-Pottery Neolithic | Phases A, B, and C | 8300–6000 |
| Pottery Neolithic/Early Chalcolithic | | 6000–5000 |
| Chalcolithic | | 5000–3800 |
| Bronze Age | Early Bronze I to III | 3800–2400 |
| | Intermediate Bronze | 2400–2000 |
| | Middle Bronze I to IIb | 2000–1550 |
| Iron Age | Late Bronze I to IIb | 1550–1200 |
| | Iron Age Ia to Ib | 1200–1000 |
| | Iron Age IIa to IIc | 1000–586 |
| Persian | | 586–332 |
| Hellenistic | | 332–63 |
| Roman and Byzantine | | 63 BCE–636 CE |
| Early Islamic | | 636–1099 CE |
| Crusader | | 1099–1291 CE |
| Late Islamic (Mameluke) | | 1291–1516 CE |
| Ottoman | | 1516–1917 CE |
| British Mandate | | 1917–1948 CE |

construction of new settlements. The active digging of wells reaching the underground water table reflects an innovative approach to water provisioning in sedentary communities.

The earliest Neolithic wells in the world currently known, dated to 10,000 BP (Pre-Pottery Neolithic B), have been discovered in two Cypriot sites: Kissonerga-Mylouthkia and Shillourocambous. In Kissonerga-Mylouthkia, on the Mediterranean coast of western Cyprus, several cylindrical, unlined shafts dug into the local Havara deposit were exposed (Peltenburg et al. 2000, 2001). Each well is about 1 m in diameter and 7–8 m deep. However, since erosion and modern quarrying have destroyed the upper part of the wells, the exact original depth is unknown.

The objective of this chapter is to summarize some of the archeological findings in Israel that may shed light on the question as to what degree of hydrological knowledge and understanding of the basic concept of groundwater table humans had in the past. The basic information about water management in ancient times in Israel is summarized in Tsuk (2011). In order to answer this specific question, this chapter deals mainly with groundwater exploitation via water wells for better understanding of the concept of aquifers, and not with springs' water and surface water systems (pipes, canals, dams, etc.). We do, however, mention water constructions or installations such as the case of the Siloam Tunnel in Jerusalem that can shed light on the research question. We also review some archeological findings associated with irrigation that employed the natural stratigraphy and hydrological conditions as in the case of the Mawasi (ancient irrigation system in coastal dune relying on a shallow groundwater) in Caesarea (Fig. 1) (Roskin et al. 2015; Shtienberg et al. 2017a, b) and the depressions in the vicinity of Yavne (Fig. 1) (Taxel et al. 2018).

Since the degree of knowledge and understanding of hydrological systems in ancient times is still debatable, we present here different views accompanied by their explanation. Below we describe the hydrogeological aspects in several selected sites according to their geographical location.

## 2 The Coastal Plain

### *2.1 The Hydrogeology of the Coastal Aquifer*

The coastal aquifer of Israel is described in detail in Chap. 4. It extends from the Gaza area to Haifa, along the coast of the eastern Mediterranean Sea (Fig. 1). Its northern part, west of the Carmel Mountain (north of Caesarea) is usually considered as a separated aquifer, named herein the Carmel coastal aquifer. Another coastal aquifer segment exists north of the Haifa Bay which is not dealt with in this chapter.

The coastal aquifer is divided vertically to several sub-aquifers, the lower of which are confined and the upper phreatic. The maximal total thickness is ~200 m near the shoreline in the southern part, decreasing to a few tens of meters in the north. The

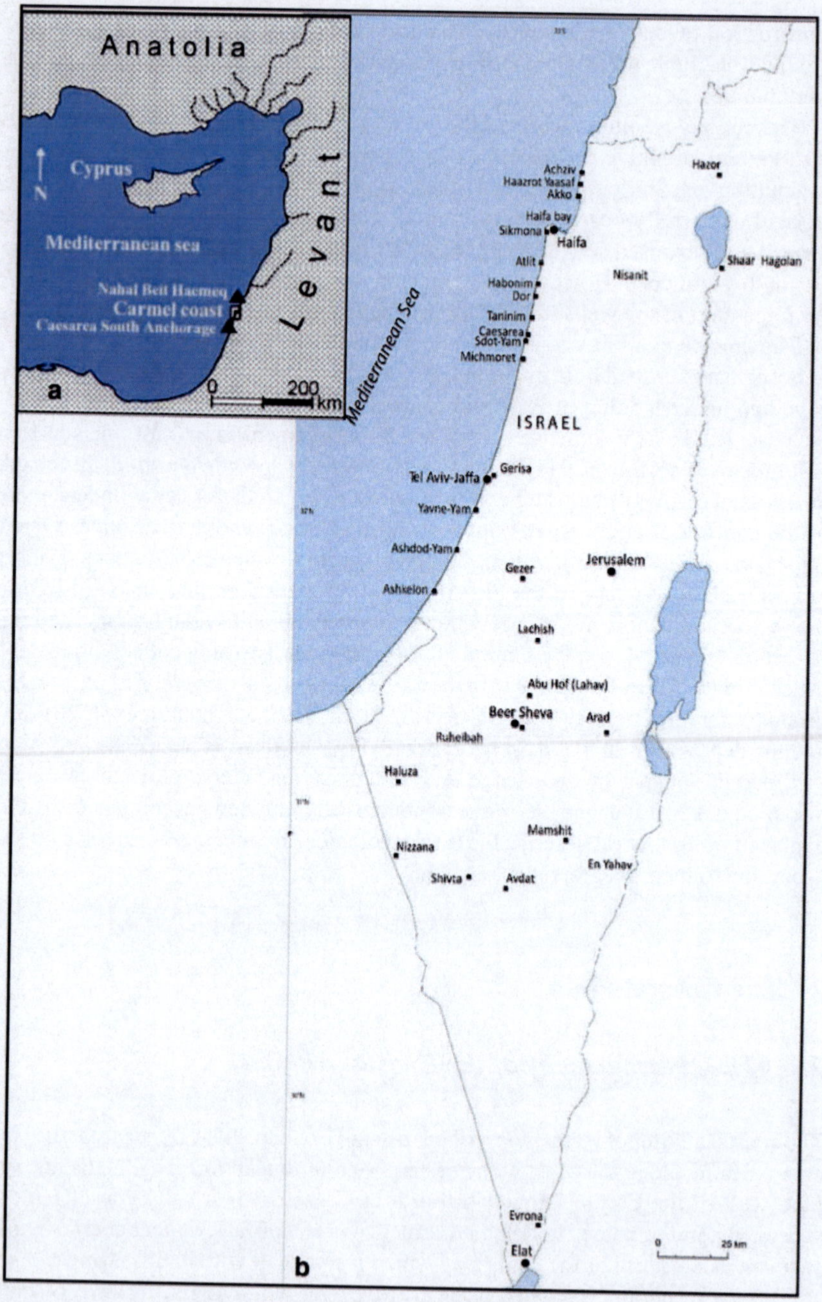


**Fig. 1** **a** Map of the Levant. **b** General map of Israel with locations of relevant archeological sites. **c** Locations of archeological sites in the Carmel area (After Galili et al. 2019)

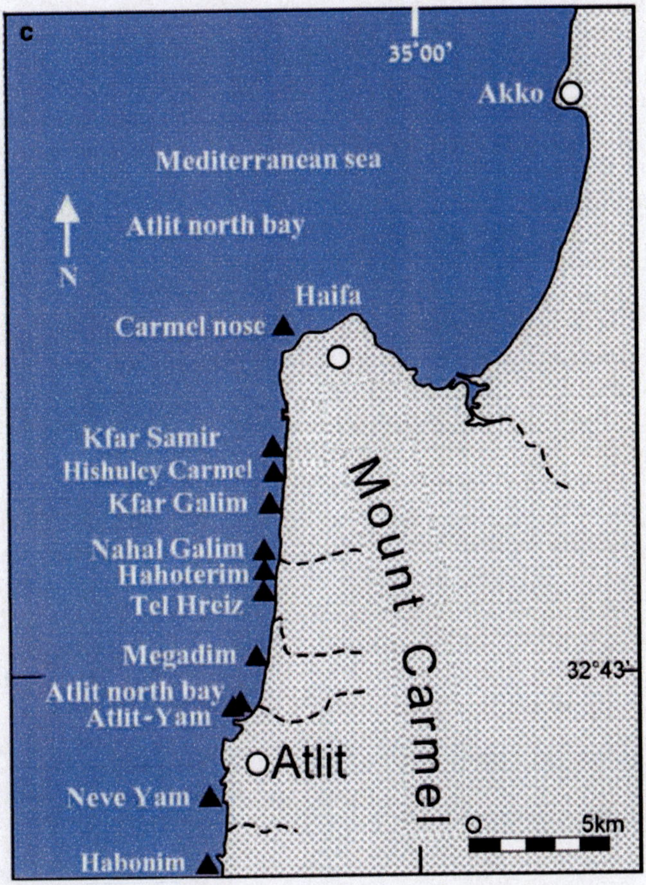


**Fig. 1** (continued)

aquifer consists of mostly calcareous sandstone (locally named "kurkar") of the Kurkar Group alternating with impermeable clay and silt units (Issar 1968).

Along the Carmel coast, the sandy unit overlies, at places, a thick clay unit overlying a calcareous sand (hereafter kurkar) unit (Cohen-Seffer et al. 2005). In many places, freshwater seepages in small amounts emerge in areas where the sand is removed. This phenomenon could have given the ancient people the idea about the local hydrology.

The important role of fresh groundwater in the coastal aquifer to the inhabitants of Israel is demonstrated not only along the coast but also offshore where currently submerged settlements existed since the Neolithic period. The following description of the different sites in the coastal area is given in a chronological order (Table 1).

## 2.2 *Submerged Neolithic Settlements off the Carmel Coast*

The present locations of human habitations on the sea bottom are due to post-glacial sea level rise that inundated prehistoric landscape and settlements situated near ancient coastlines. Their remnants were found underwater off the Carmel coast, from Haifa to Habonim (Fig. 1b, c). The prehistoric settlements, submerged off the Carmel coast belong to two main entities: The Late Pre-Pottery Neolithic (PPNC—8000–9300 BP) and the Late Ceramic Neolithic/Early Chalcolithic periods (Wadi Rabah culture—ca. 7000–8000 BP). Permanent settlements required long lasting sources of fresh water.

**Atlit-Yam PPNC wells** (Fig. 1): Northeast of the Atlit Crusader castle, about 200–400 m offshore, the Pre-Pottery Neolithic C site of Atlit-Yam is located, at a water depth of 8–12 m below present mean sea level (hereafter bsl). The site was radiocarbon dated to 9250–7970 years BP. Several stone-built circular features (0.8–1.5 m in diameter) were discovered. Three were excavated and identified as stone constructed water wells (Fig. 2; Galili and Nir 1993; Galili 2004; Galili and Rosen 2011b). The fully excavated cylindrical shaft (Well No. 11) is 5.7 m deep and 1.5 m in diameter. Its three courses superstructure protruded above the inundated land surface. The Neolithic people dug through the clay sediments and lined the shaft with ca. 28 courses of undressed stones. The lower part of the well was dug for about 1.5 m into the underlying kurkar, to reach the freshwater aquifer (Fig. 3). The fill of the well contained soft clay, kurkar and limestone fragments of various

**Fig. 2** Well (no 11) in the submerged site of Atlit-Yam (Photo I. Grinberg)

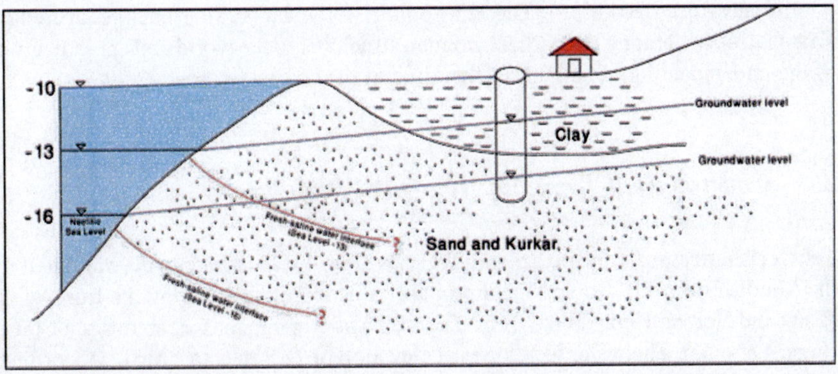


**Fig. 3** Schematic hydrogeological cross section in Atlit-Yam site (modified after Galili et al. 2005 and Galili and Rosen 2011a)

sizes, animal remains, flint, stone, bone and wood artifacts, waterlogged, and charred plant remain. The lower 2 m of the fill contained large kurkar stones and almost no animal bones. The upper fill consisted of similar components as the lower, in addition to numerous animal bones, some partly articulated, indicating that food waste was dumped there. The layers of large stones dumped in the lower parts of the fill may represent an attempt to raise the base of the well and tap into a higher level of the freshwater aquifer assumingly in response to sea level rise and saltwater seepage. The findings suggest that at some point the well stopped functioning as a freshwater source, presumably due to sea level rise and salinization, and served as a garbage pit. The radiocarbon dates of waterlogged and charred wood remain from Wells 11 and 66 range from 8540–7930 to 8977–8187 years BP, respectively, representing the final stage of their use, thus, marking the latest possible age of the well's construction (Galili 2004; Galili and Rosen 2011a, b; Galili et al. 2019). It seems that Atlit-Yam water wells, now submerged, demonstrate a good hydrogeological knowledge during that time. The Atlit-Yam wells are the most ancient stone-built water wells discovered so far.

## 2.3 *The Submerged Pottery Neolithic Wells*

The Pottery Neolithic settlements at Neve Yam, Tel Hreiz, Kefar Galim, Nahal Galim, and Kefar Samir sites (Fig. 1a) include numerous circular structures 1–1.5 m in diameter, made of undressed stones, representing storage pits and water wells (Galili and Weinstein-Evron 1985; Galili et al. 1997; 2016). At Kfar Samir, three water wells constructed of alternating courses of wooden beams and undressed stones were recovered. One of them had a rectangular opening of 1 × 0.8 m. It was excavated to a depth of 2 m without reaching the bottom. Organic materials from the well's fill and the wooden constructions were radiocarbon dated to 7964–7320 years BP

(Galili and Rosen 2011a, b). The Kfar Samir wells, and similar ones recovered at Kfar Galim, are among the earliest wooden structures in the world and are indicative of the understanding of humans of that time of the coastal hydrogeology.

## 2.4 *Coastal Wells from the Historical Periods*

**Tel Gerisa, presently in the area of Tel Aviv** (Fig. 1). This water well was dated to the Middle Bronze II (Table 1) and probably was rebuilt also during the Iron Age I (Tsuk and Herzog 1992; Tsuk 2011). The well was dug into the calcareous sandstone (Figs. 4 and 5a). The water well top is at elevation of 21.7 m above MSL. According to known hydraulic gradient in the coastal aquifer (~1‰), the water table depth in this location is expected to be at ~20 m depth and therefore the depth of the well was probably ~21 m. The well is located adjacent to the Yarkon stream. However, it is not clear whether this fact played an important role in the consideration of the residents during that time or alternatively they were aiming at the general water table of the coastal aquifer in this area (~20 m). In any case, it seems that the "engineers" of that time had an understanding of the aquifer and the ability to reach quite a deep groundwater table in sites of a relatively simple hydrogeological setup.

**Tel Ashkelon** (Fig. 1a). This site was an important city from the Early Bronze period (third millennium BCE) (Table 1) until the Mameluke period (fourteenth century CE). Most of the wells in the site are from the Roman and Byzantine periods (second–seventh centuries CE). According to a geological investigation (Lass 1989), the city was built above an underground fossil river which was active some 15 million years ago and probably deposited gravel and sand that constitutes a small aquifer in this area. Later it was covered by sand and the city was built on kurkar and sand sediments (Tsuk 2011). The dozen wells at Tel Ashkelon were dug into this restricted aquifer at depth is of ~20 m. The ancient people had to dig deep into the kurkar ridge in order to reach the aquifer within their city.

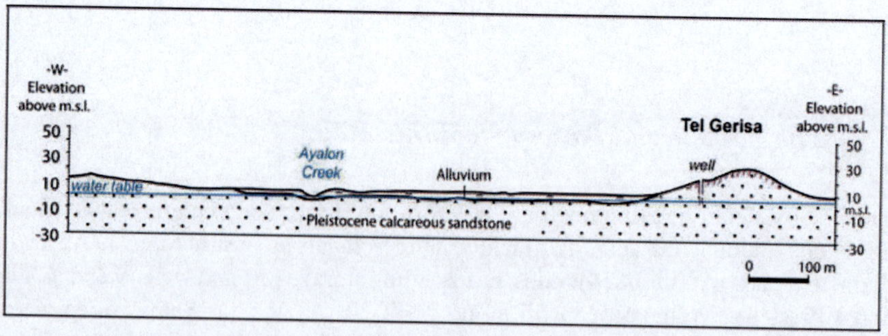


**Fig. 4** Hydrogeological cross section in the Tel Gerisa site (used by permission of Elsevier after Weinberger et al. 2008, Fig. 2)

**Caesarea, Ashdod Yam, and Yavne-Yam wells** (Fig. 1a)

From to the Persian and the Hellenistic time (sixth to second century BCE) to the end of the Ottoman period in the beginning of the twentieth century CE, several wells had been excavated along the coast of Israel.

Nir and Eldar-Nir (1987) studied Persian and Hellenistic coastal wells. The closest wells to the present coastline are the one in Mikhmoret (Fig. 1) dated to the Persian period (c. fifth century BCE), and the other in Yavne-Yam dated to the Hellenistic period (ca. third century BCE). Subsequently Sivan et al. (2004) studied 64 coastal

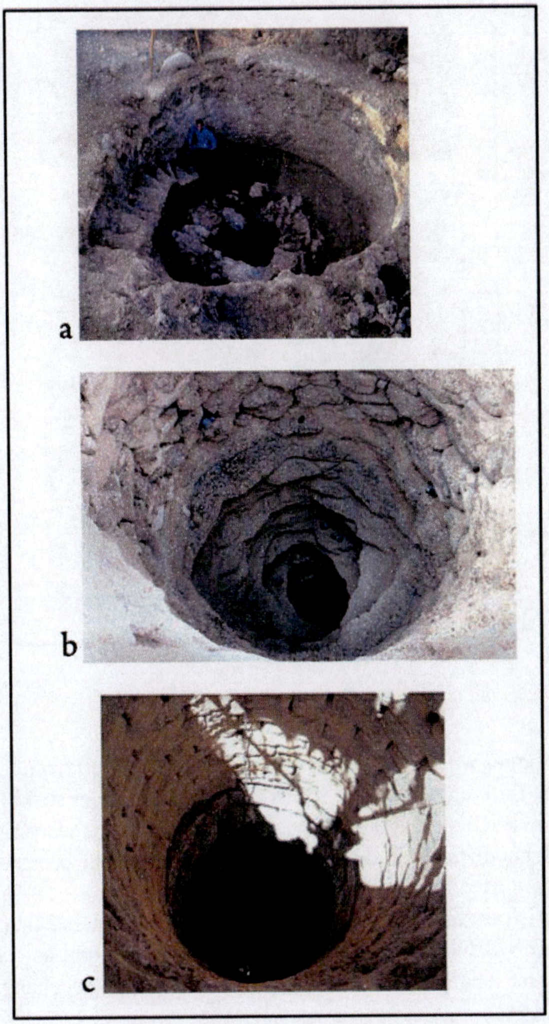


**Fig. 5** Photographs of ancients wells (after Tsuk 2018). **a** Ancient well in Tel Gerisa. **b** Ancient well in Tel Lachish. **c** Ancient well in Tel Beer Sheva. **d** The well in Shaar Hagolan

d

**Fig. 5** (continued)

wells from Caesarea, within an area of less than 1 $km^2$, dated from the first century to the end of the thirteenth century CE. In Caesarea, the most striking find was the fact that in this restricted area, elevations of the well bottoms were almost the same for same period, but different in different periods, whereby the bases of the Crusader wells eleventh to thirteenth century (CE) are the lowest of all.

Thus, this data from the last 2500 years confirm the hypothesis that people understood the coastal hydrology and dug the wells to different depths according to the changing coastline location. It seems that they reacted to the changing sea levels (Vunsh et al. 2018).

### 2.5 *The Mawasi Agriculture in Coastal Areas*

Remains of intensive agricultural were found south of Caesarea, named in Arabic "Mawasi" agriculture, close to and above the high coastal freshwater table (Shtienberg et al. 2017a, b). The inter-dune area was covered by humans by a 0.5–0.7 m thick layer of gray sediment containing pottery sherds, building refuse, and organic trash remains in order to improve the cultivation properties of the sediment, dated from the Hellenistic period (c. fourth–third centuries BCE) to the tenth century CE. Porath (1975) suggested that this agricultural system, existed in the eighth–ninth centuries and was destroyed either by the Crusader conquest of Caesarea (1101 CE) or later, after the Mamluk conquest of Caesarea (1265 CE). At Or-Aqiva, east of Caesarea, there are also remains of this system that probably existed earlier, during the Late Byzantine period or early Islamic period (fifth to seventh centuries CE) (Ad 2009).

Remains of the same type of plots, surrounded by sand berms, were found in Yavneh coastal dune field. Assemblages of pottery sherds dating to the eighth–tenth centuries CE lay on top of the sand berms. In the depressions between the berms, which are at the water table elevation, the anthropogenic more fertile gray sand was dated by optically stimulated luminescence (OSL) to the eleventh–twelfth centuries CE. In the Early Islamic period, in Tel Yavneh in the vicinity of these plots, similar pottery has been found, enabling to date it to the Early Islamic to Crusaders "plot-and-berm agroecosystem" based on top of a shallow water table as in the case of Caesarea (Taxel et al. 2018).

These two areas, dated probably from the Roman times, but certainly from Early Islamic period to the Crusader period, are indicative of the understanding of the shallow hydrological setup along the coast, which was used for irrigation of these semi-arid areas.

## 3 The Inland Zone

### 3.1 *Spring Flow Tunnels of the Iron Age II*

Long, narrow tunnels or galleries, close to several springs in the Judea Mountains, are found to cut horizontally into the rock. These tunnels are narrow (60–80 cm on the average), enabling a person to move along them easily. These tunnels were cut in a zigzagging plan which seems totally random. This was assumingly done in order to increase the rock mass conductivity and thus to increase the spring's discharge. The tunnels usually have a mild gradient outwards, leading to a catching reservoir.

Information on the recently investigated tunnel yield surprisingly reliable information on its date: The spring of 'Ein Juweiza', located about 2 km southwest of the Jerusalem neighborhood of Giloh, has a tunnel, 187.8 m long, cut into the mountain. It is the longest known tunnel of its kind (Ein-Mor and Ron 2013). Particularly outstanding is a fragment of a stone capital carved in the style known as Proto-Aeolic, which is typical to the royal architecture of the Iron Age II (ninth–eighth centuries BCE). It was found incorporated in the construction some 30 m into the tunnel (Ein-Mor and Ron 2013). This points to the fact that the first 30 m of the tunnel were cut in the Iron Age II. The inner parts could have been cut in one or more successive periods, probably following a period of drought.

## 3.2 *Inland Wells*

**Nisanit site** (Fig. 1)—this well is dated to Neolithic time (~8500 years BP) (Tsuk 2018). It was found in the western margins of the Yizreel Valley near the Nisanit springs. The depth of the well is ~8 m and its lower part was dug into the limestone (Tepper 2014). It seems to have been dug aiming to reach the groundwater table and therefore indicates a basic knowledge of hydrogeology. The well served an ancient agricultural settlement.

**Sha'ar Hagolan site** (Fig. 1)—the Neolithic well here was found in the Jordan Valley (Garfinkel et al. 2015), dated to 8400–7800 years BP. This site contains the cultural remains of the Yarmukian culture (the earliest Pottery Neolithic culture), geographically distributed in the Mediterranean belt in Israel, Jordan and Lebanon (Garfinkel 1993). The dating is based on 13 radiometric readings, seven of which derive from Sha'ar Hagolan site (Garfinkel 1999). The uppermost well's first 0.5 m was dug through the local river pebbles and the sandy sediment of the Yarmuk river delta into the underlying Lisan Formation (lacustrine marly-clayey sediments of Late Pleistocene lake, Fig. 5d). The diameter of the pit varies between 2 m (at 0–2 m depth interval), 1.2 m (at 2–2.75 m depth interval), and 0.7 m (at 2.75–4.26 m depth interval). At its base (4.26 m), the pit reached the water table, which flows through a layer of fine reddish gravel, 15 cm thick (Garfinkel et al. 2015).

**Gezer** (Fig. 1)—this site existed since the Middle Bronze Age II (eighteenth–seventeenth centuries BCE) through the Iron Age II (ninth–seventh centuries BCE) and even the Hellenistic period (fourth–second centuries BCE) (Table 1). The site has a very impressive rock-cut water system (90 m long and 1100 m$^3$ volume) from the Middle Bronze Age II (Macalister 1912; Tsuk 2011). The slanting tunnel here was excavated into the Eocene semi-permeable chalk aquitard, reaching at its bottom the clayey aquiclude of the Taqiye (Tlt) Formation. Regarding the knowledge of the ancient residents, it is unlikely that they wanted to reach the En Yarda spring, at a distance of ~500 m outside the city wall, from within the city (Fig. 6) (Yechieli et al. 2017). Did they know that groundwater can be found, perched on top of the Taqiye aquiclude? Or, alternatively, it can be speculated that they were aiming at a spring that existed at the margin of the Tel itself, during times of higher precipitation, although there is no present evidence for that. Groundwater of the springs around the Tel (Yarda, Abu Shusha springs) is of relatively good quality and was probably even better in ancient times prior to pollution by fertilizers and other sources. A preliminary study of groundwater in the reservoir at the bottom of the sloping tunnel of the water system, showed relatively good water quality, quite similar to that found

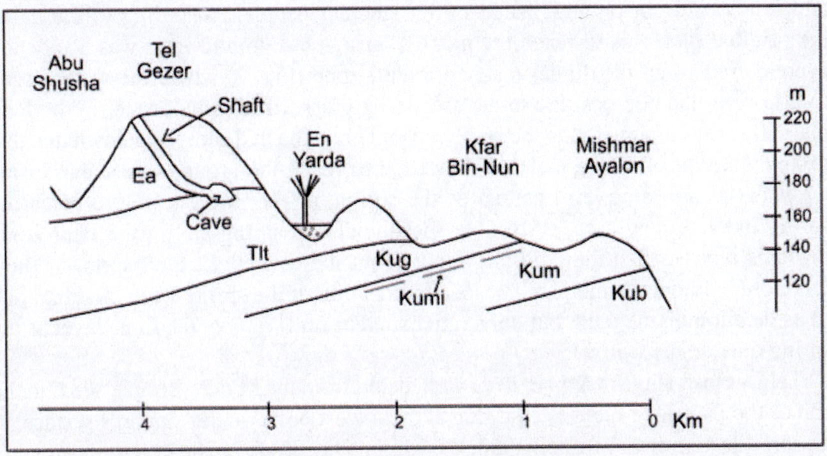


**Fig. 6** Geological cross section in the area of Tel Gezer (modified after Yechieli et al. 2017). Ea denotes Eocene era, Tlt—Tertiary clays, Kug—Upper Cretaceous chalks, Kumi—Upper Cretaceous chert, Kum—Upper Cretaceous chalks (Tlt + Kug + Kumi + Kum composed together the Mt Scopus Group) and Kub denotes the carbonates of the Judea Group

outside the Tel, with some element of local pollution (Yechieli et al. 2017). It is thus difficult to determine whether it was planned in advance, based on hydrogeological knowledge, or without such knowledge, to reach the aquifer. Moreover, some water could have been contributed from surface drainage, including through the diagonal tunnel, and not necessarily from the local Eocene aquitard. It is also possible that the climate was somewhat different in ancient times, and therefore the hydrogeological setup was different as well.

**Tel Hazor** (Fig. 1) This Iron Age site is one of the largest Tels in Israel but most debatable with regards to the hydrogeological knowledge of the ancient residents. It was studied extensively by archeologists (e.g., Shiloh 1992; Yadin 1969; Garfinkel and Greenberg 1997; Ben-Tor 1993) and hydrogeologists (Weinberger et al. 2008). The main question here is whether the ancient people found the water by chance, as suggested by Weinberger et al. (2008), or by design as claimed by the archeologists (Yadin 1969; Tsuk 2011). The water system in this site is an excellent example of Iron Age water systems (Yadin 1969; Shiloh 1992; Garfinkel and Greenberg 1997). According to Yadin (1969), the ancient "engineers" had a sound hydrogeological knowledge, since it is obvious that they anticipated to encounter the groundwater within the perimeter of the Tel at the same level as the springs outside. This was partly based on an oral opinion, at the time of excavation by U. Kafri and E. Fleisher. They assumed that the ancient inhabitants of Hazor recognized the nearby shallow perched groundwater horizon south and outside the city walls, as indicated by dense typical vegetation, and thus decided to reach the extension of this groundwater level inside the walls. On the other hand, Weinberger et al. (2008) have argued that the regional aquifer in this area is at a level tens of meters below the spring's altitude and thus could not be anticipated by the ancient people. According to the latter, the original plan was to reach the nearby springs but groundwater was accidently encountered when the digger reached a fault zone (Fig. 7) where the water began to drip from the bedrock due to the ascending pressurized groundwater in the fault plain. It is thus assumed that the ancient people had the hydrological knowledge that groundwater level is regional and they tried to reach the groundwater table inside the walls at the same level as that of the spring, or they were trying to reach the spring itself. In any case, since both the tunnel and spring are on the fault zone, it seems obvious that they did not know about the concept of fault springs. They, therefore, could not anticipate the specific location of the spring while digging, due to its location on the fault, but only anticipated to find water at the same level as the spring outside the walls.

**Tel Lachish** (Fig. 1) A 44 m deep well, dated from the Middle Bronze age II to the end of the Iron age (Table 1), was dug at the lowest point on the mound's surface, at its north-eastern edge (Tufnell 1953; Ussishkin 1982; Tsuk 2018). It was presumably intended to reach the same groundwater table as found in the alluvial gravel aquifer of the Lachish creek, running along the foot of the mound (Weinberger et al. 2008). While some water was found at the base of this well, the quantities of water seemed to be insignificant since the well penetrated the Oligocene and Eocene aquitard of low conductivity with no hydraulic connection to the gravel aquifer (Fig. 8). This supports the conclusion of Weinberger et al. (2008) that the site inhabitants did not have a sound idea about the areal hydrogeological characteristics.

**Southeastern hill-slope of Jerusalem (City of David)** (Fig. 1) The engineering mission here was to connect the city with the outside springs via shafts and tunnels with very little involvement of hydrogeological consideration as claimed by Weinberger et al. (2008). The site is dated from the Middle Bronze age II to the end of the Iron Age II (Table 1). Although this site is not determined as a hydrogeological

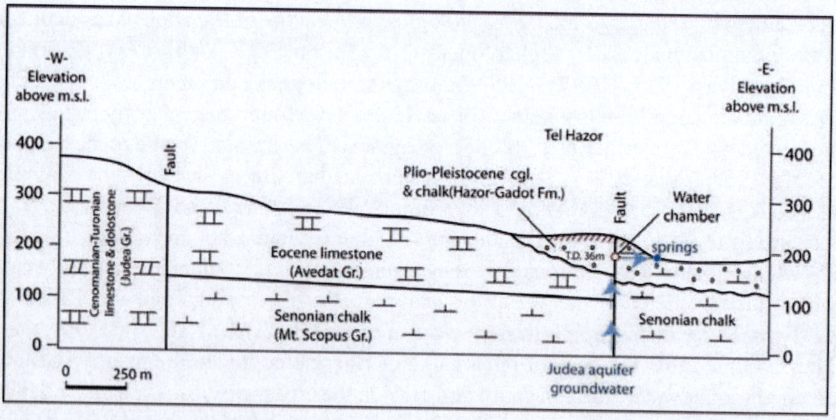


**Fig. 7** Hydrogeological cross section in the Tel Hatzor site (used by permission of Elsevier after Weinberger et al. 2008, Fig. 6)

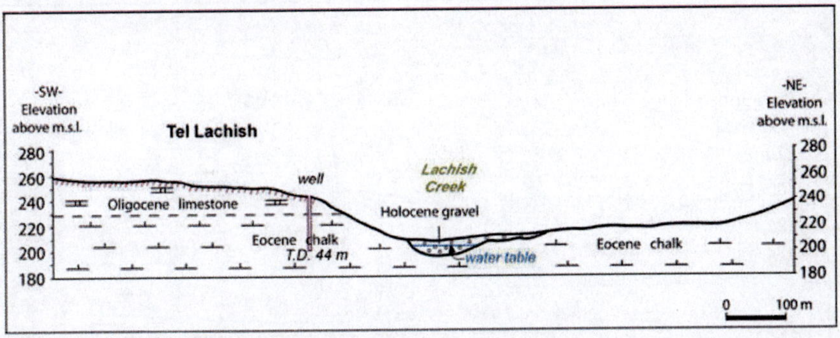


**Fig. 8** Hydrogeological cross section in the Tel Lachish site (used by permission of Elsevier after Weinberger et al. 2008, Fig. 2)

system per se, it is related to the subject due to its hosted karst system. According to Gill (1991), the diggers used the abundant natural karstic features of this carbonate sequence by enlarging them to construct the Siloam tunnel with no need for hydrological knowledge. A different view is given by Reich and Shukron (2002) who do not accept this karstic theory. They point to the facts that the tunnel's total horizontal layout, with no side tunnels, has a uniform width of 60 cm all along, and with several cutting dead-ends, all indicating that this tunnel cannot be an enlargement of an existing karstic fissure in the rock.

## 4 The Southern Part of Israel—The Negev Wells

### *4.1 Abu Hof Site (Fig. 1) (Between Kibutz Lahav and Lehavim)*

This well, dated to the Chalcolithic era (fourth millennium BCE), can also indicate some human understanding in hydrology. The well is located in a small stream channel. Its upper part was dug into Loess soil and the lower part in alluvial gravel to a total depth of 2.4 m (Alon 1988). This area has a shallow water table and until several decades ago some discharge occurred as a seepage about 500 m westward of the well (Tsuk 2011).

### *4.2 Tel Beer Sheba (Fig. 1)*

The intention at this site was probably to reach groundwater table of the alluvial aquifer in the Beer Sheba stream which gets flash floods several times in the winter,

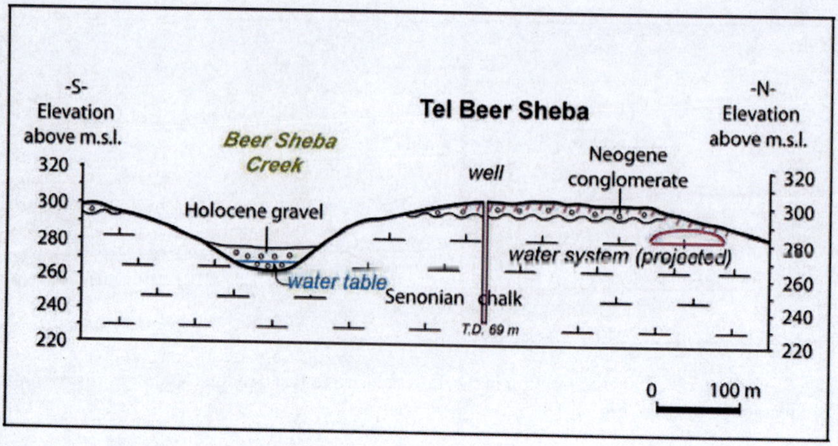


**Fig. 9** Hydrogeological cross section in the Tel Beer Sheba site (used by permission of Elsevier after Weinberger et al. 2008, Fig. 2)

by digging a 70 m deep well (Fig. 9) (Herzog 2002; Lev-Yadun et al. 1995). However, the assumption that the aquifer extends laterally underneath the mound was erroneous and the well penetrated mostly the Maastrichtian aquiclude (Weinberger et al. 2008) and yielded an insignificant amount of water.

## 4.3 *Tel Arad (Fig. 1)*

This site dates to the Early Bronze period (third millennium BCE) and a well from the Iron Age has been excavated in the Canaanite city (Amiran et al. 1985). The well is ~20 m deep and 4.6 m in diameter, whereby the lower 11 m were dug in chalk. Water was found here in 1997 (Tsuk 2011). According to Weinberger et al. (2008), the construction constitutes a well-like cistern and not a water well, since the groundwater table here today is tens of meters below the bottom of the cistern. Another interpretation was given by U. Kafri, stating that this is a groundwater well and not a cistern (Amiran et al. 1978) and the well was dug in this arid area despite the poor hydrological configuration due to the lack of alternatives. According to the latter, the well was dug to a depth of 16 m into the early Eocene chalk-limestone aquitard and reached at its bottom the Paleocene chalk aquiclude. A small water reservoir was constructed next to the well, lined at its bottom with the impervious Paleocene material which was dug from the bottom of the well, in order to clog downward water infiltration. Therefore, the latter study assumed that the ancient inhabitants of Arad had some basic understanding regarding the groundwater setup of the region. It should be noted that ancient people needed to use much less water than modern people in developed countries (~3 $m^3$ as compared to ~100 $m^3$ per person per year, respectively; e.g., Tsuk 2011).

### 4.4 *Water Wells in the Negev During the Roman and Byzantine Periods*

Much has been written about the Negev settlements that were built by the Nabateans in the Hellenistic period which flourished in Roman times and became large agricultural communities in the Byzantine period. All of the studies indicate a process of settlement development that reached its zenith during the fourth–seventh centuries CE at Ḫaluza, Ruheibeh, Nizzana, Shivta, Ovdat, Mamshit and Saʿadon (Fig. 1).

The water system of the Negev sites during the Roman and the Byzantine periods depended mainly on in-house water cisterns. Each house and church had its own cistern, but due to the desert climate and droughts the cisterns were not sufficient and a permanent water source was essential.

In Haluza, the aquifer of the Besor stream is at a depth of 8–10 m, and thus several water wells there could supply all the water demand of the city. Similarly, in Nizzana, a few wells were dug in the lower city, along the banks of Nizzana stream.

In Ovdat and Ruheibeh, the aquifer is much deeper. The Ovdat well is outside the city, next to the Roman bathhouse. The depth of this well is about 60 m. In Ruheibeh, a rock-cut well reached the groundwater aquifer in the bottom of Ruheibeh streambed (Tsafrir and Holum 1988). The well is 63 m deep, the second-deepest well in the Negev after the Beer Sheba well. Its upper part was built and lined with stones. The built portion of the well was c. 3 m in diameter and 8.8 m deep. Below the built portion, the well was rock-hewn and 54 m deep, with an average diameter of 5 m. The water table was found in May 1992 at a depth of 40 m and the water column totaled 23 m (Dahari and Sion 2017). The well provided water continually to the bathhouse which was used by all of the city's inhabitants, as well as by the pilgrims that passed through Ruheibeh, and it also supplied drinking water for the residents toward the end of summer when the household cisterns ran dry as well as during droughts. Hydrologists from the Israeli Hydrological Service estimated the flow as exceeding 20 $m^3$ per day, totaling more than 7300 $m^3$ a year.

In Mamshit and Shivta, water wells were not found (except one shallow well in Mamshit), and it seems that the only water demands were based on cisterns and open reservoirs.

## 5 Qanats in the Jordan and Arava Valleys

Qanats (called also Foggara, Aflaj, Karez, and several other names in different parts of the world) are sophisticated artificial underground water collection and transportation systems that spread across Eurasia, from Spain in the west to western China in the east. Qanats originated in Iran in the second millennium BCE and were spread from there to other regions in the Near East and North Africa mainly from the Hellenistic period onwards (Wilson 2006; Wilson and Mattingly 2003). A typical qanat was comprised of a number of parallel underground tunnels which were dug into alluvial sediments or conglomerates. These systems were designed to lead water from aquifers to agricultural areas, sometimes located several to tens of kilometers away. The construction methodology of a qanat system included the digging of vertical shafts from the ground into a horizontal tunnel (Fig. 10). These systems are highly visible from the air as the spills of earth taken out through the shafts form circular mounds on the surface, usually c. 10–20 m apart, creating direct lines of holes. Consequently, the first identification of ancient qanats in the Near East and Iran was through the development of aerial photography in the 1920s and 1930s. Looking at their global distribution, it seems clear that the spread of qanats is confined mainly to steppe areas and desert fringe zones with low precipitation, where the extensive construction of the underground systems was extended to the potential agricultural areas. Qanats are known in the Iranian plateau, in Oman, in the northern fringe of the Sahara Desert, in central Asia and western China.

The detailed study of several ancient qanat systems in the steppe areas of the Levant, and particularly, in the Arava and Jordan valleys in Israel, indicated that qanats were constructed in these regions only in the Early Islamic period, and particularly, in seventh and eighth centuries CE (Avni 2019). The spread of qanats in these regions is connected to the new agricultural and hydrological methodologies that were introduced into the region from the east, following the Arab conquest of the Near East and the opening of the borders under the new Islamic regime.

One of the best examples of incorporation of a qanat system with a farmstead newly constructed in the eighth century was found in the central and southern Arava Valley. At a farmstead in En 'Evrona, north of Eilat (Early Islamic Ayla—Whitcomb

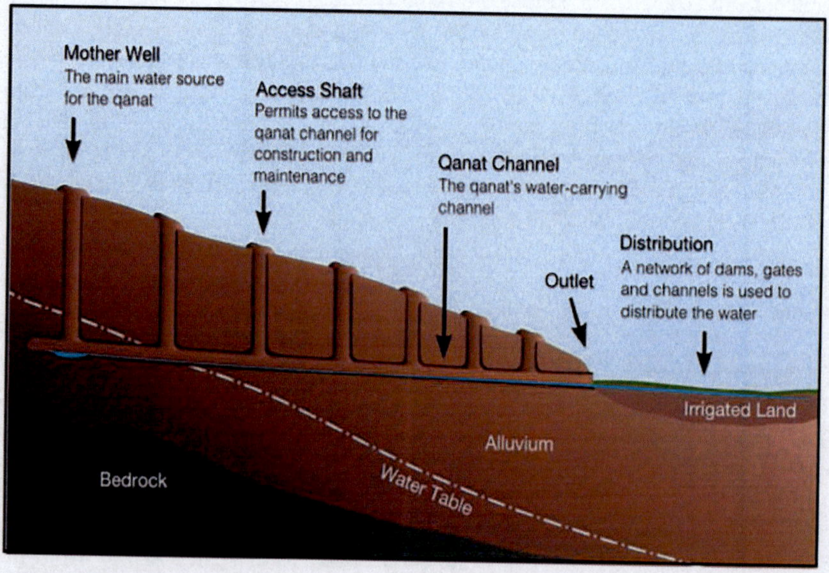


**Fig. 10** Schematic hydrogeological cross section in the qanats

1994, 2006), established in the end of the seventh and the beginning of the eighth century CE, an extensive system of qanats provided water to the farmstead from a local aquifer, irrigating the fields adjacent the site. This qanat consisted of two parallel tunnels, with shafts at distances of c. 20 m from one another. The findings from the excavations at the site show a large assemblage of pottery vessels and plant remains from the Early Islamic period (Avner 2016).

Another Early Islamic agricultural settlement was revealed in Yotvata, c. 30 km. north of Eilat. This settlement was intensified in the late seventh or eighth century CE with the construction of additional buildings at the site, including a caravanserai, serving travelers along the main road to Ayla and a large farmstead nearby (Porath 1985; Avner 2008; Davies and Magness 2015). Large agricultural fields were revealed around the site, and a system of qanats, consisting of more than 20 tunnels, was found to be constructed in conjunction to this settlement. The direct association between the qanats and the Early Islamic settlement was established through several small-scale archeological excavations (Porath 2016; Avner 2016). The main qanat in Yotvata is c. 4.5 km long, with several secondary tunnels, which were connected to the main tunnel. The qanats led the water from the aquifer to the fields, reaching the arable area to c. 400 ha (Avner 2016). These fields were cultivated only in the Early Islamic period.

As in the Arava Valley, new irrigation techniques, which included intensive use of qanats, were introduced into the Jordan Valley at the end of the seventh or early eighth century CE. The early Islamic expansion of settlement into this region included the construction of several farmsteads which were directly connected to qanats systems.

The farmsteads and their related qanats mark the introduction of a new type of settlements, which were previously unknown in the region. It is interesting to note that these agricultural sites are related to the palatial complex of the Hisham Palace (Khirbet al-Mafjar) on the outskirts of Jericho. Although no systematic survey of qanats was conducted in this region, there is evidence for the existence of other systems, some of which were re-constructed by the local population in the early twentieth century which were used for several decades (Lightfoot 2000). The continuous use of ancient qanats is known from other locations in the Near East and beyond, both for agricultural uses and for the supply of water to large cities through sophisticated qanat systems.

The evidence of qanats from the Arava and the Jordan Valley shows unequivocally that this is a new hydrological innovation that was not known in the southern Levant before the eighth century CE and was introduced by the new Islamic regime.

It seems that this innovation ameliorated and expanded the agricultural and irrigation capacities in the local villages and extended the inhabited areas well into the steppe and desert regions. While the Byzantine systems of irrigation were based mainly on runoff collection from slopes and on diversion of spring water flow through aqueducts, the new qanat systems increased the agricultural potential, relying on permanent aquifers rather than on runoff only. The finding of numerous qanat systems in the region indicates that the massive expansion of qanats in the eighth century CE, and later, was triggered by the new geopolitical situation and by the opening of the borders between east and west following the Arab conquest.

## 6 Discussion

The study of ancient water wells and water systems sheds light on the technological developments and managements of sources of water since the beginning of agriculture and sedentary settlements. Ancient water systems are of special interest not only for the details of their construction and their contents, but also as evidence that already in early periods people were able to improve the water supply by constructing artificial systems (wells, qanats, installations) to ensure permanent sources of water. In a region where water is a limiting factor, a hydrological understanding and technological development were essential and have been an important contributor to sustaining permanent settlement and improve the reliability of the subsistence economy. It should be noted that the comparison of past hydrogeological conditions to present conditions should be done with caution since climate could have been somewhat different in ancient times.

Water wells are the main man-made water supply structures that can tell us about ancient hydrological knowledge. They have been built already in Pre-Pottery Neolithic time. They are relatively simple constructions compared to water systems that required sophisticated engineering knowledge of the reservoir and the drainage systems around them. In many places, only basic hydrogeological understanding was required for constructing wells as in most cases along the coast of Israel. Humans knew where springs and shallow groundwater exist, exhibited in the field as seepages, swamps, or flooded area in events of rain. From that knowledge, it was a relatively small step to understand that digging a small hole ("Tmile") will supply a better and more constant water source. Later, deeper wells were dug in order to chase groundwater in a setup of a deeper water table. It is estimated that digging of a well in a simple hydrogeological setting (e.g., in sandstone or chalky rocks) can be done within relatively short time, at a rate of ~0.3 m per day by ~6 persons, and therefore a well of 20 m deep and 1.8 m in diameter can be dug within 60 days as estimated by Tsuk (2011).

In the PPN site of Atlit-Yam, presently located offshore, the wells indicate basic hydrogeological understanding of a confined (or roofed) aquifer in the kurkar overlain by the dark clay, unlike most other wells in the coastal aquifer which are dug into the phreatic aquifer. Therefore, the people at that time had to know that below the clay layer there is an aquifer; and they dug few meters through the confining clay layer until they reached the aquiferous layer of the kurkar. It shows more understanding than just seeing a spring or swamp and digging a shallow well in its vicinity, but rather a knowledge of the concept of a confined aquifer. This understanding could have come from noticing small freshwater "springs" emerging at the coastal zone, as at present, along the contact between the clay layer and the kurkar below along the Carmel coast. In addition, the wells required engineering knowledge for digging and supporting their walls to prevent collapse of the clayey shafts and were thus lined with courses of undressed stones. It was dug and supported by building from top to the kurkar, and cut (with no building) in the kurkar sequence.

The fact that the digging went 1.5 m into the kurkar layer, below the clay layer, indicated that the aquifer was not confined at that time, but roofed. Had it been confined, the ancient people could not have dug more than several tens of cms below the clay layer since the water pressure would cause the water to rise into the hole, preventing further digging. Therefore, at the time the PPN well of Atlit-Yam functioned, groundwater level was below the clay layer (level of ~15 m bsl) and the sea level was about 16 m bsl, as was indeed suggested by Galili and Nir (1993) and Sivan et al. (2001). It is expected that when sea level rose, groundwater level rose at the same time in the wells and therefore the wells could supply more water (a thicker water column). However, the side effect of sea level rise could adversely cause the encroachment of the fresh-saline water interface inland resulting in increasing salinity at the bottom of the well. Layers of large stones and clay recovered from the lower fill of the well suggest that the inhabitants tried to cope with the salinization by raising the base of the well in order to exploit the upper, less saline, part of the water column. Eventually, at the last phase of occupation, the well went out of use and served as a garbage pit. Later, the Atlit-Yam site was abandoned and a new set of settlements that included wells was established landward in the Pottery Neolithic and Chalcolithic periods (Galili and Weinstein-Evron 1985; Galili et al. 1988).

Along the central and southern coast, the wells were deeper and required cutting and building from top through sand and sandstone units as in the case of Tel Gerisa, dated to the Middle Bronze age and later to the Iron age (Tsuk and Herzog 1992; Tsuk 2011). A later well, from the Persian period, was found in Mikhmoret, and another, from the Hellenistic period, in Yavne-Yam (Nir and Eldar-Nir 1986; Nir 1997). All coastal wells of Ashkelon, Ashdod Yam, Yavne-Yam, Jaffa, Caesarea, and Akko, from the Roman period (Fig. 1), indicate an understanding of the changing hydrological conditions, mainly the relationship between freshwater table and the saltwater body due to the rising sea level (Sivan et al. 2001, 2004; Vunsh et al. 2018). In these later periods, people responded to sea level changes (rise and fall), as in the Crusaders period, as suggested by Toker et al. (2012). The response was to dig the wells into different depths in various periods with an awareness to the sensitive relations between fresh and salt water in the coastal zone (Sivan et al. 2004).

The study of coastal and submerged wells became a major tool in understanding sea level and coastal changes and the human response and adaptation to such changes. In the first half of the Holocene, the sea level rose from around 40 m bsl some 10,000 years ago (Galili et al. 1988, 2005; Sivan et al. 2001) attaining about the current level during the Middle Bronze Age at about 4000 years BP (Porat et al. 2008). During the occupation at Atlit-Yam, at the beginning of the Holocene, the coastline in the northern Carmel coast was some 1.5–2.5 km offshore of the present one and consisted of an indented waterline with bays and lagoons (Galili et al. 2005). The relatively rapid rise of sea level during the occupation of the Neolithic sites required active human adaptation to the changing conditions and in this case, mainly to salinization of the water wells (Galili and Nir 1993).

Consequently, Sivan et al. (2004) calculated the relative sea level (RSL) (Fig. 11) in Caesarea, based on 64 coastal wells dated from the Early Roman period (first century CE) to the end of the Crusader period (thirteenth century CE). The well's

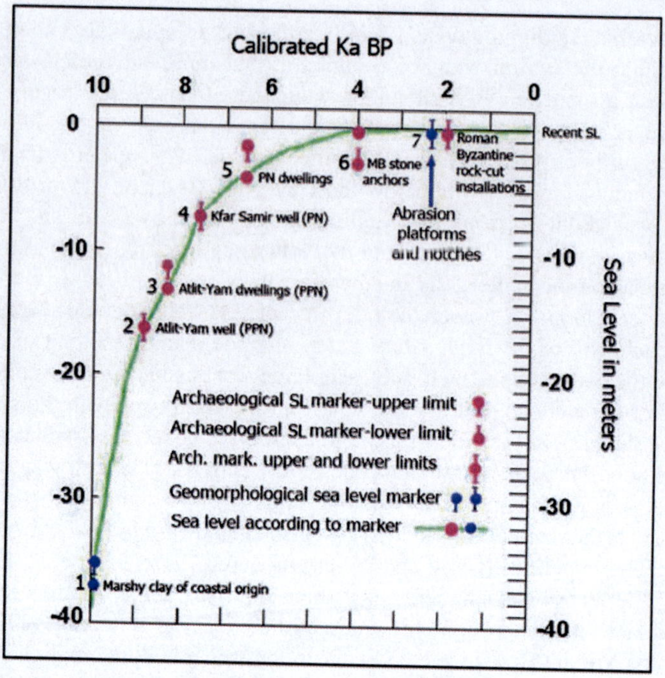


**Fig. 11** Curve depicting the sea level change in the Mount Carmel coast based on archeological and geomorphological features (modified after Galili and Rosen 2011a)

base elevation, plus 0.35 m to allow space for a water drawing jar, minus the offset between the present-day water table and sea level, measured by tide gauge, was used to evaluate the past sea level. This offset between the present-day water table and modern average sea level can be measured also either directly over a period of time by using average modern water table data, or with a model evaluating the water table as a function of the distance from the modern coastline (Levanon et al. 2013; Vunsh et al. 2018; Dean et al. 2019). The above study indicates that the elevations of the wells bottoms were almost the same for the same period, but different in various periods with the lowest bases in the Crusader period (eleventh to thirteenth centuries CE). This suggests, as mentioned before, a basic hydrological understanding.

The digging of deeper wells into phreatic aquifers seems to be a relatively easy evolutionary step forward, whereby with time, experience, and engineering advances, the well became deeper. Tel Gerisa is a good example of such deep well dated to MB II to Iron Age I (Table 1). The deep wells in the Negev desert (e.g., in Ruheiba) also indicate basic understanding in hydrogeology since a good estimation of the depth of the relatively deep water table was required. In later periods, from the Roman and for sure the Byzantine and early Islamic times (Table 1) and onward, the simple upper hydrological characteristic along the coast was used for inter-dune agriculture as seen in the Mawasi around Caesarea and Yavne. The qanats in the Arava also exhibit basic understanding of hydrogeology of alluvial systems.

In the more inland foothills areas and sites (Lachish, Beer Sheva) (Table 1), the hydrogeological setup seems to be more complicated and not fully understood, especially in large Tels that needed larger amounts of water. In these sites, the inhabitants wanted to reach from inside the Tel the same water source found some distance away in the alluvial aquifer without taking into account the lateral hydrogeological changes and resultant lower hydrological parameters and smaller yields. Thus, the lack of spatial hydrogeology understanding caused the project to be unsuccessful.

Another complicated case is exhibited in Tel Gezer, where a very impressive Middle Bronze II construction exists. Although the lower part of the diagonal tunnel reached the bottom of the local aquitard, it is not clear whether this was indeed planned or it was supposed to reach a water source outside the Tel. Since this water source is poor today and recharge from above was not observed, it is difficult to determine how usable it was in ancient times (Yechieli et al. 2017). The hydrogeological evidence in Iron Age II Tel Hazor is debatable, suggested by some that the ancient people did not understand the complexity of the system and they reached the groundwater incidentally.

Another way of looking for hydrogeological understanding is by analyzing historical sources. One such source is the Tosefta, the Jewish compilation of the religious rules which followed the Mishnah and was compiled and closed in the third or early fourth century CE. It contains information, like the present cited verse, which dates back to the Second Temple period, i.e., the first century CE. In the tractate Arachin (2, 6) we read: "Raban Shim'on ben Gamliel says: '[the spring] Shiloah' was emitting water in the [size] of an isar [=a small coin]. They said: let us enlarge it so that its waters will increase. And it was enlarged, and its waters decreased. So they closed this and it returned to be as it was in the first place." This short description shows clearly that the people in charge of the spring were not aware of its nature. They had in mind many of the springs of the Judean Mountains which feed directly from the aquifer (like the Ein Juweiza spring mentioned above). However, the spring of Shiloah, at the point where it surfaces at the lower part of the slope of the hill, which is the only spring in the vicinity of the city of Jerusalem, does not feed directly on an aquifer (Benami-Amiel et al. 2010), but is the outlet of a series of subterranean karstic fissures and fractures. For this reason, any enlargement of the exposure of the water-carrying rock is futile. Therefore, this case indicates hydrogeological ignorance.

## 7 Summary

It seems that humans in the past had some basic hydrogeological knowledge in relatively simple systems, such as the coastal area where the geological and hydrological is relatively simple. This is also seen in near alluvial river systems in the Negev, where successes in finding groundwater were recorded. Moreover, the PPN well in Atlit-Yam site indicates that humans knew to reach water even in a shallow confined (or roofed) aquifer. The knowledge in more complicated systems is somewhat debatable. For example, it seems that the effect of geological heterogeneity, between alluvial sediment in the vicinity of the river, which cut through sediments of lower permeability were not known and therefore led to failure in finding water in the dug wells. Moreover, in several more sophisticated inland sites, it appears that people dug a tunnel to reach a spring located outside the wall and encountered a good water source in an unplanned location without the hydrological understanding of the concept of fault spring.

**Acknowledgements** We wish to express our gratitude to the Israel Antiquities Authority and Haifa University for institutional support, to Care Archeological Foundation, Honor Frost Foundation, Dan David Foundation and Shelby White Levi foundations for founding the underwater research, To Howard Rosenstein and Fantasea for donating the photographic equipment, and to Itamar Grinberg and Josef Galili for the underwater photography. We are thankful to Rena Yechieli for her help in English editing.

## References

Ad U (2009) Or'Aqiva: remains of a farming complex and irrigation system from the end of the Byzantine—beginning of the early Islamic periods in the agricultural hinterland of Caesarea. Atiqot 61:49–60 (in Hebrew)

Alon D (1988) Distribution of the chalcolithic settlements in the southern Shfela and its history. In: Orman D, Stern E. (eds) Man and environment in the southern Shfela. Massada, Givataim, pp 85–86 (in Hebrew)

Amiran R, Goethert R, Ilan O (1978) The well at Arad. Biblical Archaeol Rev 13(2):40–44

Amiran R, Goethert R, Ilan O, Alon D (1985) The water-system at Arad. Qadmoniot 18:15–17

Avner U (2008) Eilat region. In: New encyclopedia of archaeological excavations in the Holy Land V. Israel Exploration Society, Jerusalem, pp 1704–1711

Avner U (2016) Desert farming in the Southern Araba Valley, Israel, 2nd century B.C. to 11th century A.D. In: Davies A, Retamero F, Schjellerup I (eds) Making agro-pastoral landscapes in pre-industrial societies: choices, stability and change. Oxbow, Oxford, pp 20–53

Avni G (2019) Early Islamic farmsteads and the spread of Qanats in Eurasia. Water Hist. https://doi.org/10.1007/s12685-018-0225-6

Bar-Yosef O, Belfer-Cohen A (1989) The origins of sedentism and farming communities in the Levant. J World Prehistory 3:447–498

Bar-Yosef O, Meadow RH (1995) The origins of agriculture in the near East. In: Price DT, Gebauer AB (eds) Last hunters, first farmers: new perspectives on the prehistoric transition to agriculture. School of American Research Press, Santa Fe, pp 39–94

Benami-Amiel R, Grodek T, Frumkin A (2010) Characterization of the hydrology of the sacred Gihon Spring, Jerusalem: a deteriorating urban karst spring. Hydrogeol J 18:1465–1479

Ben-Tor A (1993) Hazor. In: Stern E (ed) The new encyclopedia of archaeological excavations in the Holy Land, vol II, pp 604–605

Cohen-Seffer R, Greenbaum N, Sivan D, Barmeir E, Croitoru S, Inbar M (2005) Late Pleistocene—Holocene marsh episodes along the Carmel coast, Israel. Quat Int 103–120:140–141

Dahari U, Sion (2017) Ruheiba—Rehovot in the Negev as a model for a city in the desert. Kadmoniot 154:66–77

Davies G, Magness J (2015) The 2003-2007 excavations in the Late Roman Fort at Yotvata. Eisenbrauns, Winona Lake (ID)

Dean S, Horton B, Evelpidou N, Cahill N, Spada G, Sivan D (2019) Can we detect centennial sea-level variations over the last three thousand years in Israeli archaeological records? Quat Sci Rev (QSR) 210:125–135. https://doi.org/10.1016/j.quascirev.2019.02.021

Ein-Mor D, Ron Z (2013) An iron age royal tunnel spring in the region of Nahal Rephaim. In: Stiebel G, Peleg-Barkat O, Ben-Ami D, Weksler-Bdolah S, Gadot Y (eds) New studies in the archaeology of Jerusalem and its region, vol 7, pp 85–109 (in Hebrew)

Galili E (2004) Submerged settlements of the ninth to seventh millennia BP off the Carmel Coast. Dissertation, University of Tel Aviv (in Hebrew, English abstract)

Galili E, Nir Y (1993) The submerged pre-pottery Neolithic water well of Atlit-Yam, northern Israel, and its palaeoenvironmental implications. The Holocene 3:265–270

Galili E, Rosen B (2011a) Submerged Neolithic settlements off the Mediterranean coast of Israel. In: Benjamn J, Bonsall C, Pickard C, Fischer A (eds) Submerged prehistory. Oxbow Books, Oxford, pp 272–286

Galili E, Rosen B (2011b) Submerged Neolithic settlements off the Mediterranean Carmel coast of Israel and water mining in the Southern Levant. Neo-Lithics 2(10):47–52

Galili E, Weinstein-Evron M (1985) Prehistory and paleoenvironments of submerged sites along the Carmel Coast of Israel. Paléorient 11(1):37–52

Galili E, Weinstein-Evron M, Ronen A (1988) Holocene sea-level changes based on submerged archaeological sites off the Northern Carmel Coast in Israel. Quatern Res 29:36–42

Galili E, Stanley DJ, Sharvit J, Weinstein-Evron M (1997) Evidence for earliest olive-oil production in submerged settlements off the Carmel Coast, Israel. J Archaeol Sci 24:1141–1150

Galili E, Zviely D, Weinstein-Evron M (2005) Holocene sea-level changes and landscape evolution on the northern Carmel coast (Israel). Mediterranée 1(2):1–8

Galili E, Rosen B, Yaroshevich E, Orrele E, Kolska-Horwitz L (2016) Symbolic markers of the Wadi Rabah culture from the Neve-Yam submerged village and contemporary inland sites in Israel. Israel Explor J 66(2):129–150

Galili E, Kolska Horwitz L, Rosen B (2019) The Israeli model for the detection, excavation and research of submerged prehistory. TINA. Turkish Archaeological Foundation, 31–69 p

Garfinkel Y (1993) The Yarmukian culture in Israel. Paléorient 19(1):115–134

Garfinkel Y (1999) Neolithic and Chalcolithic pottery of the southern Levant. Qedem 39. Institute of Archaeology, Hebrew University, Jerusalem

Garfinkel Y, Greenberg R (1997) Area L. In: Ben-Tor A, Bonfil R, Hazor V (eds) An account of the fifth season of excavations, 1968, pp 177–294

Garfinkel Y, Vered A, Bar Yosef O (2015) The domestication of water: the Neolithic well at Sha'ar Hagolan, Jordan Valley, Israel. Antiquity 80(309):686–696. https://doi.org/10.1017/S0003598X00094138

Gill D (1991) Subterranean waterworks of biblical Jerusalem: adaptation of a karst system. Science 254:1467–1471

Herzog Z (2002) Water supply at Tel Beer Sheba in the 1st millennium B.C.E. In: Ohlig C, Peleg Y, Tsuk T (eds) Cura Aquarum of Israel. Proceedings of the 11th international water conference on the history of water management and hydraulic engineering in the Mediterranean Region, Israel, May 2001, pp 15–22

Issar A (1968) Geology of the central coastal plain of Israel. Israel J. Earth Sci 17:16–29

Lass HH (1989) Survey of wells. National Park Ashkelon, Geological Survey of Israel, Jerusalem. Report 58

Levanon E, Yechieli Y, Shalev E, Friedman V, Gvirtzman H (2013) Reliable monitoring of the transition zone between fresh and saline waters in coastal aquifers. Groundwater Monit Rem 33:101–110. https://doi.org/10.1111/gwmr.12020

Lev-Yadun S, Herzog Z, Tsuk T (1995) Conifer beams of Juniferus Phoenica in the well of Tell Beer-Sheba. In: Tel Aviv 22, No. 1, pp 128–136

Lightfoot DR (2000) The origin and diffusion of Qanats in Arabia: new evidence from the northern and southern Peninsula. Geog J 166:215–226

Macalister RAS (1912) The excavation of Gezer, vol I. London, pp 256–265

Nir Y (1997) Middle and late Holocene sea-level along the Israeli Mediterranean coast—evidence from ancient water wells. J Quat Sci 12:143–151

Nir Y, Eldar-Nir I (1986) Eustatic sea level changes and their possible influence on the Nile littoral-cell beaches. In: Raukas A (ed) Evolution and dynamics of seacoasts in conditions of relative sea level oscillation: excursion guides and abstracts, international symposium, Aug 20–26, 1986. Academy of Sciences of the Estonian SSR, Institute of Geology, Tallinn, p 63

Nir Y, Eldar-Nir I (1987) Ancient wells and their geoarchaeological significance in detecting tectonics of the Israel Mediterranean coastline region. Geology 15:3–6

Peltenburg E, Colledge S, Croft P, Jackson A, McCartney C, Murray MA (2000) Agro-pastoralist colonization of cyprus in the 10th millennium BP: initial assessment. Antiquity 74:844–853

Peltenburg E, Croft P, Jackson A, McCartney C, Murray MA (2001) Well-established colonists: Mylouthkia 1 and the Cypro-Pre-Pottery Neolithic B. In: Swiny S (ed) The earliest prehistory of Cyprus: from colonization to exploitation. American Schools of Prehistoric Research, Boston, pp 61–93

Porat N, Sivan D, Zviely D (2008) Late holocene embayment and sedimentological infill processes in Haifa Bay, SE Mediterranean. Israel J Earth Sci 57:21–23

Porath Y (1975) The gardens of Caesarea. Qadmoniot VIII:30–31 (in Hebrew)

Porath Y (1985) Ancient irrigation agriculture in the arid zones of Eretz Israel. Dissertation, Tel Aviv University (in Hebrew)

Porath Y (2016) Tunnel-well (Qanat) systems and settlements from the early Islamic period in the Arava. Atiqot 86:1–82 (in Hebrew)

Reich R, Shukron E (2002) Reconsidering the karstic theory as an explanation for the cutting of Hezekiah's tunnel in Jerusalem. Bull Am Schools Orient Res 325:75–80

Roskin J, Sivan D, Porat N, Shtienberg G, Roskin E, Bookman R (2015) Natural and human controls of the Holocene evolution of the beach, aeolian sand and dunes of Caesarea (Israel). Aeolian Res 19(A):65–85

Shiloh Y (1992) Underground water systems in the Iron age. In: Kempinski A, Reich R (eds) The architecture of ancient Israel from the prehistoric to the Persian periods. Jerusalem, pp 275–293 (in Hebrew)

Shtienberg G, Dix JK, Roskin J, Waldmann N, Bookman R, Bialik OM, Porat N, Taha N, Sivan D (2017a) New perspectives on coastal landscape reconstruction during the Late Quaternary: a test case from central Israel. Palaeogeogr Palaeoclimatol Palaeoecol 468:503–519

Shtienberg G, Dix JK, Shahack-Gross R, Yasur-Landau A, Roskin J, Bookman R, Waldmann N, Shalev S, Sivan D (2017b) Anthropogenic overprints on natural coastal aeolian sediments: a study from the periphery of ancient Caesarea, Israel. Anthropocene 19:22–34

Sivan D, Wdowinski S, Lamback K, Galili E, Raban A (2001) Holocene sea-level changes along the Mediterranean coast of Israel, based on archaeological observations and numerical model. Palaeogeogr Palaeoclimatol Palaeoecol 167:101–117

Sivan D, Lambeck K, Toueg R, Raban A, Porat Shirman B (2004) Ancient coastal wells of Caesarea Maritima, Israel, an indicator for sea level changes during the last 2000 years. Earth Planet Sci Lett 222:315–330

Stern E (1993) The new encyclopedia of archaeological excavations in the holy land volumes 1–4 ed

Stern E, Lewinson-Gilboa A Aviram J (1993) The new encyclopedia of archaeological excavations in the holy land. The Israel Explor Soc Carta, Jerusalem

Taxel I, Sivan D, Bookman R, Roskin J (2018) An early Islamic inter-settlement agroecosystem in the coastal sand of the Yavneh Dunefield, Israel. J Field Archaeol. https://doi.org/10.1080/00934690.2018.1522189

Tepper Y (2014) Enot Nisanit (Ha-Yogev Junction). Hadashot Arkheologiyot, excavations and survey in Israel, vol 126

Toker E, Sivan D, Stern E, Shirman B, Tsimplis M, Spada G (2012) Evidence for centennial scale sea level variability during the Medieval climate optimum (Crusader Period) in Israel, eastern Mediterranean. Earth Planet Sci Lett 315–316:51–61

Tsafrir Y, Holum KG (1988) Excavation at Rehovot-in-the-Negev, preliminary report. IEJ 38(3):117–127

Tsuk T (2011) Water at the end of the tunnel—touring Israel's ancient water systems, 2nd edn. Ben Yizhak Publication (in Hebrew)

Tsuk T (2018) Innovations in the study of ancient water systems. Qadmoniot 155:3–17 (in Hebrew)

Tsuk T, Herzog Z (1992) The water system of Tel Gerisa (Israel) and its contribution to the dating of underground water systems. Mitteilungen aus dem Leichtweiss-Institut fur Wasserbau, Heft 117. Braunschweig, pp 333–356

Tufnell O (1953) Lachish III (Tell Ed-Duweir), The Iron Age, London

Ussishkin D (1982) The conquest of Lachish by Sennacherib. Tel Aviv University

Vunsh R, Tal O, Yechieli Y, Dean S, Levanon E, Sivan D (2018) Evaluating ancient coastal wells as sea-level indicators from the coast of Israel. Geoarchaeology 33:403–416. https://doi.org/10.1002/gea.21663

Weinberger R, Sneh A, Shalev E (2008) Hydrogeological insights in antiquity as indicated by Canaanite and Israelite water systems. J Archeological Sci 35:3035–3042

Whitcomb D (1994) The Misr of Ayla: settlement at al-'Aqaba in the early Islamic period. In: King GRD, A Cameron A (eds) The Byzantine and Islamic Near East II: land use and settlement patterns. Darwin Press, Princeton, pp 155–170

Whitcomb D (2006) Land behind Ayla. In: Bienkowski P, Galor K (eds) Crossing the rift: resources, routes, settlement patterns and interaction in the Wadi Arabah. Oxbow, Oxford

Wilson AI (2006) The spread of Foggara-based irrigation in the ancient Sahara. In: Mattingly D et al (eds) The Libyan Desert—natural resources and cultural heritage (Society for Libyan studies monograph 6). London, pp 2015–216

Wilson AI, Mattingly DJ (2003) Irrigation technologies: foggaras, wells and field systems. In: Mattingly D (ed) The archaeology of Fazzan 1: synthesis (Society of Libyan studies monographs 5). Department of Antiquities, Tripoli, pp 235–278

Yadin Y (1969) The fifth season of excavations at Hazor, 1968–1969. Biblical Archaeologist 32:50–71

Yechieli Y, Rosensaft M, Zilberbrand M, Tsuk T (2017) The hydrogeology of Tel Gezer. Geological survey report GSI/27/2017

# Groundwater Management: An Example of a Project Related to Rehabilitation of the Southern Coastal Aquifer, as Part of the National Water System in Israel

**Israel Gev, Joshua Schwarz, Emanuel Shachnai, Svetlana Lumelsky, Sigal Oz, Shani Salomons, Adva Avital, and Ido Negev**

## 1 Introduction

The total consumption of water in Israel at present is around 2 miliard cubic meter (hereafter CM) per annum. The forecast is that the consumption will increase to 2.7 miliard CM per annum in 2030 in accordance with the domestic and agricultural demand. Until 2005, most of the consumption was based on pumping from Lake Kinneret to the National Water Carrier which diverted it to the south, as well as on the exploitation of the mountainous and coastal aquifers. Since then, Israel entered intensely into the era of seawater desalinization for domestic purposes and

I. Gev (✉)
Water Authority of Israel, Jerusalem, Israel
e-mail: israelg10@water.gov.il

J. Schwarz · E. Shachnai · S. Lumelsky · S. Oz · S. Salomons · A. Avital
TAHAL Consulting Engineers Ltd., Tel Aviv-Yafo, Israel
e-mail: schwarzj@tahal.co.il

E. Shachnai
e-mail: emmshacch@partner.net.il

S. Lumelsky
e-mail: Lumelsky.svetlana@gmail.com

S. Oz
e-mail: sigalsmail12@gmail.com

S. Salomons
e-mail: salomons.lambert.shani@gmail.com

A. Avital
e-mail: AvitalA@tahal.com

I. Negev
Mekorot Company, Tel Aviv-Yafo, Israel
e-mail: Nido@MEKOROT.CO.IL


U. Kafri and Y. Yechieli (eds.), *The Many Facets of Israel's Hydrogeology*, Springer Hydrogeology, https://doi.org/10.1007/978-3-030-51148-7_19

**Table 1** Annual replenishment and the number of production wells from the main aquifers in Israel

| Aquifer/Source name | Annual replenishment (MCM/Y) | No. of production boreholes |
|---|---|---|
| Lake Kinneret | 420 | 72 |
| Western Galilee Judea group | 120 | 102 |
| Mount Carmel Judea Group | 30 | 75 |
| Coastal | 270 | 808 |
| Central Israel Judea Group | 310 | 228 |
| Negev and Arava | 80 | 189 |
| Total | 1239 | 1474 |

sewage reclamation for agricultural irrigation. It is expected that by the year 2030 the desalinized water supply will increase from the current one of 600 million cubic meter (hereafter MCM) to 1 miliard CM. Similarly, the reclaimed water supply for agriculture is expected to increase from 500 MCM currently to around 700 MCM in 2030. Parallely, a decrease of water exploitation from Lake Kinneret and the above-mentioned aquifers is also expected, from the current volume of 1 miliard CM to some 800 MCM per annum. The expected and planned decrease of exploitation is related to the following: A decrease in precipitation due to climate changes, past over-exploitation of the aquifers, as well as urbanization and industrialization processes that decreased the natural recharge, all of which resulted in a depletion of the regional water tables. All the above also caused a deterioration of the water quality which demands a treatment of part of the exploited water.

The total estimated potential of the natural recharge of Israel is around 1.2 miliard MC per annum which is being exploited by some 1500 wells (Table 1).

As noticed above, the coastal aquifer serves as one of the main water resources of the country. It is, thus, also discussed in another chapter of the present book (Chap. 3). It extends from Mount Carmel in the north to the Gaza Strip in the south, bounded in the west by the Mediterranean base level and in the east by mountainous formations (Fig. 1). The average natural recharge to the aquifer is around 270 MCM/Y. It consists basically of calcareous sandstone, sandstone, clay and loam intercalations which subdivide the former from top to bottom to A, B, C and D sub-aquifers (Issar 1968) (Fig. 2). The aquifer is subjected to seawater encroachment from the west, mainly into its upper sub-aquifers. Salination of the aquifer occurs in places in its eastern margins through lateral flow of brackish water from the eastern aquitards. In addition, the aquifer is being salinized and contaminated by downward flowing pollutants related to anthropogenic domestic, industrial and agricultural activity (Water Authority of Israel 2001, 2002).

Due to the deterioration of the water quality (Fig. 3) and the resultant ongoing shortage of high quality water for domestic use and agriculture, measures were taken and a new operational and water management policy for this aquifer was determined.

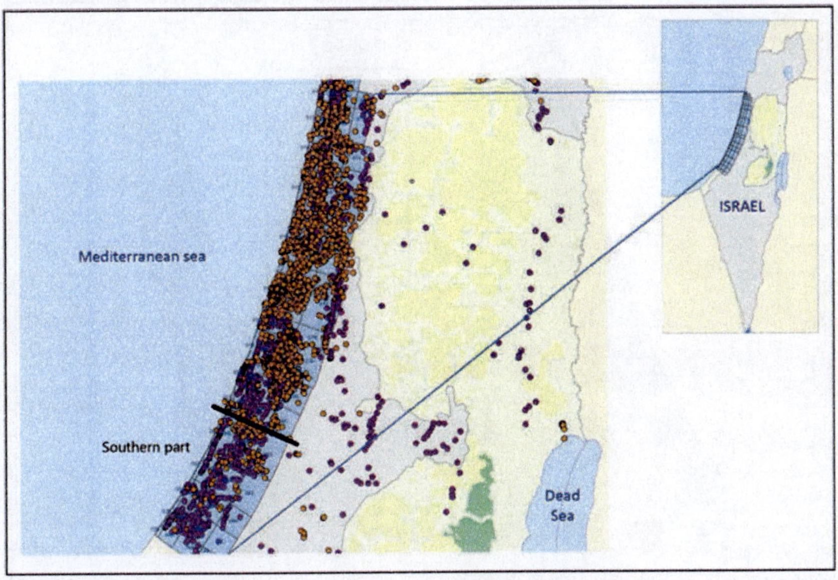


**Fig. 1** General map of Israel and the area of the coastal aquifer

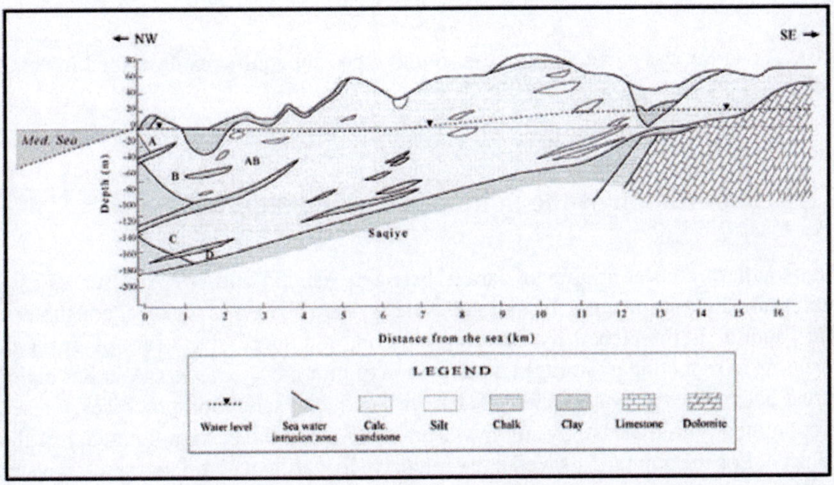


**Fig. 2** Typical hydrogeological cross section of the coastal aquifer (after Goldman and Kafri 2004)

It includes basically the following: Reduction of seawater encroachment to the aquifer from the west, massive establishment of seawater desalination plants, and prevention of salt input to the system through the brackish water flow from the aquitards in the

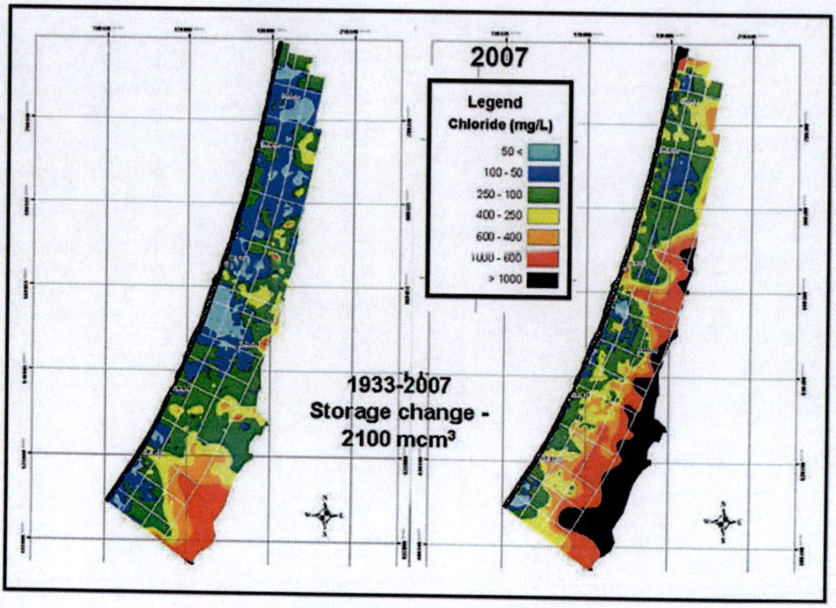


**Fig. 3** Chlorinity map of the coastal plain aquifer for the years 1933 and 2007 (after Livsitz and Tzentner 2009)

east. The latter was carried out in the southern coastal aquifer as described in detail below.

## 2 Rehabilitation of the Southern Coastal Aquifer

The southern coastal aquifer of Israel, between Ashdod and Nir-Am, serves as a major operative storage for the national water system as well as for local consumers. This aquifer is threatened by over-exploitation, and successive years of drought, resulting in reduction of storage and depletion of groundwater levels. Another major threat has to do with water quality deterioration due to salination processes, caused predominantly in the east, by saline water inflow along the eastern boundary of the adjacent Eocene aquitard, as well as by intensive irrigation with treated sewage water and by other anthropogenic activities.

In order to arrest the salination processes and allow replenishment of the aquifer, a rehabilitation plan has been designed introducing the concept of brackish water desalination as key component of aquifer rehabilitation in this area. According to Master plan for the water sector in Israel - natural water section, Gev and Shamir (2008a), and the policy document for management of the coastal aquifer, Gev and

Shamir (2008b). We suggested the following reclamation procedures for the Southern part of the aquifer.

The two main components of this plan are:

1. "The Eastern Interceptor," a battery of 33 wells located along the eastern boundary of the aquifer that pump 35 MCM/Y with chlorinities between 500 and 800 ppm, as raw water for desalination. This in turn prevents a flow of 30–35 thousand tons of salt per year into the western storage cells of the southern coastal aquifer. Desalination plants are located at Granot in the north and Lahat in the south. The desalinated fresh water of 30 MCM/Y is added to the supply system, and the residual brine, around 5 MCM/Y, is delivered to the west and discharged to the Mediterranean Sea.
2. A balanced regime pumping plan in the western part based on the annual aquifer recharge replenishment rate. The balanced pumping plan enables the storage and groundwater level restoration, both in the short and long term. Moreover, the reuse of the desalinated water in the rehabilitation area producing low salinity effluent that in turn is used for irrigation is contributing to additional elimination of salt inflow to the aquifer.

## 3 The Southern Coastal Aquifer and Its Salinity Sources

The surface area of the southern coastal plain aquifer is around 1000 km$^2$ between Ashdod and Nir-Am. It produces, annually, about 70 MCM of water. The main source of replenishment of this aquifer is through precipitation over its surface outcrops. The aquifer in this eastern area differs in facies from the aquifer in the west. Here, the sequence consists mainly of loams, rather than sandstones, and thereby the aquifer is typified by lower vertical infiltration rates and higher evapotranspiration, as compared to the west, where the salinity is low. Therefore, typical groundwater salinities in the eastern parts of the aquifer are higher with typical values of 300–500 mg/l chlorides, as compared to 30–50 mg/l in the west. Another source of salinity, as mentioned before, is lateral penetration to the aquifer by brackish groundwater from the Avdat Eocene aquitard in the east, where the water chlorinity exceeds 1000 mg/l.

Additional salinity sources to the aquifer are return flows from irrigation through the unsaturated zone, overlying the aquifer. Irrigation with National Carrier water delivered from Lake Kinneret and reclaimed sewage with a chloride content of 220–300 mg/l play also an important role as salinity sources. Artificial recharge of the National Carrier system is also a source of increasing salinity.

The chlorinity map of the coastal plain aquifer for the years 1933 and 2007 is exhibited in Fig. 3. As a result of the above, the salinity in this aquifer showed distinct salination processes particularly in its eastern parts with an average increase of chlorinity at an annual rate of 2–5 mg/l. A general map of chlorinity and contamination zones in the coastal plain aquifer are shown in Fig. 4.

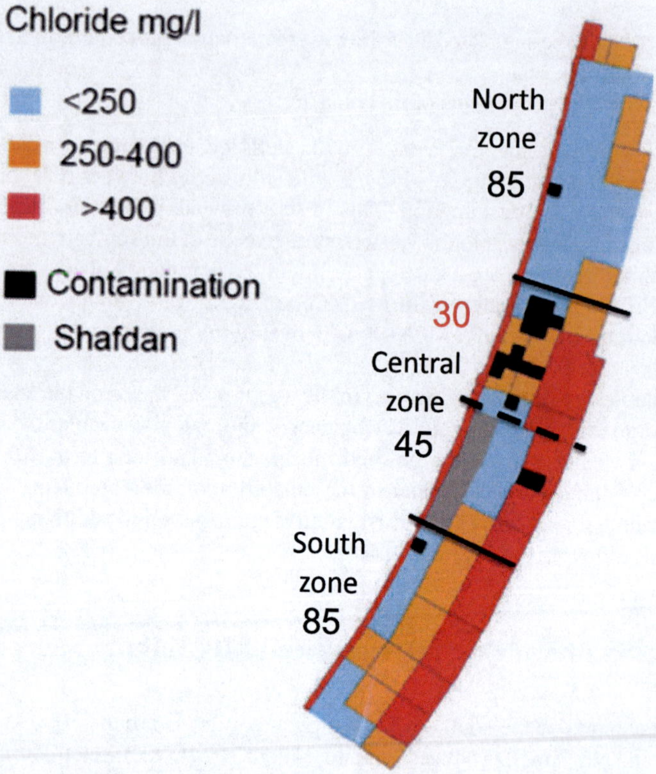


**Fig. 4** General map of chlorinity and contamination zones in the coastal plain aquifer. The number represents the average production in MCM/year

## 4 Rehabilitation Options

The rehabilitation possibilities of this aquifer have been a subject of an intensive study carried by the Israel Water Authority, and TAHAL Consulting Engineers Ltd. This study included a GMS (input output interface) model of the aquifer, including flow simulation (Modflow) and salinity conveyance (MT3D). This model was first calibrated to simulate the history of flows, water levels and salinity variations for the period 1970–2000 and then consequently used to simulate the various interventions proposed for rehabilitation.

Five main promising rehabilitation interventions were studied by the models and then further evaluated by preliminary designs and cost estimates. These are:

1. Removal and controlled disposal of point sources of domestic and industrial effluents and wastes.
2. Reduction of fertilizers and pesticides applications in agriculture.

3. Replacement of imported high salinity Kinneret water by low salinity water imported from seawater desalination plants.
4. Redistribution of production wells according to the natural replenishment, and to areas of lower salinities or to the deeper sub-aquifers of a better water quality, being protected from downward pollution.
5. Creation of a cutoff by wells in the east to intercept the higher salinity inflows from the east.

The latter intervention is the main component of the proposed rehabilitation plan.

# 5 The Eastern Interceptor Concept

The Eastern Interceptor (Fig. 5) is a battery of wells located on a line running from north to south in the eastern margin of the aquifer along the boundary of high salinity water. These wells are intended to operate uninterruptedly and continuously, aimed to prevent the flow of saline water in the aquifer to the low salinity areas of the aquifer, which will continue to operate as sources of fresh water and as a regulation facility for the national water supply system.

The water pumped by the interceptor wells is designated for desalination and further blending in the regional freshwater supply systems (Fig. 5). Alternative solutions of the residual brine disposal from desalination plants and location of production wells are considered in terms of centralizing the desalination to a single plant or decentralizing it to many single well plants or any other intermediate alternative.

## *5.1 The Eastern Interceptor Design*

The location of the interceptor line was first determined by salinity forecasts generated by the aquifer models. The 400 mg/l chloride isoline was tentatively adopted, and then studied and modified by the model simulations.

The spacing of wells was calculated by considering the flow rate in the aquifer [$m^3$/m/day] divided by the discharge capacity of a single well [$m^3$/day]. The flow rate in the aquifer was estimated by the results of aquifer simulations. Performance of existing wells enabled to estimate the anticipated discharge rates of the planned wells. These values vary from north to south and the spacing was determined accordingly. Thirty-three wells were designed with discharge rates ranging between 120 and 250 $m^3$/h, totaling 35 million $m^3$/year. Well spacing was about 500–700 m. The total length of the intercepting collector line is about 32 km.

The depth of wells and those of the perforated casings involved were designed according to the geological sequence obtained from existing well logs, reflected in geological cross sections. Pumping is designed to exploit the uppermost saline part of the aquifer.

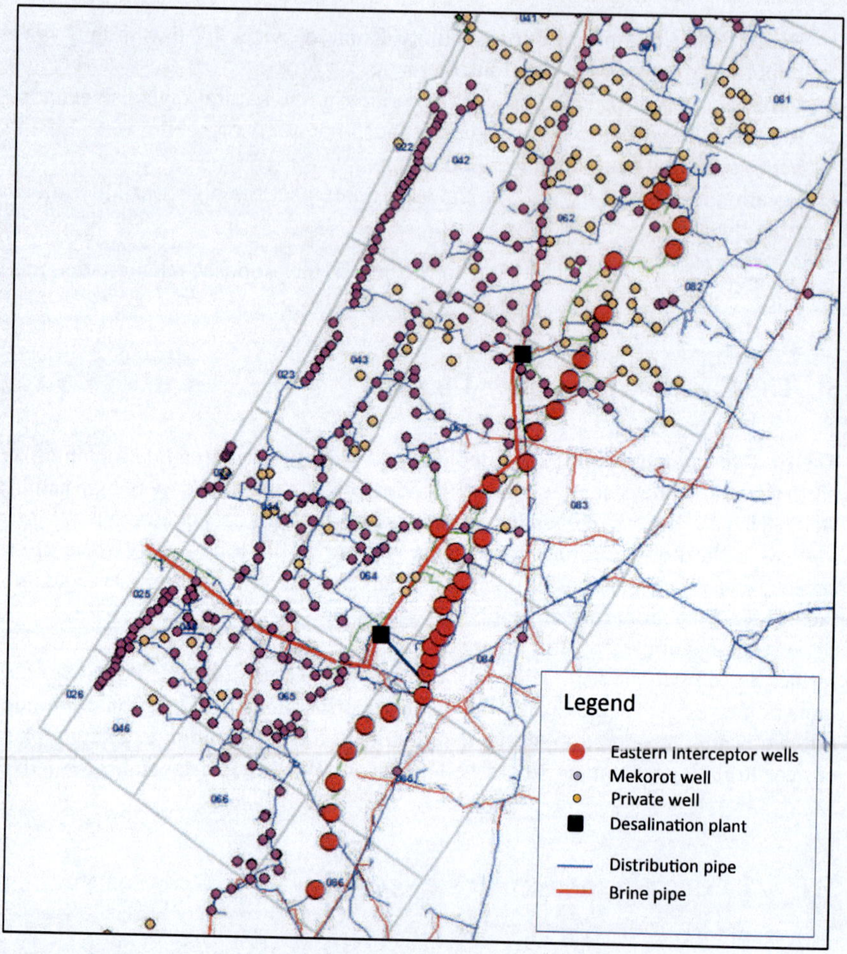


**Fig. 5** Eastern Interceptor wells and the engineering system

## 5.2 *Desalination*

Two desalination plant sites were selected, based on existing plans and following the study of various alternatives of centralization. Both desalination plants are located near the main pipelines of the regional water supply system. The first plant at Granot is designed for 2650 $m^3/h$ interceptor water in addition to 680 $m^3/h$ from other sources. The second plant at Lahat is designed for 1920 $m^3/h$ interceptor water in addition to 915 $m^3/h$ from other sources. The chlorinity of raw water is about 500–550 mg/l and the planned desalinized end product about 50 mg/l. A system of pipelines was designed to collect raw water from the interceptor wells and supply to the pair of desalination plants. The total length of the pipes is 40 km with diameters varying between 8 and 30 inches. The energy of flow is produced by the pumps of the interceptor wells. Two additional elements were required for supplying the end product water to the regional water supply system. These include product water regulation tanks of 4000 $m^3$ and pumps to elevate pressure to the regional water supply system with pumping heads of 74 and 45 m.

## 5.3 *Disposal of Desalination End Brine*

Two alternative destinations were considered for Mediterranean Sea disposal of desalination reject brines, namely the Ashdod and the Ashkelon power plants. A possibility to use both of them was also considered. The preferred plan was to use the Ashkelon site, which is planned also for brine disposal of the existing seawater desalination plant on the site. The brine disposal pipeline has a total length of 28 km and diameters ranging between 14 and 20 inches. Brine disposal also required two pumping stations in the two desalination plants.

## 5.4 *Cost*

The Eastern Interception system includes infrastructures of wells, pipes and desalination plants, producing a total capacity of 51,000 CM/day in Granot and 43,000 CM/day in Lahat (Table 2A).

The costs of the infrastructures and the desalination plants are shown in Table 2B. The total cost of water in Granot and Lahat including the cost of infrastructures and desalination plant is 0.64 $.

**Table 2** Investment and operational cost of the Eastern Interception system

| **Investments** | **Cost (million IS)** |
|---|---|
| Interceptor wells and raw water pipelines | 47.3 |
| Desalination plants and product water supply system | 88.2 |
| Brine disposal system | 8.7 |
| **Annual costs are as follows** | **(million IS/year)** |
| Capital costs | 17.4 |
| Energy costs | 11.8 |
| Operation maintenance costs | 2.4 |
| **Estimated cost of desalinated water** | **1.19 IS/$m^3$** |

**Table 3** Comparison of simulation results for the year 2025

| | Component | Reference scenario | Rehabilitation scenario |
|---|---|---|---|
| Water inflows [MCM/year] | Replenishment | 75 | 75 |
| | Eastern inflows | 22 | 22 |
| | Outflow to the sea | 10 | 10 |
| | Pumping | 93 | 45 |
| | Eastern wells pump. | 0 | 30 |
| | **Storage change** | **−6** | **12** |
| Chlorides mass balance [ton/year] | Replenishment | 19,000 | 8000 |
| | Eastern inflows | 15,000 | 15,000 |
| | Sea outflow | 3000 | 13,780 |
| | Pumping | 14,000 | 26,000 |
| | Eastern wells pump. | 15,000 | 0 |
| | **Storage change** | **6000** | **−9000** |

## 5.5 *Performance*

The rehabilitation program was first tested by including in it two components: The Eastern Interceptor aimed to reduce the entrance of salinity into the western part of the aquifer, and redistribution of the pumping in order to reduce salinity of supplied water. These coupled with the addition of fresh desalinated water was expected to considerably reduce the salinity of the supplied water.

In order to test the performance of the rehabilitation scenario, it was compared to a reference scenario assuming a continuation of the present pumping and recharge regimes. Both scenarios were simulated employing the GMS system models. Some average results, obtained for the last simulation of the year 2025 are shown in Table 3.

The tested rehabilitation program was supposed to reduce, based on the above model, the average mass of chlorinity in the aquifer by 9000 ton/year and that slowing the processes of increases clorinity of the storage. The average chlorinity of the pumped groundwater (excluding the interceptor water) is reduced from 350 mg/l in the reference scenario to 230 mg/l in the rehabilitation scenario due to pumping redistribution (Fig. 6). The salinity of water supply is further reduced in the rehabilitation scenario when considering dilution of the desalination product in the water supply system

## 5.6 *Supplementary Actions*

The proposed Eastern Interceptor project is only one component of a rehabilitation program for the southern coastal plain aquifer of Israel. It has to be supplemented by the removal, reduction and safe disposal of soil surface salinity sources, such as

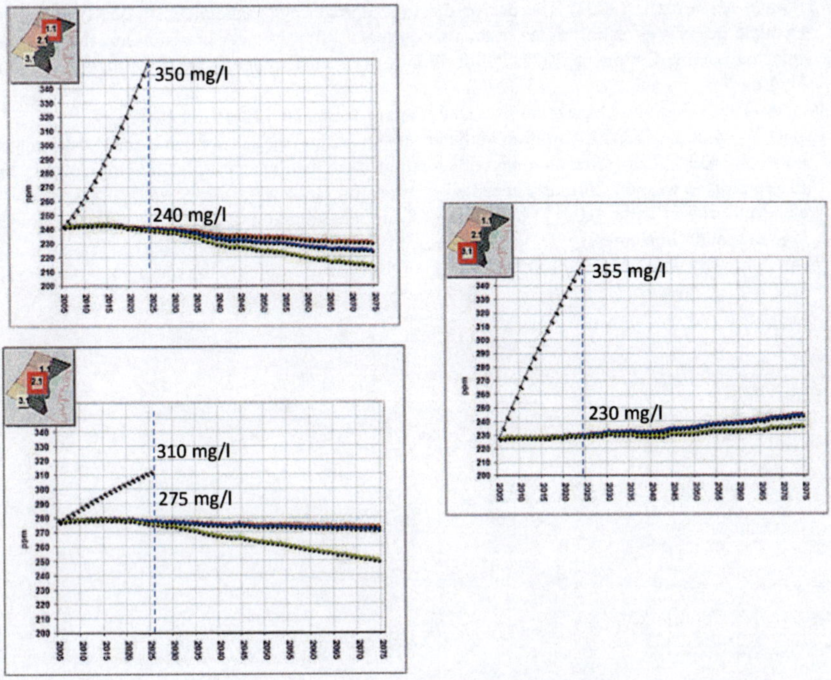


**Fig. 6** Simulation of the chlorinity concentration (ppm) to the year 2075 in the western cells of the aquifer as a result of different alternatives: black color represents the progress of salinization in reference scenario, without the Eastern Interceptor wells. The blue, brown and green colors lines represent the progress of salinization in the rehabilitation scenario, including Eastern Interceptor wells, and with different salinities of irrigation water

irrigation with high salinity water and uncontrolled waste disposal. However, this removal will have its effects on the salinity of groundwater only after a long time due to long travel times in the order of decades, between the surface and groundwater table. This removal is therefore a long term solution required only to the west of the interceptor line.

A short term solution that can be focused on particularly sensitive areas is the artificial recharge with low salinity water derived from the seawater and/or the interceptor desalination plants.

## References

Gev I, Shamir U (2008a) Master plan for the water sector—natural water section, chapter 4. Water Authority of Israel (in Hebrew)

Gev I, Shamir U (2008b) Policy document for management of the coastal aquifer. Water Authority of Israel (in Hebrew)

Goldman M, Kafri U (2004) The use of the time domain electromagnetic (TDEM) method to evaluate porosity of saline water saturated aquifers. 18th salt water intrusion meeting (SWIM) selected papers. Cartagena 2004. Publicaciones del Instituto Geologico Y Minero de Espana. Hydrog Y aguas subterraneas 15:327–340

Issar A (1968) Geology of the central coastal plain of Israel. Isr J Earth Sci 17:16–29

Livsitz Y, Tzentner E (2009) Variation of the water storage and the salt storage in the coastal aquifer between 1933 and 2007, development of a method, examination and application in selected years in comparison with the currently accepted methods (in Hebrew)

Water Authority of Israel (2001) Master plan for the rehabilitation of the coastal aquifer—hydrological model (in Hebrew)

Water Authority of Israel (2002) Rehabilitation of the coastal aquifer in the southern part of the Nir area with Ashdod (in Hebrew)

# Extended Abstracts

# Groundwater Level Changes Induced by Earthquakes

**Eyal Shalev, Hallel Lutzky, and Vladimir Lyakhovsky**

Atmospheric loading, earth tides and seismic waves are known to affect groundwater level oscillations (Cooper et al. 1965; Brodsky et al. 2003; Kitagawa et al. 2011; Wang and Manga 2009). In some cases, sustained changes of the groundwater level occur during different time spans. The duration of the short-term sustained changes (minutes) is associated with the passage of the seismic waves and it is assumed that they occur under undrained conditions. The longer time span changes (hours and days) reflect drained conditions whereby the water level in the well is affected by groundwater flow that is being triggered by the seismic waves.

The effect of groundwater behavior on seismicity and the effect of seismic events on groundwater levels were recorded in the past along the Dead Sea Transform (Kafri and Shapira 1990; Yechieli and Bein 2002). Eight wells are currently monitored in a high temporal resolution mode in Israel (Fig. 1). Six of them are artesian (Gomè 1, Meizar 2, Meizar 3, EZ 1, Sedom 2 and Neot HaKikar T-40 wells) and the remaining two wells are highly confined (Meizar 1 and Ein Hashofet 1 wells). Water pressure is monitored, using a Keller-Druck 33 PA transmitter with a set range of 0–0.1 MPa. Water pressure is recorded at a paste of 40 samples per second. In addition, barometric pressure is monitored using a Vaisala PTB110 instrument.

Co-seismic water level oscillations were recorded in several wells following many distant earthquakes as well as local events along the Dead Sea Transform (Shalev et al. 2016a). The distant Balochistan earthquake in Pakistan (2013 September 24th, Mw = 7.7) exhibits a clear waveforms and co-seismic water level oscillations, as was also observed in Gomè 1 and Meizar 1 wells, some 3000 km away from the epicenter. The

E. Shalev (✉) · H. Lutzky · V. Lyakhovsky
Geological Survey of Israel, Jerusalem, Israel
e-mail: eyal@gsi.gov.il

H. Lutzky
e-mail: hallel@gsi.gov.il

V. Lyakhovsky
e-mail: vladimir.lyakhovsky@gmail.com



U. Kafri and Y. Yechieli (eds.), *The Many Facets of Israel's Hydrogeology*,
Springer Hydrogeology, https://doi.org/10.1007/978-3-030-51148-7_20

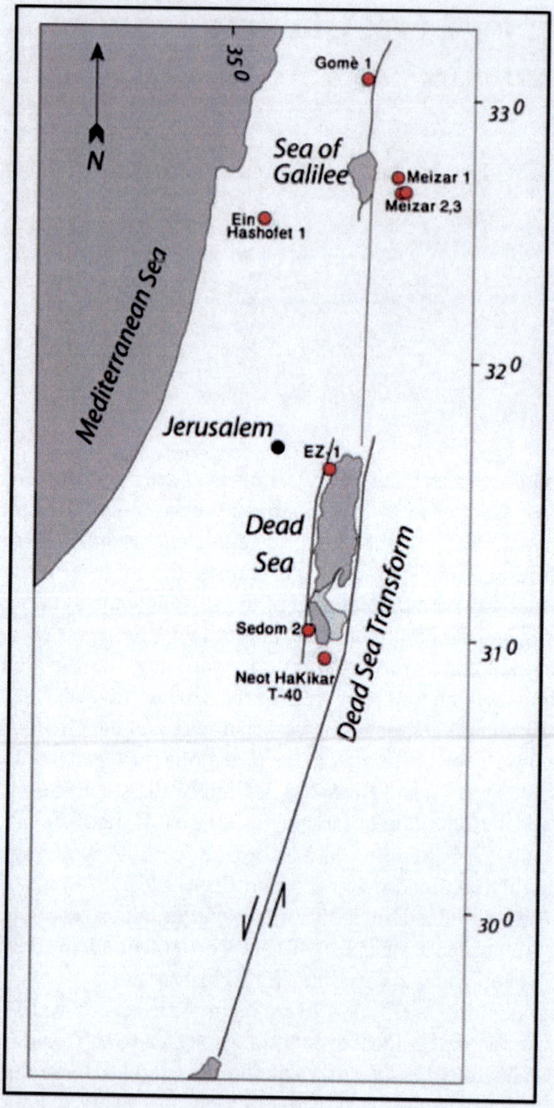


**Fig. 1** Location of eight wells with high resolution

drained response to the earthquake of the water levels in both of the wells includes a continuous decline for a few days (Fig. 2a, b). This response reflects a permeability enhancement of the nearby faults that create short-cuts between different aquifers. The high groundwater overpressure at Gomè 1 well is released through one of the faults, located only a few tens of meters away from the well. Also recognized in Gomè 1 well are the responses of the groundwater level to the largest aftershock

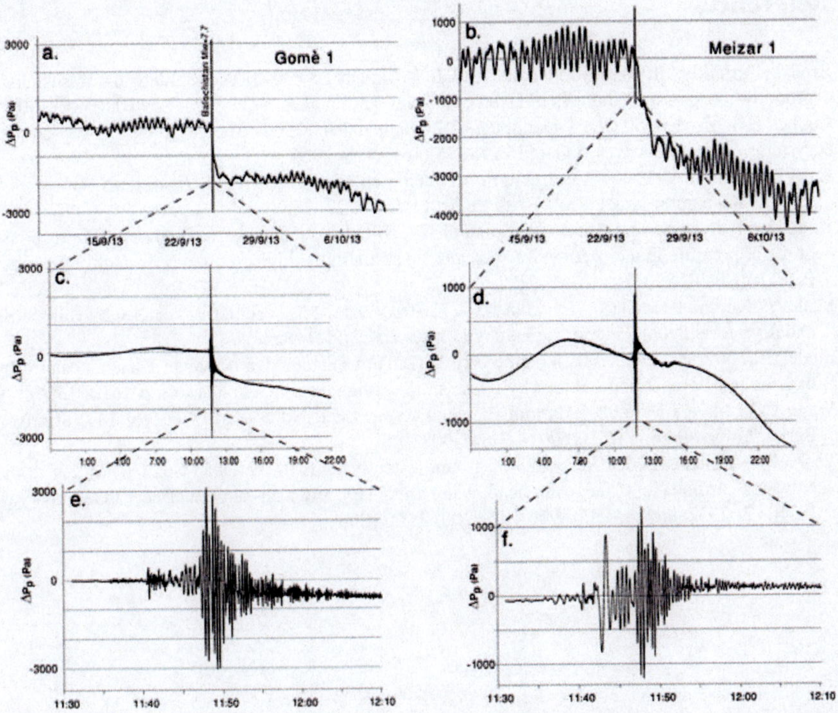


**Fig. 2** Water level responses at Gomè 1 and Meizar 1 wells to the 24 September 2013 Mw 7.7 Balochistan earthquake. **a** Drained response in Gomè 1, **b** drained response in Meizar 1, **c** undrained response in Gomè 1, **d** undrained response in Meizar 1, **e** undrained hydro-seismogram response in Gomè 1, and **f** undrained hydro-seismogram response in monitoring groundwater level in Meitzar 1

(Mw = 6.8) of the Balochistan earthquake and to the Mw = 6.4 Crete earthquake (Fig. 2a). The undrained short-term response of groundwater levels is different in the two wells, namely at the Gomè 1 well, the water pressure decreased (Fig. 2c, e), whereas at the Meizar 1 water well, the pressure increased (Fig. 2d, f).

Shalev et al. (2016b) suggested that the above described different behavior is controlled by the damage (fractures) of the aquifers. Gomè 1 well penetrated a highly fractured rock sequence, whereas the Meizar 1 well is relatively far from a fault (~5.5 km). The rate of fracture dilation strongly depends on the damage level of the rocks. At a high damage situation, rocks undergo high dilation, whereas the porosity compaction remains at the same order of magnitude. Thus, the sustained water level decline in Gomè 1 well is controlled by fracture dilation (resultant declined water level), whereas the sustained increase in Meizar 1 well is controlled by compaction (resultant increased water level).

## References

Brodsky EE, Roeloffs E, Woodcock D, Gall I, Manga M (2003) A mechanism for sustained groundwater pressure changes induced by distant earthquakes. J Geophys Res 108(B8):2390

Cooper HH, Bredehoeft JD, Papadopulos IS, Bennett RR (1965) The response of well-aquifer systems to seismic waves. J Geophys Res 70(16):3915–3926

Kafri U, Shapira A (1990) A correlation between earthquake occurrence, rainfall and water level in Lake Kinnereth, Israel. Phys Earth Planet Inter 62:277–283

Kitagawa Y, Itaba S, Matsumoto N, Koizumi N (2011) Frequency characteristics of the response of water pressure in a closed well to volumetric strain in the high-frequency domain. J Geophys Res 116(B8):B08301

Shalev E, Kurzon I, Doan ML, Lyakhovsky V (2016a) Water-level oscillations caused by volumetric and deviatoric dynamic strains. Geophys J Inter 204(2):841–851

Shalev E, Kurzon I, Doan ML, Lyakhovsky V (2016b) Sustained water level changes caused by damage and compaction induced by teleseismic earthquakes. J Geophys Res 2016JB013068

Wang C-Y, Manga M (2009) Earthquakes and water. Lect. Notes Earth Sci, vol 144. Springer, Berlin, https://doi.org/10.1007/978-3-642-00810-8

Yechieli Y, Bein A (2002) Response of groundwater systems in the Dead Sea Rift Valley to the Nuweiba earthquake: Changes in head, water chemistry, and near-surface effects J Geophys Res 107(B12):2332. https://doi.org/10.1029/2001jb001100

# The Distribution of Brines in the Central Jordan Valley and Adjacent Areas

**Eliahu Rosenthal and Peter Möller**

The thermal and pressurized Ca-chloride brines which permeate the subsurface of the Jordan—Dead Sea Transform area are a perpetual danger to all pristine water resources in the Rift. The Bet Shean Valley, which is an integral part of the Central Jordan Valley, is a regional recipient for groundwater deriving from a multi-aquifer zone which includes carbonate and basalt aquifers carrying freshwater and deep reservoirs of thermal and pressurized Ca-chloride brines. In the Valley, the natural outlets of these regional aquifers are springs, outflowing along major fault lines branching out from the Rift system (Rosenthal 1988) (Fig. 1). These faults act as connecting media between these aquifers and the upflowing brines. Until the recent decade, there was no data on the deep geology of the Valley. Starinsky (1974) was the first to propose a detailed chemical model for the formation of the regional brines. Rosenthal (1980) described in detail the regional multi-aquifer system and the relations between upflowing brines and the pristine, currently recharged water. Data attained in the last years allowed to establish the regional hydrochemical stratigraphy and information on the pressures driving the brines in the Central Jordan Valley.

In the subsurface, the deep-seated Triassic sequence includes thick beds of gypsum and anhydrite of the Mohila Formation of the upper Ramon Group (Zak 1963), with inclusions of halite, similar to the lithology in Ramallah 1 borehole (Rosenthal et al. 2006). According to Derin (2016, and personal communication 2019), the Triassic sequence in central and northern Israel is identical to the Germano-Trias type spread throughout southern Europe, between Morocco in the west to Turkey in the east, and the Mohila Formation is clearly correlated with the Keuper.

E. Rosenthal (✉)
The School of Geosciences, Tel Aviv University, Ramat Aviv, Tel Aviv, Israel
e-mail: elirose@netvision.net.il

P. Möller
Helmholtzentrum Potsdam, German Research Center for Geosciences, GFZ, Section 3, 4, Potsdam, Germany
e-mail: pemoe@gfz-potsdam.de



U. Kafri and Y. Yechieli (eds.), *The Many Facets of Israel's Hydrogeology*, Springer Hydrogeology, https://doi.org/10.1007/978-3-030-51148-7_21

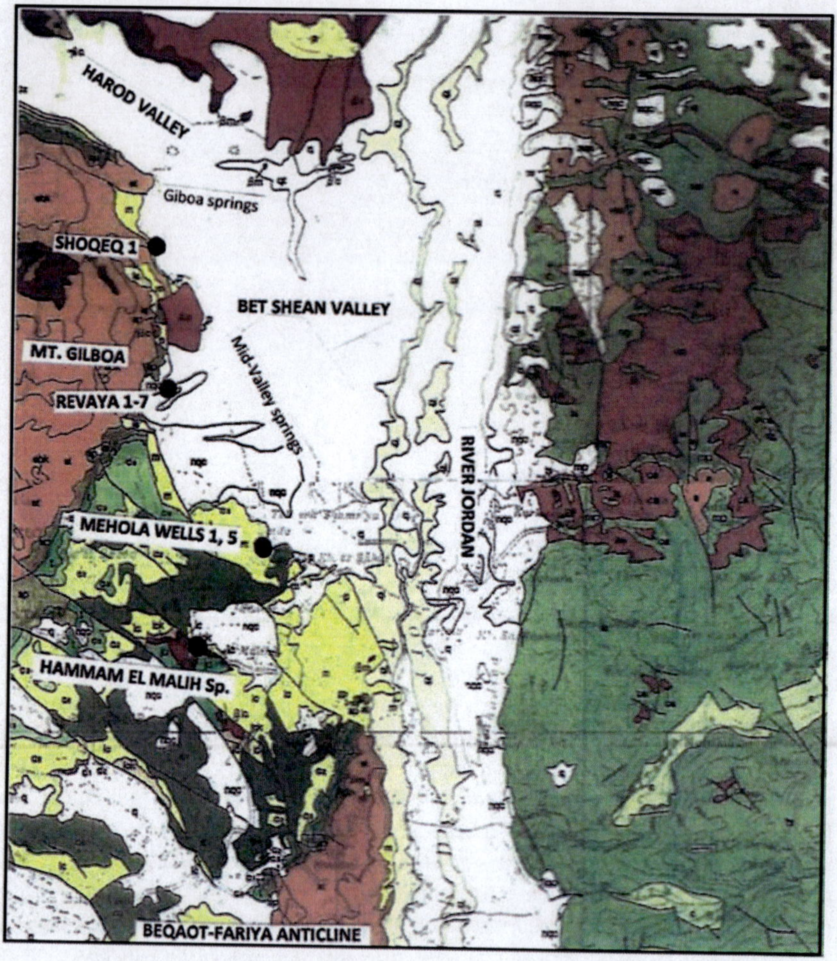


**Fig. 1** Geological location of the Bet Shean Valley with the indication of water resources

The brines in the Triassic Ramon Group in Israel have Cl concentrations in the 11–226 g/l range. Sulfate concentrations are in the range of 16–1538 mg/l. All fluids are of the Ca-chloride type with excess factors ($Q = Ca/(SO_4 + HCO_3)$(equiv) of up to 158 (in borehole Devora 2Aa). In this borehole, these brines are pressurized (480 bar) and thermal (162 °C) (Rosenthal 1988). Local occurrences of very high Cl/Br weight ratio values (>10,000) indicate local presence of halite, as found elsewhere in the Germano-Triassic zone (Alcala and Custodio 2008).

Karstic features, penetrating deeply into the Upper Triassic sequence, facilitate intraformational flow. Inverse modeling (Parkhurst and Appelo 2013) revealed that most of the brines in Jurassic formations are derivatives of Triassic brines and basaltic water which could derive from massive presence of volcanic rocks in the Ramon sequence (Derin 2016). In most cases, the Middle Jurassic Rosh Pina shales separate between the lower and upper parts of the Jurassic Arad Group sequence. The lower part contains a high amount of shales, and its permeability is much lower than that of the upper part. The average Cl content is in the range of 39–105 g/l with low concentrations of sulfate (380–1000 mg/l). These are typical Ca-chloride brines with Ca excesses of $Q = 37$–134, Na/Cl of 0.49–0.68 and low (60–127) Cl/Br weight ratios, which could have been formed as the result of seawater evaporation. In the upper part of the Arad Group, the brines are of the Na–Cl type with Na/Cl in the 0.78–1.18 range and enhanced Cl/Br ratios (111–248). Ca excess is much lower, in the 0.81–38 range. The brines could have originated from seawater which might have undergone a certain degree of evaporation without reaching the point of halite precipitation. The brackish upper water body units cannot be regarded as brine since its Cl concentration is lower than in seawater and Na/Cl and Q ratio values differ from the values characterizing Ca-chloride brines (Rosenthal 1988). Unlike the brine in Lower Jurassic and Triassic formations, this saline water body contains bicarbonate which indicates a hydrological connection with active replenishment. Of particular interest and importance is the fact that both Triassic and Jurassic brines are highly pressurized driving the brines to the surface (i.e., 480 bar in Devora 2A borehole (Rosenthal 1988) and elsewhere. This is of primary importance for the management of the regional water resources. In the southern Bet Shean area, groundwater originating from Lower Cretaceous formations was encountered in several sources. All springs are thermal and emerge along faults (Rosenthal et al. 2011) and from volcanic rocks, exposed by faulting in the core of the Faria' anticline. The Cl content reaches as high as 36 g/l. The Na/Cl (equivalent) and the Cl/Br (weight) ratios of these fluids (0.87–0.98 and 250–477, respectively) indicate that they could have evolved from seawater which did not reach the point of halite precipitation. The high Ca excess factors up to $Q = 13.3$ could have been the result of both dolomitization (Starinsky 1974) and chemical reduction of sulfates.

Summing up the evidence from Bet Shean and from the Central Jordan Valley, it occurs that the subsurface of the Rift hosts a wide range of brines (Rosenthal et al. 2009; Möller et al. 2012, 2014). According to them, the saline groundwater bodies in the Rift are as follows:

- Deep-seated basinal brines of the Ca, Mg, Chloride type.
- Neogene Sdom Sea (also termed Inland Sea) primary evaporation brine, as represented by the Ha'On saline water characterized by Mg > Ca. This saline water flows out from springs in eastern Bet Shean and seepages along the eastern and southeastern shores of Lake Tiberias, in the eastern parts of the Yarmuk Gorge and in one source along the southwestern margins of the Hula Valley. Neogene-Tertiary brines with Ca > Mg, are encountered in Tiberias Hot Springs, all along the western margins of the Rift and southward to the Arava Valley and along fault lines branching out westwards from the Rift system (Rosenthal et al. 2011).
- Neogene to recent secondary ablation brine of the Zemah 1 type with Na/Cl = 1 (Inbar 2012).

The Ha'On saline water with Mg > Ca represents the Neogene seawater which invaded the Rift and underwent 37-fold evaporation (McCaffrey et al. 1987). In spite of different Ca/Mg ratios, the similarity of Na/Cl, Br/Cl and (Ca + Mg)/Na ratios suggests that the Ha'On brine or any of its diluted equivalents was converted into the Tiberias brines outflowing along the western rims of the Rift. Under the given petrologic conditions with alkaline olivine basalt flows covering the Tertiary limestones south of Tiberias, it is reasonable to infer that the chemical evolution of the Ha'On brine into the Tiberias brines is the result of dolomitization of limestones (Starinsky 1974) and further alteration of plagioclase to albite, illite and chlorite. Uptake of halite and gypsum also plays a role. Both components are added either by leaching of evaporites or by mixing with an evaporite dissolution brine of corresponding composition. The relevant models are described in great detail in Möller et al. (2012).

The detailed knowledge on the stratigraphy of subsurface brines and their chemistry and pressures is of extreme importance for the planning and management of regional and local groundwater resources and for the further economic development of the area. These parameters facilitate now to define precisely the depths of safe groundwater exploitation and diminish the danger of further salinization.

## References

Alcala FJ, Custodio E (2008) Using the Cl/Br ratio as a tracer to identify the origin of salinity in aquifers in Spain and Portugal. J Hydrol 359:189–207

Derin B (2016) The subsurface geology of Israel-Upper Paleozoic to Upper Cretaceous, 328 p

Inbar N (2012) The evaporitic subsurface of Kinnarot basin—stratigraphy, structure, geohydrology. Ph.D. thesis, Tel Aviv University, Israel

McCaffrey MA, Lazar B, Holland HD (1987) The evaporation path of seawater and the coprecipitation of $Br^-$ and K+ with halite. J Sed Pet 57:928–937

Möller P, Siebert C, Geyer S, Inbar N, Rosenthal E, Flexer A, Zilberbrand M (2012) Relationship of brines in the Kinneret basin, Jordan-Dead Sea Rift Valley. Geofluids 12(2):166–189

Möller P, Rosenthal E, Flexer A (2014) The hydrogeochemistry of subsurface brines in and west of the Jordan-Dead Sea Transform Fault. Geofluids 14:291–309

Parkhurst D, Appelo CAJ (2013) User's guide to PHREEQC Version 2.18.3. A computer program for speciation, batch-reaction, one-dimentional transport and inverse geochemical Calculations. http//www.brr.cr.usgs.gov/projects/

Rosenthal E (1980) Hydrogeology and hydrogeochemistry of the Bet Shean, Harod and Mehola Valleys: methods for tracing groundwater flow and salinization processes in a multiple aquifer system. Ph.D. thesis, Hebrew Univ. Jerusalem (in Hebrew)

Rosenthal E (1988) Hydrochemical changes in groundwater caused by progressive upflow of Ca-chloride brines in common outlets of the Bet Shean-Harod multiple aquifer system, Israel. J Hydrol 97:107–128

Rosenthal E, Flexer A, Möller P (2006) The paleoenvironment and the evolution of brines in the Jordan-Dead Sea transform and in adjoining areas. Int J Earth Sci (Geol Rundsch) 4:725–740

Rosenthal E, Flexer A, Möller P (2009) The hydrochemical history of the Rift. In: Hoetzl H, Möller P, Rosenthal E (eds) The water of the Jordan Valley; scarcity and deterioration of groundwater and its impact on the regional development. Springer, Berlin, pp 75–82

Rosenthal E, Meiler M, Flexer A (2011) Structure controlled groundwater flow and salinization paths in the Bet Shean and Harod Valleys, Israel. J Environ Earth Sci 66(4):1071–1082

Starinsky A (1974) Relationship between Ca-chloride brines and sedimentary rocks in Israel. Ph.D. thesis, Hebrew Univ. Jerusalem (in Hebrew)

Zak I (1963) Remarks on the stratigraphy and tectonics of Triassic in Makhtesh Ramon, Israel. Isr J Earth Sci 12:87–89

# Colloid and Colloid-Facilitated Transport in Fractured Chalk

**Emily Tran and Noam Weisbrod**

Colloids are solids ranging from a few nm to a few μm in diameter (Schäfer et al. 2012) and may be composed of natural organic or inorganic matter (McCarthy and Zachara 1989). Alternatively, they may be formed by the precipitation of a dissolved species with a ligand in solution to form a stable solid (McCarthy 1998; Runde 2000; Tran et al. 2015). In saturated fractured rocks, colloids have been shown to migrate very efficiently, in many cases even faster than conservative aqueous species (Zhang et al. 2015). This increased migration velocity has been attributed to two main phenomena: (1) size exclusion, where colloids are often larger than the surrounding pore spaces in rocks, forcing them to stay within the fracture area rather than diffusing into the surrounding rock matrix, as would a dissolved species (Abdel-Salam and Chrysikopoulos 1995); and (2) charge exclusion. Most natural colloids in the subsurface are composed of mineral or organic matter which carries a negative charge. As rock surfaces also tend to be negatively charged, the colloids are repelled from the fracture surface and stay in the center of the flow path, where the flow is faster (Becker et al. 1999).

In the northern Negev Desert of southern Israel, much of the local lithology is composed of fractured chalk formations (Weisbrod et al. 2000). Previous studies have shown that the fractures in these rocks act as the primary conduits for mass and contaminant transport (Kurtzman et al. 2005) (see also Chapter "Unsaturated-Zone Hydrology in Israel: A Subjective Review"), both of dissolved and colloidal species. As the average pore throat size of this chalk is quite small (<0.4 μm, Bloomfield and Nygaard 2003), colloidal matter is likely to stay within the fracture area as previously described. In a study on the impact of colloid size on recovery and transport, it

E. Tran · N. Weisbrod (✉)
The Zuckerburg Institute for Water Research, The Jacob Blaustein Institutes for Desert Research, Ben Gurion University of the Negev, Beersheba, Israel
e-mail: weisbrod@bgu.ac.il

E. Tran
e-mail: emily.tran4298@gmail.com



U. Kafri and Y. Yechieli (eds.), *The Many Facets of Israel's Hydrogeology*,
Springer Hydrogeology, https://doi.org/10.1007/978-3-030-51148-7_22

was shown that colloids of size 0.2–1 μm exhibited faster overall transport times and greater recoveries than smaller colloids of 0.02 μm diameter (Zvikelsky and Weisbrod 2006). A later study on virus transport confirmed this, as viruses of 0.2 μm were shown to be more mobile in chalk fractures than smaller viruses of 0.02 μm, and both virus sizes were significantly more mobile than dissolved species in the same flow path (Weisbrod et al. 2013).

The mere presence of colloids in fractures is not enough to ensure that they will migrate effectively. Colloids must be stable in solution to prevent their flocculation and deposition onto fracture surfaces (Phenrat et al. 2007; Tran et al. 2018). The impact of colloid stability on mobility through chalk fractures has been shown both on a laboratory and field scale. The mobility of nano-zero valent iron (nZVI) particles, which have been used in remediation of groundwater contaminants (Crane and Scott 2012), can be manipulated by controlling their stability. As the particles naturally flocculate in solution due to the magnetic forces between them (Phenrat et al. 2007), they require a stabilizing solution to move through groundwater. The amount of stabilizing solution used can be adjusted to control nZVI stability and therefore the migration distance through chalk fractures (Cohen and Weisbrod 2018a, b).

Reactive contaminants may also sorb to mobile colloid surfaces and migrate with them through fractures (McCarthy 1998) in a phenomenon called colloid-facilitated transport. The stability of the colloids in this case not only dictates the mobility of the colloids themselves but also the potential migration of colloid-sorbed contaminants (Schäfer et al. 2012). For instance, it was shown that at low ionic strengths in fractured chalk, bentonite and humic acid colloid presence may increase the mobility of cerium, an analogue for actinides in the (III) and (IV) redox states such as Am and Pu (Tran et al. 2016). Bentonite colloids have also been shown to increase the mobility of lead through chalk fractures (Tang and Weisbrod 2009). However, at greater ionic strengths typical of those found in groundwater in the northern Negev, bentonite colloids are less likely to be stable, and sorption of radionuclides to clay colloid surfaces is anticipated to decrease (Tran et al. 2018). The role of colloid transport may therefore be minimized under such conditions (Fig. 1).

The impact of colloids on contaminant transport in fractured chalk is dependent upon colloid presence, stability, size and surface chemistry. All of these factors must be considered when predicting the mobility of colloids and colloid-bound contaminants in a fractured chalk subsurface such as that of the northern Negev Desert.

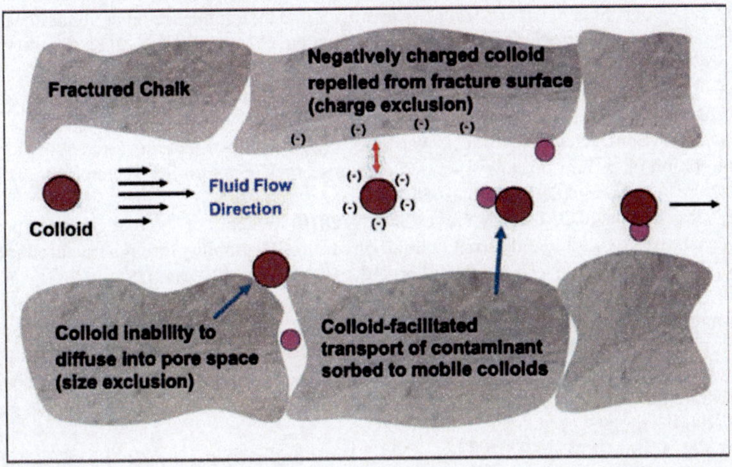


**Fig. 1** Diagram showing colloid transport in a fracture as a result of charge and size exclusion

## References

Abdel-Salam A, Chrysikopoulos CV (1995) Modeling of colloid and colloid-facilitated contaminant transport in a two-dimensional fracture with spatially variable aperture. Transp Porous Media 20(3):197–221

Becker MW, Reimus PW, Vilks P (1999) Transport and attenuation of carboxylate modified latex microspheres in fractured rock laboratory and field tracer tests. Groundwater 37(3):387–395. https://doi.org/10.1111/j.1745-6584.1999.tb01116.x

Bloomfield J, Nygaard E (2003) Overview of the geology of the northern Negev site—implications for hydrogeology, monitoring and remediation. FRACFLOW Project on Contaminant Transport, Monitoring Techniques and Remediation Strategies in Cross European Fractured Chalk. Fourth Annu Prog report I-1

Cohen M, Weisbrod N (2018a) Field scale mobility and transport manipulation of carbon-supported nanoscale zerovalent iron in fractured media. Environ Sci Technol 52(14):7849–7858. https://doi.org/10.1021/acs.est.8b01226

Cohen M, Weisbrod N (2018b) Transport of iron nanoparticles through natural discrete fractures. Water Res 129:375–383. https://doi.org/10.1016/J.WATRES.2017.11.019

Crane R, Scott T (2012) Nanoscale zero-valent iron: future prospects for an emerging water treatment technology. Elsevier 211–212:112–125. https://doi.org/10.1016/j.jhazmat.2011.11.073

Kurtzman D, Nativ R, Adar EM (2005) The conceptualization of a channel network through macroscopic analysis of pumping and tracer tests in fractured chalk. J Hydrol 309(1):241–257

McCarthy JF (1998) Colloid-facilitated transport of contaminants in groundwater: mobilization of transuranic radionuclides from disposal trenches by natural organic matter. Phys Chem Earth 23(2):171–178. https://doi.org/10.1016/S0079-1946(98)00009-3

McCarthy JF, Zachara JM (1989) Subsurface transport of contaminants. Environ Sci Technol 23(5):496–502

Phenrat T, Saleh N, Sirk K et al (2007) Aggregation and sedimentation of aqueous nanoscale zerovalent iron dispersions. Environ Sci Technol 41(1):284–290. https://doi.org/10.1021/es061349a

Runde WH (2000) The chemical interactions of actinides in the environment. Los Alamos Sci 26:392–411

Schäfer T, Huber F, Seher H et al (2012) Nanoparticles and their influence on radionuclide mobility in deep geological formations. Fundam Process Radionucl Migr Geosph 27(2):390–403. https://doi.org/10.1016/j.apgeochem.2011.09.009
Tang X-Y, Weisbrod N (2009) Colloid-facilitated transport of lead in natural discrete fractures. Environ Pollut 157(8–9):2266–2274. https://doi.org/10.1016/j.envpol.2009.03.034
Tran EL, Klein BenDavid O, Teutsch N, Weisbrod N (2015) Influence of intrinsic colloid formation on migration of cerium through fractured carbonate rock. Environ Sci Technol 49(22):13275–13282. https://doi.org/10.1021/acs.est.5b03383
Tran E, Klein BenDavid O, Teutsch N, Weisbrod N (2016) Influence of heteroaggregation processes between intrinsic colloids and carrier colloids on cerium(III) mobility through fractured carbonate rocks. Water Res 100:88–97. https://doi.org/10.1016/j.watres.2016.04.075
Tran EL, Teutsch N, Klein-BenDavid O, Weisbrod N (2018) Uranium and cesium sorption to bentonite colloids under carbonate-rich environments: implications for radionuclide transport. Sci Total Environ 643:260–269. https://doi.org/10.1016/J.SCITOTENV.2018.06.162
Weisbrod N, Nativ R, Adar EM et al (2000) Impact of coating and weathering on the properties of chalk fracture surfaces. J Geophys Res Solid Earth 105(B12):27853–27864. https://doi.org/10.1029/2000JB900295
Weisbrod N, Meron H, Walker S, Gitis V (2013) Virus transport in a discrete fracture. Water Res 47(5):1888–1898. https://doi.org/10.1016/j.watres.2013.01.009
Zhang W, Tang X-Y, Weisbrod N et al (2015) A coupled field study of subsurface fracture flow and colloid transport. J Hydrol 524:476–488. https://doi.org/10.1016/j.jhydrol.2015.03.001
Zvikelsky O, Weisbrod N (2006) Impact of particle size on colloid transport in discrete fractures. Water Resour Res 42(12):W12S08. https://doi.org/10.1029/2006wr004873

# The Use of Saline Aquifers as a Target for Deep Geologic $CO_2$ Storage in Israel

Ravid Rosenzweig, Ran Calvo, and Uri Shavit

Geologic carbon dioxide ($CO_2$) capture and storage (CCS) is considered as a promising technology for the reduction of atmospheric emissions of greenhouse gas in the short run (IPCC 2005). In CCS, $CO_2$ is separated from the gas stream of large stationary sources such as power plants, transported to the storage site and injected into suitable deep geological formations, such as saline aquifers. To prevent the buoyant $CO_2$ from migrating toward the surface, the storage formation should be sealed by an impermeable caprock. While the "structural trapping" provided by the caprock is the dominant trapping mechanism right after injection, with time, other trapping mechanisms contribute to storage security and their significance increase. These mechanisms include "capillary trapping"—the immobilization of $CO_2$ as disconnected blobs, "dissolution trapping"—the dissolution of $CO_2$ into the brine and "mineral trapping"—the incorporation of $CO_2$ in carbonate minerals as a consequence of reactions between dissolved $CO_2$ and the host rock (IPCC 2005).

As storage capacity, risk assessment, and other operational parameters greatly depend on the local geology and hydrogeological conditions, assessments of storage capacity and risks are highly site- and country-specific (Bachu 2015). In Israel, research of geologic $CO_2$ storage is being carried out in the past decade. This includes estimates of the storage potential of varying complexities (e.g., Calvo and Gvirtzman 2013; Calvo and Salamon 2016; Calvo et al. 2019); dynamic simulations of injection scenarios to elucidate storage capacity and its security (Rosenzweig and Calvo 2015; Cohen 2017); laboratory experiments to investigate various physical and geochemical

R. Rosenzweig (✉) · R. Calvo
Geological Survey of Israel, Jerusalem, Israel
e-mail: rravid@gsi.gov.il

R. Calvo
e-mail: rani.calvo@gsi.gov.il

U. Shavit
Civil and Environmental Engineering, Technion IIT, Haifa, Israel
e-mail: aguri@technion.ac.il



U. Kafri and Y. Yechieli (eds.), *The Many Facets of Israel's Hydrogeology*,
Springer Hydrogeology, https://doi.org/10.1007/978-3-030-51148-7_23

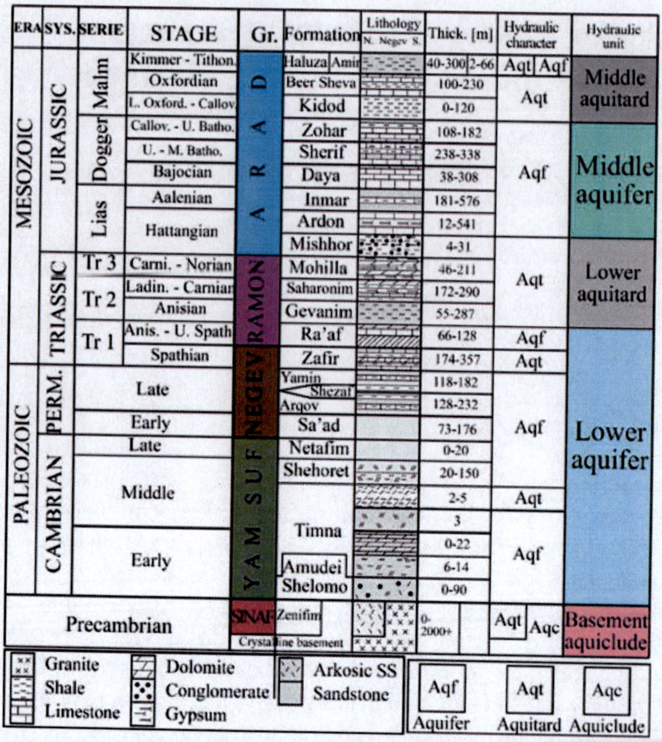


**Fig. 1** Generalized stratigraphic and hydrostratigraphic table of the Precambrian to Jurassic section in the subsurface of the Negev. Modified after Nativ et al. (1987) and Calvo and Gvirtzman (2013)

aspects of the storage process (Michel-Meyer et al. 2017; Rendel et al. 2016, 2018) as well as a small-scale injection experiment designed to in situ estimate capillary trapping (Niemi et al. 2016).

In Israel, it was found that the vast majority of the storage potential is located in a series of three saline aquifers in the Negev. These aquifers were previously studied by Nativ et al. (1987), and their hydrostratigraphic sequence was characterized. Of the three aquifers, assessments of storage potential were obtained for the lower Precambrian to late Triassic aquifer (Yam Suf and Negev groups.) and for the middle Jurassic aquifer (Arad Gr.), yielding estimated potentials of 6.2 GT and 8.7 GT $CO_2$, respectively (Fig. 1; Calvo and Gvirtzman 2013). This early assessment was followed by more detailed evaluations conducted for the middle aquifer in which the proximity to historic seismic events and to other seismo-tectonics indicators (Calvo and Salamon 2016) and the uncertainties in the aquifer thermo-physical conditions (Calvo et al. 2019) were considered. These assessments narrowed the storage potential in the middle aquifer to $7 \pm 0.9$ GT $CO_2$. A second set of studies focused on detailed evaluation of storage potential at specific structures and formations. Taragan (2018) calculated the storage potential in the sandstone-rich Negev

Group of the lower aquifer. Calvo (2014) assessed the storage capacity in the Jurassic sections of the Agur and the Qeren anticlines. Finally, Bar-Ziv (2019) evaluated the storage potential of the Jurassic section of the Zohar structure, which previously served as a natural gas reservoir and has therefore a proved containment capability. The results indicated a storage capacity that ranges from 3.4 GT $CO_2$ in the Negev Group to 1.1 GT $CO_2$ in the Qeren structure and to 200 MT and 16 MT $CO_2$ in the small Agur and Zohar structures.

While these former assessments were based on averaged properties and on empirical efficiency factors, which provide the fraction of the pore-space that can be filled with $CO_2$, for more accurate calculations and for a better evaluation of storage security and $CO_2$ migration, numerical simulations of $CO_2$ dynamics are required. Rosenzweig and Calvo (2015) and Cohen (2017) studied the behavior of $CO_2$ in the middle aquifer by modeling various injection scenarios. They found that the heterogeneity of the aquifer strongly affects the migration of $CO_2$ and its trapping. It was also found that the natural flow of the aquifer water affects the migration of the $CO_2$ plume, and to a lesser extent the dissolution trapping. Last, the simulations demonstrated that $CO_2$ leakage from the aquifer to the caprock is relatively low and does not exceed 0.1% of the injected $CO_2$.

The results of these site-specific studies led to fundamental research questions concerning various physical and geochemical aspects of the storage process. Michel-Meyer et al. (2017) studied the effect of natural horizontal flow on $CO_2$ convective dissolution using a laboratory aquifer model and a $CO_2$ analog to avoid the need to maintain supercritical conditions. While a significant effect of the horizontal flow on the convective structures was demonstrated, the horizontal flow only weakly influenced the dissolution rate.

Motivated by the possibility of gypsum precipitation and subsequent pore clogging in the vicinity of the injection well, Rendel et al. (2016, 2018) conducted laboratory experiments to study the equilibrium and kinetics of gypsum precipitation and dissolution under high-pressure conditions and in the presence of supercritical $CO_2$. Gypsum solubility was found to increase with pressure and decrease with $CO_2$ concentration. The experimental results were in a good agreement with predictions of a geochemical model allowing to assess the effect that pressure, temperature and $CO_2$ concentration have on gypsum solubility.

The most comprehensive scientific activity related to $CO_2$ storage in Israel was conducted by a European consortium led by Israeli scientists in the Heletz oil field in the south of Israel (Niemi et al. 2016 and references therein). Following a detailed characterization of the site, small-scale injection experiments were carried out during 2016 and 2017 with the purpose of assessing in situ capillary trapping within the reservoir. The experiment included several push–pull tests that were conducted before and after the injection of $CO_2$ while using several tracers having different partitioning coefficients between $CO_2$ and water. A sophisticated sampling and monitoring platform was installed on-site and was used for real-time monitoring of the conditions at the well and for collecting in situ fluid samples. Numerical simulations of the experiments were then used to elucidate the extent of residual trapping and to learn the dynamics of the injected $CO_2$ (Moghadasi et al. 2019; Basirat et al. 2019).

## References


Bachu S (2015) Review of $CO_2$ storage efficiency in deep saline aquifers. Int J Greenh Gas Control 40:188–202

Bar-Ziv O (2019) The suitability of the Zohar structure (Arad area) for geological $CO_2$ storage, Master's thesis, Haifa University, Haifa

Basirat F, Yang Z, Levchenko S, Bensabat J, Pan L, and Niemi A (2019) A coupled wellbore-reservoir model of self-producing fluids during a push-pull experiment at Heletz (Israel) pilot $CO_2$ injection site. Geophys Res Abs 21:EGU2019-14150

Calvo R (2014) A comparative study for potential $CO_2$ storage units in the subsurface Paleozoic-Mesozoic section in Southern Israel: Arad Group (Middle Aquifer and Middle Aquitard). GSI Rep. GSI/24/2014, (In Hebrew), 65 pp

Calvo R, Gvirtzman Z (2013) Assessment of $CO_2$ storage capacity in southern Israel. Int J Greenh Gas Control 14:25–38

Calvo R, Salamon A (2016) Toward a $CO_2$ sequestration project in the Jurassic saline aquifer of southern Israel: Seismotectonic characterization. Int J Greenh Gas Control 47:48–62

Calvo R, Taragan R, Rosenzweig R (2019) How large is our $CO_2$ storage capacity assessment error? Analyzing the magnitude of error in the effective capacity calculation propagated from uncertainties in the thermophysical conditions in the aquifer, the case of the Israeli Jurassic saline aquifer. Int J Greenh Gas Control 82:19–37

Cohen E (2017) The effect of petrophysical and hydrogeological properties on $CO_2$ storage in the saline Jurassic aquifer of the Negev. Master's thesis, The Hebrew University of Jerusalem (In Hebrew)

IPCC (2005) Underground geological storage. In: Metz B, Davidson O, de Coninck HC, Loose M, Meyer LA (eds) IPCC special report on carbon dioxide capture and storage. Cambridge University Press

Michel-Meyer I, Shavit U, Rosenzweig R (2017) The role of water flow and dispersion in the dissolution of $CO_2$ in deep saline aquifers. Energy Procedia 114:4994–5006

Moghadasi R, Joodaki S, Basirat F, Fagerlund F, Bensabat J, Niemi A (2019) A stochastic model for interpreting the partitioning tracer recovery from residual trapping experiment at Heletz, Israel, Pilot Injection Site. Geophys Res Abs 21:EGU2019-10178

Nativ R, Bachmat Y, Issar A (1987) Potential use of the deep aquifers in the Negev Desert, Israel—a conceptual model. J Hydrol 94:237–265

Niemi A, Bensabat J, Shtivelman V, Edlmann K, Gouze P, Luquot L, Hingerl F, Benson SM, Pezard PA, Rasmusson K, Liang T (2016) Heletz experimental site overview, characterization and data analysis for $CO_2$ injection and geological storage. Int J Greenh Gas Control 48:3–23

Rendel PM, Gavrieli I, Wolff-Boenisch D, Ganor J (2016) Gypsum solubility under pressure conditions relevant to $CO_2$ geological storage. Int J Greenh Gas Control 55:15–22

Rendel PM, Wolff-Boenisch D, Gavrieli I, Ganor J (2018) A novel experimental system for the exploration of $CO_2$-water-rock interactions under conditions relevant to $CO_2$ geological storage. Chem Eng J 334:1206–1213

Rosenzweig R, Calvo R (2015) The effect of natural groundwater flow on $CO_2$ migration and trapping: the case of the Jurassic saline aquifer of the Negev (Israel), In: Proceedings, TOUGH Symposium

Taragan R (2018) Deep geological $CO_2$ storage potential in the lower saline aquifer (Paleozoic–Triassic) in southern Israel. Master's thesis, Ben-Gurion University of the Negev, Beer-Sheva

# Groundwater Dewatering Management in Urban Areas at the Coastal Plain of Israel

**Lior Asaf**

The State of Israel has undergone extensive urbanization since the late 1950s. According to Israel's Central Bureau of Statistics (ICBS 2018), about 9,000,000 people resided in Israel at the end of 2018. However, by 2025, the population is expected to number at around 9.8 million people, and by 2035, it is estimated to reach an abounding 11 million residents. Over 91% of the population lives in cities, making Israel highly concentric and urban. Consequently, the pressure on the natural water resources is expected to rise significantly. Abounding masses of open areas and agricultural lands have turned into residential neighborhoods, urban commerce and industrial areas, with the large-scale urban development taking place along the country's coastal pain (Burmil et al. 2003). Increased impervious surface area is a consequence of urbanization, with corresponding significant effects on the hydrological cycle (Shuster et al. 2005; Grodek et al. 2011; Goldshleger et al. 2015). Carmon et al. (1997) estimated that by 2020 the built-up area above the coastal aquifer underneath the coastal plain will double as compared to the situation in 1990, reaching about 500 $km^2$.

The accelerated construction in the densely inhabited Gush Dan region along with the parking shortages in the area precipitated the construction of high-rise residential and employment towers, which include several floors of underground parking basements. In order to meet the demand for parking spaces, a large portion of the projects planned to excavate below the groundwater level in order to build several basements. Construction below the groundwater level requires dewatering systems, which lower the groundwater level in order to enable work on a dry site. In addition, the construction of the underground light rail red line began in 2014. Groundwater depletion is required for the construction of 11 underground stations, shafts and other sites where groundwater dewatering is required.

L. Asaf (✉)
LDD Advanced Technologies, 10 Gonen St., Petach Tikva, Israel
e-mail: LiorA@lddtech.com



U. Kafri and Y. Yechieli (eds.), *The Many Facets of Israel's Hydrogeology*, Springer Hydrogeology, https://doi.org/10.1007/978-3-030-51148-7_24

Multiple projects that occur simultaneously over the course of a few years in a limited area result in the creation of local gradients and new flow paths that differ from the pre-existing regional flow regime. The damages, however, do not cease at that. Along the Menachem Begin Boulevard in Tel Aviv, increasing the gradient toward the pumping centers has likely caused pollution migration from nearby highly contaminated sites, like Magen and Arvey Nahal.

Infrastructural projects often involve excavation work to subgrade depths that require aggressive groundwater control, in order to make construction feasible and safe. Construction dewatering can be achieved by designing and installing a subgrade pumping systems, which operate continuously to draw down the water table enough to allow work. Dewatering presents unique challenges since the variables, relevant to completing a project, are not easily identifiable and can often require services beyond those that were originally expected.

In many projects in Tel Aviv, there is a need to drop the water levels from a few meters to more than 15 m below the initial groundwater level for most of the construction period (1–2 years). Without suitable control measures, inflows of groundwater can flood excavations or tunnels and can lead to instability when the soils or rock around the excavation weaken and collapse—either locally or on a large scale.

The Israel Water Law determines that all the water resources in the country, above as well as below ground, are a public resource controlled by the state. The water sector in the country has been highly centralized, managed by the Water Authority. This centralized control has enabled the integrated management of water resources above and below the ground. Regulation via the planning sphere advanced considerably between 2000 and 2010 while national legislation and governance arrangements on drainage and groundwater management at urban center are still in the process of being updated (Goulden et al. 2018).

In order to issue a permit for a dewatering system, the Water Authority demands a specific hydrological groundwater model for each project. The specific parameters like the hydraulic conductivity and specific yields of recharge and abstraction wells, which are often obtained from local pumping tests as part of the design process.

Since many of the new infrastructure and high-rise residential projects at the coastal plan require large-scale dewatering plans with large pumping rates up to 500 $m^3$/h, there was a need to find a sustainable solution for the large-scale abstraction. The water authorities issued a specific abstraction rate up to 3 \$/$m^3$, making it uneconomic for the land developers to carry on dewatering project without a full recharge system. In order to decrease the influence of the recharge system on the dewatering in the upper aquifer, there was also a need to use deep recharge wells in order to recharge the water to a lower sub-aquifer.

These wells (up to 100–150 m deep) are used both for temporary dewatering purposes during the construction phase, as well as permanent sustainable storm water management solutions at a very complex high-density urban environment (Abu and Asaf 2019).

In cases where groundwater pollution is detected in the sites, above the Israeli drinking water standards, the land developer is obligated to install a treatment system on his expenses during the construction period before the water is recharged back to the aquifer, even if he was not responsible for the contamination.

The regional planning system has adopted the basic idea that groundwater below urban centers is a valuable resource. However, the transition is far from complete, and in practice, there is a need to manage the entire aquifer with multiple projects affecting both the local groundwater heads and pollution levels. The changes in the regional flow regime and the creation of local gradients and new flow paths that differ from the regional flow regime call for integrated groundwater management of dewatering operations on a larger regional scale, far beyond single project approach.

## References

Abu H, Asaf L (2019) Dan Towers, Semel South, and Semel Center dewatering design reports (Tec. reports in Hebrew)

Burmil S, Shamir U, Carmon N (2003) Urban runoff in residential neighborhoods. Technion Center for Urban and Regional Studies and the Grand Water Research Institute, Haifa, 118 p (in Hebrew)

Carmon N, Shamir U, Meiron-Pistiner S (1997) Water-sensitive urban planning: protecting groundwater. J Environ Planning Manage 40(4):413–434

Goldshleger N, Maor A, Garzuzi J, Asaf L (2015) Influence of land use on the quality of runoff along Israel's coastal strip (demonstrated in the cities of Herzliya and Ra'anana) urban runoff in Israel. Hydrol Process 29:1289–1300. https://doi.org/10.1002/hyp.10220

Goulden S et al (2018) From conventional drainage to sustainable stormwater management: beyond the technical challenges. J Environ Manage 219:37–45

Grodek T, Lange J, Lekach J, Husary S (2011) Urban hydrology in mountainous Middle Eastern cities. Hydrol Earth Syst Sci 15:953–966

Israel's Central Bureau of Statistics (ICBS) (2018) https://www.cbs.gov.il/EN/pages/default.aspx

Shuster WD, Bonta J, Thurston H, Warnemuende E, Smith DR (2005) Impacts of impervious surface on watershed hydrology: a review. Urban Water J 2(4):263–275

# The Initial Value of Radiocarbon ($^{14}C$) in the Aquifers of Israel

**Israel Carmi and Mariana Stiller**

Atmospheric $^{14}C$ enters aquifers as dissolved inorganic carbon (DIC) by dissolution of $CO_2$ in rainwater; this $CO_2$ is exhaled by plant roots into unsaturated zones (USZ) that overlie aquifers. Radiocarbon (half-life 5620 years) is of potential use in hydrology to following the flow of water in an aquifer. However, due to the interaction of DIC with carbonates that are devoid of $^{14}C$, the atmospheric $^{14}C$ value of the DIC gradually decreases in the USZ with depth. Therefore, in order to assess the age of water in an aquifer, the initial $^{14}C$ value of DIC, at the entrance of water into the aquifer, must be known.

A convenient way to estimate the initial $^{14}C$ value of DIC when it enters the aquifer, $A_{q0}$, is via the parameter $Q = A_{q0}/A_{ao}$ (Wigley 1975), where $A_{ao}$ is the atmospheric $^{14}C$ value of $CO_2$ in the rain that fed this water to the aquifer. It should be noted that root-exhaled $^{14}CO_2$ is of atmospheric origin and is subsequently dissolved as DIC into the percolating water. If $Q$ of the aquifer and $A_{ao}$ are known, $A_{q0}$ can be estimated and then used for calculating the age, $t$, of groundwater in an aquifer, from its measured $^{14}C$, $A_{qt}$, ($A_{qt} = A_{q0} \times e^{-\lambda t}$). In Israel, the unsaturated zone above aquifers is thick (more than 20 m) and it takes the rainwater a significant number of years to reach the aquifer.

The value of $Q$ has been estimated for the Mediterranean Coastal Aquifer and for the Mountain Aquifer of Israel by Carmi et al. (2006, 2009). To evaluate the $Q$ of the Coastal Aquifer, a core was taken from the USZ of the nature reserve of Nizzanim, from the surface down to the top of the aquifer. DIC was extracted from the core segments, by pumping it out of the soil water. To assess $Q$ for the Mountain Aquifer of the Judea Group, DIC was extracted from water dripping from stalactites of the

I. Carmi (✉)
Porter School of Environmental Sciences, Tel Aviv University, Tel Aviv, Israel
e-mail: carmiisr@post.tau.ac.il

M. Stiller
Geological Survey of Israel, Jerusalem, Israel



U. Kafri and Y. Yechieli (eds.), *The Many Facets of Israel's Hydrogeology*,
Springer Hydrogeology, https://doi.org/10.1007/978-3-030-51148-7_25

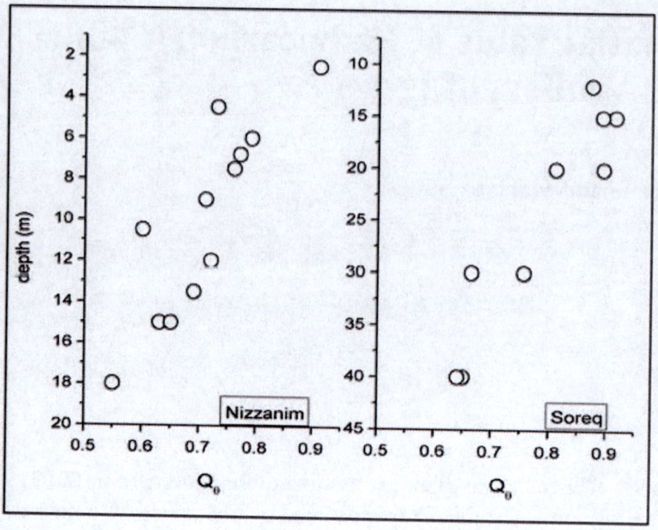


**Fig. 1** Left: $Q_\theta$ in the Nizzanim USZ (Coastal Aquifer); Right: $Q_\theta$ in the Soreq USZ (Judea Mountain Aquifer)

Soreq cave, which is situated in the foothills of the Judea Mountains. The stalactites were located below different thicknesses of the Soreq cave roof.

The $^{14}C$ activities of the DIC samples, $A_{q\theta}$, contained time-dependent atmospheric $^{14}C$ of thermonuclear origin. To calculate the time dependence of $Q_\theta$ values, $Q_\theta = A_{q\theta}/A_{a0\theta}$, the respective initial $^{14}C$ atmospheric values, $A_{a0\theta}$ and the rate of flow through the USZ are required. $A_{a0\theta}$ values were taken from the study of Levin and Kromer (1997). The rate of flow has been determined by tritium (of thermonuclear origin) as a tracer: for the vadose zones of the Judean Mountain Aquifer by Kaufman et al. (2003) and for the Coastal Aquifer by Carmi et al. (2009). Each depth in the USZ could then be referred to a specific year and its respective $A_{a0\theta}$. Finally, $A_{q\theta}$ and $A_{a0\theta}$ were used to calculate the $Q_\theta$ values. The $Q$ of the aquifer could then be estimated by extrapolation of the $Q_\theta$ values to the top of the aquifer.

The calculated $Q_\theta$ values at different depths in the unsaturated zone for the Coastal Aquifer are shown in Fig. 1 (left) and for the Mountain Aquifer in Fig. 1 (right). A clear linear decrease with depth is evident in both cases, which allowed deriving $Q$ by extrapolation of the $Q_\theta$ values to the depth of the aquifer top. The obtained results are $Q = 0.54$ (Carmi et al. 2009) for the Coastal Aquifer and $Q = 0.62$ for the Mountain Aquifer (Carmi et al. 2006).

Once the effect of the thermonuclear $^{14}C$ and tritium is over, the $^{14}C$ level in the atmosphere will become ~ constant [$A_{a0}$ ~ 100 pMC (percent of modern carbon)]. Then the initial $^{14}C$ values for the two aquifers will be $A_{q0} = 54$ pMC and $A_{q0} = 62$ pMC, respectively, for the Coastal and Mountain Aquifers of Israel.

## References


Carmi I, Stiller M, Kronfeld J, Yechieli Y, Bar-Mathews M, Ayalon A, Boaretto E (2006) Radiocarbon loss from DIC in vadose water flow above the Judea Aquifer, Israel. Radioact Environ 8:297–306

Carmi I, Yechieli Y, Boaretto E, Stiller M (2009) Carbon isotopes in pore water of the unsaturated zone and their relevance for initial $^{14}C$ activity in groundwater in the coastal aquifer of Israel. Chem Geol 268:189–196

Kaufman A, Bar-Mathews M, Ayalon A, Carmi I (2003) The vadose flow above Soreq cave, Israel. J Hydrol 273:155–163

Levin I, Kromer B (1997) Twenty years of atmospheric $^{14}CO_2$ observations at Schauinsland station, Germany. Radiocarbon 39(2):205–218

Wigley TMI (1975) $^{14}C$ dating of groundwater from closed and open systems. Water Resour Res 11:324–328

# Tracking Degradation of Organic Pollutants in Israeli Groundwater by Stable Isotope Analysis

Anat Bernstein and Faina Gelman

Gasoline stations and industrial areas are between the most significant sources of groundwater contamination by harmful organic compounds in Israel. Gasoline components, chlorinated ethylenes and nitro-explosives are among the most frequently detected groundwater pollutants (Sagi-Ben Moshe and Reshef 2018). Biochemical degradation and/or attenuation by non-destructive physical processes (e.g. absorption, dispersion, evaporation) affect the distribution of the organic pollutants in the environmental systems. Understanding of these environmental processes is important for mitigation of environmental problems caused by organic contaminants. The accepted method of following-up of the spread of the pollution through space is usually based on the monitoring the change in the pollutants' concentration and/or the concentration of the daughter products over distance and time. However, this solution is not always conclusive enough, especially in the case of heterogeneous environment systems. In the last decades, compound-specific isotope analysis (CSIA) has developed as a complementary sensitive tool to tackle the fate of organic contaminants. This methodology is based on monitoring shifts in the isotopic composition of the target compounds during the different transformation scenarios. Such shifts stem from the fact that light isotopes are converted slightly more rapidly than heavy ones (Elsner et al. 2005; Braeckevelt et al. 2012; Thullner et al. 2012). It is well accepted that shifts in the isotopic composition of a molecule are only little, if at all, influenced by processes in which the inner bonds remain intact, such as volatilization, sorption or dilution (Meckenstock et al. 2004; Braeckevelt et al. 2012). Therefore, significant isotopic shifts observed along a contamination plume

A. Bernstein
Department of Environmental Hydrology and Microbiology, Zuckerberg Institute for Water Research, Ben-Gurion University of the Negev, Sede Boqer Campus, 84990 Be'er Sheva, Israel
e-mail: anatbern@bgu.ac.il

F. Gelman (✉)
Geological Survey of Israel, 32 Yesha'yahu Leibowitz st., 9692100 Jerusalem, Israel
e-mail: faina@gsi.gov.il



U. Kafri and Y. Yechieli (eds.), *The Many Facets of Israel's Hydrogeology*, Springer Hydrogeology, https://doi.org/10.1007/978-3-030-51148-7_26

can be mostly related to degradation processes. Until now, applying CSIA in Israeli groundwater has focused mostly on gasoline additive, namely Methyl-tert-butyl ether (MTBE), trichloroethylene (TCE) and the triazine explosive compound RDX.

Methyl tertiary butyl ether (MTBE), which is a gasoline additive, is one of the common groundwater contaminants in Israel (Katz and Reshef 2018). Its occurrence in the groundwater usually results from leaks of fuel from underground storage tanks and pipes. MTBE is considered primarily a taste and odor drinking water concern and as a potential human carcinogen at high doses. In Israel, the maximal permitted level of MTBE in the drinking water is 40 μg/L. In part of the contaminated sites in Israel, the concentration of MTBE in groundwater attains up to several mg/L levels. However, a trend of decreasing MTBE concentrations, over the period 2004–2011, has been observed in ~30% of the monitoring wells (Reshef and Zhuravel 2013). In order to assess biodegradation of MTBE in situ, compound-specific isotope analysis has been applied in the field (Gelman et al. 2014; Gafni et al. 2016). The studied sites were characterized by different geochemical parameters: anaerobic with a negative redox potential, low nitrate and sulfate concentrations and high concentrations of dissolved manganese and iron as well as aerobic with high concentration of dissolved oxygen and a positive redox potential. In the anaerobic zone, a decrease in MTBE concentrations during a year, was accompanied by a significant change in $\delta^{13}C$ from −31.4 to −11.8‰ (Gafni et al. 2016; Gelman et al. 2014). Based on carbon isotopic enrichment factors ($\varepsilon$) of −17.2‰, determined in the microcosm experiments under sulfate reducing and methanogenic conditions, a half-life value of ~1 year was estimated (Gafni et al. 2016). In another study, where an active remediation by increasing oxygen level in groundwater was applied, significant decrease in MTBE concentrations was accompanied by small shifts in carbon isotope values. Based on the measured isotope composition data, the MTBE half-life was estimated as ~30 days (Gelman et al. 2014).

1,3,5-Trinitro-1,3,5-triazinane (RDX) has been detected in high concentrations up to ~2000 μg/L in adjacent to the production military site in central Israel, near the city of Ramat Hasharon. Isotopic analysis of RDX in contaminated groundwater was applied along a ca. 1.3-km long contamination plume within the Israeli Coastal Aquifer (Bernstein et al. 2010). Indigenous aerobic bacteria was isolated from the site, presenting the ability to degrade the compound (Bernstein et al. 2011). The study showed a $\delta^{15}N$ enrichment of 10.8‰ along the plume (Fig. 1). Based on geochemical data, it was suggested that the aerobic denitrification pathway of the compound is dominating, but anaerobic biodegradation was not excluded. Quantitative estimation of the degradation rates in the upper 15 m of the aquifer yielded half-life values ranging between 4.4 and 12.8 years, assuming purely aerobic biodegradation. The degradation rate presented a clear decrease with depth, which was hypothesized to be related to the lower dissolved oxygen concentration in the deep sub-surface, or the expectedly lower nutrient availability (Bernstein et al. 2010). In regard to RDX within the unsaturated soils, a non-uniform $\delta^{15}N$ composition of the compound was detected along a 46 m deep unsaturated profile (Sagi-Ben Moshe et al. 2010). The most enriched $\delta^{15}N$ value in RDX was detected in the uppermost soil layer, which was enriched by 10.6‰. This uppermost section was also characterized by high organic

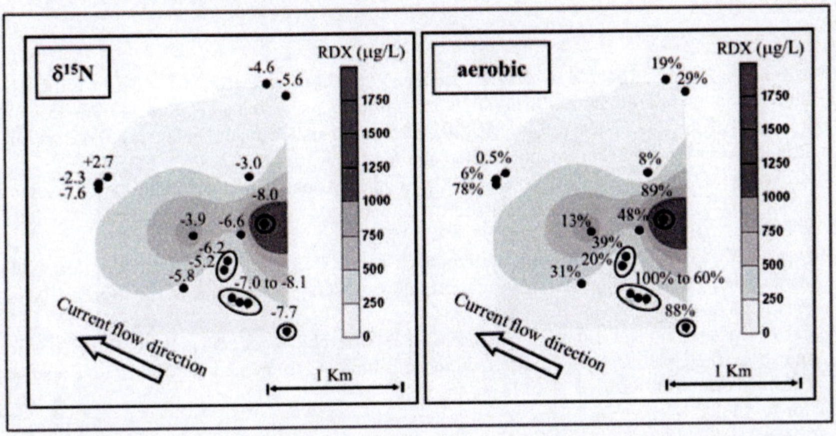


**Fig. 1** Isotope composition ($\delta^{15}N$) of RDX along *a* ≈ 1.35 km contamination plume (left), and calculated degradation extent, assuming purely aerobic degradation and $\varepsilon = -2.1‰$ (right). Modified after Bernstein et al. (2010)

content, which may have promoted the enhanced degradation of the compound. Surprisingly, the $\delta^{15}N$ values along the rest of the profile were more depleted. This contradicts the expectation that under steady-state conditions, the deeper the sample, the more enriched it would be. This discrepancy is explained by transient flow and transport conditions along the profile following changes in site's management.

Trichloroethylene (TCE) is an abundant groundwater contaminant of industrial origin. It is of the most frequently discovered industrial contaminant in the Israeli Coastal Aquifer. Microbial degradation of TCE may follow two pathways, where the most commonly detected is anaerobic reductive dechlorination. Aerobic co-metabolic oxidation is a second alternative. Demanding an auxiliary substrate, it may be questioned if this co-metabolic pathway is of high relevance. The isotope composition of TCE—both $\delta^{13}C$ and $\delta^{37}Cl$—was monitored in a wide range survey along the Israeli Coastal Aquifer, aiming to get isotopic evidence for aerobic degradation. These two processes may possibly be distinguished using the two isotopes (Gafni et al. 2018). A survey of ≈100 contaminated boreholes, from Netanya in the north to Ashkelon in the south, was conducted. No clear isotope proof was obtained for the occurrence of the aerobic pathway in the aquifer. This is despite the fact that bacteria, capable of catalyzing the process, are widely abundant (Gafni et al. 2020). Isotope data also rejects the occurrence of aerobic degradation of TCE in the Magen site (Hunkeler et al. 2011). Thus, until now isotope evidences for microbial degradation are limited to a single anaerobic site within the aquifer, and to actively remediated sites, with most enriched $\delta^{13}C$ value of −4.6‰ detected in a permanganate-treated site (Gafni et al. 2020).

## References


Bernstein A, Adar E, Ronen Z, Lowag H, Stichler W, Meckenstock RU (2010) Quantifying RDX biodegradation in groundwater using $\delta^{15}N$ isotope analysis. J Contam Hydrol 111:25–35

Bernstein A, Adar E, Nejidat A, Ronen Z (2011) Isolation and characterization of RDX-degrading Rhodococcus species from a contaminated aquifer. Biodegradation 22:997–1005

Braeckevelt M, Fischer A, Kästner M (2012) Field applicability of compound-specific isotope analysis (CSIA) for characterization and quantification of in situ contaminant degradation in aquifers. Appl Microbiol Biotechnol 94:1401–1421

Elsner M, Zwank L, Hunkeler D, Schwarzenbach RP (2005) A new concept linking observable stable isotope fractionation to transformation pathways of organic pollutants. Environ Sci Technol 39(18):6896–6916

Gafni A, Rosenzweig R, Gelman F, Ronen Z (2016) Anaerobic biodegradation of MTBE in a field site above the Israeli Coastal Aquifer: evidence from $\delta^{13}C$ compound-specific isotope analysis. J Chem Technol Biotechnol 91:1638–1645

Gafni A, Lihl C, Gelman F, Elsner M, Bernstein A (2018) $\delta^{13}C$ and $\delta^{37}Cl$ isotope fractionation to characterize aerobic vs. anaerobic degradation of trichloroethylene. Environ Sci Technol Let 5(4):202–208

Gafni A, Siebner H, Bernstein A (2020) Potential for co-metabolic oxidation of TCE and evidence for its occurrence in a large-scale aquifer survey. Water Res 171:115431

Gelman F, Resenzweig R, Ronen Z (2014) A breakdown of fuel hydrocarbons in the groundwater in Israel. Report GSI/11/2014. Geological Survey of Israel. Available from: https://www.water.gov.il

Hunkeler D, Aravena R, Shouakar-Stash O, Weisbrod N, Nasser A, Netzer L, Ronen D (2011) Carbon and chlorine isotope ratios of chlorinated ethenes migrating through a thick unsaturated zone of a sandy aquifer. Environ Sci Technol 45:8247–8253

Katz H, Reshef G (2018) Summary of activities on prevention contamination of water resources by fuel compounds (in Hebrew). Israel Water Authority. Available from: https://www.water.gov.il

Meckenstock RU, Morasch B, Griebler C, Richnow HH (2004) Stable isotope fractionation analysis as a tool to monitor biodegradation in contaminated aquifers. J Contam Hydrol 75:215–255

Reshef G, Zhuravel A (2013) Calculation of first—order decay constant for MTBE and benzene in the coastal basin (in Hebrew). Israel Water Authority. Available from: https://www.water.gov.il

Sagi-Ben Moshe S, Ronen Z, Dahan O, Bernstein A, Weisbrod N, Gelman F, Adar E (2010) Isotopic evidence and quantification assessment of in situ RDX biodegradation in the deep unsaturated zone. Soil Biol Biochem 42(8):1253–1262

Sagi-Ben Moshe S, Reshef G (2018) Summary of Activities for 2016–2017 on location, assessment and prevention of water pollution by industrial contaminants (in Hebrew). Israel Water Authority Authority. Available from: https://www.water.gov.il

Thullner M, Centler F, Richnow HH, Fischer A (2012) Quantification of organic pollutant degradation in contaminated aquifers using compound specific stable isotope analysis—review of recent developments. Org Geochem 42:1440–1460

# Modeling Groundwater Contamination for the Central Coastal Aquifer in Israel

**Alexander Yakirevich, Michael Kuznetsov, Eilon M. Adar, and Shaul Sorek**

One of the major tasks of water management is the protection of soil and groundwater resources from degradation and contamination caused by industrial, agricultural, and domestic sources. Monitoring the aquifer system, including scientifically based prediction of future changes of state, aims to control contamination processes.

Volatile organic compounds (VOCs) in the coastal plain aquifer of Israel were detected over an area of 200 $km^2$, in 82% of 90 pumping wells. The main VOCs detected in the aquifer were trichloroethylene (TCE), tetrachloroethylene (PCE), 1,1-dichloroethylene (1,1-DCE), 1,2-dichloroethylene isomers (1,2-DCE), carbon tetrachloride ($CCl_4$), 1,2-dichoroethane (1,2-DCA), and chloroform ($CHCl_3$). Exceptionally high concentrations of TCE, up to 124,000 μg/L-air in the unsaturated zone and up to 250,000 μg/L-water in the saturated zone, were measured at the saturated–unsaturated interface in a region North of Tel Aviv (Ronen et al. 2005). An integrated approach of field and model investigations was implemented to an aquifer underlying urban–industrial infrastructures and focused on the transport of TCE through the unsaturated–saturated zone (Sorek et al. 2005). Simulations addressed a three-dimensional continuum model involving three interacting mobile phases: aqueous, non-aqueous phase liquids (NAPL), and gaseous (Forsyth et al. 1988). The results indicate that VOCs are being transported also in the gas phase of the unsaturated zone

A. Yakirevich (✉) · M. Kuznetsov · E. M. Adar · S. Sorek
J. Blaustein Institutes for Desert Research, Ben Gurion University of the Negev, Sede Boqer Campus, Israel
e-mail: alexy@bgu.ac.il

M. Kuznetsov
e-mail: kmikhail@bgu.ac.il

E. M. Adar
e-mail: eilon@bgu.ac.il

S. Sorek
e-mail: sorek@bgu.ac.il


U. Kafri and Y. Yechieli (eds.), *The Many Facets of Israel's Hydrogeology*, Springer Hydrogeology, https://doi.org/10.1007/978-3-030-51148-7_27

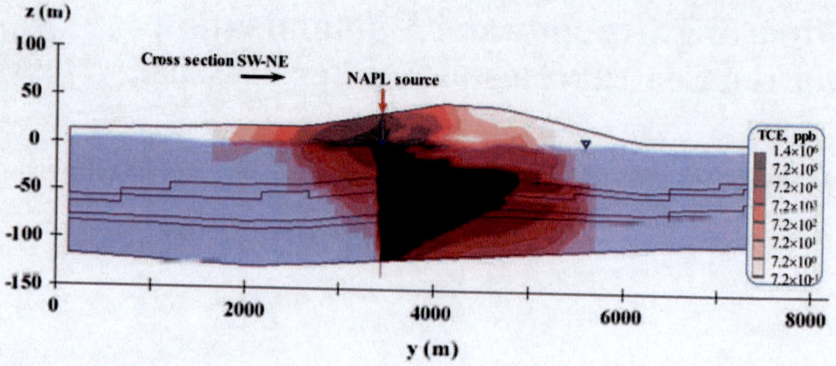


**Fig. 1** Simulated distribution of TCE in the aquifer after 50 years of NAPL percolation (modified after Ronen et al. 2005)

and that TCE vapors experience an upward flux along vertical impervious boundaries (e.g., basement walls), such that underground structures may be surrounded by VOC vapors. The TCE transport in the saturated zone is dominated by advection of groundwater toward the "Ramat-Gan" groundwater depression. After 50 years of continuous DNAPL spill, TCE migrates in the aquifer up to 2.2 km northeast, while secondary contamination of groundwater southwest from the NAPL source occurs by gaseous TCE in the unsaturated zone moving under pavement surface (Fig. 1). The subsurface NAPL migration is mainly controlled by the geological structure of the aquifer. The simulated groundwater TCE plume was significantly smaller than the actual TCE plume. This leads to the conclusion that there is more than a single source of VOC contaminants in the study area.

Industrial plants and facilities in Ramat Hasharon region were a source of organic pollutants, heavy metals, perchlorate, and nitrates that contaminated both the soil and groundwater in the area. Groundwater sampling from 89 wells in the study area indicates that significant portion of the aquifer is contaminated by perchlorate (Fig. 2a), and its concentration reaches 450,000 ppb in the vicinity of the source (Sagi-Ben Moshe 2012). The three-dimensional numerical model of water flow and contaminant transport was applied for simulating perchlorate migration in groundwater. Perchlorate was assumed as a conservative contaminant: neither sorption nor degradation reactions were taken into account. The location of the contaminant source was assigned to the effluents storage ponds, represented by a strip of 200 × 50 m. The concentration of perchlorate in percolating water was calculated as a function of time based on one-dimensional transient simulations of water flow and contaminant transport in the unsaturated layered soil profile. Simulations of perchlorate migration were carried out for the period of 2006–2030. The results indicate that the estimated total mass of approximately 280 tons of perchlorate will be washed out of the unsaturated zone into the groundwater reservoir during the coming 25 years, in addition to 1000 tons that already reached groundwater. The predicted average velocity of the plume was 15–20 m/year (Fig. 2b). It was found that the most efficient removal of

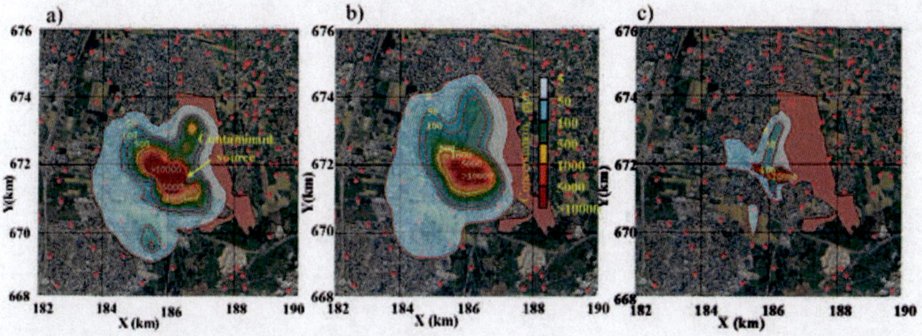


**Fig. 2** Distribution of perchlorate in groundwater: **a** observed in 2005, **b** simulated in 2030—no pumping, **c** simulated in 2030—pumping 600 $m^3$/h from four wells in the vicinity of contaminant source

the contaminant can be achieved with simultaneous groundwater abstraction at the plume center, together with pumping at the periphery of the plume, avoiding further downstream migration of the perchlorate. The contaminated groundwater plume can be almost completely reclaimed in 25 years (Fig. 2c), providing a future enforcement of a comprehensive groundwater abstraction, treatment, and management.

## References

Forsyth PA, Unger AJA, Sudicky EA (1988) Nonlinear iteration methods for nonequilibrium multiphase subsurface flow. Adv Water Resour 21:443–449

Ronen D, Graber ER, Laor Y (2005) Volatile organic compounds in the saturated–unsaturated interface region of a contaminated phreatic aquifer. Vadose Zone J 4(2):337–344

Sagi-Ben Moshe S (2012) Biodegradation and transport of explosives in sandy unsaturated zone. PhD thesis, Hebrew University of Jerusalem

Sorek SP, Kuznetsov M, Yakirevich A, Ronen D (2005) Multiphase and multi-component interactions through the unsaturated saturated Zone: field and model study. In: Nutzmann G, Viotti P, Aagaard P (eds) Reactive transport in soils and groundwater. Springer, Berlin, pp 171–185

# Groundwater Protection and Agricultural Development—The Conflict and Challenges

Ofer Dahan

## 1 Groundwater Quality and Agricultural Development

Groundwater is considered a principle source of natural drinking water worldwide. Clean groundwater is critical to the health and well-being of the environment with a direct impact on all aspects of human life, social, and economic. The history of human development all over the world was strongly dependent on the availability of fresh groundwater that could support rural and urban development year-round, regardless seasonal variability in precipitation, surface water quality, or unavoidable droughts. In recent years, water scarcity is often associated with climate changes and global warming causing abnormal long droughts. Nevertheless, on a global scale, the most common reason for water scarcity is related to continuous degradation of groundwater quality caused by man-made pollution. Apparently, groundwater pollution is an ongoing process that is responsible for immediate and future disqualification of aquifers to serve as a source of drinking water.

Groundwater pollution is attributed to a very large array of chemicals. Nevertheless, on a global scale, most drinking-water wells shut down is caused by elevated concentrations of nitrate in the aquifers (Osenbruck et al. 2006; Kurtzman and Scanlon 2011; Kourakos et al. 2012; Liao et al. 2012). Israel coastal aquifer is a major source of natural fresh water for the region. Most of the recharging zones for this aquifer underlay cultivated agricultural fields or densely populated urban zones. As such the aquifer water quality is under a continuous treat from different pollution sources, such as sea water intrusion and down-leaching of contaminants from land surface to the water table. Over recent decades, a severe degradation in quality of the coastal aquifer water has been recorded (Israel Water Authority 2014). Among all pollution potential sources, down-leaching of nitrate from cultivated soils through

O. Dahan (✉)
Zuckerberg Institute for Water Research, Jacob Blaustein Institutes for Desert Research, Ben-Gurion University of the Negev, Be'er Sheva, Israel
e-mail: odahan@bgu.ac.il



U. Kafri and Y. Yechieli (eds.), *The Many Facets of Israel's Hydrogeology*, Springer Hydrogeology, https://doi.org/10.1007/978-3-030-51148-7_28

the unsaturated zone to the groundwater is the major reason for degradation in the aquifer water quality and disqualification of wells for direct use as drinking water.

In spite of the great necessity of nitrogen fertilizer use to supply the world demand for food, many questions have been raised about the sustainability of nitrogen-based fertilizer application in soils in regard of groundwater contamination (Byrnes and Bumb 2008; Eickhout et al. 2006). As issued by the US Environmental Protection Agency (EPA), the maximum contaminant concentration of nitrate in drinking water should not excide 40 mg $L^{-1}$. At higher concentration, nitrate is harmful especially to infants, where it causes what known as the blue babies syndrome, which can lead to severe illness and even death (Fewtrell 2004; Thompson et al. 2007). Moreover, recent studies indicate that high concentration on nitrate in drinking water may increase the risk for specific cancer (Ward et al. 2018). This possess a great concern since groundwater is the major source of drinking water (Burow et al. 2010). However, the increasing concentration of nitrate in water well forces the authorities sometimes to push the bar up in terms of maximum allowed nitrate concentration in drinking water. The Israeli Water Authority has set water quality standard similar to those of the US EPA with respect to nitrate concentration. Yet, the gradual increasing concentration of nitrate in the costal aquifer has forced a new standard of 70 mg $L^{-1}$ at the year of 1994. Consequently, nitrate contamination has disqualified drinking water wells in Israel and around the world more than any other contaminant since the beginning of the twenty-first century (Russo et al. 2013).

Unfortunately, nitrate pollution in groundwater is mostly associated with agricultural development (Vitousek et al. 2009; Burow et al. 2010; Melo et al. 2012; Morari et al. 2012). The average concentration of nitrate at Israeli coastal aquifer is rising continuously by ~0.45 mg $L^{-1}$ $year^{-1}$ for the past 5 decades (Israel Water Authority 2014). Obviously, this average value represents a great special variability with a higher significance to zones that underlay agricultural fields. Today mostly the world habituated land is used for agriculture which covers major portion of the aquifers recharging zone, pointing out the dilemma between food production and water quality preservation (Roser and Hannah 2017). As such, developing efficient productive agriculture, while protecting groundwater quality from pollution hazard, is one of the most important challenges in water resources management.

Technological means to monitor and control the soil–water conditions are well developed and affordable commercially (mainly soil moisture sensors, tensiometers, and smart irrigation systems). However, monitoring and controlling the nutrient availability in the soil are still far from mitigating the required resolution for both optimal plant growth and prevention of water resources pollution. Up-to-date, fertilizer application in agriculture relies primarily on farmer's experience and expert's recommendation. All these do not provide information that is in line with the time scale of N-fertilizer mobilization, consumption, and transformation in the soil. Due to the lack of real-time information on nutrient concentration in the soil, farmers tend to implement excess fertilization regime to prevent potential nutrient deficiency and crop damage. Moreover, combining the fertilizers low cost with the plant tolerance of high N-fertilizers concentration drives farmer to use excess fertilization as a common practice. For example, In Israel, most of the intensive agricultural setups apply fertilizers in every irrigation cycle regardless the nutrient availability in the soil. As a result, major portions of the applied fertilizer are not utilized by the crop and migrate down from the root zone to the deeper sections of the unsaturated zone and ultimately to the groundwater (Dahan et al. 2007; Dudley et al. 2008). Accordingly, in situ continuous measurement of nitrate, the more mobile form of the N in the soil, is essential for both optimization of crop yield as well as reduce potential contamination of groundwater, and other related water resources such as rivers, lakes, and sea.

## 2 The Vadose Zone Role in Attenuation and Identification of Pollution Potential (See also Chapter "Unsaturated-Zone Hydrology in Israel: A Subjective Review")

Minimization of groundwater pollution is dependent on reliable and effective monitoring tools. Such monitoring tools should be capable to provide real-time information on the chemical and hydrological state of the percolating water, from the topsoil, near the land surface, to the water table below. Today, most monitoring programs aiming at preventing of groundwater pollution are based on observation wells that enable collection of hydrological and chemical information from the saturated part of the subsurface. Unfortunately, identification of pollution in a water well is clear evidence that the contaminants already crossed the entire vadose zone and accumulated in the aquifer to detectable levels. Moreover, only little can be done to fully remediate contaminated aquifers.

Nitrogen cycle in the soil transforms different forms of nitrogen into mobile and less mobile forms, depending on the soil and climate conditions. Obviously, these conditions vary seasonally and geographically. Measuring the mineral nitrogen stock in the soil profile after the growing season, as is practiced today, is not a good predictor of the amount of nitrate that may be leached during the winter season as well as of the amount that will be available to the plants in the next growing season. Therefore, effective monitoring programs must include monitoring means that provide real-time information on the chemical properties of the percolating water in the soil and deep unsaturated zone, long before contaminates reach the water table and accumulate in the aquifers water. Such monitoring programs may provide early warning for potential pollution processes that may risk groundwater quality.

## 3 Assessment of Nitrate Pollution Potential

Two main approaches are commonly practiced to characterize nitrate leaching from agricultural fields down to the groundwater: (1) Analysis of the chemical composition of soil pore water, as may be obtained by applying suction lysimeters or extracting the soil solutes in the laboratory (Feaga et al. 2010) and (2) determination of the cumulative long-term impact on groundwater as may be obtained from the chemical composition of the aquifer water through wells (Harter et al. 2002). Apparently, the chemical characteristics of the root zone pore water may vary dramatically in time scales of hours and days to seasons, according to irrigation patterns, fertilizer applications, and crop uptake, while the cumulative impact on groundwater develops in time scales of years to decades. The long time lag between the initiations of a contamination event near the land surface to its detection in the aquifer water depends on the mechanisms controlling flow and transport in the vadose zone (Billy et al. 2010; Sebilo et al. 2013). Therefore, means for in situ real-time characterization of nitrate leaching in the unsaturated zone are essential for minimization of nitrate pollution in groundwater.

The ability to characterize flow and transport processes in the vadose zone was recently improved following the development of a vadose zone monitoring system (VMS) (Fig. 1). The VMS provides real time, in-situ information on the hydraulic and chemical properties of the percolating water across the entire vadose zone. It has been implemented, in Israel as well as in other countries worldwide, in a variety of hydro-geological setups, including (1) flood water percolation (Dahan et al. 2007, 2008; Amiaz et al. 2011), (2) rain water percolation through thick unsaturated sand and clay formations (Rimon et al. 2007, 2011a; Baram et al. 2012), (3) solute transport in deep vadose zone (Dahan et al. 2009; Rimon et al. 2011b; Baram et al. 2013), (4) nitrate leaching from intensive organic farming (Dahan et al. 2014), and (5) land use impact on groundwater quality (Baram et al. 2012, 2014; Turkeltaub et al. 2014, 2015, 2016).

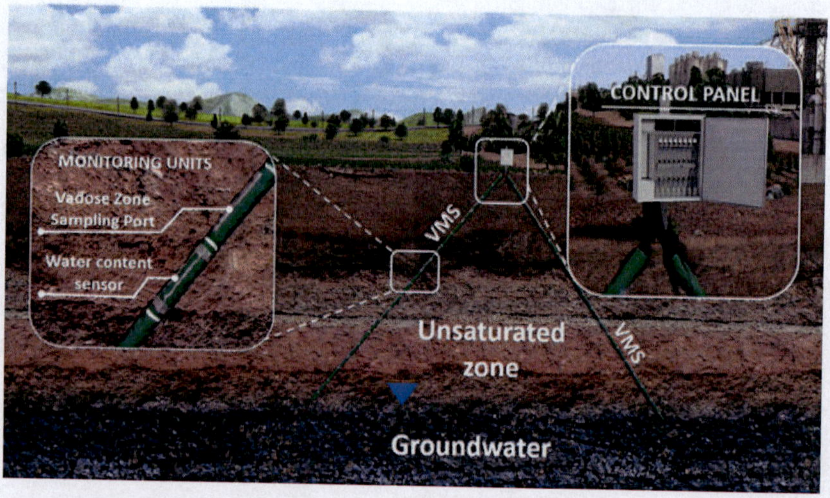


**Fig. 1** Illustration of a typical vadose zone monitoring system (VMS) installed across the unsaturated zone under and agricultural field

Several studies on the impact of different agricultural setups on groundwater quality were conducted in Israel using vadose zone monitoring systems. The monitoring systems were used to investigate the direct link between specific agricultural practices on groundwater pollution potential. For example, a study nitrate leaching from intensive organic farms showed that organic agriculture, which relay on solid organic matter as the sole fertilizer, resulted in significant down-leaching of nitrate through the unsaturated zone to the groundwater. On the other hand, similar intensive agriculture that implemented liquid fertilizer through drip irrigation, as commonly practiced in conventional agriculture, resulted in much lower rates of pollution of the vadose zone and groundwater. It has been shown that accurate fertilization methods that distribute the fertilizers through the irrigation system during the growing season dramatically reduce the potential for groundwater contamination (Dahan et al. 2014).

In another study, VMS application showed that switching crop type and consequently changing irrigation and fertilization regimes leads to alterations in deep percolation and solute concentrations of vadose zone pore water. Observations from the deep vadose zone combined with model simulations demonstrate the changes in water, chloride, and nitrate fluxes under commercial greenhouses as a consequence of a change from tomato to lettuce cropping (Turkeltaub et al. 2015).

Although the VMS enables direct access to water samples from the deep unsaturated zone, it still lacks the capability to provide continuous measurements of nitrate concentration in the resolution required to assess nitrate leaching from the root zone into the unsaturated zone.

## 4 Real-Time Monitoring of Nitrate Concentration in Soils

Developing technologies capable of carrying in situ continuous measurement of nitrate concentration in the soils is essential for optimizing fertilizers application and prevention of water resources pollution by nitrate. At present days, detecting nitrate concentration in the soil or in deeper part of the unsaturated zone requires laboratory analysis of water samples, which may be obtained from sediment extraction or suction lysimeters. Yet, this process requires intensive field and laboratory labor, which limits the required time resolution of tracking nitrate migration and transformation in the subsurface. Application of spectroscopy methods for measurements of nitrate is common in both laboratory instrumentation and water probes (Simal et al. 1985; Moorcroft 2001; Shaw et al. 2014). Nevertheless, adoption of these methodologies to the unsaturated zone imposes substantial technological challenges.

A novel monitoring system which enables in situ and continuous measurement of nitrate concentration in soil was recently developed (Yeshno et al. 2019). The monitoring system is based on absorbance spectroscopy techniques for direct measurement of nitrate concentration in soil pore water without pretreatment such as filtration, dilution, or reagent supplementation. The soil nitrate monitoring sensor combines optical flow cells with soil pore water sampling devices, which enable online continuous measurement of nitrate concentration in the soil and deep vadose zone. The

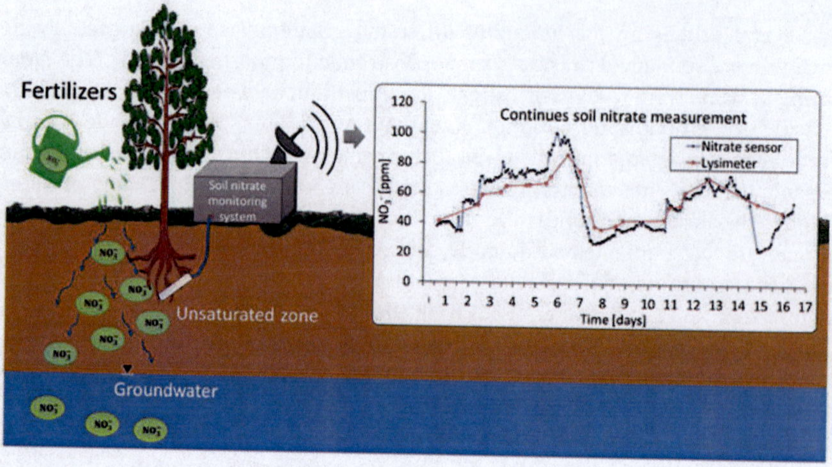


**Fig. 2** New monitoring technology for automated high-resolution measurement of nitrate concentration in soils

method uses a special analytical procedure that improves the measurement accuracy while eliminating the typical spectral interference caused by soil dissolve organic carbon. Measurement conducted by the soil nitrate sensor provided for the first time outstanding explicit data which reveals the complexities of the temporal variation in nitrate concentration in the soil with respect to irrigation cycles and fertilizers application pattern (Fig. 2). Such real-time measurements of nitrate levels in soils is crucial for optimizing fertilizers application while increasing agricultural yield and reducing the potential threat for groundwater contamination by down-leaching of nitrate from the soil to groundwater.

## References

Amiaz Y, Sorek S, Enzel Y, Dahan O (2011) Solute transport in the vadose zone and groundwater during flash floods. Water Resour Res 47:1–10. https://doi.org/10.1029/2011WR010747

Baram S, Arnon S, Ronen Z, Kurtzman D, Dahan O (2012) Infiltration mechanism controls nitrification and denitrification processes under dairy waste lagoon. J Environ Qual 41:1623–1632. https://doi.org/10.2134/jeq2012.0015

Baram S, Ronen Z, Kurtzman D, Külls C, Dahan O (2013) Desiccation-crack-induced salinization in deep clay sediment. Hydrol Earth Syst Sci 17:1533–1545. https://doi.org/10.5194/hess-17-1533-2013

Baram S, Kurtzman D, Ronen Z, Peeters A, Dahan O (2014) Assessing the impact of dairy waste lagoons on groundwater quality using a spatial analysis of vadose zone and groundwater information in a coastal phreatic aquifer. J Environ Manage 132:135–144. https://doi.org/10.1016/j.jenvman.2013.11.008

Billy C, Billen G, Sebilo M, Birgand F, Tournebize J (2010) Nitrogen isotopic composition of leached nitrate and soil organic matter as an indicator of denitrification in a sloping drained agricultural plot and adjacent uncultivated riparian buffer strips. Soil Biol Biochem 42:108–117. https://doi.org/10.1016/j.soilbio.2009.09.026

Burow R, Nolan T, Rupert G, Dubrovsky M (2010) Nitrate in groundwater of the United States, 1991–2003. Environ Sci Technol 44:4988–4997. https://doi.org/10.1021/es100546y

Byrnes B, Bumb B (2008) Population growth, food production and nutrient requirements. J Crop Prod 1(2). https://doi.org/10.1300/J144v01n02_01

Dahan O, Shani Y, Enzel Y, Yechieli Y, Yakirevich A (2007) Direct measurements of floodwater infiltration into shallow alluvial aquifers. J Hydrol 344:157–170. https://doi.org/10.1016/j.jhydrol.2007.06.033

Dahan O, Tatarsky B, Enzel Y, Kulls C, Seely M, Benito G (2008) Dynamics of flood water infiltration and ground water recharge in hyperarid desert. Ground Water 46:450–461. https://doi.org/10.1111/j.1745-6584.2007.00414.x

Dahan O, Talby R, Yechieli Y, Adar E, Lazarovitch N, Enzel Y (2009) In situ monitoring of water percolation and solute transport using a vadose zone monitoring system. Vadose Zone J 8:916–925. https://doi.org/10.2136/vzj2008.0134

Dahan O, Babad A, Lazarovitch N, Russak E, Kurtzman D (2014) Nitrate leaching from intensive organic farms to groundwater. Hydrol Earth Syst Sci 10:9915–9941. https://doi.org/10.5194/hessd-10-9915-2013

Dudley M, Ben-Gal A, Lazarovitch N (2008) Drainage water reuse: biological, physical, and technological considerations for system management. J Environ Qual 37:S25–S35. https://doi.org/10.2134/jeq2007.0314

Eickhout B, Bouwman A, van Zeijts H (2006) The role of nitrogen in world food production and environmental sustainability. Agric Ecosyst Environ 116:4–14. https://doi.org/10.1016/j.agee.2006.03.009

Feaga B, Selker J, Dick R, Hemphill D (2010) Long-term nitrate leaching under vegetable production with cover crops in the Pacific Northwest. Soil Sci Soc Am J 74:186–195. https://doi.org/10.2136/sssaj2008.0178

Fewtrell L (2004) Drinking-water nitrate, methemoglobinemia, and global burden of disease: a discussion. Environ Health Perspect 112:1371–1374. https://doi.org/10.1289/ehp.7216

Harter T, Davis H, Mathews M, Meyer R (2002) Shallow groundwater quality on dairy farms with irrigated forage crops. J Contam Hydrol 55:287–315. https://doi.org/10.1016/s0169-7722(01)00189-9

Israel Water Authority (2014) Hydrological report—Coastal Aquifer (in Hebrew)

Kourakos G, Klein F, Cortis A, Harter T (2012) A groundwater nonpoint source pollution modeling framework to evaluate long-term dynamics of pollutant exceedance probabilities in wells and other discharge locations. Water Resour Res 48(6). https://doi.org/10.1029/2011wr010813

Kurtzman D, Scanlon B (2011) Groundwater recharge through Vertisols: irrigated cropland vs. natural land Israel. Vadose Zone J 10:662–674. https://doi.org/10.2136/vzj2010.0109

Liao L, Green C, Bekins B, Bohlke J (2012) Factors controlling nitrate fluxes in groundwater in agricultural areas. Water Resour Res 48(6). https://doi.org/10.1029/2011wr011008

Melo A, Pinto E, Aguiar A, Mansilha C, Pinho O, Ferreira I (2012) Impact of intensive horticulture practices on groundwater content of nitrates, sodium, potassium, and pesticides. Environ Monit Assess 184:4539–4551. https://doi.org/10.1007/s10661-011-2283-4

Moorcroft M (2001) Detection and determination of nitrate and nitrite: a review. Talanta 54:785–803. https://doi.org/10.1016/S0039-9140(01)00323-X

Morari F, Lugato E, Polese R, Berti A, Giardini L (2012) Nitrate concentrations in groundwater under contrasting agricultural management practices in the low plains of Italy. Agric Ecosyst Environ 147:47–56. https://doi.org/10.1016/j.agee.2011.03.001

Osenbruck K, Fiedler S, Knoller K, Weise S, Sultenfuss J, Oster H, Strauch G (2006) Timescales and development of groundwater pollution by nitrate in drinking water wells of the Jahna-Aue, Saxonia, Germany. Water Resour Res 42(12). https://doi.org/10.1029/2006wr004977

Rimon Y, Dahan O, Nativ R, Geyer S (2007) Water percolation through the deep vadose zone and groundwater recharge: preliminary results based on a new vadose zone monitoring system. Water Resour Res 43(5). https://doi.org/10.1029/2006WR004855

Rimon Y, Nativ R, Dahan O (2011a) Physical and chemical evidence for pore-scale dual-domain flow in the vadose zone. Vadose Zone J 10:322–331. https://doi.org/10.2136/vzj2009.0113

Rimon Y, Nativ R, Dahan O (2011b) Vadose zone water pressure variation during infiltration events. Vadose Zone J 10:1105–1112. https://doi.org/10.2136/vzj2010.0061

Roser M, Hannah R (2017) Yields and land use in agriculture. https://ourworldindata.org/yields-and-land-use-in-agriculture#

Russo D, Laufer A, Shapira R, Kurtzman D (2013) Assessment of solute fluxes beneath an orchard irrigated with treated sewage water: a numerical study. Water Resour Res 49:657–674. https://doi.org/10.1002/wrcr.20085

Sebilo M, Mayer B, Nicolardot B, Pinay G, Mariotti A (2013) Long-term fate of nitrate fertilizer in agricultural soils. Proc Natl Acad Sci USA 110:18185–18189. https://doi.org/10.1073/pnas.1305372110

Shaw B, Wei J, Tuly A, Campbell J, Parikhd S, Buelow M, Hopmans J (2014) Analysis of ion and dissolved organic carbon interference on soil solution nitrate concentration measurements using ultraviolet absorption spectroscopy. Vadose Zone J 13(12). https://doi.org/10.2136/vzj2014.06.0071

Simal J, Lage M, Iglesias I (1985) Second derivative ultraviolet spectroscopy and sulfamic acid method for determination of nitrates in water. J Assoc Off Anal Chem 68:962–964

Thompson B, Martinez-Gaitan C, Gallardo M, Gimenez C, Fernandez D, Martínez-Gaitan C, Gallardo M, Giménez C (2007) Identification of irrigation and N management practices that contribute to nitrate leaching loss from an intensive vegetable production system by use of a comprehensive survey. Agric Water Manage 89:261–274. https://doi.org/10.1016/j.agwat.2007.01.013

Turkeltaub T, Dahan O, Kurtzman D (2014) Investigation of groundwater recharge under agricultural fields using transient deep vadose zone data. Vadose Zone J 13(4). https://doi.org/10.2136/vzj2013.10.0176

Turkeltaub T, Kurtzman D, Russak E, Dahan O (2015) Impact of switching crop type on water and solute fluxes in deep vadose zone. Water Resour Res 51(12). https://doi.org/10.1002/2015WR017612

Turkeltaub T, Kurtzman D, Dahan O (2016) Real-time monitoring of nitrate transport in the deep vadose zone under a crop field—implications for groundwater protection. Hydrol Earth Syst Sci 20:3099–3108. https://doi.org/10.5194/hess-20-3099-2016

Vitousek M, Naylor R, Crews T, David B, Drinkwater E, Holland E, Johnes J, Katzenberger J, Martinelli A, Matson A, Nziguheba G, Ojima D, Palm A, Robertson P, Sanchez A, Townsend R, Zhang S (2009) Nutrient imbalances in agricultural development. Science 80(324):1519–1520. https://doi.org/10.1126/science.1170261

Ward H, Jones R, Brender D, de Kok M, Weyer J, Nolan T, Villanueva M, van Breda G (2018) Drinking water nitrate and human health: an updated review. Int J Environ Res Public Health 15(7). pii: E1557. https://doi.org/10.3390/ijerph15071557

Yeshno E, Arnon S, Dahan O (2019) In-situ continuous monitoring of nitrate concentration in soils—a key for groundwater protection from nitrate pollution. Hydrol Earth Syst Sci 23(9):3997–4010. https://doi.org/10.5194/hess-23-3997-2019

# Occurrence, Fate and Transport of Antibiotic Residues in Groundwater in Israel

**Dror Avisar and Gefen Ronen-Eliraz**

Antibacterial substances are a sub-group of pharmaceutical compounds. Antibiotics are among the emerging micro-contaminants in water due to their potential adverse effects on the ecosystem and possibly on human health. These compounds are likely to be released into the aquatic environment via wastewater effluent and agricultural runoff as a result of (a) incomplete metabolism that may reach up to 90%, (b) improper disposal or (c) ineffective treatment removal.

Sulfamethoxazole (SMX) is a representative and typical antibacterial compound considered as an emerging micro-contaminant. Concentrations of 90–150 ng/L were detected in secondary effluents, which regularly and constantly used for agricultural irrigation about five decades, above local groundwater reservoir in the Coastal Plain Region of Israel. Consequently, groundwater samples were grabbed from the water table at a depth of 28 m of a deep phreatic, sandstone aquifer under the land of effluent irrigation. This groundwater sampled from the water table, found to contain SMX concentrations of 37 ng/L, demonstrating 16 years travel rate of SMX from surface via the unsaturated zone down to the water table. Furthermore, samples, which were taken from a nearby production well from a depth of 109 m, located downstream to the irrigated land and the local wastewater treatment plant (WWTP), contained concentrations of 20 ng/L of SMX. According to our best knowledge, this SMX groundwater contamination is the deepest pollution derived from antibiotic residues, which ever been documented in Israel (Avisar et al. 2009a).

Therefore, our findings are of concern since: (a) The contaminated production well was a major drinking water source during the period 1956–2007. Thus, it is reasonable to assume that for a considerable period of time, the local population was

D. Avisar (✉) · G. Ronen-Eliraz
Faculty of Exact Sciences, School of Environment and Earth Sciences, The Water Research Center, Tel Aviv University, Ramat Aviv, Tel Aviv, Israel
e-mail: droravi@tauex.tau.ac.il

G. Ronen-Eliraz
e-mail: gefenre@tauex.tau.ac.il


U. Kafri and Y. Yechieli (eds.), *The Many Facets of Israel's Hydrogeology*, Springer Hydrogeology, https://doi.org/10.1007/978-3-030-51148-7_29

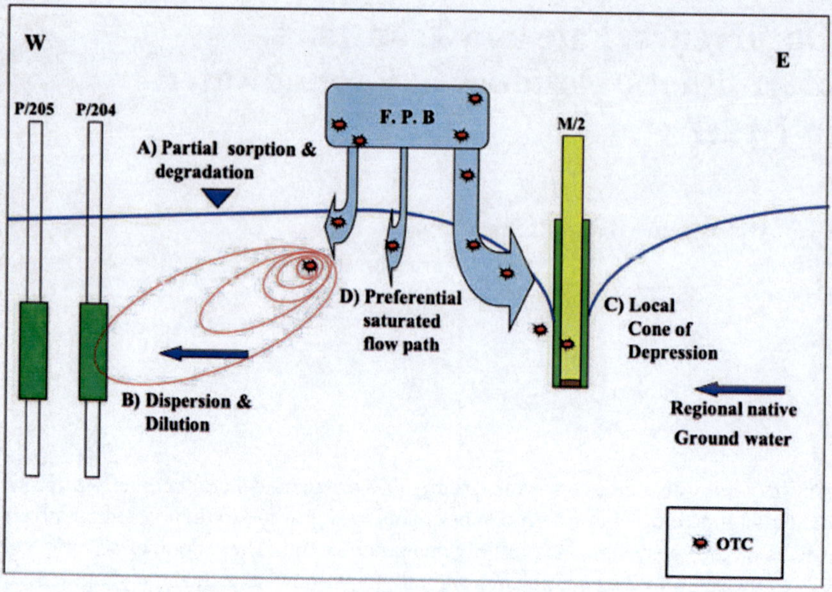


**Fig. 1** Hydrogeological models of processes contributing to OTC plume movement and detection at the studied groundwater wells

consuming routinely nanogram quantities of sulfa drugs (SMX), (b) large volumes of groundwater in the nearby regional groundwater depression, where other pumping wells are located, which were not sampled in this study and had never been monitored for antibiotics, may be presently contaminated by SMX and other antibiotic residues, and (c) wastewater effluents are used for agriculture irrigation above large areas of the phreatic Coastal Plain aquifer of Israel since the early 1960s, Thus, groundwater contamination by antibiotic residues may be a widespread phenomenon (Avisar et al. 2009a).

Oxytetracycline (OTC) is a widely used antibiotic in aquaculture, as growth promoter and a therapeutic compound. It is usually considered, according its chemical and physical characteristics, as an immobile compound, since it has the tendency to be strongly absorbed into the local sediment. In Israel, most of the aquaculture industry located above groundwater aquifers, which can severely be contaminated by leaching pollutants. Specific and unique hydrological settings were developed beneath the main pond that collected the antibiotic residues after treatment. These settings include saturated preferential flow paths, which were settled below the main fish pond along the vadose zone, enabling the adsorption-like OTC that rapidly leak and reach, by traveling through the pond floor and the aquifer toward the regional water table. Moreover, that area is characterized by a local cone of depression due to an intensive pumping of the M/2 production well (Fig. 1). Those particular and exceptional hydrological settings enable relatively rapid mobility of OTC, which conventionally considered as an immobile compound, to infiltrate from the main fishpond toward the nearby drinking water M/2 well (Avisar et al. 2009b).

These findings demonstrate that the fish pond industry may cause severe affects to the quality of groundwater. In addition to the reduction of antibiotic treatment, it is recommended to use impermeable plastic sheets to block and to prevent the leakage from fish ponds toward water resources. It is also suggested herein to avoid construction of aquaculture industry above phreatic sensitive aquifers which produce drinking water (Avisar et al. 2009b).

## References

Avisar D, Lester Y, Ronen D (2009a) Sulfamethoxazole contamination of a deep phreatic aquifer. Sci Total Environ 407:4278–4282. https://doi.org/10.1016/j.scitotenv.2009.03.032

Avisar D, Levin G Gozlan I (2009b) The processes affecting oxytetracycline contamination of groundwater in a phreatic aquifer underlying industrial fish ponds in Israel. Environ Earth Sci 59:939. https://doi.org/10.1007/s12665-009-0088-3

# Epilogue: Trends in Groundwater Development Policy in Israel Between the Jordan River and the Mediterranean Sea

**Giora Shaham**

The State of Israel is situated in the eastern Mediterranean region. In this geographical area, three main water bodies exist: 1. Lake Kinneret (the Sea of Galilee), which is being fed mostly by groundwater that issue via springs, and flowing through the Jordan River. 2. The Mountain Aquifers and 3. The Coastal Aquifer. These sources are described in detail in the previous chapters of this book. These three water sources are annually replenished by the winter precipitations at an annual average quantity of 1.2 billion $m^3$. Until about 100 years ago, there were about 800,000 inhabitants who lived in the area and consumed a very small portion of this water for drinking and agricultural purposes.

Only a hundred years have passed and population in the area has increased 20 times and is now about 15 million people between the Mediterranean Sea and the Jordan River. The current total water annual consumption in Israel for drinking, industry, and agriculture is about 2.3 billion $m^3$ (including the use of reclaimed wastewater) and is expected to almost double within 30 years.

Alongside the huge growth in the region's population, the climatic changes, that are taking place worldwide, are also causing a change in the patterns of replenishment in our region, as can be seen in the gradual drop in precipitation levels and in changes of the patterns of rainstorms. These two phenomena are gradually reducing the natural replenishment. Climate experts estimate that in about 30 years, the amount of natural replenishment will be reduced by about 20%.

The processes described above illustrate the clear, necessary, and inevitable need to be prepared for continuous production of more and more seawater desalination plants along the Mediterranean coast. It is clear to everyone that the natural water resources in this area are insufficient to meet the basic demands of the population, which is growing at a rate of 2% per year.

G. Shaham (✉)
Governmental Authority for Water and Sewage, 7 Bank Israel Street, 9095021 Jerusalem, Israel
e-mail: gioraS@water.gov.il


U. Kafri and Y. Yechieli (eds.), *The Many Facets of Israel's Hydrogeology*,
Springer Hydrogeology, https://doi.org/10.1007/978-3-030-51148-7_30

Due to the demographic impact that has been explained and illustrated above, we are facing the issue of the policy of development of the natural water sources, which are also being diminished due to climate change. The increase in the population also requires housing and employment solutions, and these factors in turn lead to the expansion of construction at the expense of open spaces, which are gradually shrinking, and as a result, the natural replenishment areas are also being reduced. It is our duty to ensure that the remaining open spaces continue to exist also as green landscapes and that the natural streams and springs will be preserved and nurtured as they feed not only the water resources, but serve as spaces for leisure and vacation, hiking, and enjoyment for the growing population.

In recent decades, hundreds of water facilities have been constructed in the area, based on thousands of wells to extract groundwater from the aquifers that, in turn, were replenished by natural precipitation. The water extraction from both Lake Kinneret and the aquifers led to the depletion of the groundwater levels to below those of the natural outlets of springs, whereby causing most of the streams and springs in the area to dry up. In the Upper Galilee, the flows of some springs are also being captured and diverted for agricultural irrigation, thus reducing the natural flow downstream.

From the picture, described above, and given the responsibility for achieving the objectives of the water sector in Israel, an obvious policy emerges:

1. Accelerate the planning and implementation of seawater desalination plants in order to meet the potable water needs of the growing population in the entire area.
2. Gradually amend the policy of groundwater development so as to enable the conservation and preservation of the natural spring water flows and ensure a minimal flow in the streams, which serve and feed the "green lungs" of Israel, used for leisure and tourism purposes.

The above policy is not a robustness policy and is not intended to completely halt the extraction of water from the natural groundwater sources and from Lake Kinneret.

Water experts in Israel have the knowledge, experience, and professional wisdom to balance the production from the natural water resources in such a manner as to enable the right management and control of the water sector, aiming to achieve the required quantitative and qualitative environmental goals. I estimate that within 20–30 years, we will be able to make the necessary changes in the redistribution of the water extraction wells and in the conveyance infrastructure of the water systems from, as well as to Lake Kinneret in order to adjust the different water sources to the various demands, as well as to restore the springs and streams in the area. This seems to be a complex but a possible task.

# Index




U. Kafri and Y. Yechieli (eds.), *The Many Facets of Israel's Hydrogeology*, Springer Hydrogeology, https://doi.org/10.1007/978-3-030-51148-7

**W**



**Y**



**Z**